Introduction to Criminal Justice

Introduction to Criminal Justice, Ninth Edition, offers a student-friendly description of the criminal justice process—outlining the decisions, practices, people, and issues involved. It provides a solid introduction to the mechanisms of the criminal justice system, with balanced coverage of the issues presented by each facet of the process, including a thorough review of practices and controversies in law enforcement, the criminal courts, and corrections.

In this revision, Edwards gives fresh sources of data, with over 600 citations of new research results. New sections include immigration policy, disparities in the justice system, Compstat and problem-oriented policing, victim services in the courts, and developments in drug policy. This edition also has expanded coverage of police use of force. Each chapter now includes a text box on a policy dilemma like cash bail or stop-and-frisk policies.

Appropriate for all U.S. Criminal Justice programs, this text offers great value for students and instructors.

Bradley D. Edwards is an Assistant Professor in the Department of Criminal Justice & Criminology at East Tennessee State University. He has previously co-authored books and book chapters in the areas of policing, corrections, and corporate misconduct. Aside from this text, recent publications with Routledge include two chapters in *Justice, Crime, and Ethics* (2017) as well as a chapter in the *Routledge Handbook on Offenders with Special Needs* (2018).

Lawrence F. Travis III is Professor Emeritus of Criminal Justice at the University of Cincinnati. His primary research interests lie in policing, criminal justice policy reform, sentencing, and corrections.

Introduction to Criminal Justice

Ninth Edition

Bradley D. Edwards
East Tennessee State University

Lawrence F. Travis III
University of Cincinnati

Routledge
Taylor & Francis Group

NEW YORK AND LONDON

Ninth edition published 2019
by Routledge
52 Vanderbilt Avenue, New York, NY 10017

and by Routledge
2 Park Square, Milton Park, Abingdon, Oxon, OX14 4RN

Routledge is an imprint of the Taylor & Francis Group, an informa business

© 2019 Taylor & Francis

The right of Bradley D. Edwards and Lawrence F. Travis III to be identified as authors of this work has been asserted by them in accordance with sections 77 and 78 of the Copyright, Designs and Patents Act 1988.

First edition published by Anderson Publishing in 2012
Eighth edition published by Routledge 2015

Library of Congress Cataloging-in-Publication Data
Names: Travis, Lawrence F., author. | Edwards, Bradley D., author.
Title: Introduction to criminal justice/Bradley D. Edwards, Lawrence F. Travis III.
Description: Ninth edition. | Abingdon, Oxon ; New York, NY : Routledge, 2019. | Lawrence Travis appears as the first named author on earlier editions. | Includes index.
Identifiers: LCCN 2018041515 (print) | LCCN 2018042425 (ebook) | ISBN 9780429426551 (Ebook) | ISBN 9781138386686 (hardback) | ISBN 9781138386723 (pbk.)
Subjects: LCSH: Criminal justice, Administration of—United States.
Classification: LCC HV9950 (ebook) | LCC HV9950 .T7 2019 (print) | DDC 364.973—dc23
LC record available at https://lccn.loc.gov/2018041515

ISBN: 978-1-138-38668-6 (hbk)
ISBN: 978-1-138-38672-3 (pbk)
ISBN: 978-0-429-42655-1 (ebk)

Typeset in Giovanni, Stone Sans, and Helvetica
by Apex CoVantage, LLC

Visit www.routledge.com/cw/edwards

Contents

Preface

For most Americans, our introduction to the criminal justice system does not occur in an academic setting. Rather, we typically experience the justice system in real life far before we ever step foot inside an institution of higher education. For some of us, that experience comes from being a victim of a crime. Others (most of us) have also been involved in the commission of at least some type of criminal or delinquent act. All of us are exposed to the media portrayals of crime and the justice system's response to those crimes. For those who are reading this text, these experiences likely motivated you to take an introductory-level criminal justice course. It is my hope that this text motivates you to continue your studies in this dynamic field.

The central purpose of this text is to provide students with a relatively brief, affordable, and comprehensive overview of the criminal justice system. As in earlier editions, the style and vocabulary are set at the reading level of the typical college freshman. I hope that reading this book is informative as well as exciting and entertaining for students. This book will introduce the reader to the systems approach to criminal justice, where law enforcement, court, and corrections personnel each make decisions that influence the operation of the overall justice system.

Many introductory students do not realize that criminal justice is a science, with an ever-increasing knowledge of the important topics covered in this text. As with previous editions, the most widespread changes in this edition of the text involved updating the references to reflect the most current knowledge available. In fact, this edition includes more than 600 new references to recent research and statistics involving crime and the criminal justice system. There is increased discussion throughout the book of important topics such as risk assessment tools and problem-oriented policing, which have become more prevalent over the past several years. In this edition, I am excited to introduce the addition of a policy dilemma box in each chapter. These boxes highlight some of the most controversial issues facing our policymakers. The policy dilemma boxes are designed to encourage the reader to think critically about these issues and facilitate classroom discussion.

Many individuals helped this book become a reality. First, I am grateful to Lawrence Travis III for creating the foundation for this book with his authorship of the first seven editions of this text. His mark will continue to be felt in this edition and any subsequent edition. I must also thank Steve Lab and John Whitehead, the original authors of Chapter 14 on the juvenile justice system.

I am also thankful to have worked with a wonderful staff at Routledge. Ellen Boyne, Pam Chester, Kate Taylor, and Michael Braswell have all been vital to this book's success. I also owe a thank-you to Hannah Medford for her help with the ancillary material. Finally, this edition would not have been completed without the love, support, and patience of my wife, Nyela. Thanks to you all.

Bradley D. Edwards

CHAPTER 1

Criminal Justice Perspectives

Imagine that you are standing on a busy street corner. You look at the people around you. What do you see?

A woman drops a postcard in a mailbox.
Across the street, a man carrying a small suitcase steps off a bus.
Several feet away from you, a couple is arguing about something.
A police car slowly passes through the intersection.
Less than half a block away, someone is jaywalking while a small child nearby reads a street sign.
A man deposits money in a nearby parking meter.
A stranger approaches and asks that you sign a petition in support of banning cell phone use by drivers.

What you probably do not see is that the mix of pedestrian and vehicular traffic is orderly. You do not notice that almost everyone watches the police car, at least briefly. You do not realize that all of these strangers at the intersection are going about their own business, apparently unaware of each other. Yet, in a well-rehearsed routine, they stop and go on cue from the traffic light. You probably do not see a crime (with the possible exception of the jaywalker).

Without realizing it, you have observed the criminal justice system in action. What you did not know is that the postcard was a monthly report the woman was sending to her probation officer. Nor was it clear that the man with the suitcase just left the state penitentiary on parole. The arguing couple may be tonight's domestic disturbance (or last night's). The slow-moving police car is searching for the small child, who is reading the street signs because he is lost. The jaywalker crossed the street to avoid walking past a group of teens gathered on the sidewalk. The man at the parking meter wants to avoid a citation. The person with the petition hopes to ensure that motorists refrain from using cell phones by making it criminal to do so.

IMPORTANT TERMS

closed system

crime control model

criminal justice

criminology

due process model

education

evidence-based policy

family model

federalism

formal social control

functions

informal social control

latent functions

local autonomy

manifest functions

open system

separation of powers

social control

system

theory

training

The workings of the criminal justice system affected the entire street corner scene just described, and all of the individuals in it. Interestingly, the individuals also directly affect the workings of that system. Should the argumentative couple become too boisterous, the shopper fail to deposit the correct coins in the parking meter, the woman not mail the postcard, and so on, you would expect some sort of official response from the justice system. Criminal justice is an integral part of our society and social living.

Sociologists often speak of the purposes of social institutions as "functions" (Parsons, 1966). **Functions** are the goals served by a social institution. For instance, schools serve the function of education. We can classify institutional functions as either manifest or latent. **Manifest functions** are the stated purposes of the institution, whereas **latent functions** are the unstated or hidden goals. Schools serve the manifest function of education through teaching students various academic subjects. They also meet the latent functions of providing child care and controlling the workforce by otherwise occupying millions of young people.

SOCIAL CONTROL

Cohen (1966:3) noted, "If human beings are to do business with one another, there must be rules, and people must be able to assume that, by and large, these rules will be observed." The making and enforcement of rules is a requirement for organized social living. **Social control** is the label given to the processes and structures that seek to limit rule-breaking behavior, or deviance.

There are a number of instruments of social control in any society, of which the law and criminal justice process are only one. Most discussions of social control attempt to classify the different means by which conformity is achieved (Black, 1976; Ross, 1926; Travis & Langworthy, 2008). These classifications focus on the procedures and processes that support conformity. The social control mechanisms in a society or a community can influence individual behavior by assigning "blame" and sanctions, or by prevention and education.

Types of Social Control

One of the most common ways of classifying social control processes is to distinguish between "formal" and "informal" social controls. **Formal social control** includes those sanctions applied by some authorized body after a public finding of fault. **Informal social control**, in contrast, refers to mechanisms that influence behavior without the need for a public finding of fault or the use of group-"authorized" sanctions. We can formally sanction a student who is disruptive in class by expulsion. In this case, the instructor, acting in his or her "official" (formal) capacity, or the educational institution itself, can apply the sanction of "banishment" on the offender. Alternatively, other members of the class can "hush" the offender by showing their disapproval without going through any formal process and punishment.

In the ideal, of course, the disruption would not have happened in the first place because the student would view the disruptive behavior as wrong or inappropriate. No matter how we achieve control over the behavior—formal sanctions, informal influence, or self-control—we stop or prevent the disruption. The social business of the class can continue with relative order and predictability. The means of control vary, but the goal of social control remains the maintenance of order in social relations. Because the goal is uniform regardless of the means, the distinction among types of social control is often artificial. Rather than being completely distinct types of control, informal and formal mechanisms lie along a continuum of controls ranging from those that are internal to the individual to those imposed on the individual externally.

Suppose the student wishes to be disruptive, but refrains from doing so because of a fear of expulsion. The student has demonstrated self-control, but the impetus for control is the threat of a formal sanction. Has social control in this case been established by informal or formal methods? In general, if the use or threat of a formal sanction is the mechanism by which we maintain social order, we call the process "formal social control." If a formal sanction is not necessary (even if such a sanction exists), then we call it "informal social control."

Criminal Justice as Social Control

The primary function of criminal justice is social control. The components of the justice process are police, courts, and corrections. These components have the manifest function of controlling different kinds of deviance defined as crime. "Crime" is only a small part of the total activities and behaviors that are the targets of social control. Most social control works through "informal" mechanisms, such as shunning or ostracizing the person who is rude, insensitive, or bothersome. Other forms of deviance defined as mental illness are handled through the mental health system. Pound (1929:4) remarked, "Law does but a part of this whole task of social control; and the criminal law does but a part of that portion which belongs to the law."

Criminal justice is the formal social institution designed to respond to deviance defined as crime. Crime control is the primary purpose of the criminal justice system, but it also serves other latent functions. Police, courts, and corrections do much more than merely fight crime. Still, our examination of the criminal justice process cannot progress until we understand this central purpose. Whatever other functions it may serve, and whatever methods it may employ, we can judge or measure the justice system as an institution of formal social control.

Focusing on the social control function of criminal justice (specifically, the control of crime) makes it easier to study and understand criminal justice practices and policies. We assess the value of a policy and procedure, or proposed changes in them, by how well they meet the objective of crime control. Theoretically, it seems easy enough to maintain an "objective" perspective, but it is often very difficult to do so in practice. Kellogg (1976:50) has observed that this perspective:

> has never made much of an impact on the administration of criminal justice, most likely because there is so little agreement as to the "objectives" of criminal justice, the purposes of punishment, and the most appropriate strategy to reduce crime.

The disagreement to which Kellogg refers concerns the means by which the justice system is expected to achieve crime control. It is not enough that criminal justice efforts control crime, those efforts must protect individual rights and otherwise be acceptable to our society. While it is true that criminal justice practices may be controversial in particular instances, the overriding interest in controlling crime is a constant goal. Although we may disagree over the use of the death penalty, wiretaps, plea bargaining, or probation, we can agree that what we want to do is reduce the incidence of crime. Unfortunately, criminal justice practices too often become focal points for debates stated in terms of the purposes of the justice system. The President's Commission on Law Enforcement and Administration of Justice (1967:70) aptly illustrated this confusion in its report:

> Any criminal justice system is an apparatus society uses to enforce the standards of conduct necessary to protect individuals and the community. It operates by apprehending, prosecuting, convicting, and sentencing those members of the community who violate the basic rules of group existence. The action taken against lawbreakers is designed to serve three purposes beyond the immediately punitive one. It removes dangerous people from the community; it deters others from criminal behavior; and it gives society an opportunity to attempt to transform lawbreakers into law-abiding citizens.

A debate may arise over whether deterring others from criminal behavior or transforming violators into law-abiding citizens is the best means of achieving the objective of social control, but no one questions the objective itself. This confusion of means and ends is not limited to disagreements over specific practices such as capital punishment but also includes ideological conflicts. People disagree not only over the appropriate forms of capital punishment (e.g., beheading, burning at the stake, electrocution, poison gas, lethal injection) but also over the appropriateness of capital punishment in general (e.g., the sanctity of life versus "an eye for an eye"). Yet, what would happen to these debates if the justice system could eliminate murder?

To further complicate an already complicated picture, the justice system is not the only social control institution in operation. The mental health system deals with many of the "rule violators" deemed inappropriate subjects for the justice system. Families, churches, schools, social organizations, and the media all serve social control purposes by informing us of what is acceptable behavior. The usefulness of the justice system must be understood within the total context of social control institutions. These other social control devices are often very effective (perhaps more effective than the criminal law), as illustrated in Box 1.1. Markowitz (2006:63) reported that when we lower the capacity of mental hospitals, the criminal justice system workload increases, writing, "In sum, public psychiatric hospital capacity is an important source of control of those whose behavior or public presence may at times threaten the social order."

That most of the pedestrians and vehicles in the illustration that opens this chapter obey the traffic lights and signs is evidence of social control. How are these individuals controlled? Some may be controlled by fear of a citation (justice system); others may react as a result of learning traffic safety at home, in school, or from the media. All of these sources of social control converge at this intersection to produce an orderly and predictable flow of traffic. How much credit for this level of conformity should go to the justice system?

In general, the criminal justice process is a formal social control mechanism. The basic social control tool available to agents of the criminal justice process is group-authorized punishment. The threat of coercive force is the ultimate sanction available for social control. The criminal justice process can be seen as the social control institution of last resort. Returning to our earlier idea of a continuum of social control, we can think of the criminal justice process as occupying the extreme end of the "formal" side of the continuum. Ideally, the individual will personally see some behavior (say, theft) as wrong, and avoid engaging in theft. If not, then the disapproval of others (family, friends, even strangers) may stop the individual from stealing. If not, then perhaps some more formal mechanism such as mental health counseling may prevent the theft. When all else fails, we can call on the criminal justice process to try to force the individual to stop stealing.

One need only consider two examples of traffic behavior to realize the complex interaction of the many sources of social control. First, compare the orderliness of most street traffic to the relative "free-for-all" chaos characteristic of most shopping mall parking lots. Second, think of the number of times you (as driver or passenger) have waited at a stoplight on a deserted street. It is clear that the presence or absence of others does not completely explain the differences in behavior. Rather, it may be the public nature of the road as opposed to the private nature of the parking lot. The criminal justice system addresses

BOX 1.1 SOCIAL CONTROL INSTITUTIONS IN AMERICA

Social control is achieved in many ways: through lessons learned by the individual about what is appropriate or inappropriate behavior, through structured opportunity that does not allow the individual the chance to deviate, through the exercise of coercive force to limit behavior. Nearly all social life affects social control, but the principal institutions in our society achieve social control in the following ways:

Lessons	Serve to teach us what behaviors are acceptable
Structures	Limit our opportunities for misbehavior
Coercion	Forces us to behave correctly, or prevents us from misbehaving

Lessons

The family	Children learn to respect others' property and opinions, and how to resolve conflict peacefully
Schools	Students learn appropriate behavior, work habits, and respect for others
Churches	Members learn rules for behavior (e.g., the Ten Commandments)
Social groups	Members learn tolerance and rules for personal relations and behavior (e.g., majority rule)
Recreation	Players learn rules and discipline, ways of behaving (e.g., fair play)
Employment	Workers learn discipline, work habits, "chain of command"
Mental health	Patients learn coping skills and ways of behaving (e.g., through token economies)
Law	Defendants and observers learn rules of behavior (laws) through their application

Structures

The family	Children are supervised, must abide by constraints on behavior (e.g., curfews)
Schools	Students follow fairly regimented academic schedules, are supervised by teachers
Churches	Members participate in legitimate activities (e.g., weekly services, Sunday school, service projects)
Social groups	Members participate in activities (e.g., formal meetings)
Recreation	Players participate in organized activities and competitions
Employment	Workers engage in defined activities and meet performance standards (e.g., production quotas)
Mental health	Patients participate in organized activities (e.g., group meetings)
Law	Statutes require certain behaviors (e.g., providing care to children, maintenance of rental property)

Coercion

The family	Children are punished for wrongdoing (e.g., "grounding," spanking)
Schools	Students who misbehave are punished (e.g., detention, suspension from school, written assignments)
Churches	Offending members are penalized (e.g., excommunication, threat of eternal damnation)
Social groups	Offending members are sanctioned (e.g., ridicule, expulsion from the group, ostracism)
Recreation	Wrongdoers are punished (e.g., game forfeiture, penalties, loss of eligibility)
Employment	Misbehavior is penalized (e.g., loss of pay, dismissal, demotion)
Mental health	Behavioral problems are controlled (e.g., passive restraint, sedation, forcible restraint)
Law	Offenders are sanctioned (e.g., fines, incarceration, execution, assessment of damages)

the issue of public social behavior, but it is not the only working mechanism of social control in those cases.

A number of factors influence people's behavior including personality, motivations, beliefs, peer pressure, and opportunities. Why people do or do not engage in crime is a complex question. Some of the explanation may be that the criminal justice process exists to punish criminal behavior, but that is not the complete answer. Flexon, Meldrum, Young, and Lehmann (2016) reported on a survey of college students, which revealed that the chances of a student committing a crime were influenced by low self-control and certain personality traits. Similarly, parental behavior has been found to be an important factor in explaining why some juveniles commit delinquent acts and other high-risk activity while others do not (Hoskins, 2014). These studies and others indicate that social control is the product of both formal and informal processes, and they are interrelated. The existence of formal controls, such as the criminal law and criminal justice process, serves to "educate" people about what is right and wrong. In this way, the law supports informal social control mechanisms, even while the law itself is a formal social control mechanism (Bianchi, 1994).

During the past few decades, the criminal justice process in the United States has undergone substantial change. Increasingly, we define the function of the justice process more broadly than crime control. As Boyd (2017:3) points out, "crime is not the problem, crime is symptomatic of much larger problems in society." As the social control institution of last resort, we have historically used the criminal justice process when all else had failed, and after crime had occurred. How much better would it be to prevent crimes from occurring in the first place? To prevent crime, the agents of the criminal justice process should work to strengthen and facilitate informal social control institutions.

Evidence of this shift in thinking abounds. Research suggests that individuals who perceive the justice system as more legitimate are more likely to comply with the law (Reisig, Wolfe, & Holtfreter, 2011). Perceived legitimacy can be shaped in many ways, such as one's upbringing, media exposure, and direct experience with the justice system (Gaeta, 2010; Mazerolle, Antrobus, Bennett, & Tyler, 2013). With the recent attention given to police-involved shootings, there is an increased focus on improving trust between the police and the communities that they serve. Police departments are now increasing the use of body-worn cameras to increase transparency, as well as training officers how to treat citizens with more respect during traffic stops and other interactions.

Another recent development involves justice reinvestment programs, which seek to reduce prison populations and divert the monetary savings into investments in high incarceration communities (Austin et al., 2013). At the same time, increased attention has been given to the collateral consequences of a criminal conviction, such as the loss of voting rights and eligibility for social benefits such as food stamps or public housing. In the past decade, most states have passed laws which mitigate the collateral consequences of being convicted of a crime (Subramanian, Moreno, & Gebreselassie, 2014). The hope is that by strengthening the communities which are impacted the most by crime, as well as reducing the lasting impacts of being convicted of a crime; the underlying societal issues which create crime can be mitigated.

Anyone studying criminal justice today will recognize the widespread use of words such as "community" and "partnership." After decades of evolution in which criminal justice professionals were increasingly isolated from the people they served, and where the "job" of criminal justice was defined narrowly as responding to crime, contemporary thinking holds that the best criminal justice is preventive, and that crime prevention is best accomplished by informal social control. As Friel (2000:15) explained, past efforts to improve public safety and reduce crime by professionalizing the criminal justice system had a downside. "The downside, however, has been growth in government bureaucracy, coupled with a tangle of laws, regulations, and red tape, which, although intended to restore the 'community' instead has removed the government from that community." Building closer links between the criminal justice process and the broader community has complicated the criminal justice picture in the United States.

Supancic and Willis (1998) studied what they called "extralegal justice" and its relationship to the formal criminal justice process. They distinguish between the two, writing (1998:193), "Legal justice includes all formal responses to crime by the police, the court system, and the corrections system. On the other hand, extralegal justice is that form of informal collective action directed against deviant and criminal conduct which is administered outside the formalized legal authority and not legally sanctioned by such authority." Extralegal justice is, they contend, an important part of a total system of social control and justice. It relates to the quality and quantity of legal justice. They suggest that an understanding of criminal justice (legal justice) cannot exist without also paying attention to informal social control in the community (extralegal justice). Martin, Wright, and Steiner (2016) studied the impact of formal social control (arrest and incarceration) in neighborhoods, and their findings support this observation. Martin et al. (2016) concluded that the overuse of formal social controls could unintentionally disrupt the informal social controls within disadvantaged neighborhoods. They suggest that police in these disadvantaged neigborhoods focus their enforcement efforts on the most dangerous suspects.

Criminal justice, as a topic of study, involves a high level of complexity. First, the study of the justice process involves the examination of social control, which itself is a complex topic. Furthermore, the justice process serves a number of conflicting—and often contradictory—purposes while achieving social control, and is characterized by a wide and expanding variety of agents, agencies, and structures. The immediate task is to develop a perspective that allows us to integrate these many components into a cohesive framework.

PERSPECTIVES ON CRIMINAL JUSTICE

Criminal justice is often distinguished from the related but distinct field of **criminology**. Criminology has been described as the scientific study of law breaking. For example, a criminologist would be interested in uncovering the nature, extent, and causes of crime (Vito & Maahs, 2015). Numerous criminological theories have been proposed through the years, becoming ever more sophisticated as time passes. The study of criminology is very important. In fact, many criminal justice academic programs require a course specializing in criminology. In contrast, criminal justice scholars focus primarily on the *response* to criminal behavior. The focus is on the progress of cases through the justice process. Which offenses and offenders are most likely to become subjects of criminal justice processing, and why? What explains why sometimes you get a traffic ticket, while at other times you receive a warning? How is it that some people get arrested, tried, convicted, and sent to prison, while others seem to "get away with murder"?

Multidisciplinary Nature of Criminal Justice

Different aspects of the criminal justice process have been the topic of study in a variety of social science disciplines. Each discipline contains at least an implicit theory of what "causes" or explains criminal justice. How one views any particular decision in the process depends partly on whether the analyst is trained as a sociologist, psychologist, lawyer, political scientist, economist, or some other profession. An arrest may be seen as an interpersonal interaction, the product of the police officer's perceptions, an exercise of legal authority, a power relation, a rational decision, or something different. In fact, most arrests probably result from a combination of these factors. The study of criminal justice operations in the United States is perhaps best described as multidisciplinary or interdisciplinary (Marenin & Worrall, 1998).

A discipline is a branch of study or learning. Thus, sociology and political science are branches of a more generic area of learning that could be called the study of "human behavior." In earlier years, the fact that programs in criminal justice at colleges and universities tended to include courses in psychology, sociology, law, political science, social work, and other disciplines illustrated the multidisciplinary nature of criminal justice study. Even today, criminal justice programs are often housed in various academic settings, including various social sciences, applied sciences, or business departments (Sloan & Buchwalter, 2017). Box 1.2 briefly describes the approaches that analysts trained in different social science disciplines might prefer when studying justice topics.

BOX 1.2 DISCIPLINARY APPROACHES TO CRIMINAL JUSTICE

Criminal justice professors (and researchers) come from a variety of disciplinary backgrounds. These backgrounds prepare them to approach justice topics and issues from different perspectives:

- *Sociologists* look to the social organization of groups and interactions among people to explain how things occur.

- *Historians* look to larger social and intellectual movements over time to explain how things occur.

- *Psychologists* look to individual motivations and perceptions to explain how things occur.

- *Political scientists* look to the processes of influence and the distribution of power to explain how things occur.

- *Lawyers* look to established legal principles, statutes, and rules to explain how things occur.

- *Economists* look to costs and benefits as an explanation of how things occur.

A full understanding of arrests, criminal penalties, or other parts of the criminal justice system is achieved through the application of several disciplinary approaches. Thus, in studying the arrest decision, the analyst should be aware of the legal, political, rational, perceptual, organizational, and personal factors in operation.

Multidisciplinary approaches remind one of the old story about the blind men meeting an elephant. Each man feels a different part of the beast and concludes that it is something different. The man touching the trunk believes it is a snake; the one with the tail believes it is a horse; and the one at the leg believes he faces a tree. The result is that there are several interpretations of the same phenomenon, each shaped by the unique perspective of the observer. Critics of the interdisciplinary approach believe that the sighted observer would describe the elephant as a large gray or brown beast with a snake-like frontal appendage, a tail, and four large legs. In other words, he would be able to describe the elephant, but not know what it is.

SYSTEMS THEORY AND THE SYSTEMS APPROACH

This book relies on the systems approach of analyzing the criminal justice system. A **system** is a set or collection of interrelated parts working together to achieve a common goal. This perspective views the criminal justice process as a whole comprising the separate, but interrelated parts of law enforcement, courts, and corrections. These parts work together to achieve the goal of crime control in our society.

Walker (1992) observed that the systems perspective is the dominant scientific paradigm of criminal justice. That is, most people studying criminal justice use a systems model to understand the process. Walker suggested that this model came to dominate thinking about criminal justice because of the American Bar Foundation survey of criminal justice in the 1950s. Those who worked on that research project assumed leadership roles in the President's Commission on Law Enforcement and Administration of Justice in the mid-1960s. After intensive study of criminal justice operations, these researchers developed a paradigm (explanatory model) for understanding criminal justice. This paradigm was based on five general observations (Walker, 1992:66–70):

1. Criminal justice is complex, involving much more than law enforcement.
2. The role of the police, as a result, is also very complex, involving more than crime control.
3. The administration of justice is largely discretionary.
4. Discretionary decisions are not well controlled by law or formal rules.
5. The agencies of criminal justice are interrelated and form a system.

These observations have directed the development of criminal justice as a field of inquiry. The focus of criminal justice study became the decision-making processes of agents and agencies of the justice process. Rather than a simple question of law enforcement or the application of rules, each decision in the process was affected, or could be affected, by a variety of forces. Furthermore, the decisions made at one point in the process (e.g., arrest) have implications for later decisions.

What emerged from this orientation was a definition of criminal justice as a complex process of social control in which decisions reflected conflicting goals and expectations. The decisions themselves were variable. Contrary to expectations, an arrest was not solely or even primarily dependent on the existence of sufficient evidence of criminality. In addition, the separate decisions of criminal justice agents and agencies were linked in a sequential fashion, so that the choices of police officers constrained prosecutors, whose choices constrained judges, and so on. The goal of reforming criminal justice processing hinged on the ability to understand and thereby control the decision-making process. Understanding this process seemed to require viewing criminal justice as a system (Conley, 1994).

Types of Systems

Systems are identified and classified in a number of ways (Sutherland, 1975). For our purposes, we need only differentiate between "closed" and "open" systems. These terms refer to the sensitivity of a system to its environment. Closed systems are relatively impervious and insensitive to the environment, whereas those that more freely interact with their environments are open.

A closed system is often self-contained. One simplistic example of a closed system is an astronaut in a space suit. Whether standing on Earth, conducting a spacewalk, or exploring another planet, the astronaut is insulated from the environment. To the degree that it functions regardless of surroundings, the life-support system of a space suit is a closed system.

An **open system** is sensitive to its environment, like a business. Among other things, changes in tax laws, wage rates, markets, environmental protection regulations, shipping rates, or costs of raw materials will affect profits. To remain profitable, a business must constantly adapt not only to internal pressures but also to external or environmental changes. Most organizations operate as open systems.

It is most accurate and useful to classify the justice system as an open system. Clearly, the justice process in American society must react to changes in the economy, population, and political components of its environment. Perhaps less clearly, it must also adapt to changes in social values, ideology, and information. We will see in later chapters how influential the environment of the justice system is in explaining the operations of the justice process. Finally, we must recognize that the criminal justice system influences the broader society.

THE ENVIRONMENT OF CRIMINAL JUSTICE

Having defined the criminal justice process as an open system, we must briefly examine the environmental factors that affect its operations. These factors have a direct impact on all aspects of criminal justice. The environment of criminal justice is both material and ideological. The material environment includes concrete resources such as money, personnel, equipment, and the like. The ideological environment comprises chiefly of values and beliefs about how the justice process should operate. Figure 1.1 illustrates the placement of criminal justice within this environment.

The Material Environment of Criminal Justice

FIGURE 1.1
The Environment of
Criminal Justice.

In simple terms, each system has three stages: (1) input, (2) throughput, and (3) output. In manufacturing, for example, input is the reception of raw materials, throughput is the production process, and output is the final product. For the criminal justice system, criminal offenses are the input, the transformation of crime suspects into convicts is the throughput, and ex-convicts are the output.

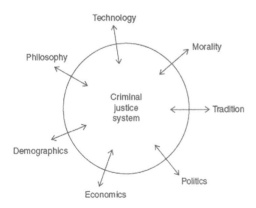

As with manufacturing, the input stage of the criminal justice system involves labor, machinery, and capital. Law enforcement officers, prosecutors, judges, defense attorneys, correctional staff, police cars, courthouses, jails, and even paper clips for reports are parts of the justice system input. In addition, the output of the justice system is not limited to "ex-convicts." Some nonguilty suspects are released at various stages in the process, as are some persons who are guilty yet not convicted of a crime. Many ex-convicts do not retain that label long before being arrested for a new crime. This illustration serves the purpose of identifying the principal material factors in the criminal justice system's environment: raw materials and the means of production.

Raw Materials

The raw material of the criminal justice system consists of criminal offenses. Thus, the system is affected by changes in the nature and distribution of crime. Suppose our "petition-pusher" from the opening part of this chapter is successful and a law is passed that prohibits all drivers from using cell phones while driving. Use of cell phones while driving becomes a new source of "raw material" for the criminal justice system. However, if all criminal laws were repealed, there would be no raw materials for criminal justice.

Unlike the manufacturing firm, the justice system has little control over the volume of raw material it receives. Imagine the effects on a manufacturing plant of deliveries of materials that far exceed the plant's capacity to produce. For example, imagine the delivery of 1 million barrels of crude oil each day to a refinery that can process only 100,000 barrels every 24 hours. Similar situations have occurred in the justice process.

In cases of large crowds or demonstrations, police officers are often instructed to overlook minor violations and concentrate on the maintenance of order. In large measure, this is because large crowds are potentially dangerous, but it is also because there may be no capacity to handle mass arrests. In one massive demonstration in Washington, DC, resulting in thousands of arrests, suspects were held in RFK Stadium until they could be processed. Although this example is perhaps the exception, the abundance of raw material is a long-recognized problem for the justice system. Increasing demands on the justice system may result in lower levels of output. Decker, Varano, and Greene (2007) report that when faced with the demands of providing additional security and dealing with an influx of some 2 million visitors during the 2002 Winter Olympics, routine police enforcement activity by the Salt Lake City Police Department decreased significantly. After the Olympics had ended, police enforcement activity returned to its previous levels. We still do not know what will be the impact of contemporary concerns about the terrorist threat that arose during the 2000s. Unless we expand the size of police agencies, adding counterterrorism responsibility will most likely detract from the size of police work.

One response to the heavy caseloads of criminal justice agencies seeks to alter this aspect of the environment. Proponents of decriminalization would remove certain categories of criminal behavior from the justice system. They suggest that the justice system devotes too many resources to the control of

FIGURE 1.2
Boston Police and Emergency Medical Services personnel stand by during a political convention. In cases of large crowds or demonstrations, police officers are often instructed to overlook minor violations and concentrate on the maintenance of order.

Ellen S. Boyne

essentially harmless or victimless crimes such as vagrancy, public intoxication, and disorderly conduct (see Luna, 2003). See Figure 1.2.

Means of Production

The means of production for the criminal justice system are the personnel, facilities, and equipment of the various justice agencies. Changes in the capacity to process criminal cases will have an effect on the entire justice system. Increases or decreases in the numbers of police, prosecutors, judges, prisons, or other components of the justice system will result in changes in the number of cases processed or in the manner in which cases are handled (Brandl, Chamlin, & Frank, 1995). Zhao, Scheider, and Thurman (2003) found that the increased number of police officers and improved police technology provided by the federal Office of Community Oriented Policing Services grants to local police agencies resulted in increased numbers of arrests over the years.

The actual outcome of changes in the means of production of the criminal justice system is not as important as the fact that the alterations will lead to adaptations in the system. For example, the recent problem of prison and jail crowding has meant that many offenders who would otherwise have been imprisoned have been released early or placed on probation instead of being sent to prison or jail. In addition, the recent trend toward "truth in sentencing" laws, which eliminate or reduce the opportunity for parole and good time credit, has been heralded as a way to "get tough" on crime and reduce sentencing discrepancies. However, these sentences also reduce the incentives for prisoners to obey prison rules, resulting in increased prisoner misconduct

(Bales & Miller, 2012). Thus, a person considering reform in one part of the justice system (more or better police, prosecutors, judges, prisons, or what have you) must be sensitive to the fact that such changes will have a ripple effect on the remainder of the justice process. Indeed, systems theory suggests that fluctuations in the environment will be met with changes in the system to limit the disruption of equilibrium. One of the most salient characteristics of the criminal justice system (and one strongly supporting the use of a systems perspective) is that it is resistant to change (Travis, 1982).

The Ideological Environment of Criminal Justice

As a social institution, and particularly one of social control, perhaps the most important aspect of the environment in which the criminal justice system operates is ideological rather than material. The criminal justice system is rife with value conflicts, political and social controversy, and inefficient organization. These attributes of criminal justice reflect our deep ambivalence about social control.

Due Process vs. Crime Control

Perhaps the most fundamental value conflict characteristic of criminal justice in the United States is that between individual freedom and social regularity. In his discussion of policing, Lundman (1980) determined that this conflict was one between liberty (freedom) and civility (order). As Culbertson (1984:vii) has observed:

> We demand that our police apprehend suspects, that our courts convict the accused, and that our correctional system, in some way, punish the convicted. We demand order. The tasks involved in insuring order would be relatively straightforward were it not for our simultaneous demand that the police, courts and correctional agencies operate within the constraints placed upon them by the law.

Perhaps the best known analysis of this controversy is Packer's (1968) models of the criminal justice system, which he called crime control and due process (see Box 1.3). The **crime control model** of criminal justice would support efficiency with an emphasis on speedy case processing. We would expect an enhancement of police powers to search and arrest, and a relaxation in the rules of evidence to allow relevant information to be presented in court. The emphasis on speed would support plea bargaining, prosecutorial discretion, and mandatory sentences as methods for hastening the disposition of cases. Although consistent with the basic tenets of a democratic society (e.g., no coerced confessions), the crime control system would operate on a presumption of guilt. In contrast, the **due process model** would vigorously protect individual rights. It would put restrictions on searches and arrests without

BOX 1.3 POLICY DILEMMA: STOP AND FRISK

New York City's stop-and-frisk policy is a prime example of the criminal justice system's conflicting goals. The policy allows the police to conduct a "pat-down" search on suspicious individuals in an attempt to find illegal weapons. Many critics of the policy view stop and frisk as a violation of one's rights and inherently discriminatory. In fact, statistics show that minorities are disproportionately targeted for these stops, and only 6% of these stops result in an arrest. Further, only half of those arrested following a frisk search are ultimately convicted of a crime and most of these convictions are for minor crimes such as graffiti and disorderly conduct (Schneiderman, 2013). New York mayor Michael Bloomberg defended the policy, claiming that it reduced crime through proactive policing. When asked about the large numbers of minorities stopped under the policy, the mayor's response was, "If we stopped people based on census numbers, we would stop many fewer criminals, recover many fewer weapons and allow many more violent crimes to take place" (Taylor, 2012).

Those individuals who advocate for a crime control model of criminal justice would likely favor the stop-and-frisk policy—as it might allow the police to effectively repress crime in a quick and efficient manner. Those favoring the due process model, however, would likely have concerns regarding the fairness of the searches.

warrants, require full trials with strict rules of evidence, and support separate sentencing hearings to protect the interests of the individual offender.

With some degree of irony, we can say that America is constitutionally unsuited for criminal justice. Our emphasis on individual liberty and constrained governmental authority requires us to tolerate a certain level of inefficiency in criminal justice. We generally do not allow potentially effective crime control practices such as random wiretaps, warrantless searches, censorship of mail, or the use of "truth serum" during interrogation. We do provide defense counsel, pretrial release, and appellate review of trials and sentences in most cases. In an effort to preserve individual liberty, we not only constrain justice agencies from engaging in many activities, but we also actively impose barriers to the agencies' swift and simple operation. Spader (1987) remarked that criminal justice practice represents a "golden zigzag" between social protection and individual rights.

The criminal justice system is designed for crime control, but the control of crime must be consistent with our social and political heritage. The justice system must achieve a balance between competing values of federalism and uniformity, vengeance and assistance, and differing political persuasions, as well as between individual actors and social regularity. Balancing these opposing forces is part of what makes the justice system so complex.

The tension between our concerns for crime control (order or safety) and due process (limited governmental power or individual liberty) is clearly visible as we struggle to respond to the threat of terrorist attacks. As Lynch (2002:2) put it, "If one examines the history of the federal government's responses to terrorism, a disturbing pattern emerges. The federal government responds to terrorist attacks on U.S. soil—such as the Oklahoma City bombing in 1995—by rushing to restrict civil liberties." From expanded wiretap authority to increased surveillance of citizens, efforts to prevent terrorism often require (or produce) increases in governmental authority and limits on individual liberty.

Value conflicts are also evident as we try to determine what to do for, with, and about criminal offenders. We generally are not content with punishment for punishment's sake (Finckenauer, 1988). Rather, we expect some ultimate "good" to arise from governmental action.

In this vein, Griffiths (1970) suggested a third model of the justice system in opposition to those described by Packer. This third type is the family model. This model assumes that the interests of society and those of the offender are the same. The net effect of criminal justice processing of an offender should be beneficial to both the offender and the society.

The term "family model" is an apt description of the conflict between vengeance and assistance. While a parent may want or need to punish a child's misbehavior, the purpose of the punishment is to correct the child's error and to restore harmony in the family. Thus, actions taken by agents of the justice system are continually compared against two standards:

Has punishment been administered?
Has the offender been "helped"?

Albanese (1996) captured the essence of this conflict in his presidential address to the Academy of Criminal Justice Sciences. As have others before him, Albanese called for a merging of the choices between punishment and rehabilitation into a choice of punishment and rehabilitation. Still, despite decades of observations that we can both punish and help at the same time, current thinking is still dominated by the conception that the two are distinct and opposite.

Depending on the political persuasion of criminal justice policymakers (i.e., liberal or conservative), radically different strategies may be adopted to control crime (Reckless & Allen, 1979). The attitudes, perceptions, and tendencies of criminal justice agents and offenders are important factors in understanding the operations of the criminal justice system. The ability of individual actors to affect criminal justice decisions and processing is known as "discretion."

Individual Discretion. Glueck (1928:480) observed that the criminal justice system was a "clumsy admixture of the oil of discretion and the water of rule." By this he meant that the rule or "law" serves to place constraints on the actions of agents of the criminal justice system, but the system relies on discretion to process cases smoothly. Regardless of the specificity of applicable law, there is always room for "judgment calls." In any specific instance, a decision to arrest, charge a suspect, impose a sentence, or grant release from prison is a judgment call. For example, the discretionary power of police officers is illustrated by our hope for a mere warning when we are caught exceeding the speed limit.

The fact of discretion in criminal justice decision making renders the explanation of specific case decisions very complex. However, an understanding of the forces at work in any given decision sheds light on the process. This chapter has attempted to illustrate these forces.

It is important to realize that discretion is not totally unfettered. Every discretionary decision is made within a context of forces operating at all levels of the justice system previously described. Therefore, the day-to-day workings of the justice system are structured by these larger and more distant factors, which must be kept in balance.

Federalism and Uniformity. One of the major criticisms of the systems approach to criminal justice is that the justice process is decentralized, disorganized, and lacks consistency. Yet, these limitations of the criminal justice system are congruent with two of our social and political values: federalism and the separation of powers.

The basic principle of governmental organization in the United States is that of **federalism**. Our nation is the result of a federation of sovereign states. The U.S. Constitution enumerates the rights and obligations of the federal government, and the Tenth Amendment includes the "reservation clause," which reads:

> The powers not delegated to the United States by the Constitution, nor prohibited by it to the states, are reserved to the states respectively, or to the people.

This amendment enables each state to pass and enforce its own criminal laws, and to create the offices and agencies necessary to perform these tasks. Thus, a federal justice system is created to deal with federal offenses (e.g., counterfeiting), and separate justice systems are created to enable each state to deal with state crimes (e.g., theft).

States, in turn, have constitutions under which they charter municipalities. These counties, cities, towns, and villages then provide for their own criminal offenses and operate offices and agencies to enforce local and state laws. This organizational structure of government ensures **local autonomy**, so that the citizens of each state and community have a fairly large degree of freedom from central control.

For the justice system, the result is thousands of police agencies at federal, state, and municipal levels; thousands of jails, courts, probation agencies, prosecutors, and defense offices; and scores of prison and parole agencies. It also results in differences in the definitions of crimes and the levels of punishments applicable to criminal behavior. Variety is central to criminal justice in the United States.

The Constitution of the United States also creates and maintains a **separation of powers** between the executive, judicial, and legislative branches of government. In simplistic terms, the legislature makes the law, the judiciary interprets the law, and the executive enforces the law. Each branch of government is checked and balanced by the other two branches. This tripartite governmental structure exists at the federal, state, and municipal levels of government.

This complex organization of the crime control function in America causes inefficiency. Yet, to preserve our interests in local autonomy and constrained governmental power, we must tolerate the inefficient organization of governmental service (Forst, 1977). Stolz (2002:52) observed:

> In the United States, governmental authority is constitutionally distributed among three levels of government—the federal, fifty state, and thousands of local governments …. Moreover, at each level of government, policy making authority is shared among three branches—the executive, legislative, and judicial …. These are the formal institutions of government and within these institutions the formal processes of government are carried out. In the criminal justice area, most policy making occurs at the local or state level but, particularly since the 1960s, the role of the federal government has expanded.

In short, a centralized, uniform system of criminal justice would be unconstitutional. Any efforts to understand the justice system and to promote consistency and simplicity in organization and in the processing of criminal cases must be sensitive to the values our society places on federalism. Effectiveness and efficiency of operation in the criminal justice system are not the only goals to be considered when analyzing its structure and operation.

CAREERS

The interdisciplinary nature of criminal justice, perhaps with the assistance of numerous crime-related television shows, has increased interest in the field of criminal justice. The ultimate goal of many reading this book will be to gain employment in the field of criminal justice. Luckily, the criminal justice field offers a great variety of employment opportunities. Employment options can be found in both the public and private sectors. Private-sector jobs include positions such as a loss prevention specialist, professional security specialist, or a fraud investigator (Tripp & Cobkit, 2013). However, most jobs within the criminal justice field exist under the umbrella of a local, state, or federal government agency. Job requirements and salaries typically vary widely depending upon the geographical location of the job as well as the particular level of government. In general, local- and state-level jobs within the criminal justice field will earn a lower salary than federal careers. However, federal agencies

BOX 1.4 SAMPLE OF CRIMINAL JUSTICE CAREER OPPORTUNITIES

Air Marshal	Detective
Arson and Fire Investigator	FBI Special Agent
ATF Agent	FBI Intelligence Analyst
Bailiff	Fish and Game Warden
Border Patrol Agent	Forensic Psychologist
Bounty Hunter	Forensic Science Technician
CIA Agent	Fraud Investigator
Correctional Treatment Specialist	Immigration Enforcement Agent
Corrections Officer	IRS Special Agent
Criminalist	Juvenile Probation Officer
Criminal Investigator	Loss Prevention Specialist
Crime Analyst	NSA Police Officer
Crime Laboratory Analyst	Park Ranger
Crime Prevention Specialist	Police Officer
Crime Scene Investigator	Private Security Officer
Crime Scene Technician	Probation Officer
Customs Agent	Secret Service Agent
DEA Agent	Security Officer

recruit from a much larger applicant pool and are thus much more selective. Employment opportunities are most often found within one of the three components of the justice system: (1) policing, (2) courts, and (3) corrections. Box 1.4 includes a sample of criminal justice employment opportunities.

Policing

Law enforcement is the most commonly cited career path among criminal justice college students. Dretsch, Moore, Campbell, and Dretsch (2014) found that 56% of criminal justice students desired a career in law enforcement. Students are most attracted to the federal law enforcement agencies, such as the Drug Enforcement Agency; Central Intelligence Agency; Alcohol, Tobacco, Firearms, and Explosives (ATF); Federal Bureau of Investigation (FBI); and U.S. Customs (Courtright & Mackey, 2004; Johnson & White, 2002). Other law enforcement career opportunities exist at the city, county, or state levels. Many local and state police agencies do not require applicants to possess a college degree, but most do require that applicants maintain a clean criminal record and often pass physical agility, psychological, and other preemployment screening.

FIGURE 1.3
An FBI Joint Terrorism Task Force member enters a home in Harrisburg, Pennsylvania, where Jalil Ibn Ameer Aziz was arrested. Aziz was eventually sentenced to 160 months of imprisonment and 12 years of supervised release for conspiracy to provide material support and resources to a designated foreign terrorist organization and transmitting a communication containing a threat to injure.

James Robinson/PennLive. com via AP

Applicants to federal law enforcement jobs typically have previous work experience and perhaps specialized experience or a proficiency in a foreign language. It is important to note that while the particular federal agencies listed above are the most popular, they are by no means the only opportunities. In fact, federal law enforcement officers are dispersed among approximately 100 federal departments, independent agencies, and subagencies (Walker, Burns, Bumgarner, & Bratina, 2008). See Figure 1.3.

Courts

Careers in the court system range from a bailiff (responsible for court security) to judgeships. Among the traditional faces of the court system are the various attorneys involved in each criminal case. As will be discussed in Chapter 8, criminal law is not one of the most desirable specializations among law school graduates. A primary reason for this lack of desirability is the relatively low income compared to other law specialties. For example, minimum entry-level salaries for assistant public defenders ranged from about $37,000 to $58,000, with a median salary of $46,000 per year (Langton & Farole, 2010). The salary of a prosecutor is generally higher, but still not great in comparison to potential private practice earnings. The median salary for chief prosecutors in all jurisdictions in 2007 was $98,000. The salary is influenced by the size of the office; the median salary was $165,700 per year in the largest jurisdictions. Many offices also employ assistant prosecutors (also known as assistant district attorneys). Assistant prosecutors with no prior experience earn between $33,000 and $51,000, depending on the size of the jurisdiction (Perry & Banks, 2011).

Judges typically earn the highest income of court employees. By mid-2015, the median salaries for judicial officials in state courts were $167,210 for chief justices of the highest court, $159,484 for intermediate appellate court justices, and $146,803 for general trial court judges (National Center for State Courts, 2016). Interestingly, recent anecdotal evidence suggests that these income levels are not high enough to either attract the best lawyers into pursuing judgeships or to retain quality judges who could make higher incomes in private practice (Hann, 2016). For example, Burbank, Plager, and Ablavsky (2012) examined the reasons that federal judges resigned from 1970 to 2009. The most common motivating factors cited by the judges were "return to private practice," "other employment," or "inadequate salary" (p. 15).

Corrections

The corrections system of the criminal justice field has a wide variety of employment opportunities. Career opportunities include being a correctional officer, correctional counselor, or a probation/parole officer. Jails and prisons often recruit individuals to help supervise inmates. In fact, correctional officers are one of the largest classifications of state employees across the country. In 2017, the median annual salary for correctional officers was $43,510 (Bureau of Labor Statistics, 2017a). Jail officers often are recruited in the same way as police officers, which is through local searches and civil service testing. Like jail officers, correctional officers in prisons are typically selected through civil service and are not particularly well paid.

Unlike police or correctional officers, it is common for a probation officer to be required to have a college degree. Recruitment of probation officers tends to be local, consistent with the court's jurisdiction. In 2017, the Bureau of Labor Statistics (2017b) reported that the median income for probation officers and correctional treatment specialists was $51,410. A parole officer is often required to have a college education and to perform duties similar to those of a probation officer, except that a parole officer typically has a smaller caseload comprised of ex-inmates. Parole officers are most often selected from statewide pools through civil service procedures.

ACADEMIC CRIMINAL JUSTICE

The study of criminal justice in an academic setting has evolved over time. Formal criminal justice education in the United States began in the 1920s and 1930s as an attempt to increase the professionalism of police departments. Many of these early criminal justice courses began and existed for many years as subspecialties of more established disciplines, such as sociology and law. These early courses resembled police academies. In fact, early criminal justice education was often much more focused on training than education (Finckenauer,

2005). As Eskridge (2003) pointed out, there are important distinctions between training and education. Training is job-specific, street-specific instruction that prepares a criminal justice professional to face his or her daily challenges. Education, on the other hand, is more of a long-term function aimed at developing a spirit of exploration, developing the academic tools (ability to read and write critically), and developing an introductory knowledge base of how the structure and process of the justice system works.

Criminal justice education began to expand and become more professional with the publication of the 1967 President's Commission on Law Enforcement and Administration of Justice. This report encouraged the increased education of criminal justice professionals as a way to improve the system. The recommendation led to government action in the form of a federal program which helped fund the education of students interested in a law enforcement field or to those currently working as a law enforcement officer but wanting to return to college (Flanagan, 2000). In 2014, *USA Today* ranked criminal justice and corrections as the sixth most popular major for college students in the United States (Stockwell, 2014).

Colleges and universities vary dramatically in the types of criminal justice programs offered. Many criminal justice programs still exist as subspecialties of other disciplines, most often sociology. However, contemporary criminal justice programs are often separate departments, with faculty specializing in distinct areas of the criminal justice system (Wrede & Featherstone, 2012). Some programs offer curriculums that are vocational-oriented, whereas others are more focused on theory and research (Birzer & Palmiotto, 2002). Regardless of the orientation of the specific program, the exposure to and application of social science research as it relates to criminal justice has been increasing. This is evident in the movement toward evidence-based policy. Proponents of evidence-based crime policy (e.g., Walker, 2015) point to the American tradition of creating policy based on what is politically popular at the time. Unfortunately, these policies are often either ineffective or actually harmful to society and the criminal justice system. Scientific research helps policymakers better understand the criminal justice system and thus make more informed policy decisions. Academic scholars, government agencies, and even nonprofit groups conduct research that provides us a clearer picture of the criminal justice process and the effectiveness of our current crime prevention policies.

EXAMINING CRIMINAL JUSTICE

So far this chapter has served as a basic introduction to the study of criminal justice in the United States. The remainder of this book will explore criminal justice practices, agents, and agencies, building on what has been described here. Three themes emerging from this introduction will guide our examination of the U.S. system of criminal justice. The first is the notion of a systems approach to understanding the operations of the justice process. Criminal justice is part of the system of social control in American society, as well as part of the larger society. Changes in the environment (materials, ideas, values, etc.) will influence the justice process. A second theme is that there is a fundamental conflict between individual liberty and collective needs for predictability. The criminal justice system, each of its decision points, and all of the decision makers involved in the system must strike a balance between the interests of the individual citizen and the interests of the community. Finally, the existence of discretion in the justice system is the third theme in our approach to studying criminal justice. In most cases, criminal justice agents (police, prosecutors, judges, corrections officials, and others) have some latitude in deciding what to do about offenses and offenders. Much of our attention will be devoted to identifying what sorts of factors help us to understand the kinds of decisions that are made.

As we progress in our examination of criminal justice in the United States, we will describe the justice system and its structure, organizations, and agents. We will investigate the range of decisions that are made in cases at each stage of the justice process, and explore the factors that are associated with different decisions. Finally, we will try to place things into the larger context of seeking a balance between due process and crime control.

See Box 1.5 for a chart that seeks to present a simple yet comprehensive view of the movement of cases through the criminal justice system.

BOX 1.5 THE CRIMINAL JUSTICE SYSTEM

Procedures in individual jurisdictions may vary from the pattern shown here. The differing weights of the lines indicate the relative volumes of cases disposed of at various points in the system. This information, however, is only suggestive because no nationwide data of this sort exist.

1. May continue until trial.

2. Administrative record of arrest. First step at which temporary release on bail may be available.

3. Before magistrate, commissioner, or justice of peace. Formal notice of charge, advice of rights. Bail set. Summary trials for petty offenses usually conducted here without further processing.

4. Preliminary testing of evidence against defendant. Charge may be reduced. No separate preliminary hearing for misdemeanors in some systems.

5. Charge filed by prosecutor on basis of information submitted by police or citizens. Alternative to grand jury indictment; often used in felonies, almost always in misdemeanors.

6. Reviews whether government evidence sufficient to justify trial. Some states have no grand jury system; others seldom use it.

7. Appearance for plea; defendant elects trial by judge or jury (if available); counsel for indigent usually appointed here in felonies. Often not at all in other cases.

8. Charge may be reduced at any time prior to trial in return for plea of guilty or for other reasons.

9. Challenge on constitutional grounds to legality of detention. May be sought at any point in the process.

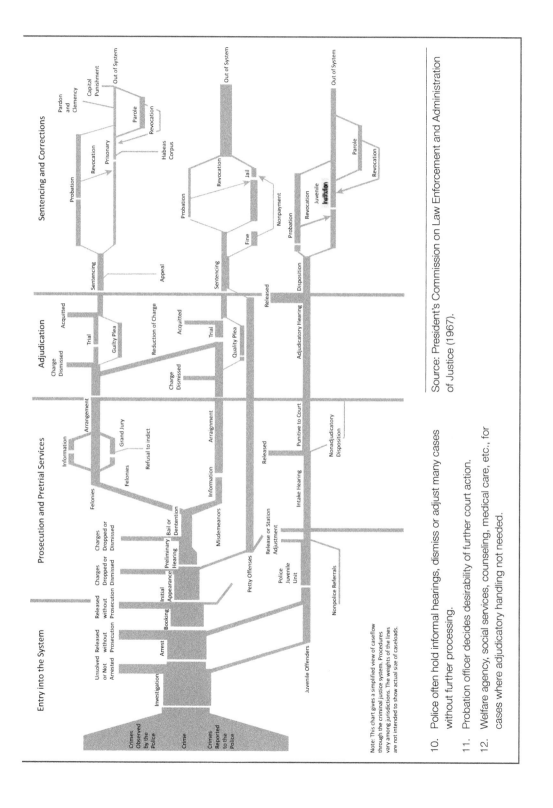

Entry into the System

Prosecution and Pretrial Services

Adjudication

Sentencing and Corrections

Note: This chart gives a simplified view of caseflow through the criminal justice system. Procedures vary among jurisdictions. The weights of the lines are not intended to show actual size of caseloads.

Source: President's Commission on Law Enforcement and Administration of Justice (1967).

10. Police often hold informal hearings, dismiss or adjust many cases without further processing.

11. Probation officer decides desirability of further court action.

12. Welfare agency, social services, counseling, medical care, etc., for cases where adjudicatory handling not needed.

PREVIEW OF FORTHCOMING CHAPTERS

In the chapters to follow, we will examine the criminal justice system. The first four chapters set the stage for analyzing criminal justice operations. This chapter has provided a foundation of criminal justice perspectives. Chapter 2 presents an overview of the operations and structure of the criminal justice system of the United States. Chapter 3 includes a discussion of law and a description of some recent changes in criminal justice, and illustrates how the system reflects changes in our thinking about crime and criminals. In Chapter 4, sources of data on the nature and extent of crime are reviewed, and an overview of the way in which cases are "selected" for justice processing is provided. The next ten chapters address the subsystems of the criminal justice system, from the detection of crime through investigation and arrest; to initial appearance in court; through formal charging, trial, and conviction; and finally to sentencing and the goals of criminal penalties. Incarceration and community-centered punishments are described and discussed, and the juvenile justice system is evaluated in a separate chapter. The last chapter presents a discussion of systemwide developments and issues and the future of criminal justice in the United States.

REVIEW QUESTIONS

1. What is the purpose of the criminal justice system?
2. What is social control?
3. What is a theory?
4. Define system and identify at least two types.
5. What components comprise the material and ideological environments of the criminal justice system?
6. What is the relationship between support for crime control and support for due process?
7. What is evidence-based policy?

REFERENCES

Albanese, J. (1996). Presidential address: Five fundamental mistakes of criminal justice. *Justice Quarterly, 13*(4), 549–565.

Austin, J., Cadora, E., Clear, T.R., Dansky, K., Greene, J., Gupta, V., et al. (2013). *Ending mass incarceration: Charting a new justice reinvestment.* Washington, DC: The Sentencing Project.

Bales, W., & Miller, C. (2012). The impact of determinant sentencing on prisoner misconduct. *Journal of Criminal Justice, 40*, 394–403.

Bianchi, H. (1994). *Justice as sanctuary: Toward a new system of crime control.* Bloomington, IN: Indiana University Press.

Birzer, M., & Palmiotto, M. (2002). Criminal justice education: Where have we been? And where are we heading? *The Justice Professional, 15*(3), 203–211.

Black, D. (1976). *The behavior of law.* New York: Academic Press.

Boyd, L. (2017, Jan.). President's message. *ACJS Today, 42*(1). Retrieved from https://c.ymcdn.com/sites/acjs.site-ym.com/resource/resmgr/acjstoday/January_ACJS_Today_2017.pdf

Brandl, S., Chamlin, M., & Frank, J. (1995). Aggregation bias and the capacity for formal crime control: The determinants of total and disaggregated police force size in Milwaukee, 1934–1987. *Justice Quarterly, 12*(3), 543–562.

Burbank, S.B., Plager, S.J., & Ablavsky, G. (2012). Leaving the bench, 1970–2009: The choices federal judges make, what influences those choices, and their consequences. *University of Pennsylvania Law Review, 161*(1), 1–102.

Bureau of Labor Statistics (2017a). Correctional officers and bailiffs. *Occupational outlook handbook.* Retrieved from www.bls.gov/ooh/protective-service/correctional-officers.htm

Bureau of Labor Statistics (2017b). Probation officers and correctional treatment specialists. *Occupational outlook handbook.* Retrieved from www.bls.gov/ooh/community-and-social-service/probation-officers-and-correctional-treatment-specialists.htm

Cohen, A. (1966). *Deviance and control.* Englewood Cliffs, NJ: Prentice Hall.

Conley, J. (Ed.). (1994). *The 1967 President's Crime Commission Report: Its impact 25 years later.* Cincinnati, OH: Anderson.

Courtright, K.E., & Mackey, D.A. (2004). Job desirability among criminal justice majors: Exploring relationships between personal characteristics and occupational attractiveness. *Journal of Criminal Justice Education, 15*(2), 311–326.

Culbertson, R.G. (Ed.). (1984). *Order under law: Readings in criminal justice* (2nd ed.). Prospect Heights, IL: Waveland Press.

Decker, S., Varano, S., & Greene, J. (2007). Routine crime in exceptional times: The impact of the 2002 winter Olympics on citizen demand for police services. *Journal of Criminal Justice, 35*(1), 89–101.

Dretsch, E., Moore, R., Campbell, J.N., & Dretsch, M.N. (2014). Does institution type predict students' desires to pursue law enforcement careers? *Journal of Criminal Justice Education, 25*(3), 304–320.

Eskridge, C. (2003). Criminal justice education and its potential impact on the sociopolitical–economic climate of central European nations: A short essay. *Journal of Criminal Justice Education, 14*(1), 105–118.

Finckenauer, J.O. (1988). Public support for the death penalty: Retribution as just deserts or retribution as revenge. *Justice Quarterly, 5*(1), 81–100.

Finckenauer, J.O. (2005). The quest for quality in criminal justice education. *Justice Quarterly, 22*(4), 413–426.

Flanagan, T. (2000). Liberal education and the criminal justice major. *Journal of Criminal Justice Education, 11*(1), 1–13.

Flexon, J.L., Meldrum, R.C., Young, J.T.N., & Lehmann, P.S. (2016). Low self-control and the dark triad: Disentangling the predictive power of personality traits on young adult substance use, offending and victimization. *Journal of Criminal Justice, 46*, 159–169.

Forst, M.L. (1977). To what extent should the criminal justice system be a system? *Crime & Delinquency, 23*(4), 403.

Friel, C. (2000). A century of changing boundaries. In C. Friel (Ed.), *Criminal justice 2000: Boundary changes in criminal justice organizations: Vol. 2.* (pp. 1–17). Washington, DC: National Institute of Justice.

Gaeta, T. (2010). "Catch" and release: Procedural unfairness on primetime television and the perceived legitimacy of the law. *Journal of Criminal Law & Criminology, 100*(2), 523–553.

Glueck, S. (1928). Principles of a rational penal code. *Harvard University Law Review, 41*, 453.

Griffiths, J. (1970). Ideology in criminal procedure or a third "model" of the criminal process. *Yale Law Journal, 79*, 359.

Hann, J. (2016). *A comparative analysis of judicial compensation in the United States and Canada: Facts, figures and comparisons.* Williamsburg, VA: National Center for State Courts.

Hoskins, D.H. (2014). Consequences of parenting on adolescent outcomes. *Societies, 4*, 506–531.

Johnson, K., & White, J. (2002). The use of multiple intelligences in criminal justice education. *Journal of Criminal Justice Education, 13*(2), 369–386.

Kellogg, F.R. (1976). Organizing the criminal justice system: A look at operative objectives. *Federal Probation, 40*(2), 50.

Langton, L., & Farole, D., Jr. (2010). *State public defender programs, 2007.* Washington, DC: Bureau of Justice Statistics.

Luna, E. (2003). Overextending the criminal law. *Cato Policy Report, 25*(6), 1, 15–16.

Lundman, R.A. (1980). *Police and policing: An introduction.* New York: Holt, Rinehart, & Winston.

Lynch, T. (2002). *Breaking the vicious cycle: Preserving our liberties while fighting terrorism: Executive summary. Policy Analysis, No. 443* (June). Washington, DC: CATO Institute.

Marenin, D., & Worrall, J. (1998). Criminal justice: Portrait of a discipline in process. *Journal of Criminal Justice, 26*(6), 465–480.

Markowitz, F. (2006). Psychiatric hospital capacity, homelessness, and crime and arrest rates. *Criminology, 44*, 45–72.

Martin, A., Wright, E.M., & Steiner, B. (2016). Formal controls, neighborhood disadvantage, and violent crime in U.S. cities: Examining (un)intended consequences. *Journal of Criminal Justice, 44*, 58–65.

Mazerolle, L., Antrobus, E., Bennett, S., & Tyler, T.R. (2013). Shaping citizen perceptions of police legitimacy: A randomized field trial of procedural justice. *Criminology, 51*(1), 33–64.

National Center for State Courts (2016). *Survey of judicial salaries.* Retrieved from www.ncsc. org/FlashMicrosites/JudicialSalaryReview/2015/resources/CurrentJudicialSalaries.pdf

Packer, H.L. (1968). *The limits of the criminal sanction.* Stanford, CA: Stanford University Press.

Parsons, T. (1966). *Societies: Evolutionary and comparative perspectives.* Englewood Cliffs, NJ: Prentice Hall.

Perry, S.W., & Banks, D. (2011). *Prosecutors in state courts, 2007– Statistical tables.* Washington, DC: Bureau of Justice Statistics.

Pound, R. (1929). *Criminal justice in America.* New York: Henry Holt.

President's Commission on Law Enforcement and Administration of Justice (1967). *The challenge of crime in a free society.* New York: Avon.

Reckless, W.C., & Allen, H.E. (1979). Developing a national crime policy: The impact of politics on crime in America. In E. Sagarin (Ed.), *Criminology: New concerns.* Beverly Hills, CA: Sage.

Reisig, M.D., Wolfe, S.E., & Holtfreter, K. (2011). Legal cynicism, legitimacy, and criminal offending: The nonconfounding effect of low self-control. *Criminal Justice and Behavior, 38*(12), 1265–1279.

Ross, E.A. (1926). *Social control: A survey of the foundations of order.* New York: Macmillan.

Schneiderman, E. (2013). *A report on arrests arising from the New York City Police Department's stop-and-frisk practices.* New York: New York State Office of the Attorney General.

Sloan, J.J., III, & Buchwalter, J.W. (2017). The state of criminal justice bachelor's degree programs in the United States: Institutional, departmental, and curricula features. *Journal of Criminal Justice Education, 28*(3), 307–334.

Spader, D.J. (1987). Individual rights vs. social utility: The search for the golden zigzag between conflicting fundamental values. *Journal of Criminal Justice, 15*(2), 121–136.

Stockwell, C. (2014, Oct. 26). Same as it ever was: Top 10 most popular college majors. *USA Today.* Retrieved from http://college.usatoday.com/2014/10/26/same-as-it-ever-was-top-10-most-popular-college-majors/

Stolz, B.A. (2002). The roles of interest groups in U.S. criminal justice policy making: Who, when, and how? *Criminal Justice, 2*(1), 69.

Subramanian, R., Moreno, R., & Gebreselassie, S. (2014). *Relief in sight?: States rethink the collateral consequences of criminal conviction, 2009–2014.* New York: Vera Institute of Justice.

Supancic, M., & Willis, C.L. (1998). Extralegal justice and crime control. *Journal of Crime and Justice, 21*(2), 191–215.

Sutherland, J.W. (1975). *Systems: Analysis, administration, and architecture.* New York: Van Nostrand Reinhold.

Taylor, K. (2012, June 10). Stop-and-frisk policy saves lives, mayor tells black congregation. *The New York Times.* Retrieved December 19, 2013, from www.nytimes.com/2012/06/ll/nyregion/at-black-church-in-brooklyn-bloomberg-defends-stop-and-frisk-policy.html

Travis, L.F., III. (1982). The politics of sentencing reform. In M.L. Forst (Ed.), *Sentencing reform* (p. 59). Beverly Hills, CA: Sage.

Travis, L., & Langworthy, R. (2008). *Policing in America: A balance of forces* (4th ed.). Upper Saddle River, NJ: Prentice Hall.

Tripp, T., & Cobkit, S. (2013). Unexpected pathways: Criminal justice career options in the private sector. *Journal of Criminal Justice Education, 24*(4), 478–494.

Vito, G.F., & Maahs, J.R. (2015). *Criminology: Theory, research, and policy* (4th ed.). Burlington, MA: Jones & Bartlett.

Walker, J., Burns, R., Bumgarner, J., & Bratina, M. (2008). Federal law enforcement careers: Laying the groundwork. *Journal of Criminal Justice Education, 19,* 110–135.

Walker, S. (1992). Origins of the contemporary criminal justice paradigm: The American Bar Foundation survey, 1953–1969. *Justice Quarterly, 3*(1), 47–76.

Walker, S. (2015). *Sense and nonsense about crime, drugs, and communities* (8th ed.). Stamford, CT: Cengage.

Wrede, C., & Featherstone, R. (2012). Striking out on its own: The divergence of criminology and criminal justice from sociology. *Journal of Criminal Justice Education, 23*(1), 103–125.

Zhao, J., Scheider, M., & Thurman, Q. (2003). A national evaluation of the effect of COPS grants on police productivity (arrests) 1995–1999. *Police Quarterly, 6*(4), 387–409.

CHAPTER 2

The Justice Process

Cases move through the justice system from the first stage of detection by police through subsequent stages to final discharge from the system. Although there are some feedback mechanisms by which a case can move back to an earlier decision point, on the whole, cases flow in one direction through the system. This processing of cases represents the "total system" of criminal justice. It includes the subsystems of law enforcement, the courts, and corrections.

In this chapter, we trace the criminal justice system of the United States. In doing so, we skip many of the details and nuances of criminal justice processing in the interests of developing an understanding of the total justice system. In other words, to some extent we ignore the trees to get a better look at the forest. Later chapters will examine the subsystems of criminal justice in more detail.

Perhaps the greatest constant of criminal justice is variety. Even things as simple as titles differ among jurisdictions. For example, prosecutors are known as state's attorneys, district attorneys, U.S. attorneys, prosecutors, and other titles. In most states, the highest court is called the state supreme court; in New York, the supreme court is a trial court, and the highest court is the New York Court of Appeals. With an appreciation that what follows here is a sketch of the justice system, we are ready to proceed.

THE DECISION POINTS OF THE CRIMINAL JUSTICE SYSTEM

The criminal justice system begins with the detection of crime, proceeds through investigation, arrest, initial appearance before the court, charging (arraignment), trial, sentencing, and possible revocation, and ends with discharge. We will examine these decision points.

Detection

As the formal social institution charged with the control of deviance that is identified as criminal, the justice system is not mobilized until a criminal

IMPORTANT TERMS

arraignment
arrest
circuits
conditional release
discharge
drug courts
initial appearance
investigation
preliminary hearing
private court
revocation
sentencing
trial
undetected crime
unfounded
unreported crime
unsolved

offense is detected. Crime that goes undetected does not influence the justice process directly. The process begins only when the justice system (usually through the police) notices a possible criminal offense.

Probably less than half of all crime is discovered by the justice system (Baumer & Lauritsen, 2010). Many crimes remain undetected because no one realizes that a crime was committed. Many others are detected but are not reported to the police, so that the justice system is not aware that criminal offenses have occurred.

Have you ever reached into your pocket or wallet for money you knew you had, only to discover that it was missing? Most of us at some time have experienced missing money. We cannot be certain that we did not spend it or lose it, but we also cannot remember when it was spent. Have we been the victims of theft? Do we report the money as stolen?

If we assume that we spent or lost the money and do not believe it was stolen, a theft may go undetected. Similarly, if we are convinced the money was stolen, we may still not report it because the sum is so small and the chance of recovery is so slim. In the latter case, a crime has gone unreported. **Undetected crime** is crime that is not known to the criminal justice system or to the victim—crimes that are not recognized as crimes. An **unreported crime** is one that victims recognize as lawbreaking behavior but is not brought to the attention of authorities.

If a person has a fight with a friend or a relative and assumes it is "personal," an assault may go undetected or at least unreported. The first decision to influence the criminal justice process is determining whether a crime may have occurred. Citizens rather than criminal justice officials most often make this decision. A second decision is reporting a crime; again, someone other than a criminal justice official most often makes this decision (Avakame, Fyfe, & McCoy, 1999). Recent surveys of crime victims indicate that only about 42% of violent victimizations are reported to the police, compared to approximately 35% of property crimes (Morgan & Kena, 2017).

Researchers have increasingly studied the reasons that influence the likelihood of crime reporting. It appears that several social and individual factors influence the likelihood of crime reporting. For example, recent studies involving identity theft (Reyns & Randa, 2017) and stalking victims (Reyns & Englebrecht, 2014) have found that the victims' perceived seriousness of the crime is one of the most influential factors determining whether victims will report the crime. Berg, Slocum, and Loeber (2013) also noted that many victims are also involved in criminal activity. They found that victims who were actively engaging in criminal behavior were less likely to report crimes to the police. Neighborhood context also appeared to influence crime reporting, as victims who lived in high-crime neighborhoods were less likely to report their victimization to the police.

Nonreporting of crime limits the ability of criminal justice agents and agencies to respond to crime. Table 2.1 presents the frequency with which different types of crimes are reported to the police, and Table 2.2 describes reasons typically given by people for not reporting violent crimes.

When a crime or suspected crime is reported to the police, the justice system is mobilized. If agents of the justice system decide that a crime has occurred, they have made the detection decision. The police respond to the report of a crime. It is then that case decision making rests with official agents of the justice process. Once the police come to believe that a crime may have been committed, it is their decision whether and how to proceed. We can say that the criminal justice system starts when justice system officials (usually the police) believe a crime has occurred. At that point, the agents of the justice system take control over the official societal response to the crime.

Investigation

Upon deciding that a crime may have been committed, the next decision is whether to investigate, and if so, how thoroughly to investigate. An investigation is the search for evidence that links a specific person to a specific crime. It is a process in which the results of initial inquiries often determine the intensity of the investigation. If, for example, someone reports a prowler, the responding officers may make a visual check of doors and windows, find nothing suspicious, and leave. Alternatively, they may note footprints near a window or find scratch marks on a door or window frame, and then intensify their investigation.

At the conclusion of the investigation, three outcomes are possible. First, police may find no evidence of criminal activity and classify the possible crime as **unfounded**, or not real. Second, evidence of possible criminal activity may support the finding that a crime was committed or attempted, but there is not sufficient evidence for an arrest. In this case, the crime will be left **unsolved** (i.e., no offender is known), and the investigation, at least theoretically, will continue. Finally, the investigation may yield evidence of both a crime and a probable guilty party. In the last outcome, we reach the next decision stage: arrest.

TABLE 2.1 Percent of Crimes Reported to Police by Type of Crime, 2016

Type of Crime	Percent Reported
Violent crimes	42.1
Rape/sexual assault	22.9
Robbery	54.0
Aggravated assault	58.5
Simple assault	37.5
Property crimes	37.5
Household burglary	49.7
Motor vehicle theft	79.9
Theft	29.7

Source: Morgan and Kena (2017).

TABLE 2.2 Reasons Given for Not Reporting Violent Crimes

Reason	Percent Giving Reason
Fear of reprisal or getting offender in trouble	13
Police would not or could not help	16
Not important enough	18
Dealt with in another way/personal matter	34
Other reasons	18

Source: Langton, Berzofsky, Krebs, and Smiley-McDonald (2012).

Arrest

Despite expectations, media portrayals, or legal mandates, police officers do not have to arrest every violator of the criminal law. The police officer makes a decision whether to arrest a suspected offender. Many factors affect the arrest decision.

Perhaps the two most important factors that determine whether police make an **arrest**—that is, take a person into custody—are (1) the seriousness of the suspected offense and (2) the quality of the evidence against the suspect. The officer can exercise tremendous discretion in this decision, especially for less serious offenses. For example, if a traffic officer stops you for speeding, a citation is not the only possible outcome, even if you actually were speeding. How often does a person give the officer excuses for his or her violation of the traffic laws? How does a person feel about the officer who issues a citation when he or she knows that the officer could have instead opted for giving a warning?

Discretionary decisions not to arrest are often the result of an officer's attempts to achieve street justice. "Street justice" is a term used to describe attempts by police to deal with problems without formal processing. For example, an officer may counsel or warn loitering juveniles, rather than arresting them. In these cases, the officer tries to solve the problem in a way that avoids the negative consequences of formal processing. As we will see in our discussion of the police, much police work is problem solving, and arrest is one of many tools police use to solve problems. Many times, however, police officers do decide to arrest a suspect. If police make an arrest, the next decision stage is reached: initial appearance.

Initial Appearance

Persons arrested for crimes are entitled to a hearing in court to determine whether they will be released pending further action. This **initial appearance** or hearing occurs relatively quickly after arrest, usually within a matter of hours. The hearing does not involve a determination of guilt, but rather an assessment of the defendant's likelihood of appearing at later proceedings. Arrested suspects are usually entitled to release before trial. With the exception of a few very serious crimes (murder, terrorism, kidnapping, etc.) specified in some statutes, arrested persons may be released while awaiting trial. Traditionally, this release happens by the posting of bail.

The primary purpose of bail is to ensure that the suspect will return to court for later hearings. The theory of bail is that a person will return to court if it would cost too much not to return. Traditional bail involves the defendant "posting bond," or leaving money on deposit at the court. If the defendant returns, the defendant gets the money back. If the defendant does not return for the next hearing, the court keeps the bail money and issues a warrant for his or her arrest.

Since the 1970s, criminal justice reforms have witnessed the rebirth of "release on recognizance" systems whereby suspects obtain pretrial release without posting bond as long as they have a job, house, family, and other ties to the community. If we expect a person to appear in court to avoid losing a few thousand dollars, it seems reasonable that he or she would also appear to keep a home, job, or family ties.

In many jurisdictions, it is possible for the prosecutor to ask for "preventive detention." In these cases, the prosecutor believes that, if released on bail, the defendant will present a danger of continued crime in the community. Upon a hearing that establishes that the defendant is indeed dangerous, the magistrate can deny pretrial release.

In many courts, bail schedules exist that link different bail amounts to different types of crime. For instance, the rate for burglary might be $5,000, but for robbery, $10,000. The bail decision, however, is not automatic. If the magistrate believes that the suspect will flee or fail to appear for later hearings, a higher bail may be set. In other cases, a lower bail than usual may be set to allow the defendant to keep his or her job or to maintain family contacts. In either case, after the initial appearance, the next decision relates to the justification for governmental (i.e., justice system) intervention in the life of the citizen.

Charging

Between the time of arrest and arraignment, the prosecutor reviews the evidence in the case and determines a formal criminal charge. The offense for which a person is arrested is not necessarily the one with which he or she will be charged. For example, the police may arrest someone for armed robbery, but be unable to prove that a weapon was used in the crime. The prosecutor may then formally charge the offender with traditional (unarmed) robbery.

Charges are brought in two principal ways: indictment by grand jury or by information. With an indictment, the prosecutor presents the case in secret to a grand jury, which decides whether the evidence is strong enough to warrant the issuance of an indictment. With an indictment, the prosecutor presents the case in open court before a magistrate, who determines if the evidence is sufficient to warrant a formal charge.

In the information process, a judge reviews the strength of the evidence against a suspect and decides if it is sufficient to have the defendant "bound over" to the felony court. Although not a determination of guilt or innocence, the **preliminary hearing** involves a judge ruling on the strength of the case against the defendant. Although the defendant ultimately may be found not guilty, if the available evidence supports probable cause to believe the defendant may be found guilty, the judge will typically order the case bound over to trial, allowing the state to continue.

About a quarter of felony arrests in large counties are dismissed before trial (Reaves, 2013). The number of cases resulting in dismissal has been dropping during the past two decades. Earlier studies of criminal prosecution (Boland,

Mahanna, & Sones, 1992) reported that nationally, 45% of felony arrests were dismissed before trial; nearly one-half were dismissed by the judge. The rates of dismissal of charges at preliminary hearings vary based on the procedures used to bring cases to court. In places where the prosecutor reviews evidence before appearing in court, the weakest cases are rejected before a preliminary hearing is held, and the number of cases dismissed by the judge is low. Where no such review occurs, the rate of dismissal at the preliminary hearing may exceed 40%.

Arraignment

At the **arraignment**, the judge notifies the defendant of the formal criminal charges against him or her and asks the defendant to plead to the charges. The arraignment is not a hearing on the facts of the case. The defendant may plead not guilty, guilty, or *nolo contendere* (no contest), or may stand silent. If the defendant pleads guilty or *nolo contendere*, the judge enters a finding of guilt. If the defendant remains silent, the judge enters a plea of not guilty and sets a trial date. The overwhelming majority of defendants plead guilty at arraignment, often as part of an agreement negotiated with the prosecutor (Devers, 2011; Reaves, 2013). In the typical plea bargain, the prosecutor drops charges or otherwise changes the seriousness of the formal charge in exchange for the certain conviction without trial. See Figure 2.1.

Trial

Whereas most cases result in a guilty plea, those that receive the most media attention and publicity involve a trial at which the defense and prosecution contest the facts and law before a neutral decision maker. Most cases that go to trial are what Walker (2015:44) termed "celebrated cases." In these cases, defendants receive full-blown trials, very often trial by jury. Much of the public believes that the jury trial is the normal operating procedure of the justice system because these cases receive the most publicity.

FIGURE 2.1
Former Michigan State University Dean William Strampel, *left*, appears during his video arraignment with Judge Richard D. Ball, *right*, regarding charges related to his oversight of Larry Nassar, who is in prison for sexually assaulting patients under the guise of treatment. Video arraignment systems allow courts to conduct the requisite arraignment process without the need to transport the accused to the courtroom.

AP Photo/Paul Sancya

At **trial**, the state (prosecutor) must prove, beyond a reasonable doubt, that the defendant committed the criminal offense with which he or she has been charged. The defense attorney seeks to discredit the state's case and, at a minimum, establish that there is some doubt as to whether the defendant committed the offense. Depending on the nature of the case, the defense can request one of two types of trials: a jury trial or a bench trial.

The jury trial is the ideal of the justice system. A panel of the defendant's peers hears all of the evidence and decides whether the defendant is guilty or not guilty. The bench trial is held before a judge alone, who hears all of the evidence and then decides whether the defendant is guilty or not guilty. If the verdict is not guilty, the justice process ends with the acquittal of the defendant. However, if the verdict is guilty (or if the defendant pleads guilty), the defendant stands convicted of the crime and reaches the next decision point in the justice system: sentencing.

Sentencing

Sentencing is the point at which officials select criminal punishments. The sentencing decision is bifurcated (i.e., has two parts). First, the judge decides the type of sentence. This can range from a fine to incarceration and covers a wide variety of alternatives, including probation, confinement in jail or prison, and combinations such as probation with a fine. In capital cases, such as murder, the type of sentence may be death. The second part of the decision involves the conditions of the sentence. These include the conditions of supervised release (probation), such as curfew, employment, and so on, as well as the length of prison term for those incarcerated.

Sentencing power is shared among the three branches of the government. The legislative branch sets limits on penalties by establishing minimum and maximum prison terms and fine amounts, by declaring some offenses ineligible for probation, and by other similar actions. The judicial branch is where the sentencing judge selects the actual type and conditions of sentence from alternatives allowed by the legislature. The executive branch has the power to pardon, to offer clemency, and, often, to authorize parole. Box 2.1 illustrates the evolution of sentencing reform in America.

BOX 2.1 AMERICAN SENTENCING STAGES

A criminal sentence is one of the most controversial aspects of the criminal justice process. Think of a recent crime that has received a lot of media attention in your area. It is likely that a quick poll among your classmates would uncover a variety of suggestions regarding the proper punishment of this crime. It is also likely that these suggestions would have been different if the poll were conducted several decades ago. This is due to the ever-evolving public opinion regarding the proper punishment of criminals. This public opinion, combined with social science research and practical realities such as budget limitations, have resulted in a shifting of sentencing policy over time. Tonry (2013) outlined four stages of sentencing that have occurred since 1930:

1. **Indeterminate sentencing** (1930–1975): The judge had primary control over the type of sentence given (such as prison, probation or fine, and the upper and lower bounds of the length of prison sentences within statutory limits), but the actual time served was determined by the parole board.

2. **Sentencing reform** (1975–1984): Jurisdictions attempted to address the public's concern that the sentences being given were inconsistent and potentially biased. Many jurisdictions created guidelines to help make judges' sentences and parole decisions more consistent. Jurisdictions also moved toward determinate sentences. With determinate sentencing, the judge sets the type of sentence and the length of prison sentences within statutory limits, but the parole board may not release prisoners before their sentences have expired, minus time off for good behavior, or "good time."

3. **Tough on crime** (1984–1996): During the tough on crime era, several punitive policies were introduced, including truth-in-sentencing laws and mandatory minimum sentences.

4. **Equilibrium** (1996–Present): Some of the most punitive punishments established during the tough on crime era have been revised to narrow their scope, and innovations such as drug courts are increasingly used as alternatives to incarceration.

Beginning in the mid-1990s, public support for the toughest sentencing practices declined (Ramirez, 2013). Still, most of these policies are still in place as legislators and correctional officials struggle to deal with the budgetary and societal impacts that such policies had on American society. Detailed descriptions of these issues will occur in later chapters.

Source: Ramirez (2013); Tonry (2013).

Keep in mind that there are far more misdemeanor crimes committed in the United States than felonies. As such, most convicted offenders are sentenced to probation or a fine and are not incarcerated. Approximately two-thirds of those convicted of felonies are sentenced to incarceration in either a prison or jail (Reaves, 2013). Those who are incarcerated most frequently gain release from prison through parole or mandatory release, and are required to live in the community under supervision and to obey conditions of release similar to those placed on probationers (Carson, 2018). Failure to obey these conditions can lead to the next possible decision point in the justice process: revocation.

Revocation

The overwhelming majority of criminal offenders sentenced to correctional custody serve some portion of their sentence under community supervision on either probation or parole. Both of these sentences are a form of **conditional release**, where the offender remains in the community if he or she abides by certain conditions, such as reporting regularly to a supervising officer, observing a curfew, or refraining from further criminal activity. Violation of the conditions of release constitutes grounds for the **revocation** of liberty. For instance, a probationer ordered not to consume alcohol can lose his or her liberty if caught drinking. The revocation process is a miniature justice system in which the probation or parole officer detects and investigates violations of conditions, and arrests and prosecutes violators who are tried by the sentencing judge (if on probation) or parole authority (if on parole). Upon conviction of violating the conditions of release, the violator may be sentenced to incarceration or continued supervision. In 2016, over 28% of all inmates admitted to state or federal prisons were parole violators (Carson, 2018).

With the exception of the death penalty, incarceration in prison is this country's most severe penalty. Convicted offenders receive this sentence either directly from the court or, more indirectly, through the revocation of conditional liberty. Despite the continued rise in life and long-term sentences (Nellis, 2017), a sentence of life imprisonment is still relatively uncommon in comparison to the total number of convicted offenders. Thus, for most offenders, a day comes when they are no longer under the control of the justice system. The last point in the justice process is discharge.

Discharge

Eventually, most criminal offenders receive a discharge from their sentences. Discharge is final release from criminal justice control or supervision. For some, this discharge will occur at the expiration of their term. For someone sentenced to a 10-year prison term, discharge will take place 10 years after the date of sentencing, whether the person was in prison for the full 10 years or received an earlier release by parole or reduction in term for good behavior.

Many states, however, have adopted procedures for early discharge. An offender serving a 10-year term may be paroled after serving 3 years, and then, after successfully completing 3 years (for example) under parole supervision, may receive an early discharge; thus, the offender may be released from sentence after serving only 6 years. Other jurisdictions in which no formal early discharge procedure exists may place similar offenders on unsupervised parole status after some time. In this case, the offender technically is still under sentence but is not supervised in the community and, for all practical purposes, has been discharged.

Upon discharge from sentence, the convicted offender becomes a member of free society again. In most cases, the record of conviction and collateral effects of conviction (limits on civil rights, employability, and the like) will haunt the ex-convict. Conviction of a crime, especially a felony, often disqualifies the offender from certain types of occupations, such as those requiring licensure or certification (teaching school, practicing law or medicine, and the like). In some cases, felony conviction leads to "civil death," that is, the offender has no rights to vote, hold public office, etc. Some states impose these penalties upon felons regardless of whether the offender was sentenced to incarceration or probation (Miller & Spillane, 2012).

Box 2.2 graphically portrays the decision points of the criminal justice system.

BOX 2.2 DECISION POINTS OF THE CRIMINAL JUSTICE SYSTEM

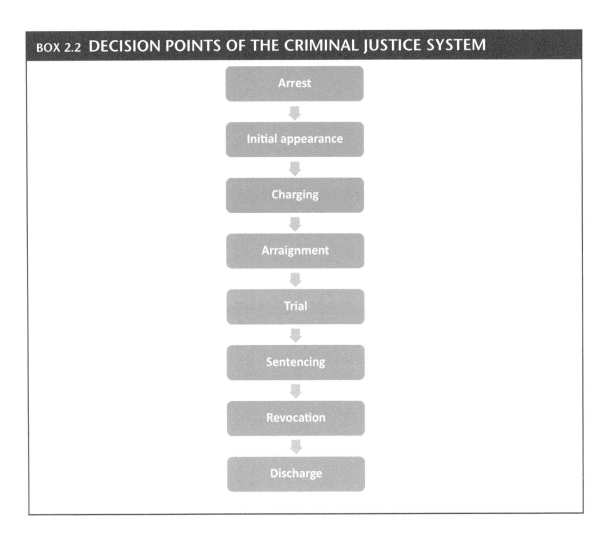

THE TOTAL CRIMINAL JUSTICE SYSTEM

As our brief description of the justice system illustrates, cases move through the various decision points on a contingency basis. If police detect a crime, an investigation may begin. If the investigation yields sufficient evidence, police may make an arrest. If police make an arrest, formal charges may be brought. The operative word is "if." Approaching this issue from the other direction, the sentence depends on the conviction, which depends on the charge, which depends on the investigation, which depends on the detection of crime. Each decision in the justice process is in large part determined by previous decisions. To a certain degree, earlier decisions depend on past practices in later points of the justice process. For example, if a county prosecutor routinely dismisses cases involving possession of minor amounts of marijuana, law enforcement officers are more inclined to stop arresting persons for possession of small amounts of that drug.

As the concept of a system implies, the components of the justice process (the decisions) are interdependent. As a result, the practices of all justice agencies affect those of every other agency to some extent. Similarly, environmental pressures will affect the operations of each justice agency to some degree. Some examples illustrate the manner in which environmental pressures and agency changes have systemwide effects: the ongoing debate regarding immigration policy, the focus on homeland security, the "war" on drugs, and the redefinition of domestic violence arrest policies.

Immigration Policy

The issue of immigration policy in the United States has received increasing attention in recent years. Policy decisions related to both legal and illegal immigration have long been controversial and complex. The public largely supports legal immigration, believing that those entering the United States strengthen the country through their hard work and talents (Jones, 2016). However, support for illegal immigration is much more divisive, with many Americans indicating a blanket opposition to illegal immigration regardless of the immigrants' personal attributes (Wright, Levy, & Citrin, 2016). Donald Trump's immigration-related rhetoric during the 2016 election reignited the public debate regarding this issue and it is still unclear exactly how his election will impact the immigration process described below.

It is important to distinguish between legal and illegal immigration. The United States permits a limited number of permanent immigrants to legally enter the country each year. In 2014, approximately 64% of newly admitted permanent immigrants entered through a visa program dedicated to relatives of U.S. citizens and existing permanent residents. Additionally, permanent legal status can be obtained for those with valuable skills or those seeking refuge or asylum from their home country. Finally, a small number of persons are admitted through a lottery system designed to favor prospective immigrants from home countries with historically low rates of immigration to the U.S. (American Immigration Council, 2016). It is important to note that a majority of lawful permanent immigrants eventually become U.S. citizens (Gonzalez-Barrera, 2017).

Enforcement of federal immigration laws is a multifaceted effort. Many immigrants are removed after being detained for attempting to illegally cross the U.S. border. In these situations, the detention, processing, and removal efforts can each be carried out by the federal government. However, immigration enforcement also relies on the cooperation of state and local governments. From 2008 to 2014, U.S. Immigration and Customs Enforcement (ICE) developed a program which asked local jurisdictions to share the fingerprints of each person arrested so that the immigration authorities could check the individual against immigration records. Immigration authorities could then request that the local agency detain the individual to allow the initiation of deportation procedures. However, lack of trust in the federal government, pro-immigrant climates in many communities, and fear of legal liability resulted in many local agencies refusing to cooperate with these detainer requests (Chen, 2016). This lack of cooperation from local jurisdictions significantly impacted federal immigration authorities' ability to enforce the immigration laws.

In late 2014, the Department of Homeland Security issued a memorandum detailing the prioritization of immigration law enforcement resources (Johnson, 2014). This memo created three priority levels which were meant to guide the enforcement of immigration laws. Priority 1 involved those undocumented immigrants who pose a danger to national security, are apprehended at the border attempting to illegally enter the country, members of street gangs, or those convicted of felonies. Priority 2 included those convicted of three or more misdemeanors, those convicted of a "significant misdemeanour" such as drug trafficking, driving under the influence (DUI), etc., and those who have recently entered the United States illegally or significantly abused the U.S. visa program. Finally, priority 3 includes those who have been issued a final order of removal after January 1, 2014. While the memo did not discourage the apprehension or removal of illegal immigrants who do not fit into one of these priorities, it made clear that the government's resources should be spent first on those who are a danger to society as identified by the priorities. In fiscal year 2016, over 93% of illegal immigrants removed by ICE were those who were clearly associated with one of these three priorities (see Table 2.3).

In the past decade, many local municipalities have passed immigration-related ordinances. Many of these ordinances were meant to discourage immigration into their communities, such as requiring employers to use the "E-verify" system which allows employers to verify the employment eligibility of their workers. At the same time, other municipalities were creating laws meant to

TABLE 2.3 ICE Removals by Priority Rank, 2016

Priority	Removals	Percentage
Priority 1	201,020	83.7
Priority 2	31,936	13.3
Priority 3	4,952	2.1
Total removals with priority	238,466	99.3
Unknown priority	1,789	0.7
Total removals	240,255	100

Source: U.S. Immigration and Customs Enforcement (2016).

benefit immigrant populations, such as not permitting government workers to ask about immigration status. These laws were meant to encourage immigrants to report criminal activity, use city services, and otherwise become better integrated into their communities (Steil & Vasi, 2014).

BOX 2.3 POLICY DILEMMA: SANCTUARY CITIES

Cities which refuse to cooperate with federal requests to keep an illegal immigrant detained pending a deportation are often called "sanctuary cities". Following the 2014 Homeland Security memo, the government began to see increased cooperation from many of these cities. However, the 2016 election of Donald Trump has put these sanctuary cities at odds with the federal government's priority to strictly enforce immigration law. Although these detainer requests are voluntary, President Trump signed an executive order threatening to withhold certain federal funds from cities that fail to abide by these requests. Do you feel that cities should be forced to abide by federal requests to keep an illegal immigrant detained? What steps could the federal government take to increase the voluntary cooperation from local municipalities?

The complex task of enforcing immigration laws demonstrates the interdependent nature of the justice system. Federal law enforcement efforts are impacted by local law enforcements' cooperation (or lack thereof) with the federal detainer orders described above. In turn, this cooperation is influenced by outside forces such as the municipalities' trust of the federal government and the political climate in each community. Finally, local and state laws may influence the immigration patterns within the country (Cebula, 2016), impacting both the political culture of a locality and the amount of resource strain that immigration enforcement might place on the local law enforcement agencies.

Homeland Security

In the wake of the terrorist attacks on the World Trade Center and the Pentagon on September 11, 2001, the United States began to wage a "war" on terror. The response of the United States to terrorism has been controversial, especially the response of local and state criminal justice agencies. Counterterrorism efforts in the United States illustrate the conflict between due process and crime control.

After the 2001 attacks the U.S. Congress took quick action to pass the USA PATRIOT Act, which broadened the powers of law enforcement and policing agencies to collect evidence and conduct surveillance on suspected terrorists, both domestic and international (see Box 2.4). In addition, billions of dollars of funding were allocated for counterterrorism activities. The federal government created a cabinet-level Department of Homeland Security that, as is described below, resulted in a major realignment of federal law enforcement organizations.

BOX 2.4 NATIONAL SECURITY LEAKS

Counterterrorism efforts since the September 11, 2001, attacks have included increased surveillance gathering within the United States and abroad. A series of recent information leaks have illustrated America's conflict between privacy and security.

In early 2010, Pfc. Chelsea Manning leaked more than 700,000 classified documents to Wikileaks, a website devoted to publishing secret information. Manning had become increasingly disillusioned with the Iraq War during her service as an intelligence analyst (Leigh & Harding, 2011). The massive leak resulted in a large public outcry

and ultimately led to Manning's conviction and sentence to 35 years in prison (Savage & Huetteman, 2013). This sentence was ultimately commuted by President Obama.

The most recent large-scale intelligence leak came when Edward Snowden revealed details of surveillance activities by the National Security Agency (NSA) to a British newspaper. Snowden has been charged with three federal crimes (Shane, 2013), but received temporary asylum by Russia (see Figure 2.2). The NSA has defended its surveillance programs, claiming that they have helped prevent numerous terrorist attacks since 9/11 (Savage, 2013).

FIGURE 2.2
Edward Snowden talks during a simulcast conversation during the SXSW Interactive Festival in March 2014, in Austin, Texas. Snowden talked with American Civil Liberties Union's principal technologist Christopher Soghoian, and answered tweeted questions.

Jack Plunkett/Invision/AP

Although some local criminal justice systems and agencies have made substantial investments in counterterrorism, most have not. Randol (2013) reported that while most local police departments had developed written plans on how they would respond to a terrorist attack, they were less likely to have made progress on terrorism prevention policies. One controversial mechanism which local police departments have used to increase their terrorism response capabilities is through the purchasing of surplus military equipment. A recent national survey of police departments found that over half of police departments had acquired some type of military surplus equipment (Johnson & Hansen, 2016). The most common types of items acquired were vehicles, night vision equipment, body armor, and protective clothing/masks.

Aside from terrorism response, it is not clear what we might expect local criminal justice agencies to do in terms of counterterrorism. The Federal Bureau of Investigation operates scores of Joint Terrorism Task Forces that involve federal, state, and local representatives. The goal of these task forces is to prevent and respond to terrorist acts in the United States. Much of the effort involved in fighting terrorism is aimed at intelligence gathering and data analysis. Most local criminal justice agencies do not have the skill or the resources to do this. Local police typically find themselves as information gatherers who report to regional or state data repositories or "fusion centers" where the information is analyzed.

Terrorist acts differ from "normal" crimes in terms of motivation. The 9/11 attacks involved kidnapping, murder, and arson. Local justice systems can respond to these offenses with existing policies, procedures, and law. Acts that fit federal definitions of terrorism, however, become federal offenses and typically are removed from local jurisdictions. Still, public concern, political pressure, the availability of additional federal funding, and other forces have made terrorism at least a nominal concern of local criminal justice agencies, especially police. At this point, the most important visible changes resulting from an emphasis on terrorism are greater cooperation between federal and local officials and increased pressure on local police to gather intelligence information. The counterterrorism tasks of local police complicate relations with the public, especially immigrants and minority group members (National Institute of Justice, 2008). Otherwise, more than a decade after the attacks, for most of American criminal justice, little has changed.

The War on Drugs

In 1973, the state of New York adopted legislation hailed as "the nation's toughest drug law" (U.S. Department of Justice, 1978). Although the most severe sanctions were finally eliminated by legislation in 2009, the law serves as a model for how the system responds to such a policy decision. The goal of this law was to "crack down" on those who sold heroin and other dangerous drugs. It had provisions for very stiff sentences and placed controls on plea bargaining. To cope with the anticipated increase in drug offense cases, it provided for the creation of more courts. The intent of the legislation was clear: to apprehend, convict, and punish those who sold heroin.

The effect of the law, however, is less clear. The officers and agencies of the justice system appear to have adapted to the changes to reduce the potentially disruptive effects on normal court operations that would result from the new law. Although there were no dramatic increases in arrests for the sale of heroin, fewer of those arrested were indicted, fewer of those indicted pleaded guilty, and fewer were convicted. For those convicted, both the rate of incarceration and the length of prison terms increased after the law took effect. However, in the final analysis, 3 years after the law was passed, the percentage of those arrested for heroin sale or possession going to prison remained stable at 11%, a figure identical to that occurring before the law was passed.

Several explanations are possible. First, a probable reason why the number of arrests did not increase was that most people already considered the sale and possession of large quantities of heroin to be serious offenses (even before the new law). The new law changed neither law enforcement nor public attitudes. The fact that fewer defendants pleaded guilty meant that prosecutors needed to be more certain of getting a guilty verdict before taking a case to trial. Indictments decreased as marginal cases were dismissed or downplayed. The increased number of trials created a backlog for the courts so that fewer cases were processed and acquittals were handed down in some cases in which previously a plea of guilty had ensured conviction.

The mandatory sentencing provisions of the legislation may account for the higher incarceration rate and more severe prison terms imposed under the new law. This suggests that there was no conscious effort to undermine the intent of the tough antidrug law, but rather, the court component of the justice process adapted to new pressures reflexively. As part of a system, the courts sought to maintain equilibrium and adapted to stresses and strains minimizing their impact.

In this example, the legislation initially and most specifically affected the criminal courts, and an effort was made to alleviate the strains through the creation of new courts. Had there been no new courts, it is likely that even more cases would have been dismissed or the backlog of cases would have been even greater. The law did not directly affect law enforcement. The effect of changes in prison sentences on corrections was not dramatic for two reasons. First, because heroin dealers are only a very small proportion of all those sentenced to prison, even large increases in their terms or rate of incarceration would not dramatically affect prisons. Second, the percentage of those arrested who were actually sentenced to prison did not change, and the effects of longer terms would not be felt until several years after those who received longer sentences had been imprisoned.

The war on drugs, having raged now for four decades in its most recent form, has produced changes in the characteristics of prison populations. For many years, drug offenders accounted for the largest part of prison population growth (Cohen & Kyckelhahn, 2010; Harrison & Beck, 2003). This trend finally appears to have slowed and even might be reversing as the public attitude toward drug crime becomes more tolerant (Carson, 2018; Swift, 2016). Nevertheless, the decades-long crackdown on drug crimes has contributed to the continued problem of prison and jail crowding and prompted the development of intermediate sanctions, specialized drug courts, and other adaptations in the criminal justice system. It has also had a disproportionately harsh impact on the poor, women, and members of minority groups (Welch, Wolff, & Bryan, 1998).

Domestic Violence Arrest Policies

The redefinition of domestic violence also illustrates the interdependency of the criminal justice system. Johnson and Sigler (2000) compared public opinion about violence against women over a 10-year period and reported that public tolerance for violence has decreased as criminalization of such behavior has become more common. All 50 states have now passed laws allowing the arrest of misdemeanor domestic violence that occurs outside the officer's presence (Zeoli, Norris, & Brenner, 2011). Again, public perceptions of offense seriousness and the severity of justice system response are related.

One of the most important policy changes in response to domestic violence has been a proliferation of preferred or mandatory arrest policies and laws. Despite these new laws, police officers arrest domestic violence offenders in less than 40% of the cases that are reported to the police (Reaves, 2017). The victim signing a criminal complaint against the offender is still very influential in the decision to make an arrest, as is the extent of injury to the victim. From 2006 to 2015, police made an arrest 89% of the time when the victim was seriously injured and signed a complaint against the offender, compared to just 35% of the time when the victim was seriously injured but did not sign a complaint. When the victim suffered minor injuries, an arrest was made 68% of the time if the victim signed a complaint against the offender, compared to 35% when no complaint was signed (Reaves, 2017). The decision to arrest appears to be influenced by several legal and extralegal factors (Lee, Zhang, & Hoover, 2013; Roark, 2016).

The impact of mandatory arrest policies is unclear. Initial research (Sherman & Berk, 1984) showed that suspects who were arrested following a domestic violence complaint exhibited fewer instances of future violence compared to those were given a warning or asked to leave. Subsequent research has shown that the impact of arrest is different for white offenders and for black offenders, and may be different for people of different economic levels (Maxwell, Garner, & Fagan, 2002). There is also evidence that mandatory arrest policies have increased the arrests of domestic violence victims (Finn & Bettis, 2006). Efforts to more aggressively prosecute domestic violence have even resulted in the jailing of some victims who fail to appear to testify in court against their abusers. Finally, some recent evidence suggests that mandatory arrest laws might increase intimate partner homicides or the premature deaths of domestic violence victims (Iyengar, 2009; Sherman & Harris, 2015). What is clear is that the adoption and implementation of policy reforms in this area have been neither easy nor trouble free (Ostrom, 2003; Whitcomb, 2002).

All of these issues have proven to be difficult for criminal justice policymakers and reformers to manage. Experience with these efforts to change criminal justice practices in dealing with illegal immigrants, counterterrorism, drug offenses, and domestic violence illustrates how the justice system interacts with its environment. In some cases, changes occur in all aspects of the justice process, such as drug enforcement, resulting in more arrests, convictions, and changes in the correctional population. In other cases, the system is able to adapt and minimize the impact of a reform by increasing rates of case dismissal or plea bargaining, or reducing the severity of sentences. All of these examples show that the criminal justice process operates as a system, adapting to change and pressure. They also indicate the complexity of evaluating the operations of the criminal justice process. This complexity becomes clearer when one examines the structure and organization of the agencies that comprise the system of criminal justice in the United States.

THE COMPONENTS OF CRIMINAL JUSTICE

It is common to divide the criminal justice system into three parts: law enforcement, courts, and corrections. Each of these three parts of the justice system itself comprises a multitude of separate agencies and actors. The organizations that make up the total criminal justice system are differently structured and funded, and draw from different personnel pools.

One of the most important distinctions among similar agencies is jurisdiction. Police departments, courts, and correctional agencies may be municipal (village, township, city, or county), state, or federal in nature. They may be specialized, like the U.S. Postal Inspectors, or they may have general duties, as does a typical police department. They may be public or private (such as security guards, many halfway houses, and other entities that provide crime control services). In this section, we examine the nature of criminal justice agencies in law enforcement, courts, and corrections.

Law Enforcement

There are so many agencies with law enforcement mandates that it is not possible to state their true number with confidence. In 1967, the President's Commission on Law Enforcement and Administration of Justice estimated that more than 40,000 police departments were in existence. Later, the U.S. Department of Justice (1980:24) reported that there were close to 20,000 state and local law enforcement agencies. This report, however, did not include townships with populations of less than 1,000 nor did it include federal or tribal law enforcement agencies. Most recently, the Bureau of Justice Statistics (Banks, Hendrix, Hickman, & Kyckelhahn, 2016) identified about 18,000 federal, state, and local police agencies.

Federal Law Enforcement

A number of federal law enforcement agencies exist. These agencies tend to be small with specific mandates, yet in total, federal law enforcement is very complex. We are all aware of the Federal Bureau of Investigation (FBI), and most of us have heard of the U.S. Marshals; the Postal Inspectors; the Drug Enforcement Administration (DEA); the Bureau of Alcohol, Tobacco, Firearms and Explosives (ATF); the Immigration and Customs Enforcement; Customs and Border Protection; the Internal Revenue Service (IRS); and the Secret Service. Yet, many are unaware of the law enforcement duties of the National Park Service, the U.S. Supreme Court Police Department, the National Gallery of Art Protection Staff, and other federal police agencies. We seldom consider the military police, the tribal police departments on Native American reservations, or the investigative duties of auditors and staff of such organizations as the Federal Trade Commission (FTC) (Travis & Langworthy, 2008).

Reaves (2012) reported that the federal government employed about 120,000 full-time officers with arrest powers who were authorized to carry firearms. The bulk of these employees worked for the U.S. Customs and Border Protection, and they included approximately 17,000 employees of the Federal Bureau of Prisons. Nineteen other federal agencies employed 500 or more such officers and agents. The most common duties performed by these federal officials were criminal investigations (37%), police response and patrol (23%), and inspections (15%). These federal employees do not include the officers of the Transportation Security Administration, created in the wake of the 9/11 terrorist attacks.

Creation of the Department of Homeland Security in 2002 resulted in organizational changes in federal law enforcement. The Department of Homeland Security is now the single largest employer of federal law enforcement officers, administering the U.S. Coast Guard, Secret Service, Federal Protective Service, and U.S. Customs and Border Protection (except for some revenue functions), and has taken over the responsibilities of the Immigration and Naturalization Service, which was abolished. With these changes, the Department of Homeland Security employs more than 45% of federal officers, and the Department of Justice employs about 33% (Reaves, 2012).

Because they serve the entire nation, these agencies recruit nationally and tend to have more stringent entry requirements than do most other police departments. The FBI, for example, requires a bachelor's degree in combination with investigatory experience or postgraduate training. Because federal law enforcement is funded at the federal level, salary and benefits for federal law enforcement officers are often higher than those paid to municipal police.

State Law Enforcement

Each state has a primary law enforcement agency which is distinct from their federal or municipal counterparts. The responsibilities and names of state agencies vary considerably. The state agency in your state may be called "state police," "highway patrol," "public safety," "state patrol," or "state troopers" (Reaves, 2011). The highway patrol typically enforces traffic laws on state and federal highways. Other forms of state police often have a broader mission and sometimes charge their state police with general law enforcement duties (Cordner, Seifert, & Ursino, 2014). In addition, several states have specialized state units to combat drug offenses, organized crime, liquor and cigarette tax violations, and the like. Finally, many states also charge their park services with law enforcement obligations. Reaves (2011) reported that state police agencies employed more than 93,000 full-time officers.

Like federal agencies, state agencies recruit from a pool of candidates that is considerably larger than that tapped by most local police departments. Entry requirements tend to be less stringent than federal law enforcement, but much more stringent than many local police departments. Moreover, in many states, the salary and benefits paid to state police officers are higher than those paid in most local departments.

Municipal Law Enforcement

Municipal or local police departments provide the bulk of law enforcement services, as shown in Figure 2.3. These include the traditional city or

FIGURE 2.3
Percentage Distribution of Police Personnel by Level of Government.

Source: Calculations by author made from U.S. Census Bureau Annual Survey of Public Employment and Payroll, 2014. Retrieved from www.census.gov//govs/apes/historical_data_2014.html

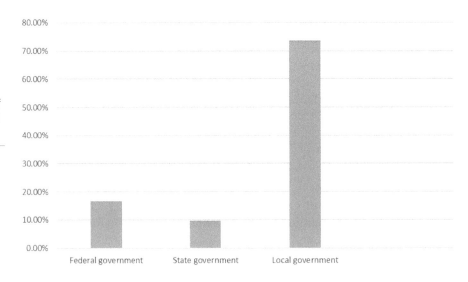

township police department, as well as the county sheriff. The majority of police departments in the United States are local ones, and most police agencies are small. In fact, more than half of local police departments employ fewer than 10 full-time officers (Reaves, 2011). Most police officers, however, work for large departments, because the relatively few large departments employ a great many officers. The largest 5% of police departments employ more than 60% of all local officers.

Municipal police departments rarely conduct national searches or recruitment drives, with the exception of a few, usually larger police departments. Most local police departments recruit locally and employ civil service testing to enlist new officers (Sanders, Hughes, & Langworthy, 1995). Sheriffs generally are elected, but many sheriff's deputies are recruited through civil service. It is common for police protection to comprise a major portion of a municipality's budget. The average starting salary for a full-time local police officer is $44,400. As might be expected, departments serving smaller communities generally pay lower salaries than those serving larger communities (Reaves, 2015). See Figure 2.4.

FIGURE 2.4
A Philadelphia bicycle cop observes the neighborhood. The bulk of law enforcement services are provided through municipal or local police departments.

Ellen S. Boyne

Private and Other Public Law Enforcement

In addition to the agencies described above, there are hundreds of special-purpose law enforcement agencies in cities and counties. Reaves (2011) identified more than 1,700 public, special-purpose police agencies including housing authority, school, airport, university, transit, and park police (see Figure 2.5). Furthermore, there are thousands of private and semipublic law enforcement agencies in the United States. For example, most factories, amusement parks, and hospitals have security staff, as do most retail chain stores. Many residential buildings and developments also have private security. Private police and private security personnel outnumber the public police by a ratio of at least three to one (Maahs & Hemmens, 1998). Additionally, the coroner or

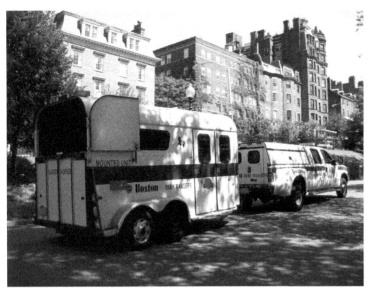

FIGURE 2.5
The job of a park ranger is not always rural in nature; urban park rangers, like the ones in Boston Common, patrol city parks and park facilities.

Ellen S. Boyne

medical examiner is often considered to be a law enforcement official because of the investigative duties of that position.

As we have seen, it may not be possible to speak accurately of law enforcement—or even of the police—in the United States. The diversity of agencies, standards, and duties is nearly mind-boggling. Because law enforcement is the largest (numerically) component of the justice process, a review of justice agencies in courts and corrections is less complicated, but only marginally so.

Courts

The picture that most Americans have of the court system is probably limited to that of a state felony court of general jurisdiction, where felony defendants are processed to determine guilt and innocence and proscribe a sentence that might include a lengthy prison term. In fact, the court system in the United States is fragmented and complicated (Malega & Cohen, 2013). Each state has a different court system, as does the federal government. There are federal, state, and municipal courts. Thousands of courts of limited jurisdiction are also in operation. These courts are divided further in terms of the types of cases they may hear and the types of decisions they may reach.

There are approximately 340 justices of the Supreme Court and other courts of last resort in the 50 states, District of Columbia, and federal systems. Approximately 990 additional justices serve in intermediate courts of appeal, with more than 27,000 judges serving in general trial courts (Malega & Cohen, 2013). Interestingly, the number of judges in general is on the increase as the caseloads of the court systems have declined over the past decade (Schauffler et al., 2016).

Federal Courts

Federal judges and justices of the U.S. Supreme Court are nominated by the president and appointed with the advice and consent of the U.S. Senate. These judges have lifetime tenure. Federal magistrates are appointed to 8-year terms by federal district judges.

The federal courts are organized by **circuits**, with 12 regional circuits covering the entire nation and foreign possessions. Within the circuits, 94 district courts are trial courts. The circuit court is the appeals court for all trial courts in its region. In addition, more than 500 federal magistrates within these districts may hear minor offenses and conduct the early stages of felony trials and more serious civil trials. See Figure 2.6.

Federal courts decide cases of federal interest, for example, charges of federal law violation. Federal appeals courts also decide federal constitutional issues, even if such issues were raised during state trials or proceedings.

FIGURE 2.6
U.S. District Court Building, Denver, Colorado.

Ellen S. Boyne

State Courts

State judicial systems are similar to the federal judiciary in structure. They generally comprise trial courts, intermediate appellate courts, and a state supreme court. State judges and justices are either appointed (as is the federal judiciary) or elected. Members of most state judiciaries are in office for specified terms of office (unlike federal judges, who have lifetime tenure). Three states provide judges tenure until they reach the age of 70 (Malega & Cohen, 2013).

Federal judges are recruited nationally (although district court judges and circuit court judges are generally selected from among candidates residing in the particular district or circuit). State court judges are elected statewide (or appointed) for statewide posts (e.g., the office of justice of the state supreme court), or from the jurisdiction of the lower court (e.g., the county of a specific county court).

Local Courts

There is a plethora of local courts in the United States. These are courts of limited jurisdiction because they are not allowed to decide felony cases, serious misdemeanors, or civil suits seeking damages above fairly low dollar amounts. Often these are known as justice of the peace courts. In many places, these limited-jurisdiction courts are known as police courts or mayor's courts. They usually decide traffic offense cases, hear violations of local ordinances and petty offenses, and make bail determinations. In states that retain the office of justice of the peace, often no formal legal training is required for this position. These limited-jurisdiction courts are not authorized to conduct jury trials, and their decisions may be appealed to courts of general jurisdiction, which are also known as trial courts. Salaries for these local courts are usually not commensurate with what an attorney could earn in the private practice of law. However, many of these courts operate on a part-time basis, and members of the bar may serve as justices of the peace.

Other Courts

Every court system has a number of special-jurisdiction courts. For example, the federal judiciary has a tax court, and states usually have a court of domestic relations and/or a juvenile court. Several jurisdictions also have bankruptcy courts and other special jurisdiction courts. A relatively recent innovation is what may be called a **private court**. In some places, offices or commissions for dispute resolution have been developed to divert cases away from the formal courts (Aaronson, Kittrie, Saari, & Cooper, 1977). Here, the parties to a dispute sit with a lay negotiator (or team of negotiators) and attempt to resolve their problem without resorting to the courts. Most of these private courts are staffed by volunteers or by paid staff whose salaries are lower than that of a judge. Examples of these types of private courts appear on television in shows like *The People's Court* or *Judge Judy*. Court specialization within the criminal justice system has also increased with the development and spread of special **drug courts** dedicated to the processing and supervision of drug cases, as illustrated in Figure 2.7. Other special courts are increasingly common. Rottman and Casey (1999) described these as "problem-solving courts" where courts

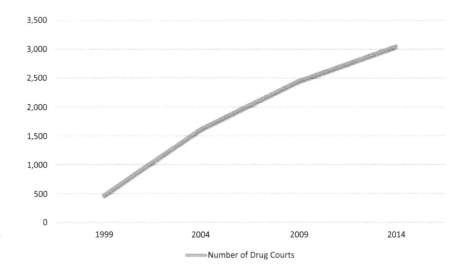

FIGURE 2.7
Growth of Drug Courts.

Source: Marlow, Hardin, and Fox (2016).

(judges, prosecutors, and the defense bar) work with offenders, victims, service providers, and the broader community to develop long-term solutions to the problems that bring cases to court.

Prosecution

At all levels of courts, from local to federal, the prosecutor represents the interests of the state (not the victim). In the federal system, the prosecutor is the U.S. attorney or the deputy U.S. attorney. These are lawyers appointed by the nomination of the president with the consent of the Senate. Local prosecutors are common in most states; for the most part, they are lawyers elected at the county level. Prosecutors have many titles, including district attorney, state's attorney, county attorney, circuit attorney, commonwealth's attorney, and solicitor (Perry & Banks, 2011). The Bureau of Justice Statistics reported the existence of more than 2,300 prosecutor's offices responsible for felony cases in state criminal courts. The average prosecutor's office closes more than 1,200 cases per year. To handle the large caseloads, these offices employed about 78,000 people, including approximately 34,000 attorneys (Perry & Banks, 2011). Most often, a prosecutor's office will have a chief prosecutor along with one or more assistant prosecutors with a large support staff.

Defense

There are three basic structures for the provision of defense counsel: private retention, public defenders, and assigned counsel. Private retention refers to the possibility of the defendant retaining his or her own attorney. Private retention is unusual because most criminal defendants cannot afford attorney fees. However, wealthy or notorious defendants often retain celebrated defense attorneys. Public defenders usually have an appointed director or an administrator who hires a sufficient staff to perform legal research, investigation, clerical, and administrative services (Strong, 2016). The most common form of criminal defense system is the public defender, but most criminal courts use two or more methods of providing defense counsel, including assigned counsel and contract systems. In the provision of assigned counsel, judges are presented either with a list of all attorneys practicing in their jurisdiction, or with a list of those attorneys willing to take on criminal defense cases. The judge then appoints an attorney for each indigent defendant from this list; he or she usually moves down the list from the first name to the last. The attorneys selected and assigned are then paid a set fee, which is usually on an hourly rate not to exceed some upper limit per case. In contract systems, the court enters an agreement with a law firm, bar association, or private attorney for indigent defense services for a specified period of time at a specified rate (Harlow, 2000).

Like prosecutors, defense attorneys employed in public defender offices (and most assigned counsel schemes) are not paid as well as judges, nor are they paid as much as they could earn in private practice as retained defense attorneys. Again, like prosecutors, young attorneys often seek this kind of work to gain trial experience.

Both prosecutors and defense attorneys recruit staff from local bar associations. Although the local nature of the recruitment is comparable to recruiting for most police officers and judges, the requirement of membership in the bar limits the pool of possible applicants.

Witnesses and Jurors

Many other persons are involved in the court process in addition to prosecutors, defense counsel, and judges. There are court support staff, such as court clerks, stenographers, bailiffs, and administrators; however, we will focus here on witnesses and jurors.

A variety of persons may serve as witnesses in a criminal case (Woodard & Anderson, 1984). Generally, the arresting officers and any investigators are called as witnesses in a criminal case. If any passersby saw the offense, they too may be called to testify. Sometimes the defendant (or a codefendant) is called to testify in criminal cases (but the defendant cannot be required to be a witness). Depending on the nature of the case, or of the defense, expert witnesses may be called. These individuals are first established as having special knowledge not commonly available to the average citizen. Experts in areas such as ballistics, forensic medicine, and psychology or psychiatry (for instance, when an insanity defense is raised) are asked to bring special knowledge to bear on issues at trial. The victim of a crime is useful only as a witness. Crimes are public wrongs; individual suffering is not at issue in criminal trials. In recent years, however, there has been an increased emphasis on using the criminal process to redress the harms suffered by individual victims. Balancing the interests of the victim with those of the defendant is a complicated task (Office for Victims of Crime, 2002).

Citizens participate directly and most strongly in the criminal justice process in the courts. Citizens make up the two types of juries used in the courts. Grand juries of citizens sit and listen to the prosecutor's case before deciding whether an indictment should be issued. Trial juries sit and listen to the criminal trial before deciding if the defendant should be convicted. In several states, and in death penalty cases, the jury also recommends a sentence to the judge after deciding to convict the defendant.

Jurors are selected from lists of residents in the court's jurisdiction. Often these lists are voter registration rolls, telephone books, or the billing records of utility companies. Trial jurors are then subjected to *voir dire*, a process by which the prosecutor and defense attorney seek to discover whether the jurors have any prejudices that could affect their decision in the trial. Attorneys may challenge a juror suspected of being unable to make an objective decision. The judge may dismiss a challenged juror.

Corrections

The general categories of incarceration and community supervision make up corrections. This general classification, however, grossly oversimplifies this complex component of the justice system. Incarceration includes both prisons and jails, whereas community supervision covers both probation and parole. With this dichotomy, it is not clear where such sanctions as halfway houses or "split sentences" fall.

Incarceration

The most frequent place of incarceration for criminal offenders and those suspected of criminal acts is the jail. There are more than 3,200 jails in the United States (Stephan & Walsh, 2011), most of which are municipal—either city or (more frequently) county jails. Most jails do not have treatment staffs of counselors, psychologists, and therapists. The major occupational group in jails is correctional officers. Many jail officers are minimally trained and low paid. Often, jail officers are members of the police department or the sheriff's department that is responsible for jail operation. Jail officers have traditionally earned less than patrol officers (Kerley & Ford, 1982).

The nation's jails supervise some 740,000 inmates on any given day, but because of the relatively short time most persons stay in jail, 10 million or more people may "do time" in jail each year (Zeng, 2018). More than 60% of those held in jail are not yet convicted and are awaiting trial. It is not possible to determine how many jail admissions are of repeat offenders.

The nation's prisons house more inmates than do jails on any given day, but because of the longer terms, fewer people serve prison time each year than jail time. Whereas jails usually are municipal, the federal or state governments operate prisons. Prisons are more apt than jails to have counselors, therapists, industries, and educational programs, partly because prisons are larger and hold inmates longer, and partly because they have a larger resource base (state taxes) than do city and county jails. Still, the most common occupational category in prisons is that of correctional officers (Stephan, 2008).

Nonincarceration

The most common form of nonincarcerative sanction (after fines, perhaps) is probation. On any day approximately 3.8 million persons are under probation supervision (Kaeble & Cowhig, 2018). Probation officers supervise these persons in the community and also write presentence investigation reports and operate other programs, depending on the jurisdiction.

Probation departments operate either at the state or local (either city or county) level. State-level probation allows more consistency among probation services across the state, whereas local-level probation allows for flexibility based on local needs. States also differ in terms of into which branch of government probation falls. Most states administer probation under the executive branch, whereas other states prefer to place probation under the judicial branch.

Parole is similar to probation, except that parole is always a function of a state agency; parole officers are, therefore, state employees. At any given time, more than 870,000 persons are under parole supervision (Kaeble & Cowhig, 2018). These persons were released early from incarceration (mostly from prison) and are supervised by parole officers. Most states have parole boards in the executive branch of government that are responsible for deciding which inmates will be granted early release, as well as what should be the proper conduct of the prisoners' parole periods.

Private-Sector Corrections

As with law enforcement and the courts, there is private involvement in corrections as well. Traditionally, many correctional practices were the province of voluntary or private initiatives. Throughout the 1980s until the present, there has been a growing movement to privatize corrections, with private companies con-

FIGURE 2.8
Neighborhood Watch programs exist across the country. Neighborhood Watch is an example of a crime prevention program that enlists the active participation of residents in cooperation with law enforcement to reduce crime.

Ellen S. Boyne

structing and operating prisons and jails in addition to providing other services on a contract basis (Travis, Latessa, & Vito, 1985). In addition to these for-profit private correctional enterprises, volunteer service is relatively common in corrections. Volunteers write to and visit prison inmates, provide services to probation and parole offices and clients, and serve on a variety of boards and commissions. The boards and commissions range from those that govern halfway houses to citizen court-watching groups. Neighborhood Watch programs (see Figure 2.8), and other citizen crime prevention projects have also increased the citizens' role in law enforcement. One of the most important trends in criminal justice during the past decade has been the resurgence of private initiative in the criminal justice system.

SYSTEMS AND CRIMINAL JUSTICE STRUCTURE

What this chapter has demonstrated is that the criminal justice system in the United States is extremely complex. The various agencies that comprise the system are organized at different levels of government, utilize different resource bases, and select differentially qualified personnel in different ways. In short, although the justice system appears too diverse to be a system, the interdependence of its parts and its sensitivity to environmental changes support a systems approach.

There are at least 52 criminal justice systems in the United States: one for each state, the federal government, and the District of Columbia. This may be an underestimate of their numbers. If city police can arrest someone for violating a city ordinance, and we convict and punish them in a city court, do we have a city justice system? Although we can argue that there are many criminal justice systems in the United States, we will continue to examine and discuss the criminal justice system as a whole.

This systems approach to the study of criminal justice seems especially appropriate. Without a prevailing approach, we might be forced to throw up our hands in despair, unable to make sense of the confusion. Why do we have so many agencies? Why do these different agencies have conflicting and sometimes competing jurisdictions and goals? The answer is because they are part of an open system. The large number of agencies and the various levels and branches of government involved are a manifestation of environmental impact on American criminal justice. Given our political and cultural values of federalism, local autonomy, and the separation of powers, we should not be surprised at the confusion in the justice system; it would be more surprising if there were no confusion. A single, well-organized, monolithic criminal justice system for the entire nation may well be "un-American."

REVIEW QUESTIONS

1. Identify the 10 decision points of the criminal justice process discussed in this chapter.

2. How does the justice process work as a directional flow of cases in the total system?

3. Give two examples of how the environment of the justice process affects the operations of all justice agencies.

4. What are the basic components of the justice process?

5. Describe the different types, levels, and staffing patterns of the components of the justice process.

6. Why is the systems approach especially appropriate to the study of American criminal justice?

REFERENCES

Aaronson, D.E., Kittrie, N.N., Saari, D.J., & Cooper, C.S. (1977). *Alternatives to conventional criminal adjudication*. Washington, DC: U.S. Government Printing Office.

American Immigration Council (2016). *How the United States immigration system works*. Retrieved from www.americanimmigrationcouncil.org/sites/default/files/research/how_the_united_states_immigration_system_works.pdf

Avakame, E.F., Fyfe, J.J., & McCoy, C. (1999). "Did You Call the Police? What Did They Do?" An empirical assessment of Black's theory of the mobilization of law. *Justice Quarterly, 16*(4), 765–792.

Banks, D., Hendrix, J., Hickman, M., & Kyckelhahn, T. (2016). *National sources of law enforcement employment data*. Washington, DC: Bureau of Justice Statistics.

Baumer, E., & Lauritsen, J. (2010). Reporting crime to the police, 1973–2005: A multivariate analysis of long-term trends in the National Crime Survey (NCS) and the National Crime Victimization Survey (NCVS). *Criminology, 48*(1), 131–186.

Berg, M.T., Slocum, L.A., & Loeber, R. (2013). Illegal behavior, neighborhood context, and police reporting by victims of violence. *Journal of Research in Crime and Delinquency, 50*(1), 75–103.

Boland, B., Mahanna, P., & Sones, R. (1992). *The prosecution of felony arrests, 1988*. Washington, DC: Bureau of Justice Statistics.

Carson, E.A. (2018). *Prisoners in 2016*. Washington, DC: Bureau of Justice Statistics.

Cebula, R. (2016). Give me sanctuary! The impact of personal freedom afforded by sanctuary cities on the 2010 undocumented immigrant settlement pattern within the U.S., 2SLS estimates. *Journal of Economics and Finance, 40*, 792–802.

Chen, M.H. (2016). Trust in immigration enforcement: State noncooperation and sanctuary cities after secure communities. *Chicago-Kent Law Review, 91*, 13–57.

Cohen, T., & Kyckelhahn, T. (2010). *Felony defendants in large urban counties, 2008*. Washington, DC: Bureau of Justice Statistics.

Cordner, G., Seifert, M.W., & Ursino, B.A. (2014). *State police and community policing*. Washington, DC: Office of Community Oriented Policing Services.

Devers, L. (2011). *Plea and charge bargaining: Research summary*. Bureau of Justice Assistance: U.S. Department of Justice.

Finn, M.A., & Bettis, P. (2006). Punitive action or gentle persuasion: Exploring police officers' justifications for using dual arrest in domestic violence cases. *Violence Against Women, 12*(3), 268–287.

Gonzalez-Barrera, A. (2017, June 29). *Recent trends in naturalization, 1995–2015*. Retrieved from www.pewhispanic.org/2017/06/29/recent-trends-in-naturalization-1995-2015/

Harlow, C. (2000). *Defense counsel in criminal cases*. Washington, DC: Bureau of Justice Statistics.

Harrison, P.M., & Beck, A.J. (2003). *Prisoners in 2002*. Washington, DC: Bureau of Justice Statistics.

Iyengar, R. (2009). Does the certainty of arrest reduce domestic violence? Evidence from mandatory and recommended arrest laws. *Journal of Public Economics, 93*, 85–98.

Johnson, I., & Sigler, R. (2000). Public perceptions: The stability of the public's endorsements of the definition and criminalization of the abuse of women. *Journal of Criminal Justice, 28*(3), 165–179.

Johnson, J.C. (2014, Nov. 20). Policies for the apprehension, detention and removal of undocumented immigrants [Memorandum]. Washington, DC: Department of Homeland Security.

Johnson, T.C., & Hansen, J.A. (2016). Law enforcement agencies' participation in the military surplus equipment program. *Policing: An International Journal of Police Strategies & Management, 39*(4), 791–806.

Jones, B. (2016, April 15). *Americans' views of immigrants marked by widening partisan, generational divides*. Pew Research Center. Retrieved from www.pewresearch.org/fact-tank/2016/04/15/americans-views-of-immigrants-marked-by-widening-partisan-generational-divides/

Kaeble, D., & Cowhig, M. (2018). *Correctional populations in the United States, 2016*. Washington, DC: Bureau of Justice Statistics.

Kerley, K.E., & Ford, F.R. (1982). *The state of our nation's jails, 1982*. Washington, DC: National Sheriffs' Association.

Langton, L., Berzofsky, M., Krebs, C., & Smiley-McDonald, H. (2012). *Victimizations not reported to the police, 2006–2010*. Washington, DC: Bureau of Justice Statistics.

Lee, J., Zhang, Y., & Hoover, L.T. (2013). Police response to domestic violence: Multilevel factors of arrest decision. *Policing: An International Journal of Police Strategies & Management, 36*(1), 157–174.

Leigh, D., & Harding, L. (2011). *Wikileaks: Inside Julian Assange's war on secrecy*. New York: Public Affairs.

Maahs, J., & Hemmens, C. (1998). Guarding the public: A statutory analysis of state regulation of security guards. *Journal of Crime and Justice, 21*(1), 119–134.

Malega, R., & Cohen, T.H. (2013). *State court organization, 2011*. Washington, DC: Bureau of Justice Statistics.

Marlowe, D.B., Hardin, C.D., & Fox, C.L. (2016). *Painting the current picture: A national report on drug courts and other problem-solving courts in the United States*. Alexandria, VA: National Drug Court Institute.

Maxwell, C.D., Garner, J.H., & Fagan, J.A. (2002). The preventive effects of arrest on intimate partner violence: Research, policy and theory. *Criminology and Public Policy, 2*(1), 51–80.

Miller, B.L., & Spillane, J.F. (2012). Civil death: An examination of ex-felon disenfranchisement and reintegration. *Punishment & Society, 14*(4), 402–428.

Morgan, R.E., & Kena, G. (2017). *Criminal victimization, 2016*. Washington, DC: Bureau of Justice Statistics.

National Institute of Justice (2008). *Policing in Arab-American communities after September 11*. Washington, DC: National Institute of Justice.

Nellis, A. (2017). *Still life: America's increasing use of life and long-term sentences*. Washington, DC: The Sentencing Project.

Office for Victims of Crime (2002). *The crime victim's right to be present. Legal series bulletin #3 (January)*. Washington, DC: Office for Victims of Crime.

Ostrom, B. (2003). Domestic violence: Editorial introduction. *Criminology and Public Policy*, 2(2), 259–262.

Perry, S., & Banks, D. (2011). *Prosecutors in state courts, 2007—Statistical tables*. Washington, DC: Bureau of Justice Statistics.

President's Commission on Law Enforcement and Administration of Justice (1967). *Task force report: The police*. Washington, DC: U.S. Government Printing Office.

Ramirez, M.D. (2013). America's changing views on crime and punishment. *Public Opinion Quarterly*, 77(4), 1006–1031.

Randol, B.M. (2013). An exploratory analysis of terrorism prevention and response preparedness efforts in municipal police departments in the United States: Which agencies participate in terrorism prevention and why? *Police Journal, 86*, 158–181.

Reaves, B.A. (2011). *Census of state and local law enforcement agencies, 2008*. Washington, DC: Bureau of Justice Statistics.

Reaves, B.A. (2012). *Federal law enforcement officers, 2008*. Washington, DC: Bureau of Justice Statistics.

Reaves, B.A. (2013). *Felony defendants in large urban counties, 2009—Statistical tables*. Washington, DC: Bureau of Justice Statistics.

Reaves, B.A. (2015). *Local police departments, 2013: Personnel, policies, and practices*. Washington, DC: Bureau of Justice Statistics.

Reaves, B.A. (2017). *Police response to domestic violence, 2006–2015*. Washington, DC: Bureau of Justice Statistics.

Reyns, B.W., & Englebrecht, C.M. (2014). Informal and formal help-seeking decisions of stalking victims in the United States. *Criminal Justice and Behavior, 41*(10), 1178–1194.

Reyns, B.W., & Randa, R. (2017). Victim reporting behaviors following identity theft victimization: Results from the National Crime Victimization Survey. *Crime & Delinquency, 63*(7), 814–838.

Roark, J. (2016). Predictors of dual arrest for offenders involved in heterosexual domestic violence arrests. *Policing: An International Journal of Policing Strategies & Management, 39*(1), 52–63.

Rottman, D., & Casey, P. (1999). Therapeutic jurisprudence and the emergence of problem-solving courts. *National Institute of Justice Journal, 240*, 12–19 (July).

Sanders, B., Hughes, T., & Langworthy, R. (1995). Police officer recruitment and selection: A survey of major police departments in the U.S. *Police Forum, 5*(4), 1–4.

Savage, C. (2013, June 18). N.S.A. chief says surveillance has stopped dozens of plots. *The New York Times*. Retrieved from www.nytimes.com/2013/06/19/us/politics/nsa-chief-says-surveillance-has-stopped-dozens-of-plots.html?pagewanted=all

Savage, C., & Huetteman, E. (2013, August 21). Manning sentenced to 35 years for a pivotal leak of U.S. files. *The New York Times*. Retrieved from www.nytimes.com/2013/08/22/us/manning-sentenced-for-leaking-government-secrets.html?pagewanted=all

Schauffler, R.Y., LaFountain, R.C., Strickland, S.M., Holt, K.A., & Genthon, K.J. (2016). *Examining the work of state courts: An overview of 2015 state court caseloads*. Washington, DC: National Center for State Courts.

Shane, S. (2013, June 21). Ex-contractor is charged in leaks on N.S.A. surveillance. *The New York Times*. www.nytimes.com/2013/06/22/us/snowden-espionage-act.html?_r=0

Sherman, L.W., & Berk, R.A. (1984). The specific deterrent effects of arrest for domestic assault. *American Sociological Review, 49*, 261–272.

Sherman, L.W., & Harris, H. (2015). Increased death rates of domestic violence victims from arresting vs. warning suspects in the Milwaukee domestic violence experiment. *Journal of Experimental Criminology, 11*, 1–20.

Steil, J.P., & Vasi, I.B. (2014). The new immigration contestation: Social movements and local immigration policy making in the United States, 2000–2011. *American Journal of Sociology, 119*(4), 1104–1155.

Stephan, J. (2008). *Census of state and federal correctional facilities, 2005*. Washington, DC: Bureau of Justice Statistics.

Stephan, J., & Walsh, G. (2011). *Census of jail facilities*. Washington, DC: Bureau of Justice Statistics.

Strong, S.M. (2016). *State-administered indigent defense systems, 2013*. Washington, DC: Bureau of Justice Statistics.

Swift, A. (2016, Oct. 19). *Support for legal marijuana use up to 60% in U.S.* Gallup Poll. Retrieved from www.gallup.com/poll/196550/support-legal-marijuana.aspx

Tonry, M. (2013). Sentencing in America, 1975–2025. *Crime and Justice, 42*, 141–198.

Travis, L.F., & Langworthy, R.H. (2008). *Policing in America: A balance of forces* (4th ed.). Englewood Cliffs, NJ: Prentice Hall.

Travis, L.F., Latessa, E.J., & Vito, G.F. (1985). Private enterprise in institutional corrections: A call for caution. *Federal Probation, 49*(4), 11–16.

U.S. Department of Justice (1978). *The nation's toughest drug law*. Washington, DC: U.S. Government Printing Office.

U.S. Department of Justice (1980). *Justice agencies in the United States*. Washington, DC: U.S. Government Printing Office.

U.S. Immigration and Customs Enforcement (2016). *Fiscal year 2016 ICE enforcement and removal operations report*. Retrieved from www.ice.gov/sites/default/files/documents/Report/2016/removal-stats-2016.pdf

Walker, S. (2015). *Sense and nonsense about crime, drugs, and communities: A policy guide* (8th ed.). Stamford, CT: Cengage.

Welch, M., Wolff, R., & Bryan, N. (1998). Decontextualizing the war on drugs: A content analysis of NIJ publications and their neglect of race and class. *Justice Quarterly, 15*(4), 719–742.

Whitcomb, D. (2002). Prosecutors, kids, and domestic violence cases. *National Institute of Justice Journal, 248*, 2–9 (March).

Woodard, P.L., & Anderson, J.R. (1984). *Victim/witness legislation: An overview*. Washington, DC: U.S. Department of Justice.

Wright, M., Levy, M., & Citrin, J. (2016). Public attitudes toward immigration policy across the legal/illegal divide: The role of categorical and attribute-based decision-making. *Political Behavior, 38,* 229–253.

Zeng, Z. (2018). *Jail inmates in 2016.* Washington, DC: Bureau of Justice Statistics.

Zeoli, A.M., Norris, A., & Brenner, H. (2011). A summary and analysis of warrantless arrest statutes for domestic violence in the United States. *Journal of Interpersonal Violence, 24*(14), 2811–2833.

CHAPTER 3

Crime and Crime Control

The business of the American criminal justice process is "crime"; yet, this does not explain much. One obstacle to the study of our system of criminal justice is our lack of precision in discussing the issue of crime. In short, what is crime?

Although it is relatively easy to provide examples of crime, it is not so easy to define it. We tend to assume a common meaning for the word "crime." The variety of actions and nuances of behavior that constitute crime is nearly infinite. If asked to name a crime, how many of us would say shoplifting, drunken driving, price fixing, or failure to register for the selective service? We are far more likely to mention murder, bank robbery, rape, or burglary. In that sense, we have a fairly clear common definition of crime, but one that is inadequate for the study of criminal justice.

These mental images of crime reflect those offenses that cause the most concern. Of the many different types of behaviors that we have defined as criminal, some types are more commonly agreed on to be criminal than others. There tends to be consensus among us about the criminality of the more serious offenses that involve actual physical harm or direct economic harm to individuals (Cullen, Link, & Polanzi, 1982). There is considerably less agreement about those offenses that do not cause such direct and potentially personal harm (Miethe, 1982; Newman & Trilling, 1975).

Similarly, we carry mental images of criminals about which there is general agreement. Most people would probably describe the average criminal as a relatively young, mean, menacing male. They seem to believe that the criminal knows—but does not care—that his or her behavior is wrong and harmful. The criminal is simply bad or lazy, preferring crime to some other more appropriate mode of earning a living or settling arguments. Yet, as with crimes, there is a wide variety of criminals. The hulking street offender, bullying rapist, calculating white-collar offender, college student selling drugs, and political terrorist are all criminals.

Faced with the wide array of crimes and criminals, we need to organize our understanding of each in order to appreciate the demands placed on the justice

IMPORTANT TERMS

actus reus

career criminal

crime

deviance

felony

"hot spot" policing

life-course criminality

mala in se

mala prohibita

mens rea

misdemeanor

problem-oriented policing

routine activities

strict liability

violation

system. Both crimes and criminals have been sorted into classes for ease of understanding. Before turning to these, however, we should try to answer the question, What is crime?

DEFINING CRIME

Crime refers in part to a set of behaviors that society deems to be wrong and in need of control. Most often, classifying a behavior as a crime includes a reference to the intent of the actor. In our society, the specification and definition of crimes are legislative functions. The legislature has the authority to declare certain behaviors criminal and describes the conditions under which a person may be said to have committed a crime. (Box 3.1 provides an example of a portion of a criminal statute.) Therefore, from a legalistic perspective, we can conclude the cause of crime in America is the legislature. Without legislative action, there would be no conduct designated as crime. **Crime** is an act or omission in violation of a law punishable by the state. In the United States, the requirement that the behavior violates the law means we need legislative action to establish crime.

BOX 3.1 SEC: 2911.12 OHIO REVISED CODE: BURGLARY

A   No person, by force, stealth, or deception, shall do any of the following:

1.   Trespass in an occupied structure or in a separately secured or separately occupied portion of an occupied structure, when another person other than an accomplice of the offender is present, with purpose to commit in the structure or in the separately secured or separately occupied portion of the structure any criminal offense.

2.   Trespass in an occupied structure or in a separately secured or separately occupied portion of an occupied structure that is a permanent or a temporary habitation of any person when any person other than an accomplice of the offender is present or likely to be present, with purpose to commit in the habitation any criminal offense.

3.   Trespass in an occupied structure or in a separately secured or separately occupied portion of an occupied structure, with purpose to commit in the structure or separately occupied portion any criminal offense.

B   No person, by force, stealth, or deception, shall trespass in a permanent or temporary habitation of any person when any person other than an accomplice of the offender is present or likely to be present.

Whoever violates division (A) in this section is guilty of burglary. A violation of division (A)(1) or (2) of this section is a felony of the second degree. A violation of division (A)(3) of this section is a felony of the third degree. Whoever violates division (B) of this section is guilty of trespass in a habitation when a person is present or likely to be present, a felony of the fourth degree.

Source: Ohio Revised Code (2010).

Although most of us do not think of legislative action as a necessary cause of crime, we understand the legislative role. If asked to define the cause of crime, most people will contend that bad companions, ignorance, poverty, psychological disturbance, or some other factor is what makes people break the law.

Yet, when we see someone doing something that we believe is wrong, we are also apt to say, "There ought to be a law." This statement reflects an understanding of the role of the legislature. No matter how wrong the behavior in question, we cannot do anything about it unless there is a law against it.

This is an important concept for understanding the criminal justice system. The justice system is constrained by the law. We generally cannot use the justice process to control behavior that is unpleasant but not criminal. There are limits, then, to what level of control agents of the criminal process can assert. Of course, the power of the justice system to control behavior often leads people to pass laws. In fact, Congress passes over 50 new laws per year, which has resulted in more than 4,500 criminal laws at the federal system alone (Koch & Holden, 2015).

Criminologists have debated the definition of crime for many years (Schwendinger & Schwendinger, 1975). Some argue that only those behaviors identified in criminal laws are crimes. Others seek a broader definition that includes actions that are socially harmful or immoral. The issue of defining crime is somewhat different for these criminologists than it is for our purposes because they are trying to explain deviance, of which crime is one type. If the focus is on deviant behavior, there is no need to consider legal status. **Deviance** is behavior that violates socially accepted standards of proper conduct.

The approach we are using in this book requires a definition of crime that identifies those behaviors on which the criminal justice system focuses. Criminologists, however, are seeking to identify a set of behaviors explained by theories of criminal behavior (Tittle, 2000). Criminologists usually seek to explain the behavior of individuals (Willis, Evans, & LaGrange, 1999), whereas we wish to understand the criminal justice system and its parts as a social institution.

The definitions of most crimes contain two components. First, there is an action (or lack of action) known as *actus reus*. Second, there is the intent or mental condition of the offender known as *mens rea*. To be considered a criminal, it is usually not enough to do something illegal; one must also intend to do what is illegal to be convicted of a crime.

Most jurisdictions define the crime of burglary as the unlawful entry of a place with the intention of committing a crime therein. To be convicted of burglary, one must unlawfully enter a place (a home, business, storage building, etc.). Simply entering, however, does not make one a burglar. The entry is the *actus reus*. To be a burglar, it is also necessary that one enter with the intent to commit a crime while inside. This intent to commit a crime is the mental state of the offender, the *mens rea*.

Neither of the following two hypothetical persons is a burglar:

> P. was invited to a party at a neighbor's home. P. did not care much for these neighbors, but cared greatly for several of their possessions. P. accepted the invitation and attended the party for the express purpose

of obtaining the property of the neighbors. P. is not a burglar, because the entry was lawful. In this case, P. is a thief.

Q. was walking home from a party at which large quantities of alcoholic refreshments were consumed (the largest quantity by Q.). Passing a clothing store, Q. blacks out from the combined effects of too many beverages and a long, tedious conversation with someone named P., a neighbor of the host. Q. falls over and crashes through the display window of the shop, landing in a huge pile of coats, on which Q. falls fast asleep. The police arrive within minutes, responding to the alarm, to discover the quietly resting Q., who is not a burglar, because there was no intent to commit a crime in the store. (Indeed, there was not even intent to enter.)

To obtain a criminal conviction, the state must prove all elements (both *actus reus* and *mens rea*) of an offense beyond a reasonable doubt. Television murder-mystery plots often include a missing victim when there is reason to believe that someone was murdered but the body cannot be found. The characters remark that it will be difficult to bring charges without the *corpus delicti*. Because of the plot, and the similarity between the words "corpus" and "corpse," audiences sometimes think that *corpus delicti* refers to the dead body. In fact, it refers to the body of the crime. The lack of a motive (also a frequent plot line) also hinders the filing of charges because, without a motive, it is difficult to establish intent, another part of the *corpus delicti*.

For some crimes, the job of the state in proving the guilt of an offender is somewhat easier. These crimes, known as **strict liability** offenses, are based simply on the action of the defendant (Samaha, 2017). In these cases, if the prosecution proves that the defendant engaged in the prohibited behavior, a conviction will occur. Regardless of the intent of the offender, she or he is strictly liable for the consequences of the behavior.

Strict liability often applies to white-collar crimes. For example, the law may presume that the head of a company is responsible for the wrongdoing of his or her employees, even if the company head is unaware of the activity. Especially with strict liability offenses, the old adage "Ignorance of the law is no excuse" is true.

This brief explanation of the definition of crimes and of the elements of an offense helps us understand the nature of crime. However, this explanation alone does not help us to obtain a perspective on crime that will be useful to our examination of the criminal justice process. The legislatures in the various U.S. criminal jurisdictions have managed to define a large number of widely divergent behaviors and mental states as crimes. The justice system must respond to all of them with limited resources and ability. To better organize and deploy the limited resources for the control of crime, criminal justice officials must sort and rank this plethora of offenses.

CLASSIFICATION OF CRIMES AND CRIMINALS

We have already mentioned one simple classification of crimes as the difference between serious and less serious offenses. This same simplistic distinction is drawn between dangerous and normal crimes. Those offenses that are most threatening to individuals are usually defined as serious (or dangerous), whereas those that are less directly threatening are classed as less serious (or normal).

Sometimes this distinction between dangerous and normal crimes is explained as the difference between offenses that are wrong in themselves (*mala in se*) and those that are wrong because they are prohibited (*mala prohibita*). That is, certain crimes appear to be obviously criminal, whereas others are apparently criminal only because we say they are wrong.

Mala in se offenses encompass traditional or street crimes that seem wrong regardless of their legality. Purposely or carelessly causing physical harm or suffering to someone, or taking the property of others, is an act that most people believe to be simply wrong. One does not need a criminal law to realize that killing a person without cause (and often with cause) is wrong. These are the very offenses about which we have the most agreement and around which most of our mental images of crime focus. As Luna (2003:1) put it, "every U.S. jurisdiction has on its books a set of crimes and punishments that are incontrovertible, involving acts and attendant mental states that must be proscribed to constitute a just society—murder, rape, robbery, arson and the like."

Mala prohibita offenses, however, are those acts that are wrong because they are defined as wrong. The use of narcotics by adults within the confines of their own homes and the refusal to pay income taxes (especially if the money might be better spent on something else) are not behaviors that are necessarily wrong (at least in a secular perspective). What makes these behaviors criminal, and therefore wrong, is that society has prohibited them and defined them as crimes. We have the least agreement on these offenses, and they raise the most serious issues of individual liberty. Luna (2003:15) wrote, "These offenses are marked by the absence of violence or coercion, with parties engaged in voluntary transactions for desired goods or services."

A second major way in which we classify criminal offenses is into categories of felony, misdemeanor, or violation. These three levels of crime reflect the different seriousness of behaviors. Before explaining this difference, it is important to remember that there are exceptions to every rule. The following are merely rules of thumb.

A **felony** is the most serious level of offense and generally is punishable by a term of more than 1 year in a state prison. A **misdemeanor** is a less serious offense, generally punishable by a term of no more than 1 year in a local jail. A **violation** is the least serious offense and typically does not carry an incarceration penalty; the penalty is limited to a fine or loss of privilege. For example, in most states, a theft of $1,000 or more is a felony and can be punished by imprisonment for a number of years, whereas theft of $50 is a misdemeanor and can be punished by a jail term of up to several months. Exceeding the speed limit is a violation and punished by a fine of less than $500 (except for repeated offenses). As these examples illustrate, the classes of crimes reflect the amount of harm caused by the criminal behavior.

Still another distinction drawn between crimes is to label them as being either ordinary (normal) or aggravated (dangerous) (Newman, 1987:28–30). It is possible to rate crimes as being "better" or "worse" than each other, within the same crime type. A burglar may not do any more damage to a home than that required for entry and theft; this is an ordinary burglary. However, the burglar may vandalize the home in addition to breaking in and stealing; this might be an aggravated burglary. We will return to this type of classification system later when we discuss sentencing.

The ordinary-versus-aggravated distinction is generally used within some less precise classification (such as "robbery") to differentiate between the seriousness of several instances of the "same" behavior. As with the other classifications of crimes, the purpose of this distinction is to clarify the response the justice system should make. Typically, the agents and agencies of the justice system are more willing to expend resources in response to aggravated felonies than in response to ordinary violations. Indeed, many people express this rational choice when stopped for a traffic violation by wondering why the officer is not out "fighting crime" rather than focusing on trivial matters.

Ample evidence suggests that the type of crime is an important consideration in criminal justice decisions. Serious crimes are more likely to be reported by victims and witnesses, and are more likely to be investigated and processed by the police (Wilson & Ruback, 2003). Criminal justice officials at all points of the system devote more attention and resources to aggravated, *mala in se* felonies than to other, less serious crimes. When trying to understand criminal justice processing then, offense seriousness is an important consideration.

When crimes are defined as felonies, they are more likely to be investigated than misdemeanors. Lytle (2013) reviewed the research on criminal justice processing across all types of criminal justice decisions from arrest through parole. After reviewing several factors, Lytle concluded that crime seriousness was the most significant predictor of decisions. For example, felonies were 2.5 times more likely to be arrested than misdemeanors. Further, the defendant's race, gender, and ethnicity played a factor in who was processed through the system. We will further examine the possibility of discrimination in the next chapter.

DEFINING CRIMINALS

Like crimes themselves, the people who commit them are of an infinite variety. Assuming that there is a preventive component to the justice system's overall mission to control crime, knowledge of the type of offender is as important as knowledge about the type of crime (Holmes, 1989). We identify certain types of offenders as deserving specific types of justice system responses. A Bureau of Justice Statistics report (1985:1) stated:

> Programs aimed at the serious, recidivistic offender require the capability to identify dangerous offenders at key decision points in the criminal justice system, such as pre-trial release and sentencing …

> These programs are designed primarily to increase the effectiveness of criminal justice by targeting resources on offenders considered most likely to recidivate and on offenders whose detention is most likely to have an incapacitative or deterrent effect.

One common method of classifying offenders is by the crime they committed—someone guilty of murder is a murderer, someone guilty of robbery is a robber, someone guilty of burglary is a burglar, and so on. This is often how the media and correctional authorities identify offenders. Police frequently improve on this simple scheme by adding details of crimes, such as the time of the crime, type of weapon used, and characteristics of the victim. These added details comprise a *modus operandi* (M.O.) file. This type of classification is limited because it does not tell much about the offender. Research has suggested that most offenders engage in a variety of criminal behavior (DeLisi & Piquero, 2011). Today's robber may have been yesterday's thief and may be tomorrow's burglar. However, some research suggests that offenders may specialize at least in terms of types of crime. Deane, Armstrong, and Felson (2005) reported that some offenders are likely to commit crimes of violence, whereas others are unlikely to be violent. Offenders may not specialize along the lines of specific crimes as much as within violent or nonviolent crimes.

Similarly, there are certain crime types that seem to "go together" and might represent something of a criminal lifestyle. Deane and his colleagues reported a link between armed robbery, other armed violence, selling drugs, and serious property crime. Criminals who engage in one of these crimes are also more likely to be involved in the others. Osgood and Schreck (2007) reported finding evidence of a specialization in violence among a sample of juvenile offenders, whereas Lo, Kim, and Cheng (2008) found modest support for specialization in violence among adults. These studies suggest that although it may not be accurate to speak of burglars or robbers, it may be accurate to classify offenders' crime types in terms of property offenders, violent offenders, or other types of offense such as drug offenders.

In contrast, Lin and Simon (2016) question the degree to which sex offenders specialize. A number of recent changes in criminal law, including such things as the Adam Walsh Child Protection and Safety Act and Megan's Law, which require sex offenders to register and limit where these offenders can live, are based on assumptions that these offenders specialize in sex offenses and will repeat their crimes. Lin and Simon found that among a sample of sex offenders only 4.5% had been arrested exclusively for sex offenses, and only 20% had more arrests for sex offenses than other types of crimes. Other research has found that rapists were less specialized than those who committed incest or child molestation (Harris, Smallbone, Dennison, & Knight, 2009). Still, even these latter offenders were more likely to commit a range of crimes than to be limited only to their sexual offenses. In short, evidence is mixed regarding the degree to which offenders specialize in the types of crime that they commit.

Another typical classification of offenders is similar to the ordinary/aggravated distinction applied to crimes. Here, offenders are identified as either "first-time" (ordinary) or "repeat" (more dangerous) offenders. Many jurisdictions have special procedures for the handling of repeat offenders, known as career criminals. The distinction drawn is one between periodic (or occasional) criminality and a criminal lifestyle. Those who lead a criminal life—that is, who routinely engage in criminal behavior—are responsible for a disproportionate share of crime committed (DeLisi, 2016).

Criminal Careers versus Career Criminals

Criminologists have studied criminal behavior through examination of the "criminal careers" of offenders (Sullivan & Piquero, 2016). Criminal careers have traditionally been studied through four dimensions (Blumstein, Cohen, Roth, & Visher, 1986). *Participation* refers to the distinction between those who engage in criminal behavior and those who do not. Researchers recognize that criminality is not always central to the personality of an offender. The average person probably has committed (or will someday commit) some type of crime. *Frequency* refers to how active the criminal is during their criminal career. Some criminals will commit many crimes in a short amount of time, while others will take breaks of months or even years between their criminal activities (Baker, Metcalfe, & Piquero, 2015). *Seriousness* refers to the relative severity of the crimes that an individual offender commits. Finally, *career length* is the total length of time that an individual is involved in criminal behavior. From a policy perspective, the ability to accurately predict an individual's criminal length would allow for punishment that protects the public during the offender's criminal career yet does not over-punish and thus "waste" prison capacity that would be better served for an offender who is actively committing criminal offenses (Blumstein, 2016).

The **career criminal** is someone for whom crime is a normal activity and for whom being a criminal is part of self-identification. This person is often called a "hardened criminal" or persistent offender. It has been suggested that these persistent offenders commit the majority of crimes (DeLisi, 2005). By attempting to focus attention on these individuals, justice officials hope to have the greatest impact on the crime rate.

Many researchers seek to identify the paths followed by offenders throughout their lives that lead them into and out of crime. In contemporary parlance, this approach to the study of criminals is known as the study of **life-course criminality**—how people engage in or refrain from crime over the course of their lives (Farrington, 2003). The life-course approach sees the decision to engage in criminal activity (and other human activities) as a combination of biological, psychological, and sociological forces which interact with and influence each other throughout a person's life (Wright, Tibbetts, & Daigle, 2008). For example, all of us have specific personality traits, are born into different family and social environments, and have a unique set of experiences and events which will occur throughout our lives. All of these factors, as well as their timing, are thought to impact behavior (Benson, 2013).

One of the best known theories of crime over the course of offenders' lives was suggested by Terrie Moffitt (1997). She suggested that some individuals engage in crime over their entire lives, whereas others experience certain periods (usually adolescence) when they have a greater risk of criminality. Most criminal activity is committed by relatively young individuals (see Figure 3.1). As each of us age, we continue to develop and make choices which are influenced by our personalities and life situations. Certain events, or turning points, such

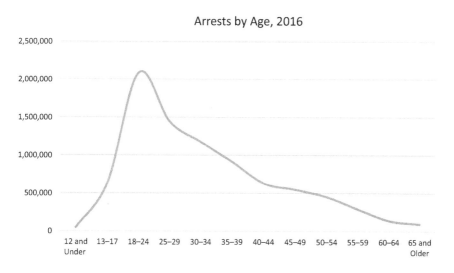

Arrests by Age, 2016

FIGURE 3.1
Arrests by Age, 2016.

Source: Federal Bureau of Investigation (2016).

as joining the military and marriage can also impact our lives and may make us less likely to engage in criminal behavior (Abeling-Judge, 2016; Sampson, Laub, & Wimer, 2006; Teachman & Tedrow, 2016). Alternatively, other life events such as being incarcerated may negatively impact our future choices, psychological well-being, and economic opportunities (Patterson & Wildeman, 2015). Life-course criminology is a subset of criminological theory that seeks to understand how people start, continue, and stop engaging in crime over the span of their lives.

Criminologists have begun to focus (in some cases, refocus) on the decision-making process of criminals. Some researchers focus on the "rational choice" theory, which treats criminals as rational decision makers who calculate the costs and benefits of crime before deciding to commit an offense (Vito & Maahs, 2017). For example, some evidence has suggested that sex offenders, who are often portrayed as impulsive and irrational, often plan out the crime and make rational decisions about their victims and the crime location to minimize the risks of apprehension (Beauregard & Leclere, 2007; Pedneault, Beauregard, Harris, & Knight, 2017). Recent evidence has also found that emotions such as fear and anger can impact the perceived costs and benefits of criminal activity (Bouffard, 2015). Understanding crime from the offender's perspective can help design and implement crime prevention strategies.

Another theme relates to place, or environmental criminology. Cohen and Felson (1979) described a **"routine activities"** theory of crime. They suggested that crime occurs when a motivated offender and a suitable target (victim or property) come together in time and space in the absence of an effective guardian. That is, there are criminals and victims in society. At some times, and in certain locations, they come into contact. Unless someone is there to prevent it (the guardian), a crime will occur. Crime, then, depends on the interaction of offenders, targets, and guardians. This theory suggests a structural approach to crime. By increasing guardianship, for example, crime can be reduced. Criminologists also have focused attention on repeat victims—those people who are frequently the victims of crime. For example, a study of female college students showed that increased partying and shopping frequency, as well as living off-campus, was significantly related to property victimization (Franklin, Franklin, Nobles, & Kercher, 2012). Another recent study of adolescent school children showed that routine activities measures were associated with the children's increased victimization by their classmates (Cho, Hong, Espelage, & Choi, 2017).

There seems to be increasing attention toward the idea that both life-course and situational factors play a role in criminal behavior. For example, Fox and Farrington (2016) examined the criminal histories of burglars in Florida. They classified burglars into five developmental categories: *young starters*, *late onset*, *low rate*, *high rate*, and *chronics*, which were based on the age at which the burglars began criminal activity, the number of years that they had persisted in criminal behavior, and the total numbers of property and violent crimes committed in that time frame. The same burglars were then placed into one of four situational categories: *opportunistic, organized, disorganized,* and *interpersonal,* which represented the level of planning that went into each crime along with other situational factors. Fox and Farrington found a significant relationship between the sets of developmental and situational categories. The organized burglaries were most likely to be committed by the chronic offenders, whereas the late onset offenders most often committed the interpersonal burglaries.

Although all of these are correctly labeled or categorized as "burglars," it is clear that they pose different levels of risk to citizens and that they require different responses by the police. The opportunistic burglar is likely to enter a property without force and grab whatever valuables can be quickly obtained and easily carried. The organized burglar is unlikely to be discovered during the crime and generally will take more high-valued items. Interpersonal burglaries most often targeted an occupied home and was motivated by anger (Fox & Farrington, 2016).

Coyne and Eck (2015) suggest that situational choice may be a more accurate term to describe the decision-making process of criminals and to bridge the gap between biosocial explanations for crime such as the life-course theory and situational approaches. They suggest that by understanding which physical environments may lead to less criminal behavior *and* addressing the way offenders and victims think about crime, the criminal justice system could develop successful crime prevention techniques.

CONTROLLING CRIME AND CRIMINALS

Francis A. Allen, former professor of law and dean of the University of Michigan Law School, was one of the first observers of American criminal justice to identify the increasing burden placed on the justice system by expansions of the criminal sanction. Allen studied the tremendous growth of criminal laws and increasing use of the justice system to deal with social problems ranging from substance abuse to health care. He considered the most important task to be the definition of what reasonably could be expected from the criminal justice system. In 1964, he wrote:

> The time has long been ripe for some sober questions to be asked. More and more it seems that the central issue may be this: What may we properly demand of a system of criminal justice? What functions may it properly serve? There is a related question: What are the obstacles and problems that must be confronted and overcome if a system of criminal justice is to be permitted to serve its own proper ends? These are broad and difficult questions, and the way in which

they are answered will affect much that is important to the community at large. (Allen, 1964:4)

At base, Allen was attempting to set priorities for the use of the criminal law. His position was that the criminal law was increasingly being applied to social welfare problems (such as public intoxication) and regulatory needs. He decried the growing reliance on criminal law to solve social problems. He urged that we decide on those behaviors that would best be the objects of criminal law, and that we limit the activities of the justice system to the control of these particular behaviors.

Other observers of American criminal justice shared this sentiment (American Friends Service Committee, 1971; National Advisory Commission, 1973). During the 1970s, a growing number of scholars and practitioners came to agree that the most sensible approach to crime control required the identification of "serious" crimes and the focusing of enforcement resources on those crimes. As the National Advisory Commission on Criminal Justice Standards and Goals (1973:84) explained, "The empire of crime is too large and diverse to be attacked on all fronts simultaneously."

Observers and agents of the justice system have long recognized the fundamental truth of this comment. Traditionally, police officers, prosecutors, judges, correctional personnel, and parole boards have adopted ad hoc strategies to maximize the effectiveness of criminal justice processing in controlling crime. In this vein, Davis (1975:1) noted:

The police make policy about what law to enforce, how much to enforce it, against whom, and on what occasions. Some law is always or almost always enforced, some is never or almost never enforced, and some is sometimes enforced and sometimes not.

With a large number of criminal laws applied to a broad variety of behavior, agents of the justice system often must choose which laws to enforce and when to enforce them (see Box 3.2). Moreover, with the wide variety of offenders, it is similarly common to enforce laws differently against different types of individuals. We are not likely to ignore the dangerous, repeat offender, regardless of the violation. In a sense, this approach to the "rationing" of justice resources seeks to maximize effectiveness. Officials use the criminal law to control the most serious offenses and offenders.

Criminal justice officials historically have devoted most of their resources to the control of more dangerous crimes and criminals. For example, police investigate suspected felonies more thoroughly than misdemeanors, and prosecutors are less willing to negotiate for guilty pleas from repeat felons. For the most part, this focus of attention on serious crimes (such as felonies and violent acts) had not been a conscious policy decision.

BOX 3.2 POLICY DILEMMA: FEDERAL MARIJUANA ENFORCEMENT

Since the late 1990s, more than half of the states have legalized some form of medicinal marijuana. This has occurred as the public's support for the legalization of marijuana is at an all-time high (Swift, 2016). However, marijuana is still considered a Schedule 1 drug (high abuse potential, no accepted medical use) according to federal law. Thus, individuals who use medical marijuana in accordance with their state laws are nonetheless violating federal law. Federal law enforcement officials generally target their enforcement efforts to those who are trafficking and distributing marijuana, rather than low-level users (Sacco & Finklea, 2014). In late 2016, the Ninth Circuit Court of Appeals (*U.S. v. McIntosh*) cited language used in a federal spending bill approved by Congess in 2014 and 2015 in its decision to prohibit federal prosecution against those who used marijuana legally in their states. Do you think that the federal government should strictly enforce the prohibition against marijuana? What impact do you think the legalization of medical marijuana in these states will have on the enforcement efforts in other states?

In the recent past, this unconscious rationing of resources (by justice officials choosing cases on which to concentrate) was exposed and adopted as formal policy in many jurisdictions. Over 30 years ago, Walker (1985:117) observed, "Career-criminal programs are the hottest fad in criminal justice these days." Programs designed to focus on career criminals existed in all three components of the criminal justice system. A brief review of the programs that flowed from these policies serves to illustrate the point and identify priorities in the justice system.

Law Enforcement Programs

Several police departments started programs that identified specific individuals as repeat offenders. The police in Washington, DC, implemented the "Repeat Offender Program" (ROP) described in Box 3.3. "Operation Cease Fire" in Chicago sought to reduce violence by targeting those identified as having a high chance of either being shot or being a shooter in the immediate future (Skogan, Hartnett, Bump, & Dubois, 2008). In each of these programs, police administrators decided to devote resources to the control of specific crimes and criminals—or to the control of crime in areas where there was reason to believe that serious crime was most likely to occur. At the same time, regular police patrol and response to calls for service in the jurisdiction continued. In the departments mentioned, and in others with similar programs, police administrators have at least tacitly decided that some crimes or criminals are more deserving of police attention.

Another type of police activity that seeks to control serious crime and criminals involves the police identifying high-risk areas where serious crimes appear likely to occur, and devoting increased patrol to those areas (see Figure 3.2). Often police departments have responded to "crime waves" by increasing police presence in a given area. This is commonly known as **"hot spot" policing** and

BOX 3.3 CAREER CRIMINALS AND THE POLICE

The Washington, DC, repeat offender program (ROP) grabbed the attention of many metropolitan police departments across the country after a 2-year evaluation of the program was completed by the Police Foundation, under a grant from the U.S. Justice Department's National Institute of Justice.

The study found that the ROP program achieved its goals in every area. For example, the program increased the likelihood of arrests of the target offenders and increased officer productivity (the officers actually made fewer arrests, but the arrests were for offenders with longer criminal histories). In addition, the program increased the likelihood that the

offender would be prosecuted and increased the chances of a conviction.

The authors of the evaluation did urge some caution. The ROP program was costly to implement, and the trade-off for the focus in crime-fighting activities appeared to be reduced order-maintenance activities by the police. Finally, the authors noted that without careful supervision, this program could lead to increased opportunities to violate a citizen's rights.

Source: Martin and Sherman (1986).

FIGURE 3.2

A patrol car passes Dealey Plaza in front of the Texas School Book Depository building in Dallas after five police officers were killed in July 2016.

AP Photo/Eric Gay

has been shown over the past two decades to be one of the most effective police strategies (Weisburd & Telep, 2014).

Police departments vary greatly in terms of their definition of a hot spot and the actions that law enforcement officers are expected to take while patrolling hot spots. Hot spot policing often means simply to increase the number of patrol officers assigned to a particular location. Commanders at one large police department applying hot spot policing described the need to "generate activity," referring to the aggressive enforcement of traffic laws and arresting lawbreakers (Haberman, 2016:498). Groff et al. (2015) found that hot spot policing was most effective when officers focused on individuals who had a long history of criminal activity. Working from a list of such offenders living in a high-crime area, officers would maintain surveillance and make frequent contact with these individuals. Groff et al. found that these tactics resulted in a 50% reduction in violent felonies.

Hot spot policing is not without its critics. Rosenbaum (2006:253) argued that the practice "can drive a wedge between the police and the community." As is evident with the many public protests involving police use of force in recent years, there seems to have been an erosion of public trust in law enforcement.

Critics fear that the increased police attention to criminal hot spots, which often are located in poor, disadvantaged communities, might have contributed to this erosion. Weisburd (2016) suggests that hot spot policing does not inevitably lead to abusive practices by law enforcement. He points out that, implemented correctly, hot spot policing is focused on particular streets in communities, instead of the entire community. His suggestion was for the police to be transparent with their communities about the location of the high-crime areas and collaborate with them regarding the proper focus of the police activities.

Court Programs

The repeat or career criminal is the subject of special treatment in at least two points of the court process: prosecution and sentencing. The goal of these special procedures is to ensure the conviction and punishment of offenders posing the greatest threat of future criminality.

Increasingly prosecutor's offices have established specialized units devoted to particular crimes. It is not uncommon for a prosecutor's office to have a domestic violence unit or a drug unit. Here again, criminal justice agents have decided to focus effort and resources on particular types of crime or criminals. At the national level, the U.S. Department of Justice has implemented Project Safe Neighborhoods, which focuses criminal justice attention on firearms offenders. The U.S. Attorney (federal prosecutor) administers the program in each of the 94 U.S. district courts. The project involves the creation of local task forces comprised of the U.S. Attorney and representatives from federal, state, and local law enforcement; prosecutors; other justice agencies; and other local leaders. The task forces identify gun crime problems and develop strategies to reduce gun crime, often involving federal prosecution for illegal gun possession and violent, gang, and drug-related offenses involving a firearm (Bureau of Justice Assistance, 2004). Evidence suggests that these measures do in fact result in a modest decline in violent crime (McGarrell, Corsaro, Hipple, & Bynum, 2010).

Differential handling of career criminals at sentencing has long been a tradition in America. Sentencing programs enacted from the late 1980s to the mid-1990s include: selective incapacitation, mandatory sentencing, and "three-strikes" laws. In each, the goal is either to control the incidence of specific crimes, or to lessen (through incarcerating offenders) the opportunity of specific criminals to commit crime in the future.

Selective incapacitation seeks to identify those offenders who are most likely to commit future crimes. This program reserves incarceration for these habitual offenders (Greenwood, 1982). Research showing the majority of crimes committed by a minority of offenders supports the idea that imprisoning those few would result in less crime. Selective incapacitation argues that because prison space is a scarce resource, reserving space for the most prolific offenders is a wise investment.

Mandatory sentencing is another strategy for controlling crime. In practice, mandatory sentencing really means mandatory incarceration. This approach relies on deterrence and does not target specific criminals, focusing instead on specific crimes. By "ensuring" that those convicted of specific crimes we believe to be dangerous will be imprisoned, the hope is that the program will deter those who might consider committing the offense.

Finally, following in this tradition, the federal government and most states began passing three-strikes laws starting in the mid-1980s. At the federal level, these laws began with the passage of the Armed Career Criminal Act in 1984, which subjected those who were convicted of certain gun crimes to a sentence of between 15 years and life if he or she had three prior qualifying convictions. Then, with the passage of the Violent Crime Control and Law Enforcement Act of 1994, offenders convicted of specific felonies for the third time were expected to be imprisoned for life (Saint-Germain & Calamia, 1996). Many states have passed three-strikes laws with similar goals, although implementation of these laws varies widely by state (Schiraldi, Colburn, & Lotke, 2004; Turner, Sundt, Applegate, & Cullen, 1995). These laws seek to identify serious, repeat criminal offenders, and ensure that they receive sentences that result in long terms of imprisonment. With the help of federal incentives, states have also moved to ensure the imprisonment of those convicted of violent crimes under what is known as Violent Offender Incarceration/Truth in Sentencing programs (Ditton & Wilson, 1999).

As with law enforcement and criminal prosecution programs, sentencing programs either target criminals (selective incapacitation) or crimes (mandatory sentencing). In both cases, a decision has been made that resources (e.g., prison) should be targeted to specific cases of criminality.

Corrections Programs

Efforts to identify and provide special services and controls for repeat offenders in correctional settings are prevalent. Correctional officials have given increased attention to classification of offenders, and have developed special "intensive supervision" programs for probationers and parolees. These efforts try to focus correctional resources on those offenders most in need of such attention.

Casey, Warren, and Elek (2011) indicated the importance of classification for the organization and delivery of correctional services. They suggested that we get the greatest return on the correctional investment through matching the available services and programs with the needs of individual offenders. The implication of classification decisions for the effectiveness of correctional treatments is only one incentive for the increasing emphasis on this process (see Figure 3.3; Bonta, 1996; Bonta & Motiuk, 1992).

In an age of prison crowding (see Figure 3.4), classification became important as a check on the efficient use of correctional resources. Clear and Cole (1986:320) noted:

> The prison crowding crisis and litigation challenging existing procedures have forced many correctional systems to reexamine their classification procedures. As space becomes a scarcer and more valuable resource, administrators feel pressured to ensure that it is used as efficiently as possible: that levels of custody are appropriate and that inmates are not held in "oversecure" facilities.

FIGURE 3.3
Inmates at the Deuel Vocational Institute in Tracy, California, are seen housed in three-tier bunks, in what was once a multipurpose recreation room. In an age of prison crowding, classification has become an important check on the efficient use of prison resources.

AP Photo/Rich Pedroncelli, File

The other side of this, of course, is that we should not hold certain dangerous prisoners in "undersecure" facilities. Among other things (such as amenability to certain types of treatment or aptitude for certain job assignments), classification also reflects risk of future criminality. A large part of the classification decision in prisons, or within probation or parole caseloads, reflects a desire to identify and control the repeat offender.

A related development in corrections involves the use of "intensive supervision" with probationers and parolees. In these programs, officials classify offenders under community supervision by risk and need. Those posing the greatest risk of future crime and those presenting the greatest needs for service are assigned to special "intensive supervision" caseloads. These caseloads are smaller than the average probation or parole caseload, and the supervising officer is expected to make more contacts with his or her clients each month. Thus, the title "intensive supervision" reflects a greater concentration of resources on offenders who are selected due to their perceived increased risk of future criminality.

The classification of prisoners is most often made with the assistance of a risk assessment tool. These tools evaluate offenders using a scoring method based on the presence or absence of several risk factors. Risk factors typically include the offender's criminal history, previous substance abuse, performance in school or

FIGURE 3.4
Prisoners under State or Federal Jurisdiction, 1990–2016.

Source: Beck and Harrison (2001); Carson (2018).

work, and the presence of antisocial personalities (Casey, Warren, & Elek, 2011). By using these tools, correctional officials hope to identify those most at risk of reoffending and provide treatment which targets their specific needs.

With many of the policies and programs being at least 20 years old, all of this discussion of career criminals and the justice system response to them may seem dated, until we realize that these programs and others like them continue today. They illustrate how criminal justice resources are devoted to those crimes and criminals defined as most serious, dangerous, or threatening. What has changed is the rate and number of persons sentenced to prison. From the early 1990s to the mid-2000s, we increasingly sentenced offenders to prison, leading to what some observers have called a policy of "mass incarceration" (Crutchfield, 2004). However, it is important to note that these "tough on crime" policies are not the cause for our selective focus on career criminals. Even those who are calling for major reforms in the nation's mass incarceration policies (e.g., Tonry, 2014) advocate for our resources to be spent proportional to the crime committed. Regardless of the specific direction that our future policies take, the focus on career criminals will likely persist.

A NEW DIRECTION FOR CRIME CONTROL

Research into the causes of crime and the effectiveness of crime prevention practices during the past 40 years has produced some changes in how criminal justice system officials seek to control crime. There is a noticeable movement today in the direction of addressing the context of crime. Recall that police are increasingly focusing on "hot spots" to control crime. While the available hot spot policing research has established its effectiveness, a question remains as to what exactly the police officers should be doing after identifying a location as a hot spot. To assist with this question, increasing attention is turning to a group of approaches based on environmental criminology. These approaches attempt to develop a greater understanding of why crimes occur at specific times and locations and then create effective intervention techniques to reduce the criminal behavior (Wortley & Townsley, 2017). Rather than viewing crime as a product of individual offenders and seeking crime prevention through control of those offenders, this perspective sees crime as a product of a social context and seeks to control crime by changing those contexts.

Returning to our earlier discussion of social control, current strategies of crime control seek to strengthen informal social control by removing obstacles to informal control and strengthening mechanisms of control other than the criminal justice process. Changing traffic patterns by blocking streets, for example, reduces the number of strangers in a neighborhood, thereby making it easier for residents to recognize those who belong in the neighborhood and those who do not belong. Using civil laws to close bars, evict tenants who sell drugs, and similar efforts reduce the chances that crimes will occur. The use of curfews to remove juveniles from the streets in the late night hours also works to reduce the likelihood of crime and misbehavior. Rather than waiting for a fight or robbery to occur and then calling the police, these strategies work to prevent the fight or robbery in the first place.

In support of this more preventive approach to dealing with problems of crime, **problem-oriented policing** has developed as a widely accepted police strategy throughout the country. This method of policing uses the SARA model to problem-solve using a situational crime prevention approach (see Box 3.4). Often used in coordination with hot spot policing, problem-oriented policing focuses on individualized responses to specific local problems that may arise in a community. These problems are often limited to a specific business, house, parking lot, or street. Other problems might involve an entire neighborhood or district (Maguire, Uchida, & Hassell, 2015). The Center for Problem-Oriented Policing (www.popcenter.org) has available more than 70 problem-specific guides and 13 response guides from which the police can find answers to their specific problems.

Interventions can take many forms. In some cases, changing the physical environment by improving lighting, altering traffic patterns, installing locks and bars, removing bushes, and the like may reduce the chances of crime (Armitage, 2017). In other cases, increased police patrols or organizing Neighborhood

---

BOX 3.4 SARA MODEL

A commonly used problem-solving method is the SARA model. The SARA model contains the following elements:

Scanning

- Identify recurring problems
- Prioritize the problems
- Develop goals

Analysis

- Identify and understand events and conditions that precede and accompany the problem
- Research what is known about the problem type
- Narrow the scope of the problem
- Develop a working hypothesis about why the problem is occurring

Response

- Brainstorm for new interventions
- Search for what other communities with similar problems have done
- Outline a response plan
- Carry out the planned activities

Assessment

- Determine whether the plan was implemented properly
- Collect pre- and post-response data
- Determine whether the broad goals and specific objectives were attained

Adapted from www.popcenter.org

Watch groups can reduce the chances of crime by increasing the risks of getting caught (Clarke, 2017). In still other instances, it may be that we need to identify and control specific high-rate offenders. In an evaluation of nearly 400 interventions coordinated by one large police department, Braga and Schnell (2013) found that nearly half of the interventions were those involving environmental-based changes (e.g., improved lighting, graffiti removal). Community outreach (e.g., holding a community event, establishing new recreational opportunities) comprised approximately 30% of the interventions. Finally, approximately 20% of the interventions conducted by this police department were enforcement-based.

Scores of programs across the nation have been implemented and assessed. Box 3.5 highlights efforts to deal with the problem of youth violence in Boston (Office of Juvenile Justice and Delinquency Prevention, 1999). This effort involved a wide range of criminal justice and community agencies. It evolved over several years and gained national recognition as an exemplary project. Recognizing that homicide was an especially pressing problem among youths, the Boston Police Department organized a Youth Violence Strike Force. The police cooperated with the federal Bureau of Alcohol, Tobacco, Firearms, and Explosives, as well as the prosecutor's office (both local and federal) and the probation department, and implemented strategies for reducing homicides. The targets of these efforts were gun traffickers (including licensed gun dealers) and gang members. Enforcement strategies involved inspections of gun dealers and targeted investigations of gun traffickers. Police officers tried to obtain gun market information from offenders charged with serious nongun charges. Gang members were identified and warned that violence (especially gun violence) would no longer be tolerated. In cooperation with prosecutors and probation personnel, the police met with gang members and warned that any violence would receive swift, severe consequences. This illustrates a focused deterrence strategy often called "pulling levers," which has been shown to be more effective than traditional unfocused policing strategies (Telep & Weisburd, 2012). In the end, these efforts seem to have reduced the incidence of homicide, and gang-related homicide in particular. The combined, coordinated efforts of police, courts, and corrections, at local, state, and federal levels, in conjunction with community service agencies, were brought to bear on the problem of juvenile homicides. One of the most remarkable things about the Boston project is not its results, but rather, the coordination of so many different agents and agencies to focus on one specific problem.

BOX 3.5 A COORDINATED APPROACH TO YOUTH VIOLENCE IN BOSTON

Beginning in 1994, the Boston Police Department identified serious violence among youths as a major problem. During the next several years, a coordinated citywide strategy emerged that involved a combination of law enforcement and crime prevention programs. Criminal justice, local government, and private social service and commercial organizations combined their efforts to address the problem. Among others, the strategy involved:

Operation Cease Fire—A collaboration between the ATF (the U.S. Bureau of Alcohol, Tobacco, Firearms, and Explosives), the Boston Police Department, the U.S. Attorney, the local prosecutor, the state probation department, the State Department of Youth Services, clergy, schools, and social service agencies to identify gang members and to warn them that violence would be met with severe sanctions.

Boston Gun Project—A collaboration between the Boston Police, the ATF, the U.S. Attorney, and the local prosecutor to reduce the number of guns available to youths, and to disrupt the illegal gun market.

Operation Night Light—A collaboration between the Boston Police and the probation department to increase surveillance and supervision of youthful offenders on probation through increased unannounced visits to probationers in their homes, schools, and workplaces during nontraditional hours (7 P.M. to midnight).

Over the past four decades, problem-oriented policing has established a strong track record of success (Scott, Eck, Knutsson, & Goldstein, 2017). For example, Weisburd, Telep, Hinkle, and Eck (2010) conducted an exhaustive review of the research involving problem-oriented policing. They found that among 45 studies that had compared crime data before and after a problem-oriented intervention, the average crime reduction was more than 40%. Studies that had used an experimental design showed a more modest but still significant reduction in crime. Some researchers have even advocated for the elimination or reduction of standard policing tactics such as random preventive patrol to focus on solutions, including problem-oriented policing and focused deterrence programs (Telep & Weisburd, 2012).

A situational or contextual approach to crime prevention holds great promise of reducing the incidence of crime, but it is not without its limits. Recall our discussion of the central conflict between due process and crime control (see Box 3.6). A juvenile curfew may prevent much crime, but it is unlikely that the juveniles affected by the curfew appreciate the limits placed on their freedom. Blocking streets to reduce through traffic may prevent some offenders from entering a neighborhood, but it also inconveniences residents who live in the area (Lasley, 1998). The current effort to go beyond traditional criminal justice practices to achieve greater crime control poses the threat of expanding crime control at the cost to individual liberty or due process.

BOX 3.6 DUE PROCESS VERSUS CRIME CONTROL: A DELICATE BALANCE

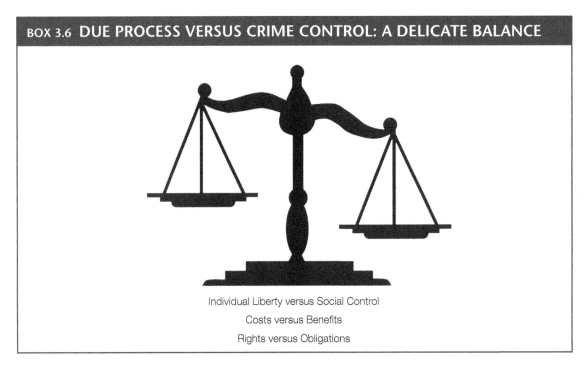

Individual Liberty versus Social Control

Costs versus Benefits

Rights versus Obligations

The new focus on the context of crime does not ignore offenders, but treats the criminal as only one of several factors that account for crime. The core logic of this argument is that offenders are only part of the crime problem, and focusing crime control efforts on offenders will then control only some of all crime. If the factors that account for crime vary, our crime control efforts should target various factors (Sherman et al., 1997; Worrall, 2008).

A final product of this broader view of crime control is the integration of criminal justice and community services. For example, training is now available to help police identify offenders who have a mental illness or intellectual disability and refer them to the appropriate services rather than placing them in jails and prisons (Edwards & Pealer, 2018). It is also now common for interagency task forces to work toward the solution of crime problems. An evaluation of Cincinnati's problem-oriented policing program showed that two-thirds of the problems required intervention by a person or entity other than law enforcement (Eck, 2014). Partnerships must be developed between criminal justice agencies and other governmental and community offices and groups not directly parts of the criminal justice system.

CRIME CONTROL IN GENERAL

This chapter began with a definition of crime and an explanation of the elements of criminal offenses. It then moved to a discussion of criminals and criminal types. We suggested throughout the discussion of these topics that the variety of behaviors and individuals included in the concepts of "crime" and "criminal," respectively, are exceedingly diverse. The justice system must respond to a large number of widely divergent instances and individuals.

To organize our perspective of this otherwise cumbersome task, we discussed the use of classes of crime or classes of offenders to allocate justice system resources. We provided descriptions of current efforts of police, court, and correctional agencies to focus attention and resources on the most serious crimes and criminals. It is important to establish priorities given the broad crime control mandate of the criminal justice system.

In practice, agents and agencies of criminal justice will respond to more serious crimes and offenders before responding to less serious incidents. For the most part, felonies and repeat offenders are more likely to attract the attention of the justice system and to receive full-scale justice processing than are misdemeanants and first offenders.

Our attention to the effect of the criminal justice process on rates and levels of crime in this chapter illustrates a core problem of criminal justice theory. Although our focus remains on the decisions in the system—how cases proceed from detection through the system—there is an equally important focus on the effect of the system on crime. Sometimes we find the explanation for why a case continues through the system is that officials expect a crime control payoff. The things that explain crime are not—and perhaps should not be—the same as those that explain criminal justice.

A redefinition of crime control that encompasses a broader range of social issues covering the context of crime promises to improve crime prevention but raises concerns about individual liberty. The task of the criminal justice system and criminal justice policymakers is to achieve a balance between crime control needs and due process requirements. Spader (1994) suggested that this balance involves weighing costs and benefits and competing values. The central questions that emerge in the conflict between due process and crime control in the American criminal justice system reflect these issues: Under what circumstances does the value of social control outweigh the value of individual liberty? How preventive should the agents and agencies of criminal justice be in their approach to crime control? How should we decide (and who should decide) which crimes and criminals deserve criminal justice attention? How much of our limited criminal justice resources should be devoted to "noncriminal" prevention activities versus detecting and apprehending criminals? We will return to these topics in later chapters.

REVIEW QUESTIONS

1. Define "crime."
2. What are the elements of a criminal offense?
3. Identify at least two ways we can classify crimes according to the level of their seriousness.
4. Define what is meant by the term "career criminal."
5. Give an example of career-criminal programs in each segment of the criminal justice system (law enforcement, courts, and corrections).
6. What is the situational approach to crime control/prevention, and how does it differ from traditional approaches?
7. What key issues emerge when agents of the justice system choose crimes and criminals on which to focus attention?

REFERENCES

Abeling-Judge, D. (2016). Different social influences and desistance from crime. *Criminal Justice and Behavior, 43*(9), 1225–1241.

Allen, F.A. (1964). *The borderland of criminal justice.* Chicago: University of Chicago Press.

American Friends Service Committee (1971). *Struggle for justice.* New York: Hill & Wang.

Armitage, R. (2017). Crime prevention through environmental design. In Wortley, R. & Townsley, M. (Eds.), *Environmental criminology and crime analysis* (pp. 259–285). New York: Routledge.

Baker, T., Metcalfe, C.F., & Piquero, A.R. (2015). Measuring the intermittency of criminal careers. *Crime & Delinquency, 61*(8), 1078–1103.

Beauregard, E., & Leclere, B. (2007). An application of the rational choice approach to the offending process of sex offenders: A closer look at the decision-making. *Sex Abuse, 19,* 115–133.

Beck, A., & Harrison, P. (2001). *Prisoners in 2000.* Washington, DC: Bureau of Justice Statistics, U.S. Department of Justice.

Benson, M.L. (2013). *Crime and the life course* (2nd ed.). New York: Routledge.

Blumstein, A. (2016). From incapacitation to career criminals. *Journal of Research in Crime and Delinquency, 53*(3), 291–305.

Blumstein, A., Cohen, J., Roth, J.A., & Visher, C.A. (1986). *Criminal careers and career criminals*. Report of the National Academy of Sciences Panel on Criminal Careers. Washington, DC: National Academy Press.

Bonta, J. (1996). Risk-needs assessment and treatment. In A. Harland (Ed.), *Choosing correctional options that work* (pp. 18–32). Thousand Oaks, CA: Sage.

Bonta, J., & Motiuk, L. (1992). Inmate classification. *Journal of Criminal Justice, 20*(4), 343–352.

Bouffard, J.A. (2015). Examining the direct and indirect effects of fear and anger on criminal decision making among known offenders. *International Journal of Offender Therapy and Comparative Criminology, 59*(13), 1385–1408.

Braga, A.A., & Schnell, C. (2013). Evaluating place-based policing strategies: Lessons learned from the smart policing initiative in Boston. *Police Quarterly, 16*(3), 339–357.

Bureau of Justice Assistance (2004). *Program brief: Project Safe Neighborhoods: America's network against gun violence*. Washington, DC: Bureau of Justice Assistance.

Bureau of Justice Statistics (1985). *Special report: Crime control and criminal records*. Washington, DC: U.S. Department of Justice.

Carson, E.A. (2018). *Prisoners in 2016*. Washington, DC: Bureau of Justice Statistics, U.S. Department of Justice.

Casey, P.M., Warren, R.K., & Elek, J.K. (2011). *Using offender risk and needs assessment information at sentencing: Guidance for courts from a national working group*. Williamsburg, VA: National Center for State Courts.

Cho, S., Hong, J.S., Espelage, D.L., & Choi, K. (2017). Applying the lifestyle routine activities theory to understand physical and nonphysical peer victimization. *Journal of Aggression, Maltreatment, & Trauma, 26*(3), 297–315.

Clarke, R.V. (2017). Situational crime prevention. In Wortley, R. & Townsley, M. (Eds.), *Environmental criminology and crime analysis* (pp. 286–303). New York: Routledge.

Clear, T.R., & Cole, G.F. (1986). *American corrections*. Monterey, CA: Brooks/Cole.

Cohen, L., & Felson, M. (1979). Social change and crime rate trends: A routine activities approach. *American Sociology Review, 44*, 588–608.

Coyne, M.A., & Eck, J.E. (2015). Situational choice and crime events. *Journal of Contemporary Criminal Justice, 3*(1), 12–29.

Crutchfield, R. (2004). Commentary: Mass incarceration, editorial introduction. *Criminology and Public Policy, 3*(2), 265–266.

Cullen, F.T., Link, B.G., & Polanzi, C.W. (1982). The seriousness of crime revisited. *Criminology, 20*(1), 83–102.

Davis, K.C. (1975). *Police discretion*. St. Paul, MN: West.

Deane, G., Armstrong, D., & Felson, R. (2005). An examination of offense specialization using marginal logit models. *Criminology, 43*(4), 955–988.

DeLisi, M. (2005). *Career criminals in society*. Thousand Oaks, CA: Sage.

DeLisi, M. (2016). Career criminals and the antisocial life course. *Child Development Perspectives, 10*, 53–58.

DeLisi, M., & Piquero, A.R. (2011). New frontiers in criminal careers research, 2000–2011: A state of the art review. *Journal of Criminal Justice, 39,* 289–301.

Ditton, P., & Wilson, J. (1999). *Truth in sentencing in state prisons.* Washington, DC: Bureau of Justice Statistics.

Eck, J.E. (2014). *The status of collaborative problem solving and community problem-oriented policing in Cincinnati.* Cincinnati, OH: School of Criminal Justice, University of Cincinnati.

Edwards, B., & Pealer, J. (2018). Policing special needs offenders: Implementing training to improve police–citizen encounters. In Dodson, K. (Ed.), *Routledge handbook on offenders with special needs.* New York: Routledge.

Farrington, D. (2003). Developmental and life-course criminology: Key theoretical and empirical issues—The 2002 Sutherland Award Address. *Criminology, 41*(2), 221–256.

Federal Bureau of Investigation (2016). *Crime in the United States, 2016.* Retrieved from https://ucr.fbi.gov/crime-in-the-u.s/2016/crime-in-the-u.s.-2016/topic-pages/tables/table-20

Fox, B.H., & Farrington, D.P. (2016). Is the development of offenders related to crime scene behaviors for burglary? Including situational influences in developmental and life-course theories of crime. *International Journal of Offender Therapy and Comparative Criminology, 60*(16), 1897–1927.

Franklin, C.A., Franklin, T.W., Nobles, M.R., & Kercher, G.A. (2012). Assessing the effect of routine activity theory and self-control on property, personal, and sexual assault victimization. *Criminal Justice and Behavior, 39*(1), 1296–1315.

Greenwood, P.W. (1982). *Selective incapacitation.* Santa Monica, CA: RAND.

Groff, E.R., Ratcliffe, J.H., Haberman, C.P., Sorg, E.T., Joyce, N.M., & Taylor, R.B. (2015). Does what police do at hot spots matter? The Philadelphia policing tactics experiment. *Criminology, 53,* 23–53.

Haberman, C.P. (2016). A view inside the black box of hot spots policing from a sample of police commanders. *Police Quarterly, 19*(4), 488–517.

Harris, D., Smallbone, S., Dennison, S., & Knight, R. (2009). Specialization and versatility in sexual offenders referred for civil commitment. *Journal of Criminal Justice, 37*(1), 37–44.

Holmes, R. (1989). *Profiling violent crimes: An investigative tool.* Beverly Hills, CA: Sage.

Koch, C.G., & Holden, M.V. (2015, Jan. 7). The overcriminalization of America. *Politico.* Retrieved from: www.politico.com/magazine/story/2015/01/overcriminalization-of-america-113991

Lasley, J. (1998). *"Designing Out" gang homicides and street assaults.* Washington, DC: National Institute of Justice.

Lin, J., & Simon, W. (2016). Examining specialization among sex offenders released from prison. *Sexual Abuse: A Journal of Research and Treatment, 28*(3), 253–267.

Lo, C., Kim, Y., & Cheng, T. (2008). Offense specialization of arrestees: An event history analysis. *Crime & Delinquency, 54,* 341–365.

Luna, E. (2003). Overextending the criminal law. *CATO Police Report, 25*(6), 1, 15–16.

Lytle, D.J. (2013). *Decision making in criminal justice revisited: Toward a general theory of criminal justice* (Unpublished doctoral dissertation). University of Cincinnati, Cincinnati, OH.

Maguire, E.R., Uchida, C.D., & Hassell, K.D. (2015). Problem-oriented policing in Colorado Springs: A content analysis of 753 cases. *Crime & Delinquency, 6*(1), 71–95.

Martin, S., & Sherman, L. (1986). *Catching career criminals: The Washington, DC repeat offender program. Police Foundation Report*. U.S. Department of Justice.

McGarrell, E., Corsaro, N., Hipple, N., & Bynum, T. (2010). Project Safe Neighborhoods and violent crime trends in US cities: Assessing violent crime impact. *Journal of Quantitative Criminology, 26*, 165–190.

Miethe, T.D. (1982). Public consensus on crime seriousness. *Criminology, 20*(3–4), 515–526.

Moffitt, T. (1997). Adolescence-limited and life-course persistent offending: A complementary pair of developmental theories. In T. Thornberry (Ed.), *Developmental theories of crime and delinquency*. New Brunswick, NJ: Transaction.

National Advisory Commission on Criminal Justice Standards and Goals (1973). *A national strategy to reduce crime*. Washington, DC: U.S. Government Printing Office.

Newman, D.J. (1987). *Introduction to criminal justice* (3rd ed.). New York: Random House.

Newman, G.R., & Trilling, C. (1975). Public perceptions of criminal behavior. *Criminal Justice & Behavior, 2*(2), 217.

Office of Juvenile Justice (1999). *Promising strategies to reduce gun violence*. Washington, DC: Office of Juvenile Justice and Delinquency Prevention.

Osgood, D., & Schreck, C. (2007). A new method for studying the extent, stability, and predictors of individual specialization in violence. *Criminology, 45*(2), 273–312.

Patterson, E.J., & Wildeman, C. (2015). Mass imprisonment and the life course revisited: Cumulative years spent imprisoned and marked for working-age black and white men. *Social Science Research, 53*, 325–337.

Pedneault, A., Beauregard, E., Harris, D.A., & Knight, R.A. (2017). Myopic decision making: An examination of crime decisions and their outcomes in sexual crimes. *Journal of Criminal Justice, 50*, 1–11.

Rosenbaum, D. (2006). The limits of hot spots policing. In Weisburd, D. & Braga, A. (Eds.), *Police innovation: Contrasting perspectives*. New York: Cambridge University Press.

Sacco, L.N., & Finklea, K. (2014). *State marijuana legalization initiatives: Implications for federal law enforcement* (CRS Report No. R43164). Washington, DC: Congressional Research Service.

Saint-Germain, M., & Calamia, R. (1996). Three strikes and you're in: A streams and windows model of incremental policy change. *Journal of Criminal Justice, 24*(1), 57–70.

Samaha, J. (2017). *Criminal law* (12th ed.). Boston: Cengage.

Sampson, R.J., Laub, J.H., & Wimer, C. (2006). Does marriage reduce crime? A counterfactual approach to within-individual causal effects. *Criminology, 44*(3), 465–508.

Schiraldi, V., Colburn, J., & Lotke, E. (2004). *Three strikes and you're out: An examination of the impact of 3-strikes laws 10 years after their enactment*. Washington, DC: Justice Policy Institute.

Schwendinger, H., & Schwendinger, J. (1975). Defenders of order or guardians of human rights? In Taylor, I., Walton, P., & Young, J. (Eds.), *Critical criminology* (pp. 113–146). London: Routledge & Kegan Paul.

Scott, M.S., Eck, J.E., Knutsson, J., & Goldstein, H. (2017). Problem-oriented policing. In Wortley, R. & Townsley, M. (Eds.), *Environmental criminology and crime analysis* (pp. 227–258). New York: Routledge.

Sherman, L., Gottfredson, D., MacKenzie, D., Eck, J., Reuter, P., & Bushway, S. (1997). *Preventing crime: What works, what doesn't, what's promising?* Washington, DC: National Institute of Justice.

Skogan, W.G., Hartnett, S.M., Bump, N., & Dubois, J. (2008). *Evaluation of Ceasefire-Chicago.* Washington, DC: U.S. Department of Justice, National Institute of Justice.

Spader, D. (1994). Teaching due process: A workable method of teaching the ethical and legal aspects. *Journal of Criminal Justice Education, 5*(1), 81–106.

Sullivan, C.J., & Piquero, A.R. (2016). The criminal career concept: Past, present, and future. *Journal of Research in Crime and Delinquency, 53*(3), 420–442.

Swift, A. (2016, Oct. 19). *Support for legal marijuana use up to 60% in U.S.* Retrieved from: www.gallup.com/poll/196550/support-legal-marijuana.aspx

Teachman, J., & Tedrow, L. (2016). Altering the life course: Military service and contact with the criminal justice system. *Social Science Research, 60,* 74–87.

Telep, C., & Weisburd, D. (2012). What is known about the effectiveness of police practices in reducing crime and disorder? *Police Quarterly, 15,* 331–357.

Tittle, C. (2000). Theoretical developments in criminology. In G. LaFree (Ed.), *The nature of crime: Continuity and change, criminology 2000: Vol. 1.* (pp. 51–101). Washington, DC: National Institute of Justice.

Tonry, M. (2014). Remodeling American sentencing: A ten-step blueprint for moving past mass incarceration. *Criminology & Public Policy, 13*(4), 503–533.

Turner, M., Sundt, J., Applegate, B., & Cullen, F. (1995). "Three strikes and you're out" legislation: A national assessment. *Federal Probation, 59*(3), 16–35.

United States v. McIntosh, 833 F. 3d 1163 (2016).

Vito, G.F., & Maahs, J.R. (2017). *Criminology: Theory, research, and policy* (4th ed.). Burlington, MA: Jones & Bartlett Learning.

Walker, S. (1985). *Sense and nonsense about crime: A policy guide.* Monterey, CA: Brooks/Cole.

Weisburd, D. (2016). Does hot spots policing inevitably lead to unfair and abusive police practices, or can we maximize both fairness and effectiveness in the new proactive policing? *University of Chicago Legal Forum, 2016,* 661–689.

Weisburd, D., & Telep, C.W. (2014). Hot spots policing: What we know and what we need to know. *Journal of Contemporary Criminal Justice, 30*(2), 200–220.

Weisburd, D., Telep, C., Hinkle, J., & Eck, J. (2010). Is problem-oriented policing effective in reducing crime and disorder? Findings from a Campbell systematic review. *Criminology & Public Policy, 9*(1), 139–172.

Willis, C., Evans, T., & LaGrange, R. (1999). "Down home" criminology: The place of indigenous theories of crime. *Journal of Criminal Justice, 27*(3), 239–247.

Wilson, M., & Ruback, R. (2003). Hate crimes in Pennsylvania, 1984–99: Case characteristics and police responses. *Justice Quarterly, 20*(2), 373–398.

Worrall, J. (2008). *Crime control in America: What works?* (2nd ed.). Boston: Pearson.

Wortley, R., & Townsley, M. (2017). Environmental criminology and crime analysis: Situating the theory, analytic approach and application. In Wortley, R. & Townsley, M. (Eds.), *Environmental criminology and crime analysis* (pp. 1–25). New York: Routledge.

Wright, J.P., Tibbetts, S.G., & Daigle, L.E. (2008). *Criminals in the making: Criminality across the life course.* Thousand Oaks, CA: Sage.

CHAPTER 4

Counting Crimes and Criminals

If we can consider the criminal justice system a business, it is a business run by people having no clear understanding of the market, the production and distribution process, or customer satisfaction. Not many commercial enterprises could succeed in such a state of ignorance. Yet, the criminal justice system does operate in ignorance.

Lack of knowledge about the types of crimes and criminals is not the only form of ignorance that hinders the criminal justice system in the United States. Not only do we not know very much about the nature of crimes and offenders, we also have difficulty determining their numbers. For decades, critics have written about the **"dark figure" of crime**; that is, the unknown amount of crime that occurs.

The dark figure represents the portion of crime about which we are ignorant. Like a half moon where part of the moon is in shadow and we can see only one-half of the lunar surface, current official crime statistics may reveal only one-half (more or less) of the actual amount of crime; the remainder is hidden in the "shadows." The dark figure is this crime in the shadows.

THE NEED FOR NUMBERS

At first we may ask, so what? How important can it be that we do not know how much crime there really is? Even when we cannot see the full moon, we know that it is there. Paradoxically, the problem is that without knowing what is in the shadows, we cannot know the importance of the crimes of which we are ignorant. Old sayings such as "Ignorance is bliss" and "What you don't know can't hurt you" do not always apply. This problem becomes clearer when we examine the uses to which we put criminal statistics.

Nettler (1984) listed four reasons for counting crime:

1. Description
2. Risk assessment

IMPORTANT TERMS

cohort studies

Crime Index

crime rate

"dark figure" of crime

defounding

forgetting

funnel effect

hierarchy rule

National Crime Victimization Survey (NCVS)

observations

official statistics

self-report studies

telescoping

unfounding

Uniform Crime Reports (UCRs)

unofficial statistics

victimization data

3. Program evaluation
4. Explanation.

Description is exactly what the term implies: painting an accurate picture of the number and distribution of criminal offenses. Such information is useful for the allocation of resources (Rich, 1995). We use knowledge of crime to determine where to concentrate police patrols and to estimate the number of prosecutors or judges needed. Descriptive measures of crime allow planners to detect changes in crime patterns over time and to adjust criminal justice operations accordingly.

Risk assessment was discussed earlier when we examined the differences between ordinary and career criminals. Accurate data about crime allow us to make estimates about the risks of people becoming offenders and of people becoming victims of crime. Not knowing how much crime actually occurs makes it impossible to predict how much crime is likely to occur. To the degree that the justice process attempts to reduce criminality, lack of knowledge about the actual level of crime hinders our ability to affect future levels. Knowing what kinds of crimes occur, and where and when they occur, enables criminal justice agents to allocate resources more efficiently (Rich, 1996). This kind of knowledge also helps individuals to avoid becoming crime victims. Often the characteristics of victims link to the crimes they experience. If victims can change their behaviors, or be more careful given those characteristics, it is possible that crime can be avoided (Fisher, Daigle, & Cullen, 2010; Vandecar-Burdin & Payne, 2010).

Program evaluation is an effort that often relies on estimates of a program's effect on crime. In the previous chapter, we discussed several programs aimed at controlling career criminals. Evaluations of these programs attempted to compare the program's effects on the criminality of offenders with the effects of "normal" criminal justice processing. Should we continue or expand these programs? We do not know. Assuming a dark figure of crime, there is no way to tell what effect these programs have had on that figure. Each year researchers report scores of evaluations of criminal justice policies and programs. These evaluations usually include a test of impact on crime. It is difficult to determine the value of criminal justice programs and policies when our knowledge about crime is limited. We typically measure crime by counting arrests, but we know that the police do not know about all crimes and do not always arrest offenders. If the data about arrests are incomplete or wrong, we must have less confidence in the conclusions of our research (Krebs, Lattimore, Cowell, & Graham, 2010; Lilley & Boba, 2009; Worrall et al., 2009).

Explanation is the most troublesome of all the reasons why we need accurate numbers about crime and criminals. Why do some people break the law? Being unaware of many criminal offenses, we can offer only partial explanations.

Many people believe that poverty causes crime, and most persons whom we know to have committed crimes are poor. However, a problem arises in that we do not know whether those who commit the dark figure of crime offenses are poor or wealthy. If these unknown offenders are not poor, then poverty explains only the criminality of those whom we can identify.

THE IMPACT OF IGNORANCE

One result of our ignorance about the nature and extent of crime is an inability to assess or predict the effects of policy changes on levels of crime. Changing our emphasis on enforcing laws against certain crimes also may have unintended and unpredictable effects. The war on drugs indicates the problem that the dark figure of crime poses for evaluations of criminal justice policy (see Box 4.1).

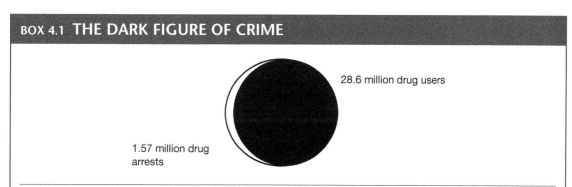

BOX 4.1 THE DARK FIGURE OF CRIME

28.6 million drug users

1.57 million drug arrests

Source: Substance Abuse and Mental Health Services Administration (2017); Federal Bureau of Investigation (2016).

In 2016, approximately 28.6 million Americans were estimated to have used illegal drugs. In the same year, only 1.57 million drug offense arrests were made.

During the 1980s, beginning with the Reagan administration, the United States declared (or more accurately, redeclared) a war on drugs, and a national anti-drug campaign developed (Albanese & Pursley, 1993:265–270). Law enforcement agencies, prosecutors, and criminal courts increased their efforts to catch, convict, and punish drug law violators. Since that time, the number of drug cases and drug offenders has grown as a proportion of the criminal justice system's "business" far more rapidly than other types of offenses and offenders. Arrests, convictions, and prison sentences for drug offenders have increased dramatically, yet we do not appear any closer to winning the "war" than we were three decades ago.

One reason we may not be able to see any progress in terms of reducing the number of drug offenses and offenders is that for years many of these crimes have been a part of the dark figure. With increased emphasis on drug crimes, agents and agencies of the justice system have uncovered offenses that have been present for years. Because drug offenses have gone unreported in the past, Zeisel (1982) noted that they present an almost limitless supply of business for the police. Changing public perceptions of the seriousness of drug offenses has supported increased drug enforcement efforts. If police ignored large numbers of casual or "small-time" drug users in the past, vast increases in arrests, convictions, and imprisonment may signal small increases in the proportion of drug offenders caught and processed.

Kraska (1992:524) observed that with drug offenders, police "can seek actively to detect drug crimes, as opposed to violent and property crimes, for which they have little choice but to react to complaints." Thus, the volume of drug offenders entering the justice system is more a product of police activity than is that of violent or property offenders. Political pressure to treat drug offenses more seriously (Linnemann, 2013), coupled with giving incentives such as profit from seizing the property of drug offenders (Baumer, 2008; Holden, 1993; Worrall & Kovandzic, 2008), spurs more aggressive police action. When applied to offenses that are largely underenforced, these activities can produce dramatic increases in criminal justice caseloads without affecting basic levels of offense behavior.

BOX 4.2 POLICY DILEMMA: CIVIL ASSET FORFEITURE

Asset forfeiture laws allow police to confiscate items (automobiles, weapons, cash, etc.) which have been used for criminal activity. Proponents of these laws argue that the seizure of assets related to criminal activity (especially drug crimes) will have a deterrent effect on crime. Additionally, the profits from these seizures are used for police and other government activities that would otherwise require taxpayer funding. However, critics point out that civil forfeiture laws often require a lower standard of proof than criminal cases (Holcomb, Kovandzic, & Williams, 2011). In other words, police can seize and keep property even if they do not have enough evidence to convict the owner of the underlying crime. Others argue that these laws create a conflict of interest where the police will focus on arresting those who have valuable assets while ignoring other types of crimes (Rothschild & Block, 2016). Do you believe that the benefits of civil asset forfeiture laws outweigh these objections?

Assuming available data are accurate, arrests for drug law violations in 2016 affected less than 6% of the population reporting use of illegal drugs that year. Doubling the number of arrests for drug violations would still leave more than 88% of users untouched. In terms of assessing the impact of doubling enforcement efforts, an evaluator would conclude that doubling efforts (and costs) would not have a significant impact on actual drug use. Indeed, between 2002 and 2008, the number of arrests for drug offenses increased by more than 15% whereas the estimated number of persons using illegal drugs increased by about 8% (Substance Abuse and Mental Health Services Administration, 2010). By 2015, the number of arrests for drug offenses had decreased by more than 12% from their 2008 levels, and reported drug use continues to rise (Substance Abuse and Mental Health Services Administration, 2015). Thus, it is unclear whether the increased enforcement of drug laws has had any impact on drug usage.

We have long recognized the inadequacies of our data about crimes and criminals. Yet, with repeated calls for improvement in the collection and use of statistics on crime, how is it that we have made so little progress? The answer lies within the complex nature of criminal justice in the United States.

Thousands of agencies in thousands of separate jurisdictions collect information. These agencies often use their own definitions of crimes and criminals, and report their data to national centers on a voluntary basis. Each agency needs different types of information for its own planning and operation. It is very difficult to follow cases through the justice process, because the police, prosecutor, criminal court, and correctional agencies use their own forms to collect the information that is useful to them, with little regard to a systemwide need for information.

Much data are available concerning the number of crimes and criminals, justice agencies, and operations of the justice system. However, we do not know exactly how much crime exists and where it occurs. The crime problem is one without clearly marked boundaries. The sources of information about crime and criminal justice are of two basic types: official statistics and unofficial statistics.

OFFICIAL STATISTICS

Official statistics are statistics provided by criminal justice agencies as official records of their activities. The most familiar of all official crime statistics is the Uniform Crime Reports (UCRs), published annually by the Federal Bureau of Investigation (FBI) from reports received by the nation's police departments. These data describe the volume of business handled by the law enforcement agencies of the country. The basic statistic of the UCR is "crimes known to the police."

Only those offenses detected by the police are crimes known to the police. If someone steals your wallet and you do not report the theft to the police, the crime is unknown to the police and not counted in the UCR. If you report the theft, or a police officer witnesses the crime, then the offense will be "known to the police." In addition, the police officer decides whether a crime has occurred and, if so, what crime it was.

In the aforementioned example, suppose you report the theft to a police officer, but the officer decides that you are not telling the truth. The process of **unfounding** occurs if the officer decides that your criminal complaint is "unfounded"; that is, the officer believes that the crime you reported is not supported by available evidence and, therefore, has reason to believe that no crime occurred. In this instance, the theft will remain "unknown" to the police because the officer considers your report untrustworthy.

A similar decision is **defounding** a crime, which occurs when a police officer decides that an offense was less serious than reported. If the criminal stole your wallet by threatening to harm you with a knife, an armed robbery occurred. If the police officer does not believe that you were actually threatened, he or she may simply record a theft of your wallet. In this instance, the police know about the crime, but the crime is reported as less serious than it actually was.

An English economist, Sir Josiah Stamp, warned of the dangers of official statistics. He stated, "The government is very keen on amassing statistics. They collect them, raise them to the nth power, take the cube root and prepare wonderful diagrams. But you must never forget that every one of these figures comes in the first instance from the village watchman, who just puts down what he damn pleases" (quoted in Platt, 1999).

The Uniform Crime Reports

The UCRs cover 29 different crimes, including eight crimes that, when totaled, are known as the **Crime Index**. We can use this total to compare levels of crime over time. The eight Crime Index offenses are homicide, forcible rape, robbery, aggravated assault, burglary, larceny-theft, auto theft, and arson. The remaining UCR crimes are considered Part II offenses and include crimes such as drug abuse violations and simple assault.

Participating police agencies voluntarily report data to the FBI. Most (but not all) police departments report to the FBI. The UCR has been published since 1930 (U.S. Department of Justice, 2014), and thus has provided information on the rate and level of crime in the United States for nearly 90 years. Nonetheless, criminologists question the value of the UCR on several grounds (Loftin & McDowall, 2010; Nolan, Haas, & Napier, 2011).

Many criminologists warn that UCR data must be used cautiously (Maltz, 1999). The data are voluntarily reported and may reflect different definitions of offenses employed by the multitude of police departments participating. Moreover, the UCR counts only non-Index crimes (simple assault, etc.) if an arrest is made in the case. As shown in Figure 4.1, less than half of Index crimes are cleared by an arrest. Thus, there is reason to believe that requiring an arrest to be made for non-Index crimes is a major limitation to the data. The data also mask the actual numbers of offenses and offenders through a reporting

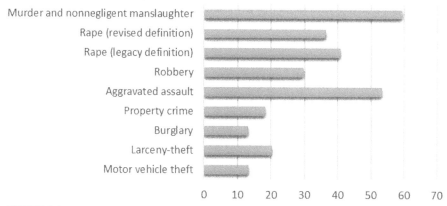

FIGURE 4.1
Percent of Offenses Cleared by Arrest or Exceptional Means, 2016.

Source: Federal Bureau of Investigation (2016).

procedure often referred to as the **hierarchy rule**. For example, if a number of crimes are committed during a single criminal episode (say a bank robber kills a teller, kidnaps a hostage, steals a car for the escape, and flees across state lines), only the most serious offense is counted (in this case, the homicide). Finally, Geis (1986) noted that changes in the crime rate may reflect police efficiency more than changes in crime. A city with an aggressive police department that proactively fights crime and clears a large percent of known crimes with an arrest will have a higher official crime rate than a city with a weaker or more passive police department, even if the true level of crime is the same for both cities.

Other criticisms have been leveled about the reporting of crimes as the crime rate in which the amount of crime is "adjusted" for population size. Using the **crime rate** formula, the number of crimes is reported as a function of population such that:

$$\frac{(\text{Crime} \times 100)}{\text{Total Population}} = \text{Crime rate}$$

In this way, the Crime Index treats crimes such as homicide and theft as equal. In addition, for many years the FBI used the decennial (10-year) census for the population total. As a result, the UCR based the 1969 crime rate on the same population as it did the 1960 rate. This caused an artificial inflation of the crime rate because the formula did not reflect the actual increase in the population (Eck & Riccio, 1979).

The purpose of the crime rate is to make fairer comparisons between jurisdictions. If we have a city with 100,000 people where 20 murders occurred last year, and a town of 5,000 people where one murder occurred last year, which is safest? The raw numbers indicate that murder is 20 times as likely to occur in the city as in the town. In fact, however, if you reside in either community, your chances of being a homicide victim are equal. The city has a homicide rate of 20 per 100,000 population. The town has a homicide rate of 1 per 5,000 (or 20 per 100,000), which is equal to the homicide rate in the city, because the city is 20 times larger than the town in population. Nolan (2004) has shown that the level of crime increases with city size so that larger cities have higher crime rates than smaller cities. This means using the crime rate to make comparisons between different-sized cities more fair is only partly successful.

Despite the problems with UCR data, it is still an important indicator of the nature and extent of crime in the nation. Rosenfeld and Decker (1999) tested the accuracy of official arrest reports as a measure of substance abuse. In comparison to public health and drug test data, they found that the arrest data were quite similar to other measures for drugs such as heroin and cocaine, but that the various measures differed for marijuana. The UCR also releases separate publications that provide some detail on the numbers of hate crimes committed in the country each year and on the numbers of police officers killed and assaulted in the line of duty. These publications are available at www.fbi.gov/stats-services/crimestats. As long as the user is aware of the limitations of the UCR and is cautious in its interpretation, it is an important source of information.

Improving the UCR

In late 1982, a task force composed of representatives of the U.S. Bureau of Justice Statistics and the FBI began to study ways of improving the UCRs (Bureau of Justice Statistics & Federal Bureau of Investigation, 1985). Although there was broad support for this revision of the crime-reporting program, the task is not yet completed. The new reporting system is the National Incident-Based Reporting System (NIBRS). In comparison to the traditional UCR, the NIBRS gathers information about crimes known to the police that are much more detailed (see Table 4.1).

The NIBRS format asks police to record information for each incident of crime, as opposed to recording information about only the most serious crime. Instead of 8 Index offenses, the NIBRS has 22 Group A offense categories and 11 Group B offenses. Finally, NIBRS records information for each incident in six segments representing data about the case (administrative), crime (offense), property, victim, offender, and arrestee. The result is a much richer and more detailed set of information about crimes known to the police.

This information enables analysts to study relationships among offenders and victims, characteristics of places and times offenses occur, and other topics (Reaves, 1993). The new reporting format, however, results in higher numbers

TABLE 4.1 Comparing the Uniform Crime Reports and National Incident-Based Reporting System

Uniform Crime Reports (UCRs)	National Incident-Based Reporting System (NIBRS)
Tracks 8 crimes	Tracks 46 crimes
Does not report arrests in specific incidents	Contains information about arrests in each incident
Gives a tally of the incidents; does not contain information on each reported incident	Contains information on each incident reported to police, including:
	Characteristics of victims and offenders. Relationship between victims and offenders. Crimes committed
	Injuries at the incident scene
	Weapons used
	Arrests made
Does not provide information about simple assault, which is the most commonly reported domestic violence offense	Provides information about simple assault
Reports only the most serious crime committed in a single incident	Requires officers to report multiple offenses, victims, and offenders

Source: Hirschel (2009:1).

of crimes reported as known to police. It is likely to be some time before the new reporting format will achieve widespread use. Well under half of the nation's law enforcement agencies currently report data in the NIBRS format (McCormack, Pattavina, & Tracy, 2017). If the NIBRS format ever becomes a true national level measure of crime, it has been suggested that the changes will go a long way toward fixing most of the problems identified with the UCR (Hirschel, 2009).

Other Official Statistics

Other criminal justice and governmental agencies collect and publish data that are relevant to measuring crime and criminals in the United States. Increasingly during the past three decades, the U.S. Bureau of Justice Statistics has gathered and published information about the operations of criminal justice agencies by surveying law enforcement, court, and correctional organizations. There are periodic censuses of the nation's jails and data on case filings in the courts of the United States. Other data are available regarding the numbers and distribution of juvenile offenders, operations of pretrial release programs, and other aspects of justice processing. Some of the more common and more important sources of official information about criminal justice operations in the United States include the Law Enforcement Management and Administrative Statistics,

National Pretrial Reporting, National Judicial Reporting, and National Corrections Reporting programs. The Bureau of Justice Statistics administers all these programs (www.ojp.usdoj.gov/bjs).

UNOFFICIAL STATISTICS

Recognizing that official statistics tell only part of the story of crime in the United States, criminologists have developed other ways of estimating how much crime occurs and who commits criminal offenses. Unofficial statistics are those measures of the rate and nature of crime that do not rely on the reporting of official agencies and agents of criminal justice. Two basic sources of crime data (collected without relying on the official reports of justice agencies) are victim surveys and self-report. These data reveal that official statistics cover only about half of all crime. They also show that persons arrested for criminal offenses are not representative of all those who commit crimes. To avoid the problems and limitations of official statistics, particularly the UCR, researchers developed these other methods of counting crimes and criminals (National Research Council, 2008; Salas & Surette, 1984). Each gives a slightly different view of the overall crime picture and thus serves as an aid to understanding the true nature and extent of crime in the United States.

Victim Surveys

In 1965, the National Opinion Research Center (NORC) of the University of Chicago conducted the first survey of crime victims. The researchers used the results to estimate the nature and extent of crime. The President's Commission on Law Enforcement and Administration of Justice (1967:96) instructed the NORC to conduct a survey of 10,000 households. The survey results indicated that crimes known to the police were only a fraction of all crimes committed.

The NORC survey contacted a spokesperson for each household surveyed and asked that person if anyone residing in the household had been the victim of certain crimes in the past year. Interviewers asked respondents whether the crime had been reported to the police and, if not, why it had not been reported. Beginning in 1971, the U.S. Bureau of the Census began collecting similar data through the National Crime Victimization Survey (NCVS).

Results of the survey over the years continue to indicate that the UCR data are incomplete. Respondents to victim surveys report that many crimes remain unreported to the police. In 2016, victims indicated that only 42.1% of violent crimes and 35.7% of property crimes had been reported to the police (Morgan & Kena, 2017). Box 4.3 compares the victim surveys of the NCVS with the UCR. Victimization data provide information about the nature and extent of crime as understood by those reporting that they have been crime victims.

BOX 4.3 COMPARISON OF THE NCVS AND THE UCR

The National Crime Victimization Survey (NCVS)

Using stable data collection methods since 1973, the NCVS has the following strengths:

- It measures both reported and unreported crimes.
- It is not affected by changes in the extent to which people report crime to police or improvements in police record-keeping technology.
- It collects information that was not available when the initial police report was made including contacts the victim has with the criminal justice system after the crime, extent and costs of medical treatment, and recovery of property.
- It collects detailed information about victims and characteristics of the victimization including who the victims are, what their relationship is to the offender, whether the crime was a part of a series of crimes occurring over a 6-month period, what self-protective measures were used and how the victims assess their effectiveness, and what the victim was doing when victimized.
- On occasion, it includes special supplements about particular topics such as school crime and the severity of crime.

The Uniform Crime Reports

The UCR program measures police workload and activity. Local police departments voluntarily report information to the FBI including the numbers of crimes reported to police, arrests made by police, and other administrative information. The UCR program has the following strengths:

- It can provide local data about states, counties, cities, and towns.
- It measures crimes affecting children under age 12, a segment of the population that experts agree cannot be reliably interviewed by the NCVS.
- It includes crimes against commercial establishments.
- It collects information about the number of arrests and who was arrested.
- It counts the number of homicides (murders and nonnegligent manslaughters), crimes that cannot be counted in a survey that interviews victims. UCRs also collect detailed information about the circumstances surrounding homicides and the characteristics of homicide victims.

Source: Bureau of Justice Statistics (1992).

The NCVS data are not directly comparable to the UCR. The two datasets do not measure identical sets of crimes and they define certain crimes in different ways (U.S. Department of Justice, 2014). The NCVS also does not report crime committed against children under 12 years old. Nonetheless, both the NCVS and UCR typically report similar year-to-year increases or decreases in the levels of both violent and property crime (Truman & Morgan, 2016). Thus, UCR data may underestimate the amount of crime, but may accurately reflect the types of crimes committed, and where and when criminal offenses are likely to occur. Taken together, the UCR and NCVS provide valuable information regarding crime trends.

Victim surveys also have their limitations (O'Brien, 1986). The victim survey's major problems may revolve around the phenomena known as "telescoping" and "forgetting." The interviewer asks the respondent if anyone has been the victim of a particular crime in the past 6 months. In **telescoping**, the respondent errs by including an offense that may have occurred, say, 7 months earlier, thereby "telescoping" it into the covered period. **Forgetting** occurs when the respondent forgets about a crime that did occur in the period under study (Schneider & Sumi, 1981). Moreover, it is always possible when interviewing a person that (for whatever reason) the person is not telling the truth in answering the questions.

Another limitation of victim surveys is that many crimes may have gone unnoticed by the respondent. The NCVS interviewer relies on one individual to have knowledge of the criminal victimizations experienced by the entire household. Finally, victim surveys do not cover some offenses (Cantor & Lynch, 2000). For example, the NCVS asks about only a small number of offenses, and it is not possible to gather data about homicide, for example, from the victims of the crime. During the past several years, the increasing cost of collecting data has meant that the size of the victimization sample and the scope of the interview have been reduced. The combination of a smaller sample size and crime rates at their lowest levels in decades during the recent past means that it is difficult to detect significant changes in rates of victimization from year to year (Catalano, 2006; National Research Council, 2008).

Improving the NCVS

In 1992, responding to many of the problems identified with the NCVS, the Bureau of Justice Statistics changed the interview used to gather victimization information. Among other changes, the survey now interviews all persons aged 12 or older residing in households included in the sample. The redesign changed some offense classifications, such as classifying all thefts as household victimizations. Interviewers now specifically ask respondents about their experiences as victims of rape or sexual assault, and the interview gathers more detailed information about victimization and some offenses such as assault (Bureau of Justice Statistics, 1996). The changes in the interview were phased into use over several years, with the final form of the new survey implemented between 1992 and 1993. The results of the changes in interview format and content included finding much higher rates of victimization. The greatest changes occurred in estimates of rape and assault (Rand, Lynch, & Cantor, 1997). In 2003, changes in federal regulations allowed survey participants to choose more than one racial category, making it difficult to compare recent victims with past victims in terms of race. Finally, the Bureau of Justice Statistics is currently in the process of identifying and testing an improved survey design to measure rape and sexual assault victimization (Truman & Morgan, 2016).

Self-Reports

Both the UCR and victim surveys attempt to describe criminal offenders. The FBI reports the characteristics of persons arrested for crimes, providing descriptions of those persons officially recognized as probably having committed crimes. The respondents to victim surveys describe the offenders involved, if possible. These data provide a description of criminal offenders as seen by the victims of crime. Both efforts at describing criminal offenders are severely limited. Moreover, although there is a fair degree of agreement between victim surveys and UCR data, it is possible that neither measure accurately reflects all crime. Thus, researchers also use a third method of counting crimes: surveys of criminal offenders.

Self-report studies attempt to measure the amount of crime committed and describe the characteristics of criminal offenders by asking people if they have committed offenses. In these studies, researchers ask a sample of the public if they have committed any crimes (Thornberry & Krohn, 2000). This crime measure yields information on the types of persons likely to commit crimes as well as another estimate of the amount of crime that is committed each year.

These studies have some obvious limitations. They share the problems of telescoping and forgetting that afflict victim surveys, and it is difficult to determine whether respondents are telling the truth. There is reason to believe that some may exaggerate to make themselves appear to be notorious and that some will be reticent, fearing that disclosure of their criminality will lead to punishment. When researchers have compared reported crimes and arrests with official records, however, they often find that respondents tell the truth (Piquero, Schubert, & Brame, 2014). For example, Peters, Kremling, and Hunt (2015) found that arrestees provided relatively truthful responses to questions of their past drug use. However, they also reported that rates of truthful responses varied among the types of people responding. For example, younger participants were less truthful than older arrestees. In addition, those who had a history of substance abuse treatment were more honest than those with no prior treatment. McElrath, Dunham, and Cromwell (1995) reported the rates of truthful responses varied by jurisdiction, race, types of people interviewed, the characteristics of the interviewer, and the type of charge for which the offender had been arrested.

Nonetheless, self-report studies indicate that almost everyone will admit to having violated some criminal law. The most important finding of self-report studies is not who does or does not break the law, but rather how often crimes are committed and how serious are those violations. Institutionalized populations of delinquents or adult criminals report more frequent and more serious law violations than do "free citizens." Males report more criminal activity than females; African Americans report more frequent and more serious offenses than Caucasians do. In general, with the exception that self-reports indicate that everyone probably breaks some law, the findings of these studies echo those of victim surveys and official reports.

Other Measures of Crime and Criminal Justice

Researchers have used several other methods to measure crime and study the criminal justice process. Information gathered from cohort studies and observations helps to better describe and explain the operations of the criminal justice process and the nature and extent of crime.

Cohort studies begin with an identifiable group (or "cohort") and trace the group's interaction with the justice system over time. The individuals studied in such research are members of a cohort. The cohort is a collection of all persons sharing a common selection characteristic. Thus, a cohort might include all entering freshmen at a university or all persons married in a given year. For criminal justice research, the selection criterion normally relates to a justice system decision (such as all those arrested in a given time period) or to an age limit (such as all those born in a specified year).

The most famous cohort study selected cases by specifying all males born in Philadelphia in the year 1948. *Delinquency in a Birth Cohort* (Wolfgang, Figlio, & Sellin, 1978) was the report of a study designed to examine the criminal careers of youths. Researchers tracked all males born in Philadelphia in 1948 for a 20-year period to determine which of them were arrested, tried, and sentenced for delinquent behavior. Additionally, they studied the distribution of delinquent offenses among the cohort and characteristics of individuals most often involved in delinquent behavior.

This study allowed the researchers to estimate the proportion of youths that would become entangled in the juvenile justice process, how serious the youths' misbehavior would be, and who among them were most likely to be delinquent. Wolfgang was involved in at least two additional cohort studies of delinquency (Navares, Wolfgang, & Tracy, 1990; Tracy, Wolfgang, & Figlio, 1990). Similar cohort analyses of persons arrested for crime could be used to estimate how the justice system processes cases from arrest to final disposition. Such data would be invaluable to an understanding of the justice process. The problem with cohort studies is one of expense. By definition, most cohort studies must be extensive; that is, they involve large numbers of cases followed over a period of several years.

For *Delinquency in a Birth Cohort*, the researchers actually identified their cohort in 1976 and backtracked through official records to estimate the subjects' involvement in juvenile delinquency. A similar study of those born this year would take 20 years to complete. Despite this limitation, cohort studies allow us to examine the operations of criminal justice agencies in a broader context than is normally possible. Cohort studies provide an estimate of the distribution of crime across an entire population over time.

Observations, as the term implies, involve researchers watching the behavior of criminals, agents of the justice system, or other samples of people. The American Bar Foundation (ABF) series mentioned in Chapter 1 is one example of an observation study. The ABF study reported observations of police, prosecutors, judges, and correctional personnel during investigation, arrest, conviction, and sentencing. Other observation studies have been conducted that seek to determine when people break the law, when they report lawbreaking, what factors justice system agents consider in their decision making, and how cases move from one stage of the justice process to the next.

Like cohort studies, observational methods can be expensive to use and, thus, often result in limited data confined to one location, or in a few decisions rather than national, systemwide descriptions. There is concern that criminal justice officials, recognizing they are being watched, will change their behavior. Spano (2007) assessed the likely effect of observers on police behavior and found that police officers do seem to react to the presence of researchers. He found that officer reactions varied and depended on characteristics of the observer and other situational factors.

Interest in and use of geographic data in the analysis of crime has increased in recent years (Johnson, 2017). Geographic data provide a different perspective from which to view crime and criminal justice, giving us a better feel for the reasons behind decisions and behaviors. Increasingly, police departments use sophisticated analyses of large databases to predict or understand the nature and distribution of crime, especially large police departments. Called "intelligence-led policing," analysts link crime data and offender information with a variety of other information about communities to identify links and predict how best to intervene to prevent crime (Bureau of Justice Assistance, 2005; Ratcliffe, 2008). These initiatives, of course, depend on the accuracy and timeliness of the data used.

SUMMARY OF CRIME STATISTICS

While each of these methods of counting crimes and criminals uses different means of gathering data and collects information from different sources, in total, the "picture" each gives us of crime is generally consistent (Nettler, 1984:98–156). Absolute numbers may vary (for instance, victim surveys may show much more crime than police reports), but the relative frequency of crimes (e.g., more thefts than robberies, more robberies than assaults, more assaults than rapes) reported by all three procedures is similar.

Crime statistics also provide a valuable insight into the types of individuals most often processed by the criminal justice system. One point that is apparent (assuming that the crime statistics are accurate) is that persons arrested and sentenced by the criminal justice system are not always representative of the general population. Poor, urban dwellers and minority group members are far more likely to be arrested and processed than their numbers in the population suggest. In addition, females are less likely to become involved in the justice process than their numbers in the population would indicate. One exception here is the juvenile justice system, where girls are often subjected to more justice processing than boys (Espinosa & Sorensen, 2016).

One of the questions raised in response to these findings concerns the fairness of the justice process. Why is it that minorities and the poor are most often arrested, convicted, and incarcerated? Why are women less likely to be arrested and convicted for crimes than are men? Is the justice system racist, sexist, and prejudiced against the poor? In short, the evidence of differential treatment of certain classes of the population has led some observers to suggest that the justice system is discriminatory.

DISPARITIES IN THE JUSTICE SYSTEM

It is apparent that certain groups are processed more frequently by the justice system. This disparity may simply be due to legal factors such as increased involvement in crime. If this were true, it would be a sign that the justice system is effective at catching and punishing those individuals responsible for criminal behavior. In other words, the disparity, in itself, does not show that the justice system is inequitable. Yet some have argued that the justice process is discriminatory and repressive because it differentially selects and processes members of disadvantaged groups, such as youth, minorities, the poor, and urban residents (Petersilia, 1983; Walker, Spohn, & DeLone, 2018). Others have suggested that the justice system is sexist because it does not subject females to equal treatment as offenders (Visher, 1983).

Age

The available data suggest that crime tends to be a young person's game. Although some researchers (e.g., Sweeten, Piquero, & Steinberg, 2013) have been able to explain many of the sociological and psychological reasons why younger people are more involved in crime, a complete explanation is elusive. The idleness, good physical condition, and lack of responsibilities that many youths enjoy may create special opportunities for them to commit crime. As one who engages in crime grows older, she or he generally risks longer prison terms as a habitual offender. Giving longer terms to older offenders means they are less likely to have the opportunities to commit the crimes that are available to the young. Finally, in a catch-22 fashion, the police actually are more likely to look for crimes among the young precisely because so many of the young have been found to be engaged in crime.

Nonetheless, the familiar age–crime curve is misleading in the fact that certain types of crimes are much more common among adults than youth (Steffensmeier, Allan, Harer, & Streifel, 1989). For example, Leal and Mier (2017) found that juveniles are much more likely to use marijuana compared to adults, while adults were more likely to use harder drugs such as cocaine and heroin. Evidence also exists that the age–crime curve is impacted by a child's socioeconomic background. Youth living in disadvantaged neighborhoods appear to start their criminal activity at a younger age and desist from crime at an older age compared to those living in more affluent neighborhoods (Fabio, Tu, Loeber, & Cohen, 2011).

Race

Neilis, Greene, and Mauer (2008:5) described four causes which are often used to explain the racial disparity in the justice system: (1) higher crime rates, (2) inequitable access to resources, (3) legislative decisions, and (4) overt racial bias. These causes are interrelated and difficult to disentangle. For example, it is possible that African Americans are overrepresented in the criminal justice system due to their increased involvement in crime. However, it is also true that African Americans reside disproportionately in lower socioeconomic neighborhoods where the conditions might not allow for the same opportunities as other population groups (Friedman & Rosenbaum, 2007). Add to this the mandatory sentencing and sentencing enhancement policies which punish behaviors that are strongly correlated with race (Schlesinger, 2011) and you can begin to see how several factors can combine to contribute to disparities in the justice system. Neighborhood disadvantage, living in poverty in places with few social services, and the presence of little informal social control all lead to greater levels of crime and greater reliance on the justice system.

A large factor appears to be urbanism. Most arrests are of young, urban offenders. Minority groups appear to be concentrated most greatly in the cities of the United States. Simply put, both the police and the young, minority males are overrepresented in the cities (Akins, 2003; Swanson, 1981). In rural and suburban areas, the population is more dispersed, and there are fewer police officers. It is more difficult for police to observe crimes and respond quickly to reports of crime. In addition, smaller communities have available more informal mechanisms of social control. People are less frequently strangers to one another and can resolve differences more easily without involving the police. It is more likely that people define disputes as "personal" and not demanding of police intervention. In the future, however, differences between rural and urban areas in terms of crime are likely to diminish. As urban sprawl continues, and with the increasing importance of a global economy and improved communications, many of the problems of cities are likely to also affect rural areas (Lee & Thomas, 2010; Weisheit & Wells, 1999).

Researchers have often examined whether police officers are biased in their enforcement of the law, as they are responsible for the discretionary decision to make an arrest and therefore begin the formal criminal justice process. Much of this research involves the most visible interaction most of us have with police: the traffic stop. Approximately 8% of U.S. drivers are stopped by the police in a given year, and around 5% of these drivers are subject to a search (Eith &

Durose, 2011). Recall from our discussion of hot spot policing (Chapter 3) that most policing efforts are often concentrated in relatively small segments of the cities. Crime tends to be concentrated in the poorer areas of a city with larger minority populations (Lofstrom & Raphael, 2016). Briggs and Keimig (2017) found that over half of the police stops in their study were conducted within the close vicinity of a hot spot but that these stops did not disproportionately include a search. Outside of those high-crime areas, however, they found that stops involving black drivers were more likely to include a search. Interestingly, Novak and Chamlin (2012) observed that white motorists were increasingly subject to searches when driving through areas in which high proportions of blacks reside.

Recall that the justice system exists in an ideological environment that contains our society's values and biases. It is possible that those who look "out of place" may arouse extra suspicions among police (Tapia, 2010). Regardless of whether the police are indeed profiling certain racial groups, evidence exists that minorities at least perceive this to be the case (Eith & Durose, 2011). Najdowski, Bottoms, and Goff (2015) found this perception could actually lead blacks to act more suspiciously when encountering a police officer. Indeed, Johnson (2006) reported that African-American and Hispanic citizens were less likely to maintain eye contact with police officers and were also likely to speak and smile differently than Caucasian citizens. These differences in behavior while dealing with police tended to make the officers more suspicious of African-American and Hispanic citizens and could then lead officers to engage in a discretionary search.

Recent attention has been given to the cumulative effect of discrimination throughout different decision points of the justice system (Kutateladze, Andiloro, Johnson, & Spohn, 2014; Stolzenberg, D'Alessio, & Eitle, 2013). Even small discrepancies in the treatment of minorities at each decision point could add up to a significant cumulative effect and result in higher rates of incarceration. This exemplifies why a careful examination of each decision point is important in determining the system's overall impact.

Gender

Females are less likely to engage in most crimes than are males, partly because of socially defined opportunities for women (both criminal and noncriminal). Additionally, the types of offenses women commit and for which police arrest women were traditionally less serious and less threatening than crimes common to males (Lauritsen, Heimer, & Lynch, 2009). We are far more likely to fully process robbers, rapists, and assaulters than we are prostitutes, thieves, and drug users. Even if they are aware of the relatively large number of female offenders, agents of the justice system are likely to concentrate their resources on the more serious offenses. Still, males appear to be at higher risk of being arrested than females (Lytle, 2014). Additionally, females appear to receive more lenient punishments at the sentencing decision point. Freiburger (2011) found that judges often account for social costs when sentencing females, and were therefore less likely to incarcerate women who lived with children or paid child support. However, this leniency toward female defendants appears to diminish among those who have significant criminal records (Koons-Witt, Sevigny, Burrow, & Hester, 2014; Tillyer, Hartley, & Ward, 2015). Finally, Bontrager, Barrick, and Stupi (2013) found that the gap in sentencing between males and females has become more balanced in recent years.

It is important to point out that females are becoming more involved in traditional crimes, and taking a more active role in the crimes they commit (Schwartz & Rookey, 2008). While arrests overall have declined for both sexes over the past decade, male arrests have declined at a faster pace than females. Among property crimes, female arrests rose over 20% from 2006 to 2015 compared to a 10% decline in the number of male arrests (Federal Bureau of Investigation, 2015). Others suggest that changes in how we treat certain crimes (e.g., drug crimes and domestic assault) have resulted in a greater likelihood of arrest for women (Schwartz, Steffensmeier, Zhong, & Ackerman, 2009).

We will return to the question of discrimination in the justice system a little later. For now, it is enough to say that the presence or absence of discrimination in criminal justice processing, the degree to which discrimination exists, and the effects of possible discrimination in the justice system are complex questions. The task of solving these questions is further complicated by the dark figure of crime and the lack of complete crime data.

UNDERSTANDING THE JUSTICE SYSTEM

Counting the number of crimes and criminals yields an estimate of the volume of "business" conducted by the criminal justice system. Other data are available that lend insight into the complex operations of criminal justice in the United States. Statistics that detail the numbers of persons arrested, prosecuted, tried, convicted, sentenced, incarcerated, or placed under community supervision, released, and discharged describe the working of the justice system.

Ideally, data obtained from all the methods discussed above would be available for answering whatever questions we might have about crime and criminal justice. Unfortunately, such data (at least on a national level) are not always available, and the student of criminal justice must rely on limited information or make inferences from what information can be obtained. Yet another problem is that the periodic revisions of how we collect information, like the development of the NIBRS and revised NCVS, often mean that it is not possible to make direct comparisons of data collected before and after the changes.

One fact readily apparent in an overview of these data is that the criminal justice system operates like a giant sieve. It continuously filters the huge volume of crimes and criminals down to the relatively small number of offenders found in the nation's prisons. By beginning with crimes known to the police and then using different data sources to track arrests, prosecutions, convictions, and sentences, it is possible (as is illustrated in Box 4.4) to observe this "funnel effect."

BOX 4.4 "FUNNEL EFFECT" OF THE CRIMINAL JUSTICE SYSTEM

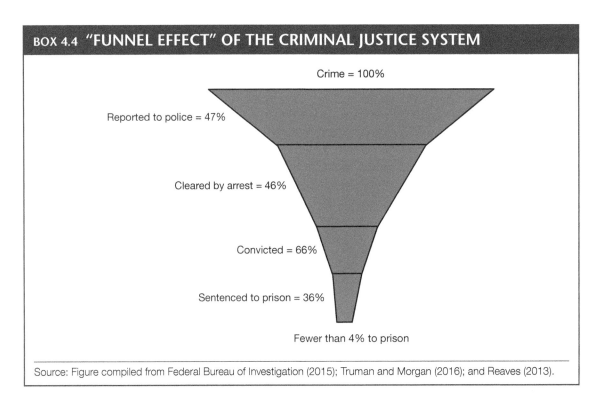

Crime = 100%

Reported to police = 47%

Cleared by arrest = 46%

Convicted = 66%

Sentenced to prison = 36%

Fewer than 4% to prison

Source: Figure compiled from Federal Bureau of Investigation (2015); Truman and Morgan (2016); and Reaves (2013).

As one moves through the criminal justice system, from the crime itself through police, courts, and correctional processing, to discharge, the volume of cases becomes progressively smaller. At each successive stage, the less serious offenses and less dangerous offenders are diverted from the justice system. "Weak cases"—those in which the evidence against the offender is less complete or less compelling—are also dropped. Some accused offenders who are innocent are removed from the process. In the end, a very select group is subjected to the full force of the criminal law.

From this perspective, it is clear that crime does indeed pay, at least in the case of an individual offense. Wilson and Abrahamse (1992:375) noted, "To someone contemplating the commission of any given crime, the answer is that it pays reasonably well." They go on, however, to observe that a career in crime does not pay. Of course, this conclusion about a criminal career's costs and benefits was based on a study of unsuccessful criminals: inmates in three state prisons. The chances of going to prison for a criminal act are slim (except regarding certain criminal acts, such as homicide, which have higher risks of imprisonment). It may be that we have too many criminals for the justice system to accommodate and that this selection process is required so that the entire justice system does not collapse under the number of cases. This is the argument most commonly raised in support of such practices as plea bargaining: that the courts could not handle the volume of criminal trials if bargaining were abolished. It is in the close observation of the filtering process that we begin to understand our society's crime control priorities. It is here, too, that we most clearly see the systemic nature of the criminal justice process. Our task is to develop an understanding of how the criminal justice system in the United States works the way it does.

Recalling our earlier discussion of systems theory, the total system of criminal justice exists within the whole "system" of American society. As an open system, criminal justice is sensitive to a variety of social values and social forces. That is, we cannot understand criminal justice processing in isolation from its place within American society. Although a comprehensive study of society is beyond our scope (and may be impossible), certain aspects of American society are relevant to our understanding of criminal justice operations. One of the most important of these is the emphasis we place on democratic values.

Criminal Justice in a Democracy

At its base, the criminal justice system is a legal system, or more accurately, part of a legal system. Black (1980) observed that legal systems can be characterized as being more or less democratic. Further, the degree of democracy shown in a legal system has implications for how the law develops, how it is applied, and what it accomplishes (Meehan & Ponder, 2002). Black concluded that more democracy in a legal system means that the law is mobilized (or applied) by the citizens more so than by governmental agents. Thus, "crimes" are brought to the attention of authorities more by citizen complaints than by police investigation. In democratic societies, the law is reactive–responsive to citizens. As a result, the law will reflect existing patterns of social stratification, and will be more responsive to changes in social morals and social structure. Thus, one effect of the high value Americans place on democracy is the creation of a legal system (criminal justice system) that reflects society in the United States.

Democracy also leads to the development of constraints on the powers of government and the ability of the criminal justice system to affect crime. Packer (1960) identified "due process" and "crime control" models of criminal justice. In the first, concerns about individual rights and liberty dominate a justice system's actions. In the latter, concerns for crime control and social order dominate. Similarly, Lundman (1980) noted that the police in American society reflect a "dynamic tension" between liberty (individual freedom) and civility (social order).

Each of these views suggests that the criminal justice system must achieve a balance between the rights of the individual and society's need for order. This balance may change over time, but the essential conflict between individual interests and social interests is central to criminal justice in the United States. How this conflict is resolved is the product of the thousands of individual decisions made in the criminal justice system each day. These decisions, in turn, reflect how that balance exists throughout American society.

With regard to the question of discrimination in the criminal justice system, the effect of democracy on the workings of the criminal justice process is evident. MacDonald (2008) argued that racial disparities which exist within the justice system do not represent discrimination. Rather, she concluded that minority group members commit more serious offenses. It is their behavior, not their race or ethnicity, which explains justice decisions. Mann (1987) argued, to the contrary, that data (especially the greater prior arrest/conviction records of minorities) represent racial discrimination. Moreover, she suggested that some of the strongest evidence of discrimination is not "quantified," like arrest statistics, but exists in the qualitative experiences of minority group members.

Who is correct? It is entirely possible that both MacDonald and Mann are correct in their assertions about the existence of discrimination in criminal justice processing. MacDonald is essentially saying that the data do not indicate, for example, that criminal justice officials consciously apply more severe treatment to minority group members. Thus, it would appear that most police, prosecutors, judges, juries, and correctional officials are not racist in their individual decisions. Mann is suggesting that minority group members are more often the targets of justice system processing because of their minority status. The aggregate data clearly indicate that this is the case.

An alternative resolution of the question might be that whereas agents of the justice system generally attempt to be fair in their decision making, society in the United States is structured so that minority group members are more likely to come under justice system control. Society may grant minority group members less access to adequate education, resulting in less employment and fewer resources. These citizens then may be less able to defend themselves from criminal charges (unable to afford bail, private defense counsel, fines, etc.) and may be less suitable for leniency in treatment (e.g., probation sentences) because of a lack of community resources such as stable residence, job, family ties, and so on. As Black's thesis would suggest, if the society in which a legal system exists is discriminatory, a democratic legal system will also discriminate.

The enforcement of drug laws provides an example of institutional or systemic discrimination. Barnes and Kingsnorth (1996) examined sentences imposed on persons convicted of drug law violations in a California court. Among other things, their findings suggested that possession of drugs favored by minority ethnic groups (crack cocaine and heroin) was defined in the law as a more serious offense (a felony) than possession of drugs favored by whites (marijuana), which was more likely to be defined as a misdemeanor offense. Defining some drugs as more dangerous than others coupled with the tendency of the police to focus on street drug markets led to disproportionate arrests of minorities. A combination of forces—including ethnic differences in drugs of choice and "drug marketing," police availability and strategies, and court decisions based on statutory definitions—together produce disproportionate rates of arrest, conviction, and incarceration for minority group members.

Beckett, Nyrop, and Pfingst (2006) studied drug enforcement in Seattle, Washington, and found that the drug crimes of minority group members and poor people were more likely to be the targets of police enforcement efforts than were the crimes of whites. For example, the police were more likely to investigate and arrest crack cocaine offenses than those involving powder cocaine, and to focus on crime in the downtown area than in suburban or other neighborhoods. In these cases, African Americans and other minorities were more at risk for arrest than were Caucasians. This has great implications in terms of sentencing, where crack cocaine has traditionally carried a more severe punishment than powder cocaine. Until 2010, the federal sentencing disparity between minimum sentences for crack and powder cocaine was 100:1. In other words, someone who had just 5 g of crack cocaine would receive a similar punishment as someone who had 500 g of powder cocaine. In 2010, Congress passed the Fair Sentencing Act, which reduced this disparity to 18:1 (Cratty, 2011).

Risk assessment tools have also recently been criticized for contributing to the age, gender, and racial disparities within the justice system (Monahan & Skeem, 2016; Tonry, 2014). These risk assessment tools are often used at several decision points in the justice system to determine which offenders will be released prior to their trial, to decide a sentence upon conviction, and to classify offenders within correctional settings. Males and younger offenders often receive a more negative score on these tools due to their statistical increased risk of reoffending. The tools can also have disproportionate effects on certain racial groups by including factors that the individual cannot control such as neighborhood context and family support system. While evaluating the impact of risk assessment tools in a juvenile justice setting, Goddard and Myers (2017:162) stated that such a tool "replicates conformist interventions, launders racial bias in the USA, and provides scientific justifications for ignoring structural causes of crime not amenable to change". Despite the criticisms, risk assessment tools have been shown to successfully predict reoffending.

This is not to say that discrimination is acceptable or inevitable. However, this observation has important implications for those who would seek to change the system. If the discrimination that exists in the justice process is rooted in the social system, changes in the justice process alone will have limited impact on levels of discrimination (Meehan & Ponder, 2002). For example, recent evidence shows that juveniles who experience discrimination (both inside and outside the justice system) also have a higher likelihood of dropping out of school and associating with delinquent peers (Unnever, Cullen, & Barnes, 2017). Changes in the social system are required to achieve either a reduction in discrimination or crime. In a democratic society, the criminal justice system is a part of—and a reflection of—the society in which it exists.

The next chapters of this book are dedicated to an in-depth examination of the practices and decisions of the agents and agencies of the criminal justice system. In this examination, we will identify the factors that seem to be most important in determining which cases remain in the system and which are diverted. Our focus will be on the decisions made in the criminal justice system and on the identification of explanations for those decisions.

REVIEW QUESTIONS

1. What is the "dark figure" of crime?
2. Give four reasons for counting crime.
3. What are official statistics? Give an example.
4. Define the terms "unfounding" and "defounding."
5. Explain how to calculate a crime rate. What does the term mean?
6. Describe two types of unofficial crime statistics.
7. What do the data reveal about discrimination in the criminal justice system?
8. Describe cohort and observation studies of crime and criminal justice.
9. Explain how the criminal justice system operates like a funnel.
10. What is the effect of democratic values on the criminal justice system?

REFERENCES

Akins, S. (2003). Racial segregation and property crime: Examining the mediating effect of police strength. *Justice Quarterly, 20*(4), 675–695.

Albanese, J., & Pursley, R. (1993). *Crime in America: Some existing and emerging issues.* Englewood Cliffs, NJ: Prentice Hall.

Barnes, C., & Kingsnorth, R. (1996). Race, drug, and criminal sentencing: Hidden effects of the criminal law. *Journal of Criminal Justice, 24*(1), 39–56.

Baumer, E. (2008). Evaluating the balance sheet of asset forfeiture laws: Toward evidence-based policy assessments. *Criminology and Public Policy, 7*(2), 245–256.

Beckett, K., Nyrop, K., & Pfingst, L. (2006). Race, drugs, and policing: Understanding disparities in drug delivery arrests. *Criminology, 44*(1), 105–137.

Black, D. (1980). *The manners and customs of the police*. New York: Academic Press.

Bontrager, S., Barrick, K., & Stupi, E. (2013). Gender and sentencing: A meta-analysis of contemporary research. *Journal of Gender, Race, and Justice, 16*(2), 349–372.

Briggs, T., & Keimig, K. (2017). The impact of police deployment on racial disparities in discretionary searches. *Race and Justice, 7*(3), 256–275.

Bureau of Justice Assistance (2005). *Intelligence-led policing: The new intelligence architecture*. Washington, DC: Author.

Bureau of Justice Statistics (1992). *Criminal victimization in the United States, 1991*. Washington, DC: U.S. Department of Justice.

Bureau of Justice Statistics (1996). *Criminal victimization in the United States, 1994*. Washington, DC: U.S. Department of Justice.

Bureau of Justice Statistics & Federal Bureau of Investigation (1985). *Blueprint for the future of the uniform crime reporting program*. Washington, DC: U.S. Department of Justice.

Cantor, D., & Lynch, J. (2000). Self-report surveys as measures of crime and criminal victimization. In Duffee, D. (Ed.), *Criminal justice 2000: Measurement and analysis of crime and justice: Vol. 4.* (pp. 85–138). Washington, DC: National Institute of Justice.

Catalano, S. (2006). *Criminal victimization, 2005*. Washington, DC: Bureau of Justice Statistics.

Cratty, C. (2011). *New rules slashing crack cocaine sentences go into effect*. CNN, Retrieved from: http://cnn.com/2011/11/01/justice/crack-cocaine-sentencing/Index.html.

Eck, J.E., & Riccio, L.J. (1979). Relationship between reported crime rates and victimization survey results: An empirical and analytical study. *Journal of Criminal Justice, 7*(4), 293–308.

Eith, C., & Durose, M.R. (2011). *Contacts between police and the public, 2008*. Washington, DC: U.S. Department of Justice.

Espinosa, E., & Sorensen, J. (2016). The influence of gender and traumatic experiences on length of time served in juvenile justice settings. *Criminal Justice and Behavior, 43*(2), 187–203.

Fabio, A., Tu, L., Loeber, R., & Cohen, J. (2011). Neighborhood socioeconomic disadvantage and the shape of the age–crime curve. *American Journal of Public Health, 101*(S1), S325–S332.

Federal Bureau of Investigation (2015). *Crime in the United States, 2015*. Retrieved from: https://ucr.fbi.gov/crime-in-the-u.s/2015/crime-in-the-u.s.-2015

Federal Bureau of Investigation (2016). *Crime in the United States, 2016*. Retrieved from https://ucr.fbi.gov/crime-in-the-u.s/2016/crime-in-the-u.s.-2016

Fisher, B., Daigle, L., & Cullen, F. (2010). What distinguishes single from recurrent sexual victims: The role of lifestyle-routine activities and first-incident characteristics. *Justice Quarterly, 27*(1), 102–129.

Freiburger, T. (2011). The impact of gender, offense type, and familial role on the decision to incarcerate. *Social Justice Research, 24*, 143–167.

Friedman, S., & Rosenbaum, E. (2007). Does suburban residence mean better neighborhood conditions for all households? Assessing the influence of nativity status and race/ethnicity. *Social Science Research, 36*, 1–27.

Geis, G. (1986). On the declining crime rate: An exegetic conference report. *Criminal Justice Policy Review, 1*(1), 16–36.

Goddard, T., & Myers, R. (2017). Against evidence-based oppression: Marginalized youth and the politics of risk-based assessment and intervention. *Theoretical Criminology, 21*(2), 151–167.

Hirschel, D. (2009). *Expanding police ability to report crime: The National Incident-Based Reporting System.* Washington, DC: National Institute of Justice.

Holcomb, J.E., Kovandzic, T.V., & Williams, M.R. (2011). Civil asset forfeiture, equitable sharing, and policing for profit in the United States. *Journal of Criminal Justice, 39,* 273–285.

Holden, R. (1993). Police and the profit-motive: A new look at asset forfeiture. *ACJS Today, 12*(2), 1, 3, 24–25.

Johnson, R. (2006). Confounding influences on police detection of suspiciousness. *Journal of Criminal Justice, 34*(4), 435–442.

Johnson, S.D. (2017). Crime mapping and spatial analysis. In Wortley, R., & Townsley, M. (Eds.), *Environmental criminology and crime analysis* (pp. 199–223). New York: Routledge.

Koons-Witt, B., Sevigny, E., Burrow, J., & Hester, R. (2014). Gender and sentencing outcomes in South Carolina: Examining the interactions with race, age, and offense type. *Criminal Justice Policy Review, 25*(3), 299–324.

Kraska, P. (1992). The processing of drug arrestees: Questioning the assumption of an ambivalent reaction. *Journal of Criminal Justice, 20*(6), 517–525.

Krebs, C., Lattimore, P., Cowell, A., & Graham, P. (2010). Evaluating the juvenile breaking the cycle program's impact on recidivism. *Journal of Criminal Justice, 38*(2), 109–117.

Kutateladze, B., Andiloro, N., Johnson, B., & Spohn, C. (2014). Cumulative disadvantage: Examining racial and ethnic disparity in prosecution and sentencing. *Criminology, 52*(3), 514–551.

Lauritsen, J., Heimer, K., & Lynch, J. (2009). Trends in the gender gap in violent offending: New evidence from the National Crime Victimization Survey. *Criminology, 47*(2), 361–400.

Leal, W., & Mier, C. (2017). What's age got to do with it? Comparing juveniles and adults on drugs and crime. *Crime & Delinquency, 63*(3), 334–352.

Lee, M., & Thomas, S. (2010). Civic community, population change, and violent crime in rural communities. *Journal of Research in Crime and Delinquency, 47,* 118–147.

Lilley, D., & Boba, R. (2009). Crime reduction outcomes associated with the state criminal alien assistance program. *Journal of Criminal Justice, 37*(3), 217–224.

Linnemann, T. (2013). Governing through meth: Local politics, drug control and the drift toward securitization. *Crime, Media, Culture, 9,* 39–61.

Lofstrom, M., & Raphael, S. (2016). Crime, the criminal justice system, and socioeconomic inequality. *Journal of Economic Perspectives, 30*(2), 103–126.

Loftin, C., & McDowall, D. (2010). The use of official records to measure crime and delinquency. *Journal of Quantitative Criminology, 26*(4), 527–532.

Lundman, R. (1980). *Police and policing: An introduction.* New York: Holt, Rinehart & Winston.

Lytle, D. (2014). The effects of suspect characteristics on arrest: A meta-analysis. *Journal of Criminal Justice, 42*, 589–597.

MacDonald, H. (2008, Spring). Is the criminal justice system racist? *City Journal*. Retrieved from: www.city-journal.org/html/criminal-justice-system-racist-13078.html

Maltz, M. (1999). *Bridging gaps in police crime data*. Washington, DC: Bureau of Justice Statistics.

Mann, C.R. (1987). Racism in the criminal justice system: Two sides of a controversy. *Criminal Justice Research Bulletin, 3*(5), 1–5.

McCormack, P.D., Pattavina, A., & Tracy, P. (2017). Assessing the coverage and representativeness of the national incident-based reporting system. *Crime & Delinquency, 63*(4), 493–516.

McElrath, K., Dunham, R., & Cromwell, P. (1995). Validity of self-reported cocaine and opiate use among arrestees in five cities. *Journal of Criminal Justice, 23*(6), 531–540.

Meehan, A., & Ponder, M. (2002). Race and place: The ecology of racial profiling African American motorists. *Justice Quarterly, 19*(3), 399–430.

Monahan, J., & Skeem, J. (2016). Risk assessment in criminal sentencing. *Annual Review of Clinical Psychology, 12*, 489–513.

Morgan, R.E., & Kena, G. (2017). *Criminal victimization, 2016*. Washington, DC: Bureau of Justice Statistics.

Najdowski, C., Bottoms, B., & Goff, P. (2015). Stereotype threat and racial differences in citizens' experiences of police encounters. *Law and Human Behavior, 39*(5), 463–477.

National Research Council (2008). *Surveying victims: Options for conducting the National Crime Victimization Survey*. Washington, DC: National Academies Press.

Navares, D., Wolfgang, M., & Tracy, P. (1990). *Delinquency in Puerto Rico: The 1970 Birth Cohort Study*. New York: Greenwood Press.

Neilis, A., Greene, J., & Mauer, M. (2008). *Reducing racial disparity in the criminal justice system: A manual for practitioners and policymakers*. Washington, DC: The Sentencing Project.

Nettler, G. (1984). *Explaining crime* (3rd ed.). New York: McGraw-Hill.

Nolan, J. (2004). Establishing the statistical relationship between population size and UCR crime rate: Its impact and implications. *Journal of Criminal Justice, 32*(6), 547–556.

Nolan, J., Haas, S.M., & Napier, J.S. (2011). Estimating the impact of classification error on the statistical accuracy of uniform crime reports. *Journal of Quantitative Criminology, 27*(4), 497–519.

Novak, K., & Chamlin, M. (2012). Racial threat, suspicion, and police behavior: The impact of race and place in traffic enforcement. *Crime & Delinquency, 58*(2), 275–300.

O'Brien, R. (1986). Rare events, sample size, and statistical problems in the analysis of NCS city surveys. *Journal of Criminal Justice, 14*(5), 441–448.

Packer, H. (1960). *The limits of the criminal sanction*. Englewood Cliffs, NJ: Prentice Hall.

Peters, R.H., Kremling, J., & Hunt, E. (2015). Accuracy of self-reported drug use among offenders: Findings from the arrestee drug abuse monitoring–II program. *Criminal Justice and Behavior, 42*(6), 623–643.

Petersilia, J. (1983). *Racial disparities in the criminal justice system*. Santa Monica, CA: RAND.

Piquero, A., Schubert, C., & Brame, R. (2014). Comparing official and self-report records of offending across gender and race/ethnicity in a longitudinal study of serious youthful offenders. *Journal of Research in Crime and Delinquency, 51*(4), 526–556.

Platt, S. (Ed.), (1999). *Respectfully quoted: A dictionary of quotations. Requested from the Congressional Research Service*. www.Bartleby.com/73, Retrieved January 26, 2005.

President's Commission on Law Enforcement and Administration of Justice (1967). *The challenge of crime in a free society*. Washington, DC: U.S. Government Printing Office.

Rand, M., Lynch, J., & Cantor, D. (1997). *Criminal victimization, 1973–95*. Washington, DC: Bureau of Justice Statistics.

Ratcliffe, J. (2008). *Intelligence-led policing*. Cullompton, Devon, UK: Willan.

Reaves, B.A. (1993). *Using NIBRS data to analyze violent crime*. Washington, DC: Bureau of Justice Statistics.

Reaves, B.A. (2013). *Felony defendants in large urban counties, 2009– Statistical tables*. Washington, DC: Bureau of Justice Statistics.

Rich, T. (1995). *The use of computerized mapping in crime control and prevention programs*. Washington, DC: National Institute of Justice.

Rich, T. (1996). *The Chicago Police Department's Information Collection for Automated Mapping (ICAM) program*. Washington, DC: National Institute of Justice.

Rosenfeld, R., & Decker, S. (1999). Are arrest statistics a valid measure of illicit drug use? The relationship between criminal justice, and public health indicators of cocaine, heroin, and marijuana use. *Justice Quarterly, 16*(3), 685–699.

Rothschild, D.Y., & Block, W.E. (2016). Don't steal; the government hates competition: The problem with civil asset forfeiture. *Journal of Private Enterprise, 31*(1), 45–56.

Salas, L., & Surette, R. (1984). The historical roots and development of criminological statistics. *Journal of Criminal Justice, 12*(5), 457–466.

Schlesinger, T. (2011). The failure of race neutral policies: How mandatory terms and sentencing enhancements contribute to mass racialized incarceration. *Crime & Delinquency, 57*(1), 56–81.

Schneider, A.L., & Sumi, D. (1981). Patterns of forgetting and telescoping: An analysis of LEAA survey victimization data. *Criminology, 23*(1), 41–50.

Schwartz, J., & Rookey, B. (2008). The narrowing gender gap in arrests: Assessing competing explanations using self-report, traffic fatality, and official data on drunk driving, 1980–2004. *Criminology, 46*(3), 637–671.

Schwartz, J., Steffensmeier, D., Zhong, J., & Ackerman, J. (2009). Trends in the gender gap in violence: Reevaluating NCVS and other evidence. *Criminology, 47*(2), 401–426.

Spano, R. (2007). How does reactivity affect police behavior? Describing and quantifying the impact of reactivity as behavioral change in a large-scale observational study of police. *Journal of Criminal Justice, 35*(4), 453–465.

Steffensmeier, D., Allan, E., Harer, M., & Streifel, C. (1989). Age and the distribution of crime. *American Journal of Sociology, 94*(4), 803–831.

Stolzenberg, L., D'Alessio, S., & Eitle, D. (2013). Race and cumulative discrimination in the prosecution of criminal defendants. *Race and Justice, 3*(4), 275–299.

Substance Abuse and Mental Health Services Administration (2010). *Results of the 2008 National Survey on Drug Use and Health: National findings*. Rockville, MD: Author. Retrieved from: www.oas.samhsa.gov/p0000016.htm#2k5

Substance Abuse and Mental Health Services Administration (2015). *Results of the 2015 National Survey on Drug Use and Health: Detailed tables*. Rockville, MD: Substance Abuse and Mental Health Services Administration. Retrieved from: www.samhsa.gov/data/sites/default/files/NSDUH-DetTabs-2015/NSDUH-DetTabs-2015/NSDUH-Det-Tabs-2015.pdf

Substance Abuse and Mental Health Services Administration (2017). *Key substance use and mental health indicators in the United States: Results from the 2016 National Survey on Drug Use and Health*. Rockville, MD: Center for Behavioral Health Statistics and Quality, Substance Abuse and Mental Health Services Administration.

Swanson, C.R. (1981). Rural and agricultural crime. *Journal of Criminal Justice, 9*(1), 19–28.

Sweeten, G., Piquedro, A., & Steinberg, L. (2013). Age and the explanation of crime, revisited. *Journal of Youth Adolescence, 42*, 921–938.

Tapia, M. (2010). Untangling race and class effects on juvenile arrests. *Journal of Criminal Justice, 38*, 255–265.

Thornberry, T., & Krohn, M. (2000). The self-report method for measuring delinquency and crime. In Duffee, D. (Ed.), *Criminal Justice 2000: Measurement and analysis of crime and justice: Vol. 4.* (pp. 33–84). Washington, DC: National Institute of Justice.

Tillyer, R., Hartley, R., & Ward, J. (2015). Does criminal history moderate the effect of gender on sentence length in federal narcotics cases? *Criminal Justice and Behavior, 42*(7), 703–721.

Tonry, M. (2014). Legal and ethical issues in the prediction of recidivism. *Federal Sentencing Reporter, 26*(3), 167–176.

Tracy, P., Wolfgang, M., & Figlio, R. (1990). *Delinquency careers in two birth cohorts*. New York: Plenum Press.

Truman, J.L., & Morgan, R.E. (2016). *Criminal victimization, 2015*. Washington, DC: Bureau of Justice Statistics.

Unnever, J., Cullen, F., & Barns, J. (2017). Racial discrimination and pathways to delinquency: Testing a theory of African American offending. *Race and Justice, 7*(4), 350–373.

U.S. Department of Justice (2014). *The nation's two crime measures*. Washington, DC: Bureau of Justice Statistics.

Vandecar-Burdin, T., & Payne, B. (2010). Risk factors for victimization of younger and older persons: Assessing differences in isolation, intra-individual characteristics, and health factors. *Journal of Criminal Justice, 38*(2), 109–117.

Visher, C.A. (1983). Gender, police arrest decisions and notions of chivalry. *Criminology, 21*(1), 5–28.

Walker, S., Spohn, C., & Delone, M. (2018). *The color of justice: Race, ethnicity, and crime in America* (6th ed.). Boston: Cengage.

Weisheit, R., & Wells, L. (1999). The future of crime in rural America. *Journal of Crime and Justice, 22*(1), 139–172.

Wilson, J., & Abrahamse, A. (1992). Does crime pay? *Justice Quarterly, 9*(3), 359–377.

Wolfgang, M.E., Figlio, R.M., & Sellin, T. (1978). *Delinquency in a birth cohort*. Chicago: University of Chicago Press.

Worrall, J., & Kovandzic, T. (2008). Is policing for profit? Answers from asset forfeiture. *Criminology & Public Policy, 7*(2), 219–244.

Worrall, J., Hiromoto, S., Merritt, N., Du, D., Jacobson, J., & Iguchi, M. (2009). Crime trends and the effect of mandated drug treatment: Evidence from California's Substance Abuse and Crime Prevention Act. *Journal of Criminal Justice, 37*(2), 109–113.

Zeisel, H. (1982). *The limits of law enforcement.* Chicago: University of Chicago Press.

CHAPTER 5

Police and Policing

The most visible image of the formal control applied by the criminal justice system comes in the form of the various police agencies across the United States. A variety of federal, state, municipal, special jurisdiction (housing authority, transit authority, etc.), and private agencies provide law enforcement services. The Bureau of Justice Statistics identified roughly 18,000 local and state police agencies. Combined with federal law enforcement, these agencies combine to employ approximately 885,000 sworn law enforcement officers across the country (Reaves, 2011, 2012).

Approximately 73% of the law enforcement agencies are relatively small, employing 24 or fewer officers (see Figure 5.1). The largest police agencies, however, employ the most officers, meaning that although the typical police organization is small, the typical police officer works in a large agency. Regardless of agency size, most sworn police personnel are assigned to patrol and similar field operations duties. Patrol officers do the bulk of police work in the United States, and the work of patrol officers is varied.

Police agencies provide a variety of services to the communities they serve, ranging from travelers' aid through ambulance service. Yet, we continue to think of them as law enforcement. Indeed, enforcing criminal laws is, at most, a part-time activity for most police departments and police officers. The police are what Wilson (1968) called the "agency of last resort." Although we focus here on the law-enforcing duties of police agencies because of our interest in criminal justice, it is also important to remember that other demands are placed on police.

Because police are available 24 hours each day, are mobile, and carry authority, we call on them to resolve many issues and problems. Most of these problems are not, strictly speaking, law-enforcement-oriented in nature (Johnson & Rhodes, 2009). The criminal law is only one of many tools available to police officers and police agencies in their efforts to keep peace in our communities. The police are responsible for dealing with stray children and dogs,

IMPORTANT TERMS

community-oriented policing

CompStat

constable

cynicism

legalistic style

order maintenance

paramilitary structure

police specialization

proactive

reactive

service style

shire reeve

watchman style

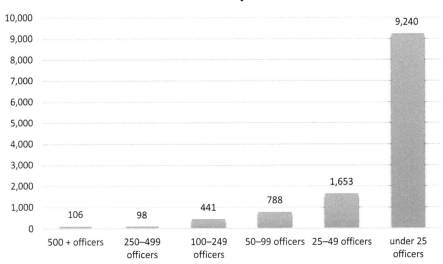

FIGURE 5.1
Distribution of Police Departments, by Size of Department.

Source: Reaves (2015).

lost travelers, injured persons, stranded motorists, traffic accidents, parades, domestic disputes, and crime. It seems that almost any disruptive event can be resolved by "calling the cops" (Payne, Berg, & Sun, 2005). In many cases, people have reported fires to the police first, rather than to the fire department. The police are often called to deal with abandoned cars, but the reason they are called is not because the car may have been involved in a crime. People call the police because the police are the first agency that comes to mind.

People also call the police because they have come to know that the police can (and probably will) do something about the problem. Bittner (1990) observed that the police have a monopoly on the legitimate use of coercive force. The ability to use force and to make people behave in certain ways defines the police role in society. The police are called to deal with a variety of problems that may require force. For example, the police may be called to intervene with a mentally disturbed person who has become disruptive but who may not have committed a crime (Wood, Watson, & Fulambarker, 2017). The police may have no way to help cure the mental disorder, but are nonetheless expected to act in a way that temporarily solves the problem.

Given the scope of police responsibilities (some assigned, some assumed, and some simply evolved), it is clear that law enforcement is only a small part of police duties. Studies of police tasks have revealed that actual crimes consume a small portion of police resources and comprise a small percentage of police tasks. Depending on the definitions used by researchers, the majority of police time is devoted to general patrol, service calls, and paperwork (Cathey & Guerin, 2009; Huey & Ricciardelli, 2015; Famega, 2005). Nonetheless, the public, the media, and the police themselves continue to define policing principally as crime fighting (Surette, 2015). This conception of the police as "crime fighters" developed historically as a response to difficulties encountered in the police role.

THE DEVELOPMENT OF AMERICAN POLICING

The idea of police controlling the behavior of individuals is a relatively recent addition to society. The American colonists did not employ police, and no police forces were created in the United States until the 1840s. In colonial times, law enforcement was the duty of every citizen, and no specialized occupational group had responsibility to ensure public order. As with so much of our justice system, we can trace the origins of public police to our English tradition.

Law Enforcement in England

Most of England had no specialized police force or public office charged with maintaining public order until after the Norman Conquest in 1066 (Critchley, 1972; Stewart-Brown, 1936). At that time, the Normans, having gained control over England and occupying a hostile population, created a centralized governmental structure based on feudalism. High-ranking officers and nobles of William the Conqueror's army received control of large parcels of England. In return, these nobles were to provide a percentage of the production from these lands to the king as taxes, and to supply soldiers in time of war. They also were required to obey the commands of the king and to ensure the "king's peace" in their lands.

To accomplish this administrative task, these nobles further subdivided their lands to lesser officers and nobles, and required them to remit a portion of their profits and to supply a number of soldiers when necessary. The subdivision continued, with each successive rank being required to pay a larger portion of taxes but supply a smaller number of soldiers. The increased taxes represented the need to meet the demands of the king and the higher nobility for income. The lower number of soldiers reflected the increased number of officers who were granted lands. If a duke promised to provide 10,000 soldiers and 5% of profits to the king, he could do it by dividing the land among ten barons. Each baron might be required to provide 1,000 soldiers (10,000 total) and 8% of profits (5% for the king; 3% for the duke). Feudalism provided a structure for government in medieval times.

Essentially rural and agrarian, England was divided into ten family units called tythings. Each tything was responsible for its own tax collection and order maintenance. With the advent of Norman control, new units of 100 families, called shires, were created. Being an occupied country, maintenance of the Norman king's law was problematic, and tax collection was difficult. The office of shire reeve emerged.

The **shire reeve** was responsible for the collection of taxes and the maintenance of the king's peace within the shire, which was usually an area similar to a county or parish in the United States today. The reeve was elected from a list of candidates approved by the lord of the manor. Over time, the shire reeve became known as the sheriff. We can see the resistance of the English to this new structure in the tales of Robin Hood, whose nemesis was the Sheriff of Nottingham.

Each manor also operated a manorial court. To assist in the day-to-day operation of the court functions, they created the office of constable. The lord of the manor selected the **constable** from among qualified property holders. The constable performed the clerical duties of the court and housed prisoners awaiting trial. Over time, these functions expanded to include general peacekeeping.

As villages and towns developed, this rural order-maintenance apparatus proved to be inadequate to the task of law enforcement in congested areas. Traditionally, every citizen was responsible for order maintenance. Drawing on this tradition, town constables were empowered to draft citizens for a watch system. In this system, citizens were required to provide unpaid watch service (typically at night) to patrol for fires and breaches of the peace. As towns grew larger and cities developed, it became increasingly difficult to find either adequate watchmen or persons willing to take the role of constable.

By the early 1800s, English towns and cities were crowded and unruly. The nobility and wealthy citizens traveled with hired guards (footmen) and avoided the more dangerous sections of town. Several experiments with paid watches, "private" law enforcement, and rejuvenated constabulary offices had all failed to provide adequately for order maintenance. One of the most famous of these experiments was the organization of the Bow Street Runners by Henry Fielding. Also known as "thief-takers," this group of men was organized to provide police protection in the Bow Street area of London. There was clearly a need to create a specialized body charged with maintaining the peace.

During the 1820s, Sir Robert Peel, British home secretary, proposed the creation of a police force for England (see Figure 5.2). This force would comprise paid, uniformed, armed, and disciplined officers whose job would be the enforcement of the law, the maintenance of order, and the prevention of crime. Crime prevention was expected to result from the presence of police on the streets. Patrolling police officers would act to deter criminal offenders. Parliament, fearing the effects of such an armed force on the "rights of Englishmen," resisted Peel's idea. In democratic societies, and especially in the Anglo-American tradition, there has always been a tension or conflict between liberty and civility (Lundman, 1980). Our need for order (civility) is in direct conflict with our desire for personal freedom (liberty). The English Parliament had to be convinced that the new police would not unduly threaten English liberty. Peel revised his proposal and, in 1829, Parliament agreed to "experiment" with a Metropolitan Police Force in London. This force would be unarmed, but in every other respect, it would mirror Peel's original plan. If it worked in London, the idea might be expanded. The Metropolitan Police Force of London, created in 1829, was the first modern police force.

The Colonial and Early American Experience

As in many other areas of social life, the American colonists relied on their traditions and experiences from England in developing a social control system. Colonial villages and towns normally had the offices of constable and/or sheriff. The duties of each were similar to those of its English counterpart (Jones & Johnstone, 2012). As towns and cities grew, the Americans experimented with watch systems. In time, it became clear that these less formal systems for order maintenance were inadequate.

Unlike England, however, there was no strong central government in America, and weapons and violence were more commonplace in the New World. Although the American police followed the same general pattern of development, these differences resulted in a modification of the English police structure to suit the situation in America. If anything, Americans were more strongly concerned with liberty, and distrust of centralized governmental power meant that any police would be limited.

By the 1840s, waves of immigrants began arriving on the shores of America, and industrialization was beginning. The population of American cities swelled, and the cities became unruly and dangerous places. The upper and middle classes viewed the urban poor, especially immigrants, as prerevolutionary. It became common to speak of the immigrant poor in America's congested cities as the "dangerous classes."

FIGURE 5.2
A member of the professional police force organized by Sir Robert Peel. Initially referred to as "Peelers," they later came to be known as "Bobbies."

Metropolitan Police, New Scotland Yard, London

To control these "dangerous" people, and to bring order and stability to the cities, Americans began to consider the creation of police forces. On many occasions, the militia or the army was used to quell riots or to break strikes (and would still be used for these purposes in the future), but these were extraordinary circumstances. Many people believed a more permanent solution to the problems of day-to-day disorder was required. The importance of ethnic and cultural conflict to the emergence and development of policing in America cannot be overstated (Barlow & Barlow, 2000).

Knowing about the English experiments and developments, many reformers began to advocate the creation of police forces for the cities. In 1845, New York City created a police force modeled after the Metropolitan Police Force of London, but with several significant differences (Jones & Johnstone, 2012:233–236). The mayor appointed police officers from among candidates recommended by political ward leaders. A Board of Police Commissioners administered the police force. Each officer was to be a resident of the ward in which he would work. Officers objected to wearing uniforms and wanted to be armed.

The New York City Police were created and funded locally. This new American police department had a weak central administration, a municipal organization, and direct political involvement. Over the years, the issues of uniforms and arms were resolved, so that municipal police were both armed and uniformed. Each city created its own police force and organizational structure. In the next 10 years, the New York City Police were followed by the creation of police in most major American cities (see Box 5.1). Monkonnen (1981) noted that local police in America were first modeled on the police of London, but that American police agencies quickly grew independent of this original model. It is important to note that the police in America developed in specific cities and towns, with no effort to create a national police force. This ensured that the American police would be under local control and direction.

BOX 5.1 MILESTONES IN AMERICAN LAW ENFORCEMENT

1748	Bow Street Runners organized by Henry Fielding
1829	Metropolitan Police Force created in London, England
1838	Boston police created with nine officers
1844	New York City police created with 800 officers
1852	Cincinnati and New Orleans police created
1854	Philadelphia and Boston police establish formal patrol
1855	Orleans police created
1857	Baltimore police created

1893	Organization of the National Chiefs of Police Union
1905	August Vollmer elected marshal of Berkeley, California
1908	Bureau of Investigation (later FBI) created; Berkeley Police School started
1924	J. Edgar Hoover named director of FBI
1930	*Uniform Crime Reports* first published
1931	Police science program started at San Jose State University
1935	FBI National Police Academy opened
1960	O.W. Wilson named Chicago police commissioner
1974	Kansas City preventive patrol experiment initiated
1982	Broken windows theory appears.
1994	Violent Crime Control and Law Enforcement Act passed (created the Office of Community Policing Services)
2001	Terrorists attack the World Trade Center and Pentagon on September 11

The American Police

Similar to New York's, most early police departments in America were not centrally organized and did not have strong leadership. This was in response to a fear of the effect of police on the exercise of rights by individuals (similar to the English Parliament's fears regarding the Metropolitan Police). It was common to employ a "police commission" to govern the department so that no single individual could gain too much power. With the growth of political machines in the cities, however, this weak administrative structure left city police forces open to manipulation and corruption.

Initially, the principal duty of the police was to maintain order. The success of a police officer was most easily established by the absence of disorder on the patrol beat. The fact that officers were recruited from the neighborhoods they were to patrol and were sometimes unwilling to arrest their friends and acquaintances meant that officers tolerated much "deviance," which the upper and middle classes found frightening. Drinking of alcohol, for example, was viewed with suspicion by many city leaders, yet tolerated (even shared) by many officers. The failure of the police to remain free of political influence and corruption, coupled with neighborhood enforcement styles, led to an early call to reform the police.

In the latter part of the nineteenth century, a reform effort to enhance police accountability and to professionalize the police emerged (Travis & Langworthy, 2008; Walker, 1977). Vollmer, perhaps the foremost proponent of police professionalism, led the reform. Vollmer sought higher personnel standards and stronger police leadership. The push for police professionalism continued well into the twentieth century and is still felt (Vogel & Adams, 1983). The police were given strong central administration and a clear crime control mandate. The focus on law enforcement was supposed to circumvent the difficulties that accompanied the more general role of maintaining order (Johnson, 1981). According to many policing historians, the movement to professionalize the police was an effort by middle-class, native-born citizens to gain control over the police and to restrain the growing immigrant population (Toch, 1997).

With the crime control mandate came an equally important definition of the police as serving a crime prevention function. The police were not only expected to detect and apprehend offenders, but police presence on the streets of a city was expected to deter others from committing crimes. From this came the tradition of preventive patrol. The uniformed officer on patrol would not only be better able to detect crime, but the patrol presence would prevent crime as well. Although the technology may have changed from foot patrol to motorized patrol (and perhaps back again), the idea is essentially the same.

The tradition of police as peacekeepers (order maintenance), however, has also remained (Kappeler, 1996). Inasmuch as the definition of the police became (and remains) one of a crime control force, the functions of a modern police department are far broader. The police are the most visible representatives of government in a community, and they represent the legitimate authority of the law. Because they are always (theoretically) present and available, the police have become responsible for handling all sorts of social and legal problems.

THE FUNCTIONS OF POLICE

The role of the police has been broadly classified into three categories. Wilson (1968) suggested that the police are responsible for law enforcement, order maintenance, and service. Further, he argued that, of the three, order maintenance is at once both the most important and the most troublesome. Order maintenance is the main purpose of police. If they do nothing else, the police must ensure that the citizens can go about their daily business safely and efficiently.

Order-maintenance activities include settling disputes, dispersing crowds, keeping sidewalks and streets clear and traffic flowing smoothly, and other important activities. These are troublesome responsibilities because the officer often must operate in the "gray areas" of the law and must choose whether to intervene and, if so, how and with whom (Eck & Spelman, 1987; Wood, Watson, & Fulambarker, 2017).

The service functions of police have evolved over time out of necessity. Citizens ask police to provide a variety of services, from giving directions to travelers to finding missing children. Partly because of a potential link to criminal behavior, police also investigate traffic accidents, provide first aid to victims, and, often, transport the injured to medical facilities. Whatever the reasons, modern police provide a wide variety of services that do not strictly conform to the role of crime control (Rossler & Terrill, 2012).

Law enforcement activities are those that relate directly to the detection and apprehension of criminal offenders. Responding to alarms and citizen complaints of crime, investigating suspicious persons and circumstances, and arresting suspected offenders are all law enforcement activities. Our focus is on the crime control function of police, although it does not comprise the bulk of police tasks. Before turning our attention to the crime control activities of the police, however, we must more fully explore the diverse obligations of contemporary police.

The Police as a Human Services Agency

To understand the role of the police, it is helpful to describe what it is that the police do. Ideally, perhaps, it would be possible to determine what it is that police are supposed to do and, from that, develop a definition of the police role. The problem is that it is not clear what it is that we want the police to do. Historically, the police acquired responsibilities because no other agency existed to perform particular tasks. Today, a variety of social, political, legal, and administrative factors shape the police role.

Whatever it is that the police do, research shows that most of what they do is not criminal law enforcement. Christine Famega (2005) reviewed studies of police officer activity and concluded that only about one-quarter of a patrol officer's time is spent responding to calls for service or dispatches. If you think of your own interactions with the police over the course of your lifetime, how often have you dealt with police officers in nonlaw-enforcement situations? The investigation of traffic accidents and crowd control at parades, demonstrations, sporting events, and the like are important, nonlaw-enforcement services provided by the police.

Many local police agencies, particularly small departments, are consolidated with emergency medical services (EMT) and fire services to create a "public safety" agency (Wilson, Weiss, & Grammich, 2016). Many police departments also serve special functions such as search and rescue, bomb disposal, underwater recovery, and even animal control (Hickman & Reaves, 2006). The driving force behind this movement is the opportunity to provide services more efficiently and the flexibility to better coordinate their responses to emergency situations (Morley & Hadley, 2013). Conversely, some agencies are opting to deconsolidate due to the management difficulties inherent with many consolidated departments (Wilson & Grammich, 2015).

The police serve as a referral center for people in trouble, linking victims, the ill, and others in need of service with available community resources. The police are available all day and every day throughout their jurisdiction. Whether crime, injury, illness, or disputes cause the police to intervene, police problems generally represent problems in living. The sheer scope and variety of situations for which people call the police are evidence of their generic helping activities. Finally, the police, like all governmental agencies, are ultimately responsible to the public (their clients) and the courts.

Historically, as a human services agency, the most important role of the police was that of first aid. Whether the problem is a lost child, domestic disturbance, landlord/tenant dispute, public intoxication, or traffic accident, the police are normally first on the scene. In this role, the police provide first aid by taking charge of the situation, providing immediate help and counseling, and giving referrals for further care (Fritsch, Caeti, Tobolowsky, & Taylor, 2004). It is common, especially in larger police departments, to give officers a directory of social service agencies to which the officers can refer citizens. The fact that the police do not always provide complete human services to resolve the living problems of those with whom they come into contact does not negate the important role of the police in the first stages of human service (Das, 1987).

The Police as a Crime Control Agency

The police control crime in one of two basic ways: reactively or proactively (Black, 1972). These types of policing represent ideal types. A **reactive** police department would only respond (or react) to crime. A **proactive** police department would use its own initiative in aggressively seeking out crimes and criminals. The strictest type of reactive police department would remain at the police station, watching television, cleaning cruisers, and the like, until a complaint of a crime was received or an alarm sounded, when the officers would rush to their cruisers and speed to the scene of the crime. Having investigated or made an arrest, the officers then would return to the station to await their next call.

In contrast, a fully proactive police department would resemble the vice squad. All officers would be in the field seeking out crime. Undercover officers would perform much police work because marked cars and uniforms would forewarn offenders. Traffic officers would establish speed traps rather than patrol stretches of highway. Decoy teams and sting operations would be prevalent.

In reality, of course, one does not find either ideal type. Rather, it is possible to classify police departments as being more or less proactive or reactive. The democratic nature of our society and the municipal organization of the majority of our police forces make police in the United States more reactive than proactive (Travis & Langworthy, 2008). As a society, we prefer to set policing priorities through our complaints and calls. The alternative is for the police to set priorities through deciding how and when to combat crime. Proactive police seeking to prevent crime are required to act on their own initiative to avoid citizen complaints about crime.

The Police as a Peacekeeping Agency

In television and motion picture Westerns, the marshal has responsibility for keeping the peace. At base, peacekeeping means the maintenance of order. The police control disruptions such as fights and riots. They maintain traffic flow and ensure a general level of satisfaction with living in the community. Police protect and enhance orderly social interactions. Wilson (1968:16) argued that order maintenance is at the core of the police task. He wrote:

> The patrolman's role is defined more by his responsibility for maintaining order than by his responsibility for enforcing the law. By "order" is meant the absence of disorder, and by disorder is meant behavior that either disturbs or threatens to disturb the public peace or that involves face-to-face conflict among two or more persons. Disorder, in short, involves a dispute over what is "right" or "seemly" conduct or over who is to blame for conduct that is agreed to be wrong or unseemly.

As Wilson's explanation implies, disorder is often noncriminal. Order-maintenance problems usually involve questions of propriety rather than questions of legality. Youths loitering on a street corner, a neighbor who plays her stereo too loudly, homeless persons congregating in a park, and other noncriminal events are frequently the basis for order-maintenance calls to the police. In these situations, people expect the responding officer to resolve the conflict and thereby restore order.

Most order-maintenance problems fall into a gray area of the law where frequently the officer has no clear legal authority to act. As a matter of practicality, the officer is compelled to do something. Order maintenance is the most common activity of police officers. These tasks often expose officers to physical danger and involve the exercise of discretion by the officer. Order maintenance is the least "consensual" part of the police task (Wilson, 1968). For these reasons, order maintenance is perhaps the most difficult aspect of policing. Jiao (1998) suggested that different communities hold different expectations of their police. He urges police to develop policies and practices based, in large part, on the expectations of the community. Other research has indicated that some police decisions reflect community characteristics. Jackson and Wade (2005) observed that the police are more proactive in their enforcement efforts in neighborhoods characterized by high rates of crime than in low-crime areas. Further, Kane and Cronin (2013) found that order-maintenance policing was more effective at reducing violent crime in areas where informal social controls are lowest. It seems the likelihood of criminal activity influences the decisions of officers to intervene with citizens, as well as the effectiveness of those interventions.

Although there may be a consensus that robbery is intolerable, that police should arrest robbers, that accident victims deserve help, and that the police should help them, there often is no such consensus for order-maintenance questions. In the case of the neighbor with the loud stereo, it is clear that dissension or disagreement exists; one party believes that the stereo is too loud and the other thinks that it is set at an acceptable level. Into this conflict steps the officer. Regardless of the outcome, at least one (if not all) of the parties will be dissatisfied.

Police are also increasingly likely to be a first responder in situations involving mentally ill or developmentally disabled individuals who have either committed a low-level crime or are exhibiting troubling behavior which the public expects to be taken care of (Reuland, Schwarzfeld, & Draper, 2009). These encounters are especially difficult for officers because there is often a lack of legal justification for an arrest or hospitalization. Police departments are increasingly collaborating with community mental health professionals to develop strategies for handling such situations (Edwards & Pealer, 2018).

Difficult as it is, order maintenance is a critical component of policing. Left unattended, minor disputes can escalate into criminal acts (such as assault or vandalism if the complaining neighbor takes the matter of the loud stereo into his or her own hands). Further, the police and the entire justice system must serve the major function of social control. We should be able to go about our daily lives in a relatively smooth and predictable fashion. More than any other police function, order maintenance ensures the routine functioning of society.

As you can see, police officers perform a variety of roles (See Box 5.2). In reality, each of the police functions work together in an effort to control crime. The police can work to control crime by enforcement of laws and by prevention. Police efforts to maintain order often have the effect of preventing crime (Braga, Welsh, & Schnell, 2015). Likewise, collaboration with mental health officials can divert mentally ill and developmentally disabled suspects from the criminal justice system so that they can receive the treatment needed to reduce future incidents. Further, as we will discuss in more detail later, the police can (and do) use criminal law to achieve order-maintenance and service functions.

BOX 5.2 POLICY DILEMMA: SCHOOL RESOURCE OFFICER

With the increased media attention on school violence, many jurisdictions have placed police officers in the K-12 school setting as a way to maintain a safe environment. These school resource officers are expected to perform many roles, including law enforcement, teacher, counselor, and order maintenance (Lynch, Gainey, & Chappell, 2016; Schlosser, 2014). Johnson (1999) found that the school resource officers did, in fact, lead to a decrease in school violence and disciplinary infractions. Though these officers may help maintain order in the school, critics argue that they, along with the increase in zero-tolerance policies, also contribute to an increased number of juveniles being processed by the criminal justice system (Price, 2009). The fear is that situations which could have previously been handled internally by the school administrators is now resulting in more formal, punitive actions which could lead a juvenile toward further criminal involvement. Are school resource officers placed in the schools in your community? If so, did you experience the school resource officers' involvement in multiple roles?

POLICE MANAGEMENT

Throughout the past 40 years, research has challenged, and in some cases redefined, the role of police. Walsh and Vito (2004) indicated that police operate within one of three existing models: (1) the rational–legal bureaucratic model, (2) the community, problem-solving model, or (3) the CompStat model. Organizational options were limited when police departments came into being in the mid-1800s. The only organizational model available that allowed control of large groups of personnel was the military model. For this reason above any other, the police often adopted a **paramilitary structure** with ranks and a chain of command. The trappings of a military organization are still a part of American policing in most places (Chappell & Lanza-Kaduce, 2010; King, 2003).

Departments organize patrol units by geography into precincts or districts. In the military model, information flows up the chain of command from the street officers to the police administrator. Orders and commands flow down the chain to the street officers. In theory, the police administrator controls the actions of the officers on the street. Organizationally, this structure enables the police to meet their conflicting functions in a routine manner. Traditional police functions include quickly responding to calls for service, randomly patrolling neighborhoods looking for criminal activity, and investigating crimes which become known to the police.

In practice, the structure of policing is different from the military model. Wilson (1968) noted that, unlike other organizations, street officers have more discretion than administrators do. The reality of police work is that officers on the street must react to a variety of situations. It is not practical for police officers to report every call and await instructions from their superiors.

As policing entered the twentieth century, the advent of the automobile as well as changes in American cities affected the structure of police departments. Although the paramilitary model remained, policing became increasingly bureaucratic. The radio and the telephone allowed more communication between officers and supervisors. In addition, the new communication tools increased demand for police service because a citizen only had to pick up a telephone to request help. However, police began to become overwhelmed by the calls for service generated by the 9-1-1 system. As the former commissioner of the New York City Police Department wrote, "chasing after those thousands of 9-1-1 calls meant putting bandages on the symptoms of the problems generating the calls" (Bratton, 1997:32). At the same time, research began to question the effectiveness of police strategies such as random police patrol and improved response time (Walker, 2015). This research, combined with the increasing crime rate, resulted in the two most widely discussed reforms in policing history: community-oriented policing and CompStat.

Community Policing

Community-oriented policing is an approach to policing that relies on community definitions of police functions and a partnership between the police and the community in the production of public safety. Each police officer is expected to have responsibility for his or her beat area. The officers work with residents to identify problems and implement solutions. This direct link between the officer and the community naturally lessens the bureaucratic control and "chain of command" characteristic of the more formal, specialized police agency (see Box 5.3).

BOX 5.3 COMMUNITY-ORIENTED POLICING

Programmatic elements

- Foot patrol
- Bicycle patrol
- Mounted (horse) patrol
- Neighborhood ministations
- Citizen police academies
- Neighborhood Watch
- School resource officers
- D.A.R.E. programs

- Citizen ride-along programs
- Neighborhood police officers
- Prioritizing calls for service
- Geographic mapping

Although community-oriented policing is often discussed as a fundamental philosophical shift in the way police agencies do business, departments reporting the existence of community policing initiatives often point to the implementation of special programs such as these as evidence that they are "doing community policing."

Source: Hickman and Reaves (2006:19, 21).

This trend seeks to actively engage the public, ultimately leading to more effective informal social controls. As Fielding (2005b) described, community policing is a "style of policing in which the police are close to the public, know their concerns from regular everyday contacts, and act on them in accord with the community's wishes" (p. 460). Community-oriented policing, however, is still elusive of definition. Many agencies report that they practice community-oriented policing, but each agency seems to describe something different with this term. Still, community-oriented policing involves an expansion of the police role from reactive crime fighters to proactive problem solvers. In partnership with the community, the police identify community problems that contribute to crime and seek solutions designed to alleviate those problems. Community policing is now a topic of basic police training in most police academies in the United States (Reaves, 2016). See Table 5.1 for some statistics on police academy training.

The rapid acceptance of this new role definition of policing once led some commentators to suggest that community-oriented policing was the new "orthodoxy," or commonly accepted purpose for the police. The police, under community-oriented philosophies, are supposed to be more closely tied to the communities they police, more accessible to the public, and less bureaucratic. The range of issues that are now defined as legitimate police problems is enormous, including street lighting, sanitation removal, recreational and health programming, housing, and almost every other public problem (Fielding, 2005a).

In practice, community-oriented policing has supported decentralized organization in police agencies, alternative patrol strategies and resource allocation, and increased police involvement in civic issues. In theory, the police should be involved in this wide range of problems because they are broadly responsible for public order, and because they are perhaps best organized (jurisdiction-wide, 24-hour availability) to learn about problems, develop solutions, and monitor outcomes. In most places adopting formal community-oriented policing programs, departments assign at least one neighborhood officer to each crime-ridden neighborhood. Permanent assignment to a specific beat is expected to help the officers develop a better understanding of the

TABLE 5.1 Percent of Police Academies with Training in Community Policing and other Special Topics		
	Percent of Academies with Training	Average Number of Hours Required
Domestic Violence	98	13
Cultural Diversity/Human Relations	95	12
Mental Illness	95	10
Sexual Assault	92	6
Domestic Preparedness/Terrorism	85	9
Community Partnership Building	82	10
Conflict Management	82	9
Problem-Solving Approaches	80	12

Source: Reaves (2016).

FIGURE 5.3

The Manchester, New Hampshire, Police Department touts its commitment to Community-Oriented Policing (C.O.P.S.) with signs posted around the city.

Ellen S. Boyne

area and its residents, and to take greater responsibility for public safety and problem solving in the beat. See Figure 5.3.

Kane (2000) found that within a few weeks, officers who were permanently assigned to an area "took ownership" of the beat and began to engage in more proactive, problem-solving efforts. Residents quickly come to appreciate having their own officer. Evidence suggests that community-oriented policing improves citizens' satisfaction with the police (Gill, Weisburd, Telep, Vitter & Bennett, 2014).

The implementation of community policing has not been easy in all departments and has rarely been implemented completely. Most often, departments have adopted the most visible and easily attainable aspects of community policing without making the fundamental organizational changes needed for community policing to truly become the new policing orthodoxy which it was once envisioned (Magers, 2004). Chappell (2009) identified three primary obstacles to the implementation of community policing: (1) lack of resources, (2) time constraints, and (3) cultural resistance from within the police department. Likewise, Liederbach, Fritsch, Carter, and Bannister (2007) found that whereas citizens valued the community policing in their neighborhoods, many officers still resisted the movement. Smaller police departments seem to have a particularly difficult transition to community policing, as the low turnover rates and infrequent hiring of new officers creates an environment that tends to value traditional techniques over more radical changes (Morabito, 2010).

CompStat

First implemented in New York City, **CompStat** represents a compromise between the decentralized community policing model and the traditional paramilitary-style bureaucracy (Dabney, 2010). This strategy gives the decision-making authority to the middle-level managers such as the district commanders and then redirects resources as needed to assist in solving each district's specific problems (Weisburd, Greenspan, Mastrofski, & Willis, 2008). At its core, CompStat has four components (Police Executive Research Forum, 2013:2):

1. Timely and accurate information and intelligence,
2. Rapid deployment of resources,
3. Effective tactics, and
4. Relentless follow up.

A core component of CompStat is the use of increasingly sophisticated crime analysis and mapping technology (Johnson, 2017; Santos, 2014). These tools allow police managers to have access to important information regarding crime in their community. Police executives then discuss these crime problems at weekly or monthly meetings in an attempt to develop solutions. Middle-level police commanders are then held accountable for implementing the crime reduction strategies to the problems in their districts.

It is important to note that CompStat is, at base, simply a method used to develop solutions to crime problems (Police Executive Research Forum, 2013). Police often use innovative strategies associated with hot spot policing, problem-oriented policing, or community policing to reduce the criminal activity. However, these strategies can be hindered by departmental organizational or cultural issues. For example, a hierarchical culture has been shown to reduce the active participation of low-ranking officers during the CompStat meetings (Bond & Braga, 2015; Willis, Mastrofski, & Weisburd, 2004). Likewise, Dabney (2010) found that the strategies developed during the CompStat meetings were being miscommunicated as they were passed down the chain of command to the street-level officers.

COUNTERTERRORISM AND HOMELAND SECURITY

In the aftermath of the terrorist attacks of September 11, 2001, there has been some concern that the increased expectations in terms of security would overshadow the emphasis on community policing. In fact, the use of community policing did decline in the years after the terrorist attacks. However, research shows that many police agencies view community policing and homeland security as complementary strategies (Randol, 2012; Roberts, Roberts, & Liedka, 2012). Community policing can improve community relations, which in turn can improve security. Even so, community policing and terrorism preparedness do at least compete for resources (Lee, 2010).

Several federal and state initiatives involve police in counterterrorism activities. The roles of police and law enforcement agencies in counterterrorism differ by size and type of agency. Many federal law enforcement agencies now operate within the Department of Homeland Security. Some local police departments,

FIGURE 5.4
A New York City Counterterrorism Officer and bomb-sniffing dog patrol the perimeter of a public viewing area.

Associated Press

especially the largest ones, have counterterrorism units or include counterterrorism efforts within expanding intelligence units. Many observers argue that the most important role of local police in counterterrorism efforts is as producers and consumers of intelligence (O'Connell, 2008). See Figure 5.4.

Perhaps the biggest change in policing since 9/11 has been the growth of intelligence-led policing—the idea of basing police activity on criminal intelligence (Ratcliff, 2016). Great effort has been directed to increase intelligence gathering and information sharing among police departments at various levels. Schaible and Sheffield (2012) found that over 90% of state agencies have allocated more resources to intelligence efforts since 9/11. Much of this effort has been directed toward the increased use of FBI-sponsored Joint Terrorism Task Forces (JTTFs), which allow state and local police agencies to collaborate with federal agencies. Approximately one-third of local law enforcement agencies collaborate with these task forces (Riley, Treverton, Wilson, & Davis, 2005). The creation of state-level fusion centers has perhaps been a more significant development. These centers serve as a clearinghouse of information for the Department of Homeland Security and provide opportunities for agencies at all levels to share information about terrorist or other criminal threats (Chermak, Carter, Carter, McGarrell, & Drew, 2013).

The role of police agencies in counterterrorism preparedness largely depends on the size of the agency. Schafer, Burruss, and Giblin (2009) reported on a study of small municipal police agencies in Illinois. They found that smaller agencies had done relatively little in response to concerns about terrorism. Agencies located close to Chicago or other large cities and larger police departments had done more preparation. Those police chiefs who had done more preparation naturally felt more prepared to respond to acts of terror. Finally, those agencies that perceived a terrorist threat were more likely to prepare to respond.

William Pelfrey (2007) reported a case study of early responses in South Carolina. His research indicated, again, that most local police agencies had done relatively little specifically to prepare for a terrorist attack. Pelfrey found that factors such as the size of the police agency, its technological capacity, and whether it had a Special Weapons and Tactics (SWAT) team predicted its level of preparation. He noted that perhaps not every police agency needed to prepare to combat terrorism.

The small size and limited jurisdiction of most local police departments militate against their having a major role in counterterrorism. It is not clear what we can reasonably expect of the local police concerning terrorism. Reaves (2016) reported that nearly every law enforcement training academy provided some basic training on terrorism to new police recruits. Although there is widespread concern about homeland security in the United States, it has never been a core part of the local police role. The structure of American police reflects their predominantly local focus.

THE STRUCTURE OF AMERICAN POLICING

The structure and organization of law enforcement agencies in the United States show the influences of historical development and the conflicting tasks expected of them. Unlike the police agencies of other countries, most American police agencies exist at a local level. This fragmentation of police service supports the value we place on federalism and local autonomy. Americans do not want national police, and we insist on maintaining a civilian police force that is distinct from the military (Kraska, 1994; Moore, 1987).

The focus of police organization and administration is to standardize the use of civil force (Alpert & Dunham, 1988). The police must weigh the mandate to control behavior against the requirement that they respect individual rights. This double responsibility places a premium on controlling and directing the actions of individual police officers. The bureaucratic response to this concern was the creation of rules and procedures for officers to follow. These departmental policies, or standard operating procedures (SOPs), became a factor determining the actions of individual officers. Although not perfect, the rules affect how officers decide to handle cases (LaFrance & Day, 2013; Mastrofski, Ritti, & Hoffmaster, 1987). Like other organizations, police agencies change slowly in response to external pressures (King, 1999).

Early police officers were sworn into office, issued uniforms and weapons, and sent to the streets. Most states now require training and certification of recruits prior to being assigned to patrol. Beginning in 1972, many police agencies developed field-training programs to evaluate how well new officers apply laws and departmental policies to field situations (McCampbell, 1986). These programs help ensure that police officers know and follow the rules of the police bureaucracy. Today field training is a component of basic police training in the overwhelming majority of American police departments. Over a third (37%) of all police academies require recruits to complete field training as part of the academy program. When not part of the basic academy training, the agency that hired the recruit usually provides field training (Reaves, 2016).

Another development in policing during the twentieth century was specialization. Large police departments, especially those in our biggest cities, use task specialization to assign officers. **Police specialization** divides tasks into special units or divisions. Whereas departments differ among themselves in how they separate and name tasks, Alpert and Dunham (1988:59) identified four basic elements: (1) administration, (2) communication, (3) patrol, and (4) internal review. They observed that many police agencies use divisions that are more precise. Patrol and investigation are two units that include the crime control function of the police. A traffic division and community relations unit may combine with the patrol unit to provide service and order maintenance. Administrative services and internal affairs units assist police administrators in running and controlling the department.

A typical large police department may have several divisions. The investigation unit, for example, may be further divided into homicide, robbery, fraud, vice, and other squads. Regardless of the complexity of the police bureaucracy, most retain the military ranks and chain of command. Figure 5.5 presents a model organizational chart for a specialized police department. This chart shows an organization that has a well-defined hierarchy—there are many steps between the individual patrol officer and the chief of police. Most police departments, however, are small and do not have such a detailed and specialized organization. Still, even the smallest of police departments has a chief of police. An alternative organization, depicted in Figure 5.6, "flattens" this hierarchy, giving patrol officers easier input into policy development and shortened lines of communication. This flatter organizational structure is more consistent with contemporary calls for community-oriented policing (Maguire, Shin, Zhao, & Hassell, 2003; Zhao, Ren, & Lovrich, 2010).

So far, most police organizations have not shown substantial change in their organizational structure. King (2005) suggested that a rigid rank structure is only one part of police hierarchy and that if officers have discretion to make decisions, the police organization can operate in a decentralized fashion while keeping the traditional rank structure.

We must remember that even in the most specialized city police department, patrol officers are generalists. Observers have called patrol the "backbone of policing." Patrol officers still provide the majority of police services. Nearly 70% of officers in local police departments have assignments to patrol (Reaves, 2015). It is unlikely that the decentralized performance of police service by patrol officers will change in the future. What may change is the amount of direction and supervision these generalist patrol officers receive from the central administration.

The degree of specialization in any police department is at least partly the result of the size of the department (Reaves, 2015). Larger departments are

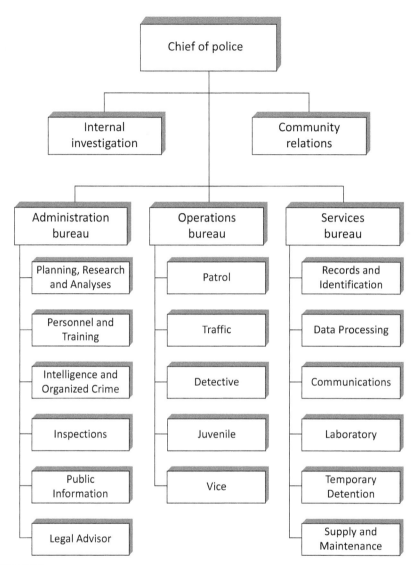

FIGURE 5.5
Organizational Chart for a Police Department.

Source: President's Commission on Law Enforcement and Administration of Justice (1967), Task Force Report: Police (Washington, DC: U.S. Government Printing Office):47.

more likely to be specialized than are smaller ones. Recall that the majority of police officers work in large, bureaucratic departments, but that most departments are small. This fact of police organization means duplication and inefficiency are part of American policing. Our police serve communities as much

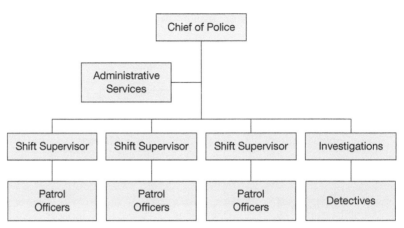

FIGURE 5.6
A Flattened Organizational Structure.

as (or more than) they serve in the enforcement of the criminal law. The price we pay for local control of police is inefficiency. To ensure that we have police who are responsive to local needs, we must be willing to tolerate multiple jurisdictions and thousands of separate police agencies.

UNDERSTANDING POLICE

Because we are examining the criminal justice system, we will proceed to discuss policing almost exclusively in terms of law enforcement. In doing so, it is easy to forget the other complex demands placed on the police. The purpose of the previous discussion was to recognize and highlight the fact that enforcing the criminal law is only one part of the police function. In dealing with many social problems, the police must develop workable, immediate responses. The multiple goals of policing complicate an analysis of the crime control actions of police departments. Nonetheless, we shall endeavor to focus on the role of the police in the criminal justice system—and that role includes crime control. Several factors influence the practice of policing, but three of the most important are the characteristics of the community, the police organization, and those of the police officer.

Police and Communities

Most policing in the United States happens at a local level. The police serve their various communities. As Wilson (1968) reported, there are varieties of police behavior. It is our goal to understand and explain police behavior as it relates to crime control. We cannot accomplish that goal if we ignore the contributions of community to the decisions of police officers.

Wilson (1968) used the frequency and formality with which officers intervened in the lives of citizens to classify police department styles. Formality refers to the use of the criminal law, whereas frequency refers to the rate at which police interact with citizens. He identified three basic styles of policing: (1) legalistic, (2) service, and (3) watchman. The **legalistic style** involved relatively frequent and formal interventions. In the **service style**, police intervened frequently, but informally (there was little law enforcement). In contrast, in the **watchman style**, police intervened infrequently. For example, in the case of a curfew violation, legalistic policing would involve the police stopping and issuing a citation (formal intervention); the service style would involve a stop followed by a warning (informal intervention); and the watchman style would predict that the officers were likely to ignore the violation (no intervention).

Because these are ideal types, we can expect variety in practice. Sometimes officers in a legalistic-style department will ignore the violation, those in a service-style department will issue a citation, and those in a watchman-style department will stop the citizen. However, the police department as a whole develops a style of policing that is generally characteristic of the organization. Wilson (1968) suggested that these different styles reflect differences in the communities served by the police. Thus, police departments develop policing styles that are appropriate to the desires and needs of the communities in which they work. Although it is difficult to determine exactly what characteristics of communities explain variation in police department styles, it seems clear that local police do exhibit different styles (Liederbach & Travis, 2008). It is also true that neighborhood or community characteristics help explain public perceptions of the police (Luo, Ren, & Zhao, 2017). For example, McNeeley and Grothoff (2016) found that racial tension in certain neighborhoods impacted residents' perception of the police.

Other research has shown that police react to neighborhood or community characteristics. Varano, Schafer, Cancino, and Swatt (2009) found that police were likely to discount the seriousness of several types of crimes depending on the neighborhood in which they occurred. Although police usually report some crimes, such as robbery, as serious crimes, police in lower-class and crime-ridden neighborhoods did not record other crimes, including drug offenses and some assaults. Police tend to identify more problems in disadvantaged neighborhoods and see more crime problems there than do the residents (Sun & Triplett, 2008). Similarly, Sabol, Wu, and Sun (2013) found that police acted more aggressively toward suspects in disadvantaged neighborhoods.

A long tradition of policing research has reported community and neighborhood effects on police actions. It is clear that community characteristics are an important part of the police equation and that explanations of police behavior must include reference to community and neighborhood factors. These community influences are likely to resist change. Zhao and colleagues (2010) found that "incrementalism" best explains community financial support for local police. A tradition of high or low support is a better explanation of current support than is any other. Communities that have supported the police in the past are likely to continue to do so, whereas those that were less supportive will continue to be less supportive. The link between communities and their police is traditional and enduring.

Police Organizations

Beyond recognizing the multiple functions served by police, we must also remain aware of the structure and organization of policing. The actions of police officers and police departments reflect different patterns of organization (Holmes, 1997). Departmental policies and procedures, as well as recruit training, serve as boundaries on police behavior. The diversity of organizational sizes, structures, and policies ensures variety in the practice of policing.

Wilson's (1968) model of police styles explains differences in police organizations resulting from different community characteristics. These organizational differences produce different officer behaviors. In this model, the organization mediates (or transmits) community characteristics to officer behavior. The greatest influence on what officers do is the product of the police organization.

Phillips and Varano (2008) reported that police officer charging decisions reflect organizational influences. The number and types of charges made in domestic violence cases by local police officers differed from those of state police. The local police decisions reflected their awareness of conditions in the local courts that affected case prosecutions. Nowacki (2015) found officers' use of lethal force related to department size and administrative policy. Smaller departments and those with tighter regulations regarding officer discretion in use of force experienced less lethal force incidents by their officers.

Not surprisingly, the police organization can control and direct the decisions of police officers. Training, regulations, supervision, and discipline all work to constrain the behavior of officers. Beyond these direct controls, much evidence supports the notion that informal understandings and rules exist that structure the decisions and behaviors of officers. Even though individual officers still exercise discretion, their choices are constrained by the organization for which they work.

Selection of Police Officers

The police officer is a very important component of the justice process. In some ways, he or she is the most important component: Police officers decide whom to subject to justice processing, what crimes to investigate, and how vigorously to enforce laws. As the most numerous and most visible agents of criminal justice, police officers are disproportionately responsible for citizens' opinions about the entire justice system. Ultimately, police officers are ordinary people entrusted with extraordinary powers, and they are charged with what some have called "an impossible mandate" (Manning, 1978).

The discretionary nature of the decisions officers make concerning the criminal justice system means that the forces that influence their decisions also influence the workings of the justice system (Brooks, 1989). If a police officer is prejudiced, the system will be prejudiced to some extent. Thus, the characteristics of officers, as well as their relationship to the organization and the citizenry, are important influences on criminal justice system operations. The characteristics of individual officers, including demographic factors (age, gender, race, ethnicity) and experience (education, rank, years on the job, etc.), have been linked to officer decisions (Fridell & Binder, 1992; Harris, 2009; Hassell & Brandi, 2009; LaFrance & Day, 2013; Riksheim & Chermak, 1993).

Because it is believed that individual officer characteristics influence police decisions, the selection and training of police officers is an important topic. Police selection typically involves a number of steps that are designed to produce the best qualified police officers. Research into the police selection process, however, indicates that it is not always effective. Metchik (1999) suggested that our reliance on a "screening out" model aimed at dismissing unsuitable candidates is not as effective as one that seeks to identify and retain good candidates. Similarly, Gaines and Falkenberg (1998) argued that traditional written examinations may not adequately identify the best candidates, and may have the negative effect of excluding female and minority candidates. Henson, Reyns, Klahm, and Frank (2010) reported that selection tests and performance in the academy may be unrelated to performance as a police officer. Likewise, selecting police officers largely based on their education might not have as substantial of an impact as once assumed. Although officers who have a college education might be less likely to resort to the use of force than less educated officers, education does not seem to have an impact on other key discretionary decisions or the frequency of citizen complaints (Bruns & Bruns, 2015; Rydberg & Terrill, 2010; Terrill & Ingram, 2016). Only 15% of local law enforcement agencies have some type of college requirement for entry-level officers (Reaves, 2015).

One of the recent changes in American policing has been the increased hiring of women and minority group members. Some observers feared that women would be unable to endure the danger and rigors of police work (Hale & Wyland, 1993; Potts, 1983). Others believed that the inclusion of women and minorities would improve community relations. Thus far, the data show that female and minority group members make good police officers (Lundman, 2009; Poteyeva & Sun, 2009; Schuck, 2014b). However, only about 12% of sworn officers are female, and about 27% are members of ethnic minority groups (see Figure 5.7). Generally, larger police agencies have greater levels of diversity (Federal Bureau of Investigation, 2016).

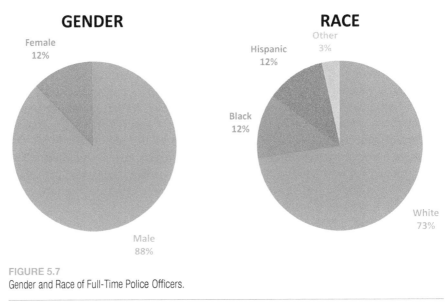

FIGURE 5.7
Gender and Race of Full-Time Police Officers.

Source: Federal Bureau of Investigation (2016).

Challenges to affirmative action policies threaten to limit diversity among police personnel. Cordner and Cordner (2011) surveyed female officers and police chiefs in several police departments in which the female representation was well below the national average. Both groups indicated that the low level of female employment was primarily due to the lack of female applicants. It appears that this was, at least in part, due to the gender-neutral recruitment process. Cordner and Cordner suggested that agencies may need to recruit in a manner that actively encourages female applicants. Overall, only 21% of law enforcement agencies have such recruitment efforts targeting females and minorities (Reaves, 2012). At the same time, there is evidence that traditional officer selection practices such as physical ability testing still work to exclude qualified female and minority candidates (Schuck, 2014a; Lonsway, 2003).

Beyond selecting police officers, training and experience also produce differences among officers, and training frequently is used as a vehicle to change officer attitudes and practices (Caro, 2011). Ingram and Lee (2015) have suggested that officer attitudes are related to job satisfaction. Johnson (2012) found that officers reported lower job satisfaction when they experienced role conflict, defined as discomfort when the required job-related behaviors are inconsistent with the officer's personal beliefs. In short, part of the task in police officer selection and training is to develop a good match between the skills and orientations of officers and the tasks and goals of the police organization.

Police Culture

Many observers have identified a distinct police culture in which officers share a widely accepted set of attitudes, values, and occupational norms. For example, police often isolate themselves from other members of society and maintain loyalty to each other when questioned by outsiders (Paoline, 2003). Officers also tend to be more conservative and conformist, as compared to the rest of society (Ellwanger, 2017). Some have suggested that the policing occupation attracts certain types of people who share similar personality traits. Others argue that the nature of the job changes an average person, thus forming a police personality. Whatever the causes for the police to have this unique culture, it is important to understand the complex set of forces that affects the policing activities of the individual officer.

Just as Wilson (1968) suggested that there are types of police organizations, others have suggested that there are types of police officers. Muir (1977), for example, argued that officers differ in terms of their willingness to use coercive force and their ability to empathize with or understand citizens. These differences result in four distinct types of officers differing in the ways they do policing. Muir used the term "enforcers" to describe those officers who are willing to use force but unable to empathize with citizens. Those comfortable with force and able to empathize he called "professionals." Officers who were unwilling to use force were either "reciprocators," who could empathize with citizens, or "avoiders," who did not understand citizens. More recently, Paoline (2004) identified seven typographies of police officers. Regardless of the specific terminology used, it is reasonable to conclude that police officers probably differ among themselves, and that these differences are relevant to an understanding of their criminal justice decisions (Travis & Langworthy, 2008).

The complexity of the job facing the individual officer is multiplied several times when we analyze the complexity of policing in society on an organizational level. The police officer is a member of an organization and, as such, must be careful to serve the ends of the organization. She or he is also a bureaucrat, and must abide by the rules and regulations of the bureaucracy. Finally, the officer is a member of an occupational group that is larger than his or her individual department. Thus, in any given situation, the behavior of the officer reflects the limits imposed by (1) the department, (2) the goals of the organization, and (3) the prestige of the occupational group. The officer is (at least subliminally) aware that his or her actions will be judged against all of these criteria. Ingram, Paoline, and Terrill (2013) found that officers who work together in the same shifts within a department develop their own police culture.

To this mix of standards, we must add the social context of the situation in which the officer is operating (Corsaro, Frank, & Ozer, 2015; Stein &

Griffith, 2017). Many times the goals or expectations of the audiences of a police department conflict with each other and with those of the department. The victim/witness has a set of expectations about how the officer will behave. The officer has expectations about himself or herself and about the victim/witness. Finally, onlookers also have their expectations. The officer is aware of many of these expectations, and they affect his or her behavior. It is increasingly likely that the officer's actions will be captured on some type of video (e.g., body camera, cell phone) for public scrutiny. Although not all calls to which the police respond are dangerous or difficult, there are no simple calls.

Studies of the police show that officers frequently act as if they are cynical (Osborne, 2014). **Cynicism** entails distrust or suspicion. In terms of police, the term refers to a perception or belief that citizens, department leaders, politicians, and other criminal justice officials are not truthful and honest in their dealings with officers. From a police officer's perspective, laws are often created by legislatures who have no background in criminal justice. Further, the officer's job is criticized by a public who have no understanding of the nature of police work (Caplan, 2003). In this sense, cynicism can create a barrier between the officer and the public for which they serve.

Police Officer Stress

The existence and nature of police officer stress has been an important but elusive topic of research for several decades. The evidence about the level of stress experienced by police officers in comparison to those in other occupations is unclear. Policing, as an occupation, contains some unique stressors, such as the chance of killing someone, the threat of attack, and the need to deal with human misery and crime. Police officers also face many of the same stressors faced by other workers, such as rotating shifts, organizational insensitivity, red tape, and the like (Ma et al., 2015; Vila & Kenney, 2002). Investigations of police stress indicate that causes and levels of stress may differ among officers (Gachter, Savage, & Torgler, 2011). Officers whose personality does not fit into the prevailing police culture seem to experience more stress than those who are part of the police culture (Rose & Unnithan, 2015) Research reveals that officers experience stress differently and develop different coping patterns based on their gender, race, level of family support, and length of service (Hassell & Brandi, 2009; Kurtz, 2008; Menard & Arter, 2013). Although police may experience stress as do other people, not all police experience it to the same degree, nor cope with it as well. Some officers respond to stress by leaving the profession (McCarty & Skogan, 2012). Burnout and turnover represent costly problems for police organizations, because replacing officers requires extensive and expensive selection and training processes.

POLICING IN THE WHOLE SYSTEM

Returning to our discussion of the location of the criminal justice system within the whole system of American society, Wilson's observations about the police indicate the influence of environmental factors. He saw police agency styles as related to community characteristics. The size, structure, and composition of the community, Wilson (1968) argued, created a "political culture" in which the police exist. Part of this culture defined the limits of police actions that were acceptable in the community. Service-style agencies existed in suburban, middle-class communities. Watchman-style and legalistic agencies operated in heterogeneous (mixed population in terms of social class and ethnicity) cities. The distinction between whether the police would demonstrate a watchman or legalistic style seemed to be related to the type of government in the cities. Those with "professional" governments (city managers) were legalistic; those in "political" governments (strong mayor) were watchman style.

Wilson (1968) did not contend that city governments directly influenced policing practices, but that differences in communities created different environments and that police styles reflected those environments. Thus, the job the police do and the way in which they do it reflects the values and structure of the community in which they exist (Travis & Langworthy, 2008). As a subsystem of the criminal justice and social systems, the police department adapts to its environment, supporting the existing equilibrium. Consider the likely public reaction to a watchman style of policing in a homogeneous, middle-class, suburban community. It is likely that citizens would complain about the failure of the police to enforce the laws, and would demand police reform. On the other hand, a legalistic style would also generate complaints about overzealous enforcement. The service style is most likely to fit the desires and expectations of the citizens in such a community.

Finally, given the essentially democratic nature of government (and policing in particular) in the United States, it is logical that the police reflect the desires of the citizens. After all, if citizens define the police job through their calls for service, the citizens direct the police. The spread of community-oriented policing promises to increase variety in policing. Each large police department can potentially become composed of a large number of distinct "neighborhood" or "community" police. The types of issues that attract police attention, and the ways in which police react to problems, may begin to vary not just between departments but within departments between neighborhoods.

The next chapter examines the law enforcement role in the criminal justice system. It continues our discussion of the forces that affect discretionary decisions by police officers. The final part of the chapter examines some contemporary issues in law enforcement. Each of these topics must be understood within the context of police history, functions, and organization.

REVIEW QUESTIONS

1. What is meant by calling the police the "agency of last resort"?
2. Briefly trace the development of policing in the United States.
3. Describe how the police can be considered to be a human services agency.
4. How can community policing be considered a revised role for the police?
5. Differentiate between reactive and proactive policing.
6. How do the multiple tasks expected of the police affect them as part of the justice system?
7. What factors influence the decisions and behavior of police officers?

REFERENCES

Alpert, G., & Dunham, R. (1988). *Policing urban America*. Prospect Heights, IL: Waveland.

Barlow, D., & Barlow, M. (2000). *Police in a multicultural society: An American story*. Prospect Heights, IL: Waveland.

Bittner, E. (1990). *Aspects of police work*. Boston: Northeastern University Press.

Black, D.J. (1972). The mobilization of law. *Journal of Legal Studies, 2*(1), 125–149.

Bond, B.J., & Braga, A.A. (2015). Rethinking the CompStat process to enhance problem-solving responses: Insights from a randomized field experiment. *Police Practice and Research, 16*, 22–35.

Braga, A.A., Welsh, B.C., & Schnell, C. (2015). Can policing disorder reduce crime? A systematic review and meta-analysis. *Journal of Research in Crime and Delinquency, 52*(4), 567–588.

Bratton, W.J. (1997). Crime is down in New York City: Blame the police. In Norman, D. (Ed.), *Zero tolerance: Policing a free society* (pp. 29–43). London: Institute of Economic Affairs.

Brooks, L. (1989). Police discretionary behavior: A study of style. In Dunham, R., & Alpert, G. (Eds.), *Critical issues in policing: Contemporary readings* (pp. 121–145). Prospect Heights, IL: Waveland.

Bruns, D.L., & Bruns, J.W. (2015). Assessing the worth of the college degree on self-perceived police performance. *Journal of Criminal Justice Education, 26*(2), 121–146.

Caplan, J. (2003). Police cynicism: Police survival tool? *The Police Journal, 7*, 304–313.

Caro, C.A. (2011). Predicting state police officer performance in the field training officer program: What can we learn from the cadet's performance in the training academy? *American Journal of Criminal Justice, 36*, 357–370.

Cathey, D., & Guerin, P. (2009). *Analyzing calls for service to the Albuquerque police department*. Albuquerque, NM: Institute for Social Research.

Chappell, A. (2009). The philosophical versus actual adoption of community policing: A case study. *Criminal Justice Review, 34*, 5–38.

Chappell, A., & Lanza-Kaduce, L. (2010). Police academy socialization: Understanding the lessons learned in a paramilitary-bureaucratic organization. *Journal of Contemporary Ethnography, 39*(2), 187–214.

Chermak, S., Carter, J., Carter, D., McGarrell, E., & Drew, J. (2013). Law enforcement's information sharing infrastructure: A national assessment. *Police Quarterly, 16*(2), 211–244.

Cordner, G., & Cordner, A. (2011). Stuck on a plateau? Obstacles to recruitment, selection, and retention of women police. *Police Quarterly, 14*(3), 295–310.

Corsaro, N., Frank, J., & Ozer, M. (2015). Perceptions of police practice, cynicism of police performance, and persistent neighborhood violence: An intersecting relationship. *Journal of Criminal Justice, 43*, 1–11.

Critchley, T. (1972). *A history of police in England and Wales* (2nd ed.). Montclair, NJ: Patterson Smith.

Dabney, D. (2010). Observations regarding key operational realities in a CompStat model of policing. *Justice Quarterly, 27*, 28–51.

Das, D. (1987). *Understanding police human relations.* Metuchen, NJ: Scarecrow Press.

Eck, J., & Spelman, W. (1987). Who you gonna call? The police as problem busters. *Crime & Delinquency, 33*(1), 31–52.

Edwards, B., & Pealer, J. (2018). Policing special needs offenders: Implementing training to improve police–citizen encounters. In Dodson, K.D. (Ed.), *Handbook of offenders with special needs.* New York: Routledge.

Ellwanger, S.J. (2017). Learning police ethics. In Braswell, M., McCarthy, B., & McCarthy, B. (Eds.). *Justice, crime, and ethics* (9th ed.). New York: Routledge.

Famega, C. (2005). Variation in officer downtime: A review of the research. *Policing: An International Journal of Police Strategies & Management, 28*(3), 388–414.

Federal Bureau of Investigation (2016). *Crime in the United States, 2015.* Retrieved from: https://ucr.fbi.gov/crime-in-the-u.s/2015/crime-in-the-u.s.-2015/tables/table-74

Fielding, N. (2005a). *The police and social conflict* (2nd ed.). London: Glasshouse Press.

Fielding, N. (2005b). Concepts and theory in community policing. *The Howard Journal, 44*(5), 460–472.

Fridell, L., & Binder, A. (1992). Police officer decision making in potentially violent confrontations. *Journal of Criminal Justice, 20*(5), 385–399.

Fritsch, E., Caeti, T., Tobolowsky, P., & Taylor, R. (2004). Police referrals of crime victims to compensation sources: An empirical analysis of attitudinal and structural impediments. *Police Quarterly, 7*(3), 372–393.

Gachter, M., Savage, D., & Torgler, B. (2011). Gender variations of physiological strain amongst police officers. *Gender Issues, 28*, 66–93.

Gaines, L., & Falkenberg, S. (1998). An evaluation of the written selection test: Effectiveness and alternatives. *Journal of Criminal Justice, 26*(2), 175–183.

Gill, C., Weisburd, D. Telep, C.W., Vitter, Z., & Bennett, T. (2014). Community-oriented policing to reduce crime, disorder and fear and increase satisfaction and legitimacy among citizens: A systematic review. *Journal of Experimental Criminology, 10*, 399–428.

Hale, D., & Wyland, S. (1993). Dragons and dinosaurs: The plight of patrol women. *Police Forum, 3*(2), 1–6.

Harris, C. (2009). Exploring the relationship between experience and problem behaviors: A longitudinal analysis of officers from a large cohort. *Police Quarterly, 12*(2), 192–213.

Hassell, K., & Brandl, S. (2009). An examination of the workplace experiences of police patrol officers: The role of race, sex, and sexual orientation. *Police Quarterly, 12*(4), 408–430.

Henson, B., Reyns, B., Klahm, C., & Frank, J. (2010). Do good recruits make good cops? Problems predicting and measuring academy and street-level success. *Police Quarterly, 13*(1), 5–26.

Hickman, M., & Reaves, B. (2006). *Local police departments, 2003.* Washington, DC: Bureau of Justice Statistics.

Holmes, S. (1997). *The occupational definition of police use of excessive force* (Unpublished doctoral dissertation). Cincinnati, OH: University of Cincinnati Division of Criminal Justice.

Huey, L., & Ricciardelli, R. (2015). This isn't what I signed up for: When police officer role expectations conflict with the realities of general duty police work in remote communities. *International Journal of Police Science & Management, 17*(3), 194–203.

Ingram, J.R., & Lee, S.U. (2015). The effect of first-line supervision on patrol officer job satisfaction. *Police Quarterly, 18*(2), 193–219.

Ingram, J.R., Paoline, E.A., III., & Terrill, W. (2013). A multilevel framework for understanding police culture: The role of the workgroup. *Criminology, 51*(2), 365–397.

Jackson, A., & Wade, J. (2005). Police perceptions of social capital and responsibility: An explanation of proactive policing. *Policing: An International Journal of Police Strategies & Management, 28*(1), 49–68.

Jiao, A. (1998). Community-oriented policing and policing-oriented community. *Journal of Crime and Justice, 20*(1), 135–158.

Johnson, D.R. (1981). *American law enforcement: A history.* St. Louis, MO: Forum Press.

Johnson, I.M. (1999). School violence: The effectiveness of a school resource officer program in a southern city. *Journal of Criminal Justice, 27*(2), 173–192.

Johnson, R. (2012). Police officer job satisfaction: A multidimensional analysis. *Police Quarterly, 15*(2), 157–176.

Johnson, R. & Rhodes, T. (2009). Urban and small town comparison of citizen demand for police services. *International Journal of Police Science and Management, 11*(1), 27–38.

Johnson, S.D. (2017). Crime mapping and spatial analysis. In Wortley, R., & Townsley, M. (Eds.), *Environmental criminology and crime analysis* (pp. 199–223). New York: Routledge.

Jones, M., & Johnstone, P. (2012). *History of criminal justice* (5th ed.). Boston: Elsevier (Anderson Publishing).

Kane, R. (2000). Permanent beat assignments in association with community policing: Assessing the impact on police officers' field activity. *Justice Quarterly, 17*(2), 259–280.

Kane, R., & Cronin, S. (2013). Associations between order maintenance policing and violent crime: Considering the mediating effects of residential context. *Crime & Delinquency, 59,* 910–929.

Kappeler, V. (1996). Making police history in light of modernity: A sign of the times? *Police Forum*, 6(3), 1–6.

King, W. (1999). Time, constancy, and change in American Municipal Police Organizations. *Police Quarterly*, 2(3), 338–364.

King, W. (2003). Bending granite revisited: The command rank structure of American police organizations. *Policing: An International Journal of Police Strategies & Management*, 26(2), 208–230.

King, W. (2005). Toward a better understanding of the hierarchical nature of police organizations: Conception and measurement. *Journal of Criminal Justice*, 33(1), 97–109.

Kraska, P. (1994). The police and the military in the post–cold war era: Streamlining the state's use of force entities in the drug war. *Police Forum*, 4(1), 1–8.

Kurtz, D. (2008). Controlled burn: The gendering of stress and burnout in modern policing. *Feminist Criminology*, 3(3), 216–238.

LaFrance, C., & Day, J. (2013). The role of experience in prioritizing adherence to SOPs in police agencies. *Public Organization Review*, 13, 37–48.

Lee, J. (2010). Policing after 9/11: Community policing in an age of homeland security. *Police Quarterly*, 13(4), 347–366.

Liederbach, J., & Travis, L. (2008). Wilson redux: Another look at varieties of police behavior. *Police Quarterly*, 11(4), 447–467.

Liederbach, J., Fritsch, E., Carter, D., & Bannister, A. (2007). Exploring the limits of collaboration in community policing: A direct comparison of police and citizen views. *Policing: An International Journal of Police Strategies & Management*, 31(2), 271–291.

Lonsway, K. (2003). Tearing down the wall: Problems with consistency, validity, and adverse impact of physical agility testing in police selection. *Police Quarterly*, 6(3), 237–277.

Lundman, R. (1980). *Police and policing: An introduction*. New York: Holt, Rinehart & Winston.

Lundman, R. (2009). Officer gender and traffic ticket decisions: Police blue or women too? *Journal of Criminal Justice*, 37(4), 342–352.

Luo, F., Ren, L. & Zhao, J. (2017). The effect of micro-level disorder incidents on public attitudes toward the police. *Policing: An International Journal*, 40(2), 395–409.

Lynch, C.G., Gainey, R.R., & Chappell, A.T. (2016). The effects of social and educational disadvantage on the roles and functions of school resource officers. *Policing: An International Journal of Police Strategies & Management*, 39(3), 521–535.

Ma, C.C., Andrew, M.E., Fekedulegn, D., Gu, J.K., Hartley, T.A., Charles, L.E., et al. (2015). Shift work and occupational stress in police officers. *Safety and Health at Work*, 6, 25–29.

Magers, J.S. (2004). CompStat: A new paradigm for policing or a repudiation of community policing? *Journal of Contemporary Criminal Justice*, 20, 70–79.

Maguire, E., Shin, Y., Zhao, J., & Hassell, K. (2003). Structural change in large police agencies during the 1990s. *Policing: An International Journal of Police Strategies & Management*, 26(2), 251–275.

Manning, P.K. (1978). The police: Mandate, strategies and appearances. In Manning, P.K., & Van Maanen, J. (Eds.), *Policing: A view from the street* (pp. 7–31). Santa Monica, CA: Goodyear.

Mastrofski, S.D., Ritti, R.R., & Hoffmaster, D. (1987). Organizational determinants of police discretion: The case of drinking-driving. *Journal of Criminal Justice, 15*(5), 387–402.

McCampbell, M.S. (1986). *Field training for police: State of the art.* Washington, DC: National Institute of Justice.

McCarty, W.P., & Skogan, W.G. (2012). Job-related burnout among civilian and sworn police personnel. *Police Quarterly, 16,* 66–84.

McNeeley, S., & Grothoff, G. (2016). A multilevel examination of the relationship between racial tension and attitudes toward the police. *American Journal of Criminal Justice, 41,* 383–401.

Menard, K.S., & Arter, M.L. (2013). Police officer alcohol use and trauma symptoms: Associations with critical incidents, coping, and social stressors. *International Journal of Stress Management, 20,* 37–56.

Metchik, E. (1999). An analysis of the "screening out" model of police officer selection. *Police Quarterly, 2*(1), 79–95.

Monkonnen, E. (1981). *Police in urban America: 1860–1920.* Cambridge: Cambridge University Press.

Moore, R. (1987). *Posse comitatus* revisited: The use of the military in civil law enforcement. *Journal of Criminal Justice, 15*(5), 375–386.

Morabito, M. (2010). Understanding community policing as an innovation: Patterns of adoption. *Crime & Delinquency, 56*(4), 564–587.

Morley, B.S., & Hadley, J.M. (2013). Public safety consolidation: Does it make sense? *FBI Law Enforcement Bulletin.*

Muir, W. (1977). *Police: Streetcorner politicians.* Chicago: University of Chicago Press.

Nowacki, J.S. (2015). Organizational-level police discretion: An application for police use of lethal force. *Crime & Delinquency, 61*(5), 643–668.

O'Connell, P. (2008). The chess master's game: A model for incorporating local police agencies in the fight against global terrorism. *Policing: An International Journal of Police Strategies & Management, 31*(3), 456–465.

Osborne, R.E. (2014). Observations on police cynicism: Some preliminary findings. *North American Journal of Psychology, 16*(3), 607–628.

Paoline, E.A., III. (2003). Taking stock: Toward a richer understanding of police culture. *Journal of Criminal Justice, 31,* 199–214.

Paoline, E.A., III. (2004). Shedding light on police culture: An examination of officers' occupational attitudes. *Police Quarterly, 7*(2), 205–236.

Payne, B.K., Berg, B.L., & Sun, I.Y. (2005). Policing in small town America: Dogs, drunks, disorder, and dysfunction. *Journal of Criminal Justice, 33,* 31–41.

Pelfrey, W. (2007). Local law enforcement terrorism efforts: A state level case study. *Journal of Criminal Justice, 35*(3), 483–518.

Phillips, S., & Varano, S. (2008). Police criminal charging decisions: An examination of post-arrest decision-making. *Journal of Criminal Justice, 36*(4), 307–315.

Police Executive Research Forum (2013). *CompStat: Its origins, evolution, and future in law enforcement agencies.* Washington, DC: Bureau of Justice Assistance.

Poteyeva, M., & Sun, I. (2009). Gender differences in police officers' attitudes: Assessing current empirical evidence. *Journal of Criminal Justice, 37*(5), 512–522.

Potts, L. (1983). Equal employment opportunity and female employment in police agencies. *Journal of Criminal Justice, 11*(6), 505–524.

President's Commission on Law Enforcement and Administration of Justice (1967). *Task Force Report: Police.* Washington, DC: U.S. Government Printing Office.

Price, P. (2009). When is a police officer an officer of the law? The status of police officers in schools. *Journal of Criminal Law & Criminology, 99,* 541–570.

Randol, B.M. (2012). The organizational correlates of terrorism response preparedness in local police departments. *Criminal Justice Policy Review, 23*(3), 304–326.

Ratcliffe, J.H. (2016). *Intelligence-led policing* (2nd ed.). New York: Routledge.

Reaves, B. (2011). *Census of state and local law enforcement agencies, 2008.* Washington, DC: Bureau of Justice Statistics.

Reaves, B. (2012). *Hiring and retention of state and local law enforcement officers, 2008– Statistical tables.* Washington, DC: Bureau of Justice Statistics.

Reaves, B. (2015). *Local police departments, 2013: Personnel, policies, and practices.* Washington, DC: Bureau of Justice Statistics.

Reaves, B. (2016). *State and local law enforcement training academies, 2013.* Washington, DC: Bureau of Justice Statistics.

Reuland, M., Schwarzfeld, M., & Draper, L. (2009). *Law enforcement responses to people with mental illnesses: A guide to research informed policy and practice.* New York: Council of State Governments Justice Center.

Riksheim, E., & Chermak, S. (1993). Causes of police behavior revisited. *Journal of Criminal Justice, 21*(4), 353–382.

Riley, J.K., Treverton, G., Wilson, J., & Davis, L. (2005). *State and local intelligence in the war on terrorism.* Santa Monica, CA: Rand Corporation.

Roberts, A., Roberts, J., & Liedka, R. (2012). Elements of terrorism preparedness in local police agencies, 2003–2007: Impart of vulnerability, organizational characteristics, and contagion in the post-9/11 era. *Crime & Delinquency, 58,* 720–747.

Rose, T., & Unnithan, P. (2015). In or out of the group? Police subcultures and occupational stress. *Policing: An International Journal of Police Strategies & Management, 38*(2), 279–294.

Rossler, M., & Terrill, W. (2012). Police responsiveness to service-related requests. *Police Quarterly, 15,* 3–24.

Rydberg, J., & Terrill, W. (2010). The effect of higher education on police behavior. *Police Quarterly, 13,* 92–120.

Sabol, J.J., Wu, Y., & Sun, I. (2013). Neighborhood context and police vigor: A multilevel analysis. *Crime & Delinquency, 59*(3), 344–368.

Santos, R.B. (2014). The effectiveness of crime analysis for crime reduction: Cure or diagnosis? *Journal of Contemporary Criminal Justice, 30*(2), 147–168.

Schafer, J., Burruss, G., & Giblin, M. (2009). Measuring homeland security innovation in small municipal agencies: Policing in a post-9/11 world. *Police Quarterly, 12*(3), 263–288.

Schaible, L.M., & Sheffield, J. (2012). Intelligence-led policing and change in state law enforcement agencies. *Policing: An International Journal of Police Strategies & Management, 35*(4), 761–784.

Schlosser, M.D. (2014). Multiple roles and potential role conflict of a school resource officer: A case study of the Midwest police department's school resource officer program in the United States. *International Journal of Criminal Justice Sciences, 9*, 131–142.

Schuck, A.M. (2014a). Female representation in law enforcement: The influence of screening, unions, incentives, community policing, CALEA, and size. *Police Quarterly, 17*, 54–78.

Schuck, A.M. (2014b). Gender differences in policing: Testing hypotheses from the performance and disruption perspectives. *Feminist Criminology, 9*(2), 160–185.

Stein, R.E., & Griffith, C. (2017). Resident and police perceptions of the neighborhood: Implications for community policing. *Criminal Justice Policy Review, 28*(2), 139–154.

Stewart-Brown, R. (1936). *The Serjeants of the peace in medieval England and Wales.* Manchester, UK: Manchester University Press.

Sun, I., & Triplett, R. (2008). Differential perceptions of neighborhood problems by police and residents: The impact of neighborhood-level characteristics. *Policing: An International Journal of Police Strategies & Management, 31*(3), 435–455.

Surette, R. (2015). *Media, crime, and criminal justice: Images, realities, and policies* (5th ed.). Stamford, CT: Cengage.

Terrill, W., & Ingram, J.R. (2016). Citizen complaints against the police: An eight city examination. *Police Quarterly, 19*(2), 150–179.

Toch, H. (1997). The democratization of policing in the United States: 1895–1973. *Police Forum, 7*(2), 1–8.

Travis, L., & Langworthy, R. (2008). *Policing in America: A balance of forces* (4th ed.). Englewood Cliffs, NJ: Prentice Hall.

Varano, S., Schafer, J., Cancino, J., & Swatt, M. (2009). Constructing crime: Neighborhood characteristics and police recording behavior. *Journal of Criminal Justice, 37*(6), 533–563.

Vila, B., & Kenney, D. (2002). Tired cops: The prevalence and potential consequences of police fatigue. *National Institute of Justice Journal, 248*, 16–221.

Vogel, R., & Adams, R. (1983). Police professionalism: A longitudinal cohort study. *Journal of Police Science and Administration, 11*(4), 474–484.

Walker, S. (1977). *A critical history of police reform: The emergence of professionalism.* Lexington, MA: Lexington Books.

Walker, S. (2015). *Sense and nonsense about crime, drugs, and communities* (8th ed.). Stamford, CT: Cengage.

Walsh, W.F., & Vito, G.F. (2004). The meaning of CompStat: Analysis and response. *Journal of Contemporary Criminal Justice, 20*, 51–69.

Weisburd, D., Greenspan, R., Mastrofski, S., & Willis, J.J. (2008). *CompStat and organizational change: A national assessment.* Washington, DC: U.S. Department of Justice.

Willis, J.J., Mastrofski, S.D., & Weisburd, D. (2004). CompStat and bureaucracy: A case study of challenges and opportunities for change. *Justice Quarterly, 21*(3), 463–496.

Wilson, J.M., & Grammich, C.A. (2015). Deconsolidation of public-safety agencies providing police and fire services. *International Criminal Justice Review, 25*(4), 361–378.

Wilson, J.M., Weiss, A., & Grammich, C. (2016). *Public safety consolidation: A multiple case study assessment of implementation and outcome.* Washington, DC: Office of Community Oriented Policing Services.

Wilson, J.Q. (1968). *Varieties of police behavior.* Cambridge, MA: Harvard University Press.

Wood, J.D., Watson, A.C., & Fulambarker, A.J. (2017). The "gray zone" of police work during mental health encounters: Findings from an observational study in Chicago. *Police Quarterly, 20*, 81–105.

Zhao, J., Ren, L., & Lovrich, N. (2010). Police organizational structures during the 1990s: An application of contingency theory. *Police Quarterly, 13*(2), 209–232.

CHAPTER 6

Law Enforcement in the Criminal Justice System

Three principal decision points of the criminal justice system occur in the law enforcement segment of the process: (1) detection of crime, (2) investigation, and (3) arrest. As decisions, the police have choices about what, when, how, and whom to detect, investigate, or arrest. The existence of choices means that the police have discretion in these decisions. These activities comprise the scope of police crime control, with the exception of preventive practices such as uniformed patrol and proactive problem-solving efforts. It is these three decision points to which we will devote most of our attention in this chapter.

DETECTION

Detection is the discovery of crime or probable crime by the police. Once the police come to believe that a crime has been committed, the justice process begins. Detection hinges on many factors, ranging from the seriousness of the alleged crime and the observation powers of the officer, to the credibility of the complainant or witness. As indicated by our earlier discussion of "unfounding," sometimes the officer decides that a reported crime did not actually occur. In this case, she or he would label the complaint "unfounded," and would proceed as if nothing out of the ordinary had happened (Spohn, White, & Tellis, 2014).

Detection of crime by police takes place in two major ways, as described in Chapter 5: reactively or proactively. The most common way in which police come to learn about crime is reactively, through the receipt of citizen complaints. Most crimes are brought to the attention of the police rather than discovered by them. Some, however, are detected proactively, through undercover operations or through the observations of officers on patrol. Still other crimes are discovered through the actions of related agents, such as investigative grand juries, legislative committees, and the like. These latter means of detection can best be categorized under the heading of "reactive detection," because in these cases police gain knowledge of the existence of crimes from complaints.

IMPORTANT TERMS

arrest

automobile search

booking

cold case squads

consent search

crackdown

custodial interrogation

detection

entrapment

exclusionary rule

frisk

good faith exception

hot pursuit

interrogation

inventory search

investigation

lineups

low visibility

Miranda warnings

plain view doctrine

probable cause

procedural justice

search

search incident to
lawful arrest

sting operation

street sense

throw-downs

working rules

We refer to detection as a decision process because often what appears to be a quick decision, made on the spur of the moment, is actually the culmination of months or years of training or effort. This is particularly true of many of those crimes discovered by the police. What seems to be a nearly arbitrary decision to stop and question someone, or to check the license of a particular automobile, is in fact the result of a long process of learning and intelligence gathering. What the officer does is decide that further investigation is warranted. Stroshine, Alpert, and Dunham (2008) found that officers develop "working rules" that guide their detection decisions. **Working rules** are guidelines for behavior that identify the circumstances that justify different police actions. Such rules might have the officer check cars parked in business areas during nonbusiness hours.

Smith, Makarios, and Alpert (2006) reported on a study of police officer decision making in Miami. The results suggested that police officers form suspicion based on prior experience. For instance, if the officer has primarily encountered crime among males, the officer will be more suspicious of men. In the end, this means officers will be more observant of males, more likely to search and question males, and more likely to arrest males. The researchers noted that this may lead to a sort of "self-fulfilling prophecy" over time because police will look for crime where they expect to find it—among males, for example. In turn, they will find more crime among males because that is where they look, and that will encourage them to continue to be more suspicious of men than of women. Of course, other data (self-report and victimization) suggest that men are more likely to be criminal, so increased police suspicion of males may simply be reasonable.

Sacks (1978) described **street sense**, which he called an "incongruity procedure." Street sense is the ability of experienced police officers to "know" who is likely to be a criminal or to be dangerous. Sacks's term is appropriate because what the police officer relies on is that something about the individual or circumstances is incongruous or "not right"; that is, something does not fit. What appears normal to the average citizen may appear strange to the experienced officer. Police officers learn to observe clues, such as overcoats worn on warm days, mud splatters on rear license plates, and the like. The average citizen either does not notice such things or does not interpret them as possibly being crime related. After several years of experience, most officers become quite adept at the use of the incongruity procedure. The detection of crime then often rests on the police officer's perception or interpretation of the incongruous circumstances. Stroshine et al. (2008) reported that one common "working rule" focused on people or things that were "out of place."

Over time, and with training, police officers become sensitive to crime and potentially criminal circumstances. Police officers see (and seek) evidence of potential crime in circumstances that the typical citizen would not view as suspicious. Several efforts have been made to improve officer accuracy in identifying criminal situations. One of the most controversial of these has been the use of profiles in the enforcement of drug laws. Based on data taken from arrests, some jurisdictions developed "profiles" of people who are likely to be engaged in certain types of crimes. For example, young, male, minority group members driving late-model automobiles are targeted for traffic stops and preliminary investigations as potential drug couriers. These profiles understandably antagonize law-abiding citizens whom they target and contribute to poor public relations with the police (Reitzel & Piquero, 2006; Schuck, Rosenbaum, & Hawkins, 2008).

INVESTIGATION

Investigation is a process that continues throughout the law enforcement segment of the criminal justice system, and that often continues into the court segment as well. The decision involved in investigation has two parts: (1) whether to investigate a suspected crime and (2) how best to proceed with the investigation if one is initiated.

An investigation is the accumulation of information and evidence that links a particular person or group of persons to a particular crime or set of crimes (Blair & Rossmo, 2010). It is the process by which formal criminal charges can be brought against identified individuals. As an evidence-gathering activity, the principal tools of investigation are search and interrogation. Other tools and skills employed depend on the nature of the offense and the resources that are available to the police, for example, forensic analyses, lineups, and surveillance (Palmiotto, 1984).

Search

Search involves the seeking out of evidence of a crime or the location of a suspect. It entails the physical inspection of papers, premises, and possessions by the police. The U.S. Constitution provides that searches be conducted only on the issuance of a warrant based on probable cause. In practice, however, the warrant requirement has proven impractical and problematic. The U.S. Supreme Court has recognized a number of exceptions to the warrant requirement. Most of these depend on a determination of reasonable behavior by the police under the circumstances. If the police have behaved reasonably, the Court often rules the search to be valid.

Police can obtain a search warrant from a judge or magistrate if they provide the court with information that establishes the existence of **probable cause** (evidence that leads a reasonable person to conclude that a crime has occurred and evidence of the crime may be found) to support the search. In this case, the court has the chance to review the police decision and protect the right of the suspect to be free from unreasonable intervention by the police.

To control unreasonable or improper police behavior in searches, the Supreme Court has adopted the **exclusionary rule**. This rule states that police cannot use illegally or improperly obtained evidence in a trial. The logic of the rule is that the police conduct a search to obtain evidence of criminal behavior, and excluding this evidence from trial defeats the purpose of the search. Thus, theoretically, if the police cannot conduct a legal search, they will not conduct any search.

The Supreme Court developed and imposed an exclusionary rule on federal law enforcement in 1914, in *Weeks v. United States*, but refused to apply it to the states, hoping instead that the various state systems would arrive at a better solution to the problem of illegal searches. Finally, in 1961, the Court applied the rule to the states in the case of *Mapp v. Ohio*.

Acting on a tip that a suspect (wanted in connection with a bombing) and gambling equipment were at the home of Dolree Mapp, Cleveland police went there and asked for admission, which Mapp refused. Three hours later, more officers arrived and again the police asked to enter. When Mapp refused, one of the officers displayed papers that he claimed to be a search warrant and the officers entered the home, searched the premises, and discovered pornographic materials in a trunk in the basement of the house. The police arrested Mapp, charging her with possession of obscene materials.

The Supreme Court ruled that the police behavior in this case was unacceptable and ordered the exclusion of evidence (the obscene materials) from any future trial of Mapp. Weighing heavily in the Court's decision were the facts that the police had time to secure a warrant but did not do so, that the officers claimed to have had a warrant (but never introduced one in Mapp's trial), and that Mapp was apparently not guilty of the crimes of which she was suspected. In the end, the Court held that the exclusionary rule used to support the Fourth Amendment of the U.S. Constitution was applicable to state cases through the due process clause of the Fourteenth Amendment.

The *Mapp* case and the *Miranda* decision (discussed below) led to charges that the Supreme Court was "handcuffing" the police. Critics of the Court launched sharp attacks on this interpretation of the Constitution. They believed that an "overly liberal" stance characterized these and similar decisions of the Court (Leo & Thomas, 1998). Nonetheless, the Supreme Court had interpreted the Constitution, and the exclusionary rule became the law of the land (Wilson, 1988).

Given that the Court would distinguish between the types of searches that would yield admissible evidence and those that would not, it was possible to determine exceptions to the warrant requirement by learning the circumstances under which the Court had ruled evidence was admissible in warrantless searches. Generally, the Court has identified the following circumstances to be exempt from the warrant requirement:

- Limited protective searches (frisks)
- Searches incident to lawful arrest
- Searches conducted in emergency situations or in hot pursuit of a suspect
- Border searches, on the consent of the person being subjected to the search
- Searches of automobiles
- Inventory searches
- Seizure of evidence "in plain view" of the officer.

The frisk is a traditional police practice that the Supreme Court has upheld when conducted under specific circumstances. The Court outlined the requirements for a valid frisk in 1968 when it decided the case of *Terry v. Ohio*. In this case, an experienced Cleveland police officer conducted a limited "pat down" of the outer garments of suspected robbers to ensure his own safety as well as that of innocent bystanders.

In the *Terry* case, a 39-year veteran officer observed two suspects repeatedly walk past and look into a store. The officer concluded that the men were "casing" the store for a robbery, and he approached them to investigate. When the two stopped to confer with another man, the officer approached and asked for their names. When they mumbled in response, he turned one of them around and patted him down, finding a revolver. He frisked the other man, found another gun, and placed both men under arrest.

The Supreme Court ruled that the officer had acted reasonably under the circumstances and that he took the evidence (guns) in a reasonable search. Therefore, they allowed the use of the evidence at trial. If an officer has a reasonable suspicion that criminal activity is occurring, the officer has the authority to stop persons and to request identification and information. Moreover, if the officer has a reasonable belief that these individuals may be armed, and if nothing in the initial investigation dispels this reasonable fear, an officer may conduct a "frisk" to protect him- or herself and others.

The ability for the police to conduct a frisk is not without limits. For example, a federal judge (*Floyd v. New York*, 2013) rejected a long-standing "stop and frisk" policy by the New York City Police Department, calling it a form of indirect racial profiling. At its peak in 2011, the NYPD conducted nearly 700,000 stop and frisks against mostly innocent people (Figure 6.1). Police were most often initiating a stop and frisk because of factors such as being in a high crime area or the time of day rather than actual suspicious activity (Avdija, 2014). The policy did appear to impact the number of gun-related crimes in New York (Bellin, 2014). However, the arbitrary nature of the stops and the disproportionate

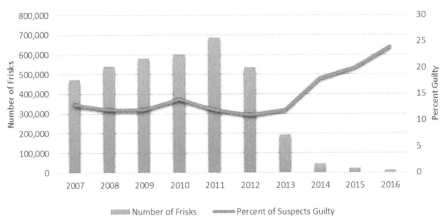

FIGURE 6.1
Stop and Frisk: New York City.

Source: New York Civil Liberties Union (2018).

impact on certain demographic groups made this practice unconstitutional. As shown in Figure 6.1, the likelihood of finding a weapon has increased as the stop and frisks have become more limited, and constitutional, in scope.

The frisk is a protective search limited in scope to only that which is required to ensure that the person with whom the officer is dealing is not capable of injuring the officer. Upon a lawful arrest, however, the officer's authority to search is broader. Whereas frisking is limited to a "pat down" of the outer garments to discover weapons, a full search by the officer is permissible in the case of an arrested person. The Court allowed **search incident to lawful arrest** to protect officers through discovery of any weapons, and to secure any evidence of the crime that the offender might otherwise be able to destroy. This type of search is limited to the area within the immediate control of the offender (*Chimel v. California*, 1969). In April 2009, the U.S. Supreme Court decided *Arizona v. Gant*. Police stopped Gant for a traffic violation and arrested him for driving with a suspended license. After handcuffing Gant and placing him in the police cruiser, the officer searched Gant's car and found drugs. The Court ruled the police could not introduce the drugs at trial because there was no reason to search the car once Gant was in custody. In these circumstances, there was no chance Gant could obtain a weapon or destroy evidence in the car. Search incident to lawful arrest only applies to the area immediately in the reach or control of the suspect.

The Supreme Court recognized a **hot pursuit** exception to the warrant requirement in the case of *Warden v. Hayden* (1967). While the Court generally prefers that police officers obtain warrants before conducting a search, it recognizes that at times this is impractical. In the *Hayden* case, the police learned of the crime (a robbery) from a cab driver who followed the suspect to a building. The police arrived at the scene within minutes after the suspect entered the building and the cab driver stated that the suspect had gone inside. The police entered the building and searched the premises for evidence of the robbery and for the suspect described to them by the cab driver. The Court decided that, under the circumstances, although the police could have cordoned off the building and awaited a warrant, the time lag involved might have allowed the suspect to destroy the evidence of the crime or to escape. Thus, when in hot pursuit, officers may follow and search for the suspect. Likewise, officers can conduct a search in an emergency situation without obtaining a search warrant. For example, police can enter a home to protect an occupant from imminent injury.

The law regarding search at international borders is different from the law of searching residences, according to the Court. Given the volume of traffic that crosses our borders and faces the U.S. Customs and Border Protection Service, as well as the potential for smuggling, the Court allows searches of persons and possessions at the nation's borders without the requirement of warrants. Indeed, the U.S. Customs and Border Protection Service periodically conducts random searches at border points. Recently, border patrol officials have increased the frequency of warrantless searches of electronic devices such as cell phones and laptops (U.S. Customs and Border Patrol, 2018; Waddell, 2017), leading many privacy advocates to raise concern about the intrusive nature of such searches. This will likely be the source of continued debate and future legal decisions.

It is possible for a competent adult to waive his or her right of protection from unreasonable searches. In a **consent search**, the suspect consents or agrees to the search so any evidence found will be admissible. In the *Mapp* case, for example, if Mapp had voluntarily allowed the police to enter her home and search for the bombing suspect, she would not have been successful in her efforts to have the evidence excluded from trial. Anyone with a right to grant consent may do so. In *United States v. Matlock* (1974), the Court ruled that a co-occupant of an apartment could grant consent to a search. In a later case, *Georgia v. Randolph* (2006), the court ruled that police must observe the objections of an occupant even if a co-occupant has given consent for the search. More recently, the Court ruled that an objector must be physically present to prevent a search from taking place. In this case, *Fernandez v. California* (2014), the defendant initially refused to allow a search, but was then arrested in connection with a crime. After the defendant was taken into custody, his co-tenant consented to a search.

Because of the mobility of automobiles, the Court treats **automobile searches** differently from searches of houses or buildings. In *United States v. Ross* (1982), the Court ruled that police officers could conduct a full search of an automobile that the officers had legitimately stopped, as long as they had probable cause to believe that the automobile contained contraband. The Court has applied the warrant standard of probable cause to the search of automobiles. Thus, as long as the police have probable cause to believe that an automobile contains contraband or evidence, they may search the vehicle and any containers in it. In 2012, the Supreme Court expanded the definition of an automobile search to include Global Positioning System (GPS) tracking devices in the case of *United States v. Jones*. These devices have been used by law enforcement to track the movement of suspected criminals, and the decision clarifies that the police must obtain a search warrant before GPS tracking is authorized.

A related issue surrounding automobiles in particular is the **inventory search**. If the police seize an automobile or other item, they generally search it to determine the contents for the purpose of inventory. They do not want the owner to claim that valuable property is missing after retrieving the car. In *Chambers v. Maroney* (1970), the Court ruled that a search of an automobile shortly after it was seized was permissible. In *Coolidge v. New Hampshire* (1981), however, the Court held that a search of an automobile required that the officers first obtain a warrant. The fact that the police had held the car for a long period before the search negated the possibility that the *Coolidge* case involved an inventory search. Because the car was in police custody and could not be moved, the probable cause standard of *Ross* was inapplicable because the officers could have obtained a warrant.

Not only may officers search an automobile they have legitimately stopped, they can also order the driver and passengers from the car and frisk those individuals (*Maryland v. Wilson*, 1997). Beyond this, in the event the police legitimately arrest the driver, they are authorized to search the passenger compartment of the automobile (*Thornton v. United States*, 2004).

The **plain view doctrine** does not pertain to a type of search. What it means is that if an officer sees criminal evidence in plain view (i.e., the evidence did not have to be searched for), the officer may seize that evidence. Thus, if you approach an officer for directions, and you are openly carrying a controlled substance, the officer may seize the substance with no other justification than that it was in plain view. In *New York v. Class* (1986), an officer trying to read the VIN (vehicle identification numbers) through the windshield of a stopped automobile reached into the car to clear material from the number. While doing so, the officer saw a gun under the seat. The Court ruled that the officer had a right to check the VIN and thus the discovery of the gun "in plain view" was not violation of the warrant requirement. A related issue is that of "open fields": the issue of whether or when police can search for evidence of crime in fields that are fenced but visible. The Supreme Court has ruled that police may make special arrangements to fly over these areas to look for evidence and as long as the police view the evidence from publicly accessible areas, the search and seizure is acceptable (*Florida v. Riley*, 1989). With the increased availability of unmanned drones, privacy advocates are encouraging states to pass legislation clarifying and restricting law enforcement's use of these drones to conduct warrantless searches (McNeal, 2014).

The legal authority to conduct a search often requires an understanding of the exceptions above and can be complex, as evidenced by the case of *Kentucky v. King* (2011). In this case, the police were involved in a chase of a suspected drug trafficker. The police lost sight of the suspect, but knew that the suspect had entered either of two adjacent apartments. As the police came to the apartments, a strong smell of marijuana coming from one of the doors led police to believe that the suspect was in that particular apartment. After knocking on the door and yelling "police, police, police," the officers heard noises behind the door that led them to believe that evidence was being destroyed. At this point, the police knocked down the door and arrested three people who were using drugs in plain view of the officers. Interestingly, the three defendants were not the initial suspect; the police later found the trafficker in the adjacent apartment. The Court ruled that the police acted properly. The officers had only entered the property after reasonable belief that evidence was being destroyed. Once the officers were inside the apartment, the drugs were in plain view.

The purpose of the exclusionary rule and the general requirement of a warrant are to protect the liberty of the individual citizen. The fact that there are exceptions indicates a consideration for order and crime control. The police must walk a fine line between what is necessary for the protection of the public and the control of crime, and what the protection of individual liberties requires. The warrant process, whereby a magistrate reviews the evidence and either confirms the judgment of the police (issues a warrant) or rejects it, and court rulings as to the admissibility of evidence are examples of how the justice system attempts to check the discretionary powers of the police.

Decisions whether to search, how to search, and what to search are governed by evidentiary standards such as "reasonable suspicion" or "probable cause," but in practice are discretionary in nature. The officer on the street must decide whether to take investigatory steps. Although this decision reflects the evidentiary standards of the courts, many other factors come into play when making this decision.

Most crimes reported to the police are never solved. The investigation of crimes must consider the available resources and the likelihood that the case can be solved. Police detectives often experience heavy workloads which can impact the effectiveness of their investigations (Roberts & Roberts, 2016). If there is enough evidence to suggest that the case can be solved, detectives are more likely to devote scarce investigation resources in solving it (Roberts, 2008). Cases which have the availability of DNA evidence, witnesses, and motives are more likely to be solved than those which lack these factors. Given equal probabilities of solving cases, however, other concerns exert an effect, including the seriousness of the crime. As with other criminal justice decisions, police exert more effort responding to cases that are seen as more serious (involving physical harm, threats, or substantial property loss). Police detectives, especially those with limited resources, will understandably prioritize homicide and rape investigations ahead of misdemeanor thefts.

Other factors also can affect the decisions of police officers when detecting crime. A rash of crimes, negative media pressure, the importance or status of the victim, political interference, and similar factors all can play roles in the decision to investigate. Increasingly, police agencies are instituting what are known as "cold case squads." These are teams of detectives dedicated to pursuing unsolved cases lacking significant leads. As cold case squads proliferate, the traditional wisdom that police devote attention to new cases with the greatest chance for successful resolution may have to change. Here too detectives select cases based on the availability of evidence. They most often select cases where new witnesses have come forward with evidence, when new DNA evidence is available, or when new DNA technology becomes available to test old evidence (Davis, Jensen, Kuykendall, & Gallagher, 2015).

Interrogation

Interrogation of suspects and witnesses has long been a mainstay of criminal investigation. We are all familiar with entertainment-media portrayals of police investigations in which detectives "grill" the suspect or continually return to the witness to extract details of the crime. Interrogation is a "search" for evidence through seeking testimony or responses to questions put to the suspect.

Just as the Fourth Amendment to the U.S. Constitution protects the homes, papers, and possessions of the citizenry from unreasonable searches, the Fifth Amendment protects citizens from overzealous interrogation by the police. The Fifth Amendment states that no one can be compelled to give testimony against oneself. *Miranda v. Arizona* (1966), perhaps the most famous of the Court's exclusionary rule cases, involved the Fifth Amendment.

In the *Miranda* case, as in many other cases preceding it, the police arrested the suspect and held him in custody for several hours while questioning him about the crime. At the conclusion of the interrogation, the suspect had confessed to the crime. Miranda's attorneys appealed the conviction stating Miranda was not aware that he did not have to speak during the interrogation and that he had a right to an attorney during questioning. Thus, the attorneys contended that police obtained the confession improperly and it should not have been allowed as evidence at the trial. The Supreme Court agreed, ruling that when police have a suspect in custody, they must advise the suspect that he or she may remain silent, that what is said may be used against the suspect in court, and that the suspect has the right to either a retained (hired) or appointed attorney during questioning (see Box 6.1).

The *Miranda* decision revolutionized police interrogation practices. One of the reasons the case is so well known is that the Court required that police provide suspects with warnings as to their rights at interrogation, and these required warnings became known as the *Miranda* warnings. Traditional practices of "incommunicado" (not permitting the suspect to see or speak with anyone except the investigating officers), as well as the psychological advantage held by the police in such interrogations, were abolished.

Within 2 years of the decision, the U.S. Congress included language in the Omnibus Crime Control Act of 1968 that redefined the conditions making a confession voluntary. The U.S. Department of Justice, however, never sought to implement that statute (*Dickerson v. United States*, 2000). As long as the confession was voluntary, the statute did not require that the police give the *Miranda* warning. In 2000, the U.S. Supreme Court reviewed this issue and held that the requirements of the *Miranda* ruling still apply. Essentially, the Court ruled that absent the warnings it would not be possible to conclude that a suspect voluntarily confessed because the warnings ensure that the suspect knows his or her rights and that he or she need not cooperate with the police.

Since the original *Miranda* decision, numerous cases have been heard by the Court regarding interrogation, particularly regarding the issue of a time lapse between interrogation sessions. The Court has generally been consistent in ruling that when a suspect requests the assistance of counsel during an interrogation, any future waiver of *Miranda* rights is invalid (Ross & Myers, 2010). Finally, in 2010 the Supreme Court held that a break in custody of two or more weeks is all that is needed to allow police to attempt to question a suspect again. In *Maryland v. Shatzer* (2010), Shatzer, who was in prison, refused to answer questions about alleged child sexual abuse. More than 2 years later, the police reopened the case and interrogated Shatzer. This time he answered questions, and the police used his statements to convict him of the child abuse crime. The court ruled that although Shatzer originally invoked his Fifth Amendment privilege, the long break in custody meant that the second questioning did not violate Shatzer's original refusal. Because he was warned again before the second questioning, Shatzer's testimony was admissible.

The *Miranda* decision, most of all, led some commentators to charge that the Supreme Court was "handcuffing the police." These critics expressed fears that,

BOX 6.1 *MIRANDA* WARNINGS

- You have the right to remain silent.
- Anything you say can and will be used against you in a court of law.
- You have the right to an attorney during questioning.
- If you cannot afford an attorney, one will be appointed for you by the court.

because of this decision, the police would be unable to obtain confessions, and thereby convict far fewer offenders (Cassell, 1996). However, most evaluations of the effects of the *Miranda* warnings on conviction rates and the recovery of stolen property have failed to support these earlier criticisms. Leo (2001) concluded that the *Miranda* decision has had very little impact on the criminal justice system. As Leo pointed out, most offenders in modern society have knowledge of these rights through exposure to popular television programs. In addition, the overwhelming majority of suspects voluntarily waive their *Miranda* rights, thus not creating a problem for law enforcement personnel. These arguments have been backed up by surveys of police chiefs. Only 1% of police chiefs nationwide believe that too many offenders get off easy as a result of *Miranda* warnings (Zalman & Smith, 2007). In fact, many police chiefs report that one outcome of the *Miranda* ruling has been a proliferation of more professional and effective police officers (Time & Payne, 2002). In part, the critics were wrong in their contention because they probably did not consider the system qualities of the justice process and the ability of systems to resist change and to maintain equilibrium. Several practices developed that may have served to blunt the effect of *Miranda*.

A practice that enabled police to get around the *Miranda* requirement was "psychological warfare." One effect of the *Miranda* decision may have been to increase the skill of police interrogators. Davis (2008) called the modern interrogation process an extended "anti-*Miranda* warning," where the interrogator works to "convince the suspect that everything he says—preferably including a detailed confession to the crime at hand—can and will work to his benefit, whereas denial or failure to talk to his interrogator can and will be held against him" (p. 1). Suspects can be motivated to confess either by (1) increasing stress and discomfort to motivate confession as a method of escaping the situation or (2) leading the suspect to the rational conclusion that confessing is in his or her best interests. Interrogators have developed many tactics to facilitate confessions through one or both of these paths (Davis, 2008). One such interrogation technique involves officers telling suspects that they have physical evidence or witnesses when no such evidence exists. Officers believe guilty suspects will confess when told the police have other evidence (Alpert & Noble, 2009). Other techniques include role playing, exaggerating, or minimizing the severity of the crime, and the use of implied promises of leniency (Ellwanger, 2017).

It is important to note that *Miranda* warnings are only necessary whenever there is a **custodial interrogation** (Ellwanger, 2017). In other words, a suspect is only entitled to the warning in situations where they are under arrest *and* being interrogated by officers. Therefore, officers can arrest a suspect and simply listen to him or her talk. If the suspect provides any incriminating statements, these are generally allowable in court. Likewise, the warnings are not required when an officer is engaged in a simple interview, which is often conducted by the police at the early stages of an investigation when they do not have a particular suspect and are primarily concerned with gaining general information about the crime. An interrogation begins when, for example, the officers ask accusatory questions to a specific individual.

The "Good Faith" Exception

In 1967, the U.S. Supreme Court decided the case of *Katz v. United States*. In this case, federal investigators who suspected Katz of involvement in gambling placed an electronic eavesdropping device on the exterior of a telephone booth they believed Katz used to transmit betting information. Katz was convicted of transmitting betting information by telephone in violation of a federal law. He appealed, stating that the federal agents had violated the Fourth Amendment by not securing a warrant for the eavesdropping. Prior to this case, the Court interpreted the Fourth Amendment as designed to protect places. By locating the device on the exterior of a public telephone booth, the agents believed they had not violated Katz's rights. The Supreme Court, however, reversed the conviction, ruling that the Fourth Amendment protected people, not places. The decision turned on Katz's expectation of privacy. In this case, although law enforcement officials took pains to comply with the Constitution and believed they were behaving appropriately, a guilty defendant was set free.

The *Katz* case illustrates one problem with the exclusionary rule. The rule seeks to deter police officials from violating citizens' rights. What happens when the police believe they are obeying the law? If the police make an honest mistake, their behavior cannot be deterred. What we need, many observers believed, was a "good faith" exception to the exclusionary rule. If the police work in good faith in the belief that they are not violating the Constitution, then any evidence they obtain should be admissible.

In 1984, the Court established a good faith exception in the case of *United States v. Leon*. In this case, officers relied on information from a confidential informant to obtain a search warrant. At trial, Leon's attorneys argued that the information on which the warrant was based was insufficient to establish probable cause, and thus the evidence from the search should be suppressed (excluded). The judge agreed. Here, officers sought and obtained a warrant, executed the search, found large quantities of drugs, and yet the evidence was still suppressed because of an error by the magistrate issuing the warrant. The Supreme Court ruled that because the exclusionary rule seeks to deter police misconduct rather than punish errors of judges or magistrates, the state could use the evidence. If the police conduct a search believing in good faith that the search is permissible, then the evidence can be used at trial, making a **good faith exception**.

Since then, the Court has expanded on the good faith exception. In *Arizona v. Evans* (1995), the Court ruled that a police search of Evans based on a mistaken belief that he was a fugitive was legal. Police stopped Evans for a traffic violation, and the subsequent computer check revealed an outstanding arrest warrant. The officer arrested Evans based on the warrant, and then searched the car, finding marijuana. The arrest warrant, however, had been removed two weeks earlier. There was no warrant for Evans's arrest at the time of the search. Still, the officer believed, in good faith, that Evans was wanted and that the arrest and subsequent search were constitutional. The Supreme Court ruled that the exclusionary rule is not designed to prevent clerical errors (the computer record of the warrant had not been updated), and thus the officer was acting reasonably and the evidence was admissible. In *Utah v. Strieff* (2016), the police illegally stopped a man who had left a house which police had suspected of being involved in dealing of narcotics and learned of an outstanding arrest warrant during that stop. Upon arresting the suspect, police found methamphetamine and drug paraphernalia. The Court ruled that this evidence could be used against the suspect because the valid arrest warrant was unconnected to the illegal stop. Once the officer discovered the arrest warrant, he was required to make an arrest and thus conducted a constitutional search upon lawful arrest.

The USA PATRIOT (Uniting and Strengthening America by Providing Appropriate Tools Required to Intercept and Obstruct Terrorism) Act of 2001 has raised concerns about potential governmental intrusion on individual rights. The act allows investigators to execute search warrants without notifying the owner or occupant of the location. It has eased restrictions on the use of wiretaps and other electronic surveillance. Although aimed at counterterrorism and national security, the powers given law enforcement officers to detect and prevent terrorism can and will also be used to detect and apprehend more traditional criminal offenders.

Other Investigatory Practices

In addition to search and interrogation, police obtain evidence in a number of ways. The offender's age, race, sex, and size can be determined from hairs found at the scene of the crime. Police can identify clothes worn by the offender from fibers collected at the crime scene. The police can ascertain the weapon used in the crime from ballistics examinations. All these techniques are staples of a forensic scientist's investigation. While rarely employed because of costs and a lack of necessity, the techniques of a forensic scientist sometimes provide the answers to investigators (Peterson, Sommers, Baskin, & Johnson, 2010). Forensic science is becoming increasingly important in criminal investigations, as can be seen in the increased attention paid to DNA typing. The Combined DNA Index System (CODIS) serves as a database allowing for comparisons between crime scene DNA samples and existing profiles of convicted offenders. While still relatively rarely used, the application of DNA analysis has experienced amazing advances in the past several years (Dale, Greenspan, & Orokos, 2006). Still, despite television dramas, forensic examinations are not common to most investigations, and DNA evidence is rarely used to solve cases (Schroeder & White, 2009). Television dramas based on crime scene investigators have popularized forensic science and crime scene analysts, often leading to unrealistic public expectations (Ley, Jankowski, & Brewer, 2012; Rhineberger-Dunn, Briggs, & Rader, 2016).

Identification Techniques: Throw-Downs and Lineups

More common is the identification of offenders through throw-downs and lineups, whereby the police seek the identity of a suspect from a pool of possibilities by having witnesses examine photographs (**throw-downs**) or observe possible offenders (**lineups**). The major concern of the courts regarding these practices is that the police not be too suggestive in their behavior (Cutler, Duaherty, Babu, Hodges, & Wallendael, 2009). For example, the identification of a white male suspect from a lineup comprised of the man and six black females would be too suggestive. To protect against the possibility that the police would encourage a false identification, the Supreme Court has held that suspects in custody have the right to have an attorney present during lineups (*United States v. Wade*, 1967). Unless the victim or witness knows the offender, it is unlikely that these procedures by themselves will yield the offender. More often, police use these procedures after they have enough evidence to conclude that the suspect is probably guilty of the offense. Thus, these identifications tend to constitute supporting evidence and are not the heart of the case.

Surveillance

Still another investigation tool is surveillance. Police use both physical and electronic surveillance techniques to gather evidence of criminal activity. Wiretaps, hidden microphones, cameras, and other forms of surreptitious surveillance such as GPS tracking devices generally require warrants and are therefore used infrequently. The point of most surveillance is to gather evidence on persons suspected of crimes, but against whom insufficient evidence exists to obtain a search warrant. Usually, surveillance does not involve tapping telephone lines or watching individual residences. The bulk of surveillance conducted by police agencies is physical surveillance, by which officers watch a certain location or follow a suspect to gather evidence (Nunn, Quinet, Rowe, & Christ, 2006). One exception is the use of video surveillance of public areas. The British have been using video surveillance of public areas for years, and the practice is becoming more common in the United States.

Surveillance techniques yield evidence of specific criminality and information (or "intelligence") on suspicious persons and places. In the former case, officers generally have reason to believe that a particular person is engaged in criminal activity, or that crimes are occurring at a particular place, and the surveillance seeks to provide evidence about those specific offenses. In intelligence gathering, police suspect an individual or location and conduct surveillance in hopes of obtaining further evidence or information to confirm or reject their suspicions.

Technological advances are constantly improving the intelligence potential of surveillance methods. Police use license plate readers, both at static locations and mounted on police vehicles, to help locate stolen cars or suspect vehicles. Police can review computerized records of the license plate scanning effort later to determine if a particular vehicle was at the scene. Face-recognition technology coupled with video surveillance helps police to determine if a particular person was observed at some place. Cameras are also increasingly being placed at red lights or along busy streets to catch motorists who violate traffic laws. We will discuss technology more in Chapter 15. It is enough to recognize that technological advances greatly increase the surveillance capabilities of the police.

Informants

Informants are another investigation tool used by the police. To some, the use of informants seems inappropriate because it frequently requires the police to join forces with criminals to enforce the law. Nonetheless, informants are extremely useful to police in intelligence gathering. Usually neither criminals nor the police respect the informant, in keeping with a norm that one should not be a "rat." However, informants are often able to learn and observe things that an undercover officer would find difficult to discover even after months of work—and that a uniformed officer would find impossible to obtain. Police can use information from informants, even anonymous tips, but they must be able to convince the magistrate that the informant's information is credible. In *Illinois v. Gates* (1983), the Supreme Court allowed police to rely on an anonymous tip about Gates's involvement in drug trafficking as long as the police had taken reasonable steps to verify the accuracy of the tip.

Informants do not make particularly good witnesses. Typically, their credibility is questioned either because of their own criminal pasts, or because their information is provided with the expectation of a reward and not as a matter of civic responsibility. Add to this the fact that many informants are not willing to testify in open court, and this valuable source of information for police becomes less than ideal for solving crimes. To combat these weaknesses, many police departments employ undercover police officers. These officers are assigned to work in "plain clothes" and are directed to mix with the general public so that their identities as police officers will not be readily apparent.

Crackdowns

Yet another police response to crime that combines investigation with detection and arrest is the **crackdown** (Sherman, 1990). In crackdowns, the police devote increased attention and resources to either specific types of crime (drug sales, prostitution, street robberies, etc.) or to crime in particular places (downtown areas, parking lots, schools, etc.). Crackdowns often involve practices of saturation patrol, in which a large number of uniformed officers are used to flood a locale with police, or policy directives, in which officers are directed to arrest in certain crime situations whenever possible. Crackdowns allow the police to show that they are "doing something" about crime, and are often used as a strategy to implement hot spot or problem-oriented policing.

Undercover Operations

Certain types of crimes, such as the "victimless" offenses associated with vice enforcement, are not likely to yield complainants. The customer who solicits a prostitute or purchases drugs is unlikely to complain to the police. In so doing, of course, the complainant would implicate himself or herself in criminal behavior. Thus, the police must aggressively (proactively) seek evidence of these crimes. Almost every major police department has a vice unit comprised of officers who attempt to uncover instances of victimless crimes. These officers pose as either potential consumers or potential providers of the illicit goods and services, and then wait for would-be offenders to approach them.

A "sting" is a form of undercover operation in which police present an opportunity for offenders to commit a crime. The **sting operation** gets its name from the slang of confidence artists. These "con artists" gain the trust of their victims and then take the victim's money. The actual taking of the money is the "sting." Because the police pretend to be criminals to gain the trust of offenders until they have sufficient evidence to make an arrest, the undercover operations also are called stings. Police departments conducting sting operations often use deception techniques such as disguises, fake storefronts, false advertising (increasingly using the internet), or decoy cars to "bait" the offender into committing the crime (Newman, 2007). When the criminal strikes, the officers will either make an arrest or use that intelligence to identify higher level criminals.

A common police sting involves the establishment of a "fencing" operation; that is, police officers pose as dealers of stolen goods. Police photograph burglars and thieves when they come to sell (fence) goods they have stolen. Later, police arrest these offenders based on the evidence obtained in the fake fence operation. Stings are also used to identify and arrest those engaging in prostitution, drug dealing, and child pornography (Dodge, Starr-Gimeno, & Williams, 2005; Newman, 2007).

There are two commonly raised criticisms of undercover operations. First, there are those who argue that these operations, especially sting operations, create crime because they provide an easily identifiable outlet for stolen property, which encourages people to steal (Lyman, 2017). Second, and perhaps more important, there are those who suggest that such operations ensnare the innocent through entrapment.

The **entrapment** defense to criminal charges is applicable when the entrapped offender did not have the inclination to commit a crime absent police enticement. If the police entice an otherwise innocent individual to commit a crime, that individual may have a defense of entrapment (Levanon, 2016). Assuming that you would not generally think of committing a crime, how would you respond if I offered you $1 million to transport some drugs across the country for me?

For many people, this offer is too good to refuse. If you were to say yes to my offer and I then wished to arrest you for transporting drugs, you could say you were entrapped. You only committed the crime because I came to you with the idea and offered you an inducement too great to refuse. The same logic applies if an officer suggests a price and activity to a suspected prostitute. This is true regardless of the price offered; the law does not question judgment and worth as much as it does motive.

It is important to distinguish between entrapment and encouragement. Unlike entrapment, it is permissible for the police to "encourage" someone to complete a crime the person is already contemplating. An officer who approaches a prostitute and offers to pay for a service may be placing the idea in the mind of the prostitute. Someone contemplating an act of prostitution, however, already has the criminal idea. The fact that an officer allows a person to make a proposition is not entrapment; it is encouragement.

The investigation of crimes, through whatever means the police employ, provides a basis for the next decision of importance in the law enforcement segment of the justice process: Arrest. When police have probable cause to believe that an individual has committed a crime, we expect them to place that person under arrest. Yet, as with investigation and detection, arrest is a decision process that depends on a number of factors.

ARREST

Arrest is the act of taking a person into custody. Once arrested, a person is no longer free to leave. The authority to arrest offenders is but one of a number of tools available to the police officer in his or her efforts to maintain order. There are times when this authority is the best or most appropriate tool; at other times it is inappropriate. The decision regarding the appropriateness of arrest is often a discretionary one for the officer.

Goldstein (1960) discussed police discretion in deciding whether to employ the criminal law. He identified the decision not to arrest as "noninvocation discretion" (decision not to invoke the law). He termed the process "low-visibility decision making." If an officer decides not to make an arrest when an arrest is justified, who has knowledge about the decision? Generally, only the officer and the offender are aware of the failure to arrest. Thus, this decision has **low visibility**, meaning that most people (including the police administration) do not see it. If an officer decides not to issue a traffic citation to you, no one will know unless the officer or you report this decision. How many times have you reported to the police station that an officer did not issue a citation to you?

The other face of noninvocation discretion is, logically, "invocation discretion" (the decision to invoke the law). This decision is more visible, because the officer must report the arrest, the suspect will most likely have legal counsel, and eventually the case may get into the courts. Yet, the decision to arrest someone is often as discretionary as the decision not to arrest. Officers or departments follow rules of thumb when deciding to invoke the criminal law. Occasionally, an officer will opt to enforce the law when the rule of thumb would suggest noninvocation. Exemplifying this are "tolerance limits" regarding excessive speed. Most departments attempt to avoid bad public relations (and close court decisions) in traffic cases by suggesting that offenders not be cited if they are traveling at a rate within a specified difference from the speed limit. If the speed limit is 55 mph, we might expect traffic officers to cite only those motorists traveling in excess of 60 mph. However, a motorist stopped for traveling 58 mph could be cited instead of warned (regardless of the tolerance limit) if the officer feels the motorist's behavior is inappropriate. Schafer and Mastrofski (2005) described traffic enforcement in one community in which the policy of the department supported leniency by officers, but also noted that citizen behavior was associated with the use of more formal and more severe sanctions. Sun, Payne, and Wu (2008) reported that a number of situational and neighborhood factors influence officer decisions.

A responding officer might decide that arrest is necessary to separate combatants in a dispute, or that such an arrest might prove more harmful than beneficial in a given situation. An officer might arrest an ill child to secure needed medical treatment in the absence of parents, or might fail to arrest a known offender in return for information about other offenses or offenders. In all these situations, it is apparent that arrest is a tool for the officers involved, or a means to an end. To understand the exercise of the arrest power in a particular instance, one must understand the intent of the officer involved.

A number of researchers have investigated the arrest decision by police officers. Many would be surprised at how often the police do not make an arrest. In fact, Bonner (2015) found that police failed to make an arrest in the vast majority of dispute cases that she examined. Factors predicting arrest include the seriousness of the crime, the age of the involved parties, the suspect and victim's demeanor, signs of alcohol and other drug abuse, and prior encounters with an offender (Terrill & Paoline, 2007). In domestic violence cases where many agencies have mandatory or preferred arrest policies, police still often fail to make arrests (Phillips & Sobol, 2010). Nationwide, police make an arrest during their initial response to domestic violence less than a quarter of the time (Reaves, 2017). Circumstances that influence arrest in intimate partner violence include the race, age, and marital status of the victim and the offender, as well as the type of violence involved and the neighborhood characteristics of the complaint (Dichter, Marcus, Morabito, & Rhodes, 2011; Lee, Zhang, & Hoover, 2013). Clearly, the available research indicates that police officers decide between arrest and nonarrest in many cases.

When police do make an arrest, the justice process becomes fully involved. The arresting officer must file reports, and the offender/suspect must be transported to a detention facility and booked (where police make an official record of the arrest). Shortly after the arrest, the suspect must be given the opportunity to contact an attorney. The suspect has the right to be considered for pretrial release. These procedures quickly move the case from the law enforcement segment of the justice system into the courts.

As is evident in the *Miranda* ruling, the behavior of police officers is much more tightly controlled after arrest than before it. Once the suspect is in custody, the procedures designed to protect the liberty of individuals are initiated. Police officers often complain that suspects are returned to freedom on the streets through bail before the officers have completed their paperwork. Typically, these officers are expressing frustration with the justice system because it seems to be "stacked" in favor of the suspect. In fact, they are correct. The criminal justice system is intentionally constrained by the Constitution to protect the due process rights of the accused who are facing a potential punishment from the criminal justice system.

Booking, the official public recording of an arrest, seeks to prevent the police from holding an offender/suspect incommunicado; anyone can check the booking records to determine whether a person is in police custody. Bail or pretrial release is a provision of the Eighth Amendment of the U.S. Constitution, which is designed to prevent a suspect from unnecessary confinement prior to conviction and to allow the defendant to be free to cooperate in his or her own defense. A third stage, the preliminary hearing (which will be discussed in Chapter 7), is a review of the evidence and the arrest decision by a magistrate or judge. The hearing is held to determine whether the police had sufficient grounds for placing a person in custody. All these procedures illustrate the care with which we protect the rights of the individual in conformity with the presumption of innocence. Frustration arises because, although the law requires police and other actors in the justice system (at least theoretically) to presume that the suspect is innocent, to do their jobs, they often have to believe in the probable guilt of suspects.

DUE PROCESS, CRIME CONTROL, AND THE POLICE

Even if the core mission of the police is crime control, crime control is not the only criterion on which we assess the police and police actions. The police must not only strive to reduce, prevent, and respond to crime, they must do so in a constrained fashion. We require the police to enforce the law and protect the public, but to observe and protect individual rights in the process (Bayley, 2002). Our discussion of entrapment illustrates the tension that exists between due process and crime control in policing. It is not enough for the police to establish that someone broke the law; they must do so without unduly influencing the person to engage in crime. The requirements of due process support a reactive style of policing by restricting what police can do to detect and arrest potential criminals.

The exclusionary rule also illustrates the tension between due process and crime control. The exclusionary rule applies to only those cases in which the police find evidence that crime has occurred. If the police search a home without a warrant and find no criminal evidence, there is no evidence to exclude. The exclusionary rule applies only when there is evidence that someone is guilty of a crime. The exclusionary rule works to release the guilty. In doing so, the rule implies that due process concerns outweigh crime control concerns. The courts would rather let a criminal go free than allow the police to violate constitutional rights.

The good faith exception to the exclusionary rule represents the balancing act that characterizes the entire American system of criminal justice. It seeks to define when, and under what conditions, our interests in individual liberty outweigh our interests in crime control. When do we trust the police to act in our best interests, and when are we suspicious of police powers?

The movement toward community-oriented or problem-oriented policing also demonstrates this balance. As the police become more proactive in seeking to prevent and control crime, the ability of citizens to control police behavior through calls for service is reduced. When the police are encouraged to identify and select problems that seem to produce crime, and when they are urged to focus on disorder at least as much as crime, the role of the police expands. Rather than responding to citizen calls for service, the police themselves determine when and how they will intervene.

This discretion can cause problems between the police and the community. If there is disagreement among members of the community about what is acceptable behavior, the police end up taking sides when they decide whether to act to control such behavior (Kelling, 1999). As important, when the behavior in question relates to order more than crime, the authority of the police to act is unclear. If citizens believe that they are not violating any laws, they may be resistant to police efforts to control them. Facing resistance, the officers often must rely on force, or the threat of force, to make citizens obey. This, in turn, leads to citizen views of the police as brutal and authoritarian. Many researchers have studied citizen attitudes toward the police (e.g., Frank, Smith, & Novak, 2005; Gau, 2010; Warren, 2011). In general, citizens who have had negative experiences with the police, or who have friends, family, and acquaintances who report negative experiences with the police, are less likely to report positive feelings toward the police themselves. This negative opinion of the police can, in turn, affect the suspect's demeanor in subsequent encounters with the police (Rosenbaum, Schuck, Costello, Hawkins, & Ring, 2005). In a sense, citizen attitudes toward police can be a self-fulfilled prophecy. If the citizen expects a negative interaction with the police, their disrespectful demeanor will result in a negative reaction by the officer, which then serves to reinforce the suspect's original view of the police (Dunham & Alpert, 2009). This raises several important questions. When, and for what types of disorder, should the police direct citizen behavior? If a behavior is not criminal in itself, should the police be able to intervene to prevent future crime?

These issues have become even more important as local law enforcement has been called on to take a more active role in homeland security. Recent federal legislation and growing public concern about the threat of terrorist attacks have resulted in increased demands that police take steps to prevent terrorist activity (Johnson & Hunter, 2017). Preventive actions, of course, are in some sense less democratic than reactive, enforcement actions, because the police work under their own initiative. Critics of our reaction to the terrorist threat point out that efforts to prevent future terrorist acts that negate our freedoms are, in the end, proof that the terrorist attacks have been successful (Lynch, 2002). To the extent that concerns about the harm of terrorist attacks outweigh our concerns about individual liberty, we can expect a shift in police practice toward a greater emphasis on crime control and a reduced concern with due process.

See Table 6.1 for a brief summary of some important cases in law enforcement.

TABLE 6.1 **Selected Court Cases in Law Enforcement**	
Weeks v. United States 232 U.S. 383 (1914)	Evidence illegally obtained by law enforcement officers is not admissible in federal criminal prosecutions. This is referred to as the "exclusionary rule"
Mapp v. Ohio 367 U.S. 643 (1961)	The exclusionary rule, which prohibits the use of evidence obtained as the result of unreasonable searches and seizures, is applicable in state criminal proceedings (See *Weeks v. United States* for the exclusionary rule)
Miranda v. Arizona 384 U.S. 436 (1966)	Police must advise a suspect in custody of his or her right to remain silent, the fact that what is said may be used against him or her in court, and of the right to the counsel of a hired or appointed attorney during questioning
United States v. Wade 388 U.S. 218 (1967)	A suspect in custody has the right to an attorney during postindictment lineups
Warden v. Hayden 387 U.S. 294 (1967)	A warrantless search may be valid under circumstances of "hot pursuit"
Terry v. Ohio 392 U.S. 1 (1968)	Stop and frisk is valid on "reasonable suspicion" in order to ensure the safety of police and bystanders
Chimel v. California 395 U.S. 752 (1969)	When making a valid arrest, police may search the area of the arrestee's "immediate control," whether the arrest is with or without a warrant
Chambers v. Maroney 453 U.S. 42 (1970)	Police may perform a warrantless search of an automobile shortly after it is seized
Brewer v. Williams 430 U.S. 387 (1977)	Conversations with or appeals to a suspect that may induce a confession constitute an interrogation that requires both *Miranda* warnings and the right to counsel

Coolidge v. New Hampshire 403 U.S. 443 (1981)	A search is valid only if the warrant is issued by a neutral and detached magistrate. Search of a car after it is in police possession for a period of time does not constitute a valid inventory search
United States v. Ross 456 U.S. 798 (1982)	Police can conduct a full search of an automobile as long as they have probable cause to believe that the automobile contains contraband
Illinois v. Gates 462 U.S. 213 (1983)	The two-pronged test established under *Aguilar* and *Spinelli* is abandoned in favor of a "totality of circumstances" approach. The task of an issuing magistrate is to make a practical decision whether, given all the circumstances, there is a fair probability that the evidence of a crime will be found in a particular place
United States v. Leon 484 U.S. 897 (1984)	Even if the warrant relied on is eventually found invalid, the exclusionary rule allows the use of evidence obtained by officers acting in reasonable reliance on the validity of the warrant
Dickerson v. United States 530 U.S. 428 (2000)	*Miranda* waivers are still required when police officers are interrogating suspects who are in custody. This supersedes the provision of the Omnibus Crime Control Act of 1968, which provides that a defendant's statement is admissible in federal court if given voluntarily, regardless of whether *Miranda* rights have been waived
Arizona v. Gant 556 U.S. 332 (2009)	The Fourth Amendment requires law enforcement officers to demonstrate an actual and continuing threat to their safety posed by an arrestee, or a need to preserve evidence related to the crime of arrest from tampering by the arrestee, to justify a warrantless vehicular search incident to arrest conducted after the vehicle's recent occupants have been arrested and secured
Maryland v. Shatzer 559 U.S. 98 (2010)	Police may reopen questioning of a suspect who previously asked for counsel if there has been a 14-day or more break in custody
United States v. Jones 565 U.S. 400 (2012)	Installing a GPS tracking device on a vehicle constitutes a search under the Fourth Amendment, requiring the police to obtain a search warrant
Utah v. Strieff 136 S. Ct. 2056 (2016)	Evidence seized after an arrest from a preexisting warrant is admissible even if the officer's original stop of the individual was illegal

Source: Portions of this table were adapted from del Carmen and Walker (2012).

ISSUES IN LAW ENFORCEMENT

It is difficult to identify a limited set of important issues in contemporary policing. Almost everything about the police is problematic, and given the dilemma inherent in policing a free society, all these problems are important. However, it is necessary to narrow the field and examine only a few of the problems. Police misconduct, for example, is a traditional problem in policing, as people often suspect that the police act inappropriately. Additionally, the use of force—especially deadly force—has been an issue in policing since the first use of force by police. The relationship between the police and the community also has traditionally been a source of conflict. Finally, the role of the police in controlling crime remains unclear.

Police Misconduct

Police misconduct is an especially vexing problem. Law enforcement officers are held to a higher standard of moral and legal conduct than the general public at the same time that they encounter stressful and dangerous situations on a daily basis (Bishopp, Worrall, & Piquero, 2016). Actions that can be considered misconduct range from improper treatment of citizens (e.g., excessive force, harassment) to the violation of internal departmental rules such as the abuse of sick time or driving the department's car in a reckless manner. More so now than ever before, acts of police misconduct are likely to be substantiated with video evidence taken from either a private citizen's cell phone or the departments' dashboard and/or body cameras. (See Box 6.2.)

One of the most notable forms of misconduct involves corruption. The police have broad powers, and yet they operate with decentralized patrols. The discretionary authority of police officers is generally well known (i.e., we all hope for a "break," even if we do not get it). This combination of power and lack of accountability is what makes corruption possible. In the end, police corruption is difficult to define. Withrow and Dailey (2004) suggested that corruption lies on a continuum that includes lawful activity. Police behavior is corrupt, they said, when the police officer comes to expect or demand special treatment or reward.

"Police corruption" is a term that almost everyone understands, but that has a variety of meanings to different people. Clearly, most people see taking bribes not to enforce the law, stealing from crime scenes, and similar acts of crime as corrupt. There is less agreement about other actions in which police officers may engage. Officers who conduct background checks on persons for their friends, or who use their access to criminal and driving records for personal information, may be corrupt. Often, what people call police corruption is something that is "not right," but is not a violation of the law. Some police misconduct consists of violations of departmental regulations and can be sanctioned, whereas other improper police behavior is not covered by any rules. In an exploratory

BOX 6.2 POLICY DILEMMA: BODY WORN POLICE CAMERAS

The recent media attention to police use-of-force situations has ignited a debate as to whether police officers should wear body worn cameras. Proponents advocate for the use of these cameras as a way to increase the perceived legitimacy of police as well as provide objective evidence when accusations of police misconduct are made. Recent surveys have suggested that both the public and police are generally supportive of these cameras (Crow, Snyder, Crichlow, & Smykla, 2017; Jennings, Fridell, & Lynch, 2014; Smykla, Crow, Crichlow, & Snyder, 2016). Research has also shown that the presence of body worn cameras can reduce the level of citizen complaints against the police (Ariel, Farrar, & Sutherland, 2015; Hedberg, Katz, & Choate, 2016). These findings seem to support the increased usage of police worn body cameras. However, some observers have urged caution. The potential challenges for departments wishing to implement body worn cameras include privacy concerns, the police unions' concern that police may hesitate to use force when necessary due to fear of being on camera, and the vast costs associated with infrastructure and data management (Sousa, Coldren, Rodriguez, & Braga, 2016).

study of police officer sexual violence, Maher (2003) discovered a continuum of police misbehavior with regard to sexual behavior. This continuum ranged from nonsexual contacts, such as making a traffic stop for the purpose of getting a closer look at the driver, viewing unsuspecting lovers having sex in their car, and unnecessary callbacks or visits with an attractive crime victim, to serious criminal acts such as rape and sexual shakedowns. The continuum illustrates one problem with police misconduct: Some misbehavior is clearly illegal (rape), whereas other behavior is inappropriate but not yet defined as a punishable wrong.

Research suggests that misconduct most often occurs among officers who are relatively new to the job before gradually decreasing over the course of an officer's career (Harris, 2012, 2014, 2016). It also appears that a small number of officers are often involved in repeated acts of misconduct (Worden, Harris, & McLean, 2014). For example, Rabe-Hemp and Braithwaite (2012) found that nearly 40% of officers involved in sexual misconduct investigations are repeat offenders, averaging four victims over a 3-year span of offending. Similarly, Stinson et al. (2013) found that nearly half of officers arrested for drug-related misconduct were charged with multiple types of misconduct.

Punch (2009) points out that police misconduct, to some extent, represents a failure within the organization as well as the individual. For example, departments which enact policies that increase officers' stress or are viewed by officers as unfair may have more problems with misconduct. Many departments are also implementing early intervention systems which can identify early signs of problems (e.g., citizens' complaints, vehicle accidents, failure to complete training) among officers and refer those officers to counseling or increased training (Worden, McLean, Paoline, & Krupa, 2015). It is thought that these behaviors may serve as a warning sign that the identified officers may become involved in more serious misconduct in the future. Unfortunately, these early intervention systems have not been shown to reduce officer misconduct and are not always taken seriously by the departments (Shjarback, 2015). It may be that these early intervention systems are simply a way for departments to mitigate their civil liability in the case of a complaint or lawsuit.

Use of Force

Similar fears surround the issue of justified use of force by the police. Among all police powers, the ability to employ physical force legitimately to secure compliance with police orders, or with the law, is one of the most problematic issues in law enforcement (Prenzler, Porter, & Alpert, 2013). Police officers resent having their decisions to employ force second-guessed by civilians or by the internal affairs unit of the department. Citizens fear the unbridled use of force by the police. The issue revolves around the definition of excessive force, that is, force that is greater than what would be required to achieve the lawful aims of the officer.

Few people seriously question an officer's ability and duty to use force to protect self or others from death or serious harm. Police officers often respond to inherently dangerous situations in which they or others are at risk of being harmed. In fact, approximately 10% of police officers are assaulted or killed each year (see Figure 6.2). Many believe that we should allow police officers to use force to gain compliance from resisting suspects or to prevent the escape of dangerous offenders who are likely to injure people again. Questions arise, however, when a police officer strikes an offender "too many times" or when an officer shoots and kills an innocent or unarmed person.

The actual use of substantial force by police officers against citizens is a relatively rare event. In fact, less than 2% of those who have contact with the police experience the threat or use of force (Hyland, Langton, & Davis, 2015). The amount of force used by officers typically represents a continuum ranging from verbal commands to physical force such as grabbing a suspect, using pepper spray, or pointing their service gun at the suspect (Terrill & Paoline, 2013). It appears that most use of force is at a lower level, such as threatening force or shouting (see Figure 6.3). The level of force used depends on situational factors such as the level of resistance provided by the suspect. Lee, Vaughn, and Lim (2014) noted that police use of force is also linked to the level of violent crime in a community. That is, the police are most likely to use force in places that violent crime is also most likely to occur.

Perhaps not surprisingly, the perceived justification for using force often varies between the police and those who force is being used upon. Hyland et al. (2015) found that approximately 75% of those who were subject to verbal or physical force later described the police actions as excessive (Table 6.2). In a given interaction, the police might justify the use of force based on a threat to

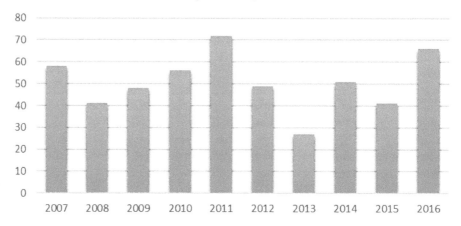

Number of Law Enforcement Officers Feloniously Killed, 2007–2016

FIGURE 6.2
Number of Officers Killed, 2007–2016.

Source: Federal Bureau of Investigation (2016).

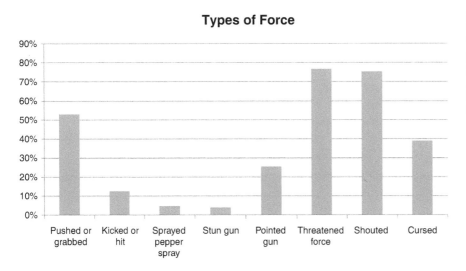

Types of Force

FIGURE 6.3
Types of Force Used or
Threatened by Police,
among Residents who
Reported Force Used.

Source: Eith and Durose
(2011).

TABLE 6.2 **Perceptions of Police Behavior during Contacts Involving Threat or Use of Force, 2002–2011**		
Race	Average Annual Number of Contacts with Police with Threat or Use of Force	Percent Viewing Police Behavior as Appropriate
All races	715,500	13.3
White	445,000	15.8
Black	159,100	8.7
Hispanic	90,100	10.6
Source: Hyland, Langton, and Davis (2015).		

officer safety, whereas a suspect might focus on claims of procedural injustice or having their actions misinterpreted (Rojek, Alpert, & Smith, 2012).

Fears of excessive police force are particularly clear in cases where the officer used deadly force. Illustrating this concern are police use-of-force cases such as the recent instances of Michael Brown and Eric Garner. To some, these are clear cases of police brutality. African Americans are not the only targets of police use of force, but they are subjected to police assaults disproportionately. Because of the discretionary nature of decisions to employ force, citizens are suspicious of officer motives in these cases. Many people feel that the Michael Brown and Eric Garner cases would not have occurred if the victims had been white.

Determining a definitive answer regarding why African Americans are disproportionately subject to police force is a difficult task, especially given the lack of reliable statistics on this issue. Currently, the best available data concerning the use of deadly force originate from media organizations which have compiled databases from news reports of police shootings. With regards to fatal police shootings, the available data show that the rate of police shootings has held steady in the years following the Michael Brown incident (*The Washington Post*, 2018). While African Americans are indeed killed by police at a higher rate than other races, Shane, Lawton, and Swenson (2017) pointed out that disproportionality, in itself, does not account for the situational and contextual factors that an officer faces when using deadly force. In other words, these data do not prove or disprove the concern that police officers are disproportionately using deadly force on African Americans due to racial bias.

Ultimately, the outcome of each police–citizen encounter is situational. Mears, Craig, Stewart, and Warren (2017) argued that implicit biases and spur of the moment decisions from both the police officers and citizens during an encounter could contribute to increased use of force. Imagine a scenario where a police officer comes into contact with a young black man walking along a sidewalk late at night in a high-crime neighborhood. Due to preexisting biases, the officer may perceive this man as dangerous and wish to conduct a pat down search for weapons. The young man, who also has a preexisting negative view of the police, might act disrespectfully to the officer, further raising the officer's suspicions. It should be easy to see how these biases could result in the unnecessary use of force. It is common for the public and the police to not recognize their respective biases toward each other. Nonetheless, these biases affect our social interactions on a daily basis.

The core of the problem is that we have granted our police the power to inflict death and injury without a clear understanding of the fact that the officer must decide when to use this power. The decisional aspect of police use of force is what troubles us most (Terrill, 2005). When the evidence reveals that certain people are more likely to be victims than others, we fear discrimination based on factors other than the threat posed to the officer or to the law. In addition, research has also found that a small proportion of officers are responsible for a disproportionate number of incidents involving force (Brandl & Stroshine, 2003). If we could control the decisions of officers to employ force, we would be less troubled by the issue.

Most police agencies attempt to constrain the use of force by officers through training and policy. Terrill and Paoline (2013) reported wide variation in the comprehensiveness and thoroughness of use-of-force policies. Most agencies used a use-of-force continuum, but policies varied widely by department. For example, some agencies instructed their officers to use chemical sprays and conducted energy devices (stun guns) at earlier stages in the progression of force than other agencies. Moreover, some departments take factors such as the level of resistance by suspects into consideration, whereas others do not give explicit guidance to officers relative to the level of resistance encountered. Increasingly, police agencies are turning to technological solutions such as less lethal weapons as a means to overcome concerns about the use of force. While

FIGURE 6.4
A police officer demonstrates the use of a Taser, an electric stun gun used as a "pain compliance" device by law enforcement. Taser use has been the subject of controversy in several lawsuits, and at least one appeals court has ruled that using a Taser can constitute excessive force.

Gareth Fuller/PA Wire
URN: 29376626/Press
Association via AP Images

there is still some concern about the safety of electronic control devices (e.g., stun guns like the Taser), most police departments are issuing less lethal weapons to their officers (Reaves, 2015) (see Figure 6.4).

Community Relations

Community relations is a long-standing issue that has received much attention in recent years. However, the problem of police and community relations has always plagued the police departments of the United States. To what extent should the community determine policing policy and priorities? In the early days of American policing, police officers served in their own neighborhoods, and the officers very clearly reflected community standards and norms in the enforcement of the laws. Later, policing became more centralized and more impersonal. Communities now desire closer control of their police departments and more input in the setting of police policy. The problem arises when we define the police as experts about policing matters but still want them to observe our wishes. The police oppose what they term "political interference," but must maintain community support.

Less than one-fourth of U.S. residents aged 16 or older have a face-to-face contact with the police in a given year (Berzofsky, Ewing, & DeMichele, 2017). Therefore, much of one's perception of the police is a result of friends or family members who may have had contact with the police or the media portrayal of police activities. Police and citizen contact is most common among younger, white, male citizens and most typically occurs during traffic stops. This limited amount of citizen contact with the police contributes to strained relations between the police and the community. Not only is there little interaction, most of it takes place in stressful circumstances. Police officers typically assume a professional, detached role in dealing with citizens, suggesting that the police are separate from the communities they serve.

In reality, the police are dependent on, and responsible to, the community. At the same time, the community has charged the police with controlling the behaviors of community members while complaining about the fact that the police control behavior. The dilemma is that the community has hired the police to do many things, and some of these are things that the community does not really want to see done. Community relations problems center on those issues in which it is not clear who (the police or the community) should directly define the police role (Hunter, Barker, & de Guzman, 2018).

The movement to community-oriented policing sought to increase the role of citizens in setting police policy. In this model, the police work with the community to identify problems and select solutions. The community defines problems, suggests and/or approves solutions, and cooperates in police efforts to improve the quality of life in the community. A problem arises when the community has a negative attitude toward the police. While Americans generally report positive attitudes toward the police, several factors have been shown to negatively influence these views.

Support and satisfaction with police appear linked to feelings of personal safety. Those who feel safest in their communities are most likely to be supportive of the police (Haberman, Groff, Ratcliffe, & Sorg, 2016; Wentz & Schlimgen, 2012; Wu, Sun, & Triplett, 2009). Those persons least in need of police assistance are most happy with police service. Research has also shown that individuals who live in racially tense neighborhoods or experience greater social distance with police officers hold a more negative attitude toward police (Lee & Gibbs, 2015; McNeeley & Grothoff, 2016). As police seek to prevent disorder in these neighborhoods so as to improve community conditions and prevent serious crime, the targets of disorder control are community members the police are seeking to assist. When the police crack down on loitering, for example, the persons confronted by the police tend to be those living in the neighborhood (McNamara, Crawford, & Burns, 2013). Despite these challenges, many police departments which focus on community-oriented policing tactics have been able to enjoy higher levels of community support (Gill, Weisburd, Telep, Vitter, & Bennett, 2014).

Police Crime Control

A final issue for our consideration is the role of the police in the control of crime. The police do not define what behaviors are criminal, do not control their own budget, and do not control the social and psychological forces that lead individuals to commit crimes, yet they are responsible for the control and prevention of crime. Manning (1978) called crime control an "impossible mandate" for police. The police are responsible for the control of something over which they have no control.

To control crime, police have adopted a number of strategies throughout the years that theoretically appear to serve the function of crime control. Many of these, however, are of questionable value (Walker, 2015). Rapid response and preventive patrol are but two of these strategies.

For decades, police leaders distributed the typical police force over the entire jurisdiction for which it is responsible. They based this distribution on the belief that officers would be able to respond more quickly to calls for aid, and that quicker responses would prove more successful in apprehending offenders. Research indicates that this is only partially true. While rapid police response can help apprehend offenders if the crime is in progress (Cihan, Zhang, & Hoover, 2012), complainants often do not notify the police soon enough that an offense has occurred. By the time most victims call the police, responding within a minute of the call normally is not important (Cordner, Greene, & Bynum, 1983).

Another reason for placing officers out in the community is to provide a "police presence" to deter potential violators and reassure law-abiding citizens. When they are not responding to calls for help, we expect the police to patrol and to prevent crime. Research on police questions the assumption that patrolling prevents crime. There is little reason to believe that preventive patrol is the crime-reduction strategy it is thought to be (Kleck & Barnes, 2014; Telep & Weisburd, 2012). While the effectiveness of preventive patrol requires further study, the general effectiveness of the police in controlling crime has come into question.

The primary issue involved in the question of police effectiveness appears to be one of preventing crime. While there is interest in how to improve police efficiency in apprehending and processing offenders, and there is hope of success in this area, the greatest questions revolve around the ability of the police to prevent crimes. As described in Chapter 3, the increasing emphasis on proactive, problem-solving approaches to policing and hot spot policing provide renewed hope that police can impact crime rates. Indeed, crime rates have declined in recent years as these innovations have been implemented. As we have found, however, these approaches raise questions about citizen support for the police. Overly zealous enforcement efforts may cause citizens to question whether the police are concerned about helping citizens (Tyler, 2005). If people believe they are simply the targets of police enforcement and that police are not concerned about their welfare, they may begin to question the legitimacy of the police, and of the law more generally. In the end, public perceptions of unjust and uncaring enforcement can lead to "rebellion," with higher rates of crime and less citizen cooperation with police. The control and prevention of crime by the police requires a balance between the needs for crime control and requirements of due process.

Many police departments have increased the level of attention given to the ways in which they interact with citizens (Fischer, 2014; Murphy & Tyler, 2017). The perspective of procedural justice suggests that people oppose rules when they believe their treatment is unfair. An outcome (whether the officer writes a ticket or arrests us) is less important than how the officer treats us in terms of our assessment of the legitimacy of the officer's actions. If we believe the process is fair, we tend to support it regardless of outcomes (Sunshine & Tyler, 2003). The reverse of this, of course, is that we resist police authority if we think our treatment is unfair. We are less likely to obey the law, cooperate with the police, or report crime when we see police treatment of us as unfair. Preventive efforts lead to police intervention in minor disturbances or before crimes occur. Citizens subjected to police control in these "noncriminal" situations can question the fairness of police action. Balancing prevention with procedural justice is a difficult task (Gau & Brunson, 2010), but one that is essential if the police are to reach their full capacity to control crime.

THEORY AND THE POLICE

How can we explain the actions of police? Why do the police arrest some people and not others? How is it that police decide to use force against some individuals and the types of force they will use? What explains the decision of police officers to defound a reported crime, or to conduct an investigation? Theories of criminal justice can explain these questions.

We have seen that the law explains much police behavior. The more serious the crime or suspected crime, the more seriously the police respond. The characteristics of the community in which the police work are also important explanatory factors. The police are asked to do different things, and do the same things differently, based on community characteristics. The police organization, its rules, supervision, size, resources, and other characteristics explain much officer behavior. So too, the characteristics of individual officers are related to what they do and how they do it.

As we explored the police, we identified many explanations for police behavior reported in the research literature. We will see that similar explanations exist for court and correctional decisions. We may not have well-established "grand theories" of criminal justice, but we do have a great deal of consistency in what research we find to be important explanations of criminal justice decisions.

REVIEW QUESTIONS

1. Identify and explain the three principal justice system decision points contained in the law enforcement component of the system.

2. Under what circumstances has the U.S. Supreme Court ruled that police need not obtain a warrant prior to conducting a search?

3. What is meant by the "exclusionary rule," and how does it reflect the conflict between due process and crime control?

4. What is the significance of the *Miranda* ruling, and what does it require of the police?

5. What is the good faith exception to the exclusionary rule?

6. Define what is meant by the term "entrapment."

7. What pressures influence the decisions and behavior of police officers?

8. Identify three contemporary issues in American law enforcement.

REFERENCES

Alpert, G., & Noble, J. (2009). Lies, true lies, and conscious deception: Police officers and the truth. *Police Quarterly*, *13*(2), 237–254.

Ariel, B., Farrar, W.A., & Sutherland, A. (2015). The effect of police body-worn cameras on use of force and citizens' complaints against the police: A randomized controlled trial. *Journal of Quantitative Criminology*, *31*, 509–535.

Avdija, A.S. (2014). Police stop-and-frisk practices: An examination of factors that affect officers' decisions to initiate a stop-and-frisk police procedure. *International Journal of Police Science & Management*, *16*(1), 26–35.

Bayley, D. (2002). Law enforcement and the rule of law: Is there a tradeoff? *Criminology & Public Policy*, *2*(1), 133–154.

Bellin, J. (2014). The inverse relationship between the constitutionality and effectiveness of New York City "Stop and Frisk". *Boston University Law Review*, *94*, 1495–1550.

Berzofsky, M., Ewing, G., & DeMichele, M. (2017). *Police–public contact survey: Assessment and recommendations for producing trend estimates after 2011 questionnaire redesign*. Washington, DC: Bureau of Justice Statistics.

Bishopp, S.A., Worrall, J., & Piquero, N.L. (2016). General strain and police misconduct: The role of organizational influence. *Policing: An International Journal of Police Strategies & Management*, *39*(4), 635–651.

Blair, J., & Rossmo, K. (2010). Evidence in context: Bayes' theorem and investigation. *Police Quarterly*, *13*(2), 123–135.

Bonner, H.S. (2015). Police officer decision-making in dispute encounters: Digging deeper into the "black box". *American Journal of Criminal Justice*, *40*, 493–522.

Brandl, S., & Stroshine, M. (2003). Toward an understanding of the physical hazards of police work. *Police Quarterly*, *6*(2), 172–191.

Cassell, P. (1996). *Miranda*'s social costs: An empirical reassessment. *Northwestern University Law Review*, *90*, 387–499.

Cihan, A., Zhang, Y., & Hoover, L. (2012). Police response time to in-progress burglary: Multilevel analysis. *Police Quarterly*, *15*(3), 308–327.

Cordner, G.W., Greene, J.R., & Bynum, T.S. (1983). The sooner the better: Some effects of police response time. In Bennett, R.R. (Ed.), *Police at work: Policy issues and analysis* (pp. 145–164). Beverly Hills, CA: Sage.

Crow, M.S., Snyder, J.A., Crichlow, V.J., & Smykla, J.O. (2017). Community perceptions of police body-worn cameras: The impact of views on fairness, fear, performance, and privacy. *Criminal Justice and Behavior*, *44*(4), 589–610.

Cutler, B., Duaherty, B., Babu, S., Hodges, L., & Wallendael, L. (2009). Creating blind photo arrays using virtual human technology: A feasibility test. *Police Quarterly*, *12*(3), 289–300.

Dale, W., Greenspan, W.M., & Orokos, D. (2006). *DNA forensics: Expanding uses and information sharing*. Sacramento, CA: SEARCH: National Consortium for Justice Information and Statistics.

Davis, D. (2008). Selling confession: The interrogator, the con man, and their weapons of influence. *Wisconsin Defender, 16*(1), 1–16.

Davis, R.C., Jensen, C., Kuykendall, L., & Gallagher, K. (2015). Policies and practices in cold cases: An exploratory study. *Policing: An International Journal of Police Strategies & Management, 38*(4), 610–630.

del Carmen, R.V., & Walker, J.T. (2012). *Briefs of leading cases in law enforcement* (8th ed.). Boston: Elsevier (Anderson Publishing).

Dichter, M., Marcus, S., Morabito, M., & Rhodes, K. (2011). Explaining the IPV arrest decision: Incident, agency, and community factors. *Criminal Justice Review, 36*, 22–39.

Dodge, M., Starr-Gimeno, D., & Williams, T. (2005). Puttin' on the sting: Women police officers' perspectives on reverse prostitution assignments. *International Journal of Police Science & Management, 7*(2), 71–85.

Dunham, R., & Alpert, G. (2009). Officer and suspect demeanor: A qualitative analysis of change. *Police Quarterly, 12*, 6–21.

Eith, C., & Durose, M. (2011). *Contacts between police and the public, 2008*. Washington, DC: Bureau of Justice Statistics.

Ellwanger, S. (2017). Deception in police interrogation. In Braswell, M., McCarthy, B., & McCarthy, B. (Eds.), *Justice, crime and ethics* (9th ed.). Burlington, MA: Elsevier.

Federal Bureau of Investigation (2016). Law enforcement officers killed & assaulted, 2016. Retrieved from https://ucr.fbi.gov/leoka/2016

Fischer, C. (2014). *Legitimacy and procedural justice: A new element of police leadership*. Washington, DC: Bureau of Justice Assistance.

Frank, J., Smith, B., & Novak, K. (2005). Exploring the basis of citizens' attitudes toward the police. *Police Quarterly, 8*(2), 206–228.

Gau, J.M. (2010). A longitudinal analysis of citizen's attitudes about police. *Policing: An International Journal of Police Strategies & Management, 33*(2), 236–252.

Gau, J., & Brunson, R. (2010). Procedural justice and order maintenance policing: A study of inner-city young men's perceptions of police legitimacy. *Justice Quarterly, 27*(2), 255–279.

Gill, C., Weisburd, D., Telep, C.W., Vitter, Z., & Bennett, T. (2014). Community-oriented policing to reduce crime, disorder and fear and increase satisfaction and legitimacy among citizens: A systematic review. *Journal of Experimental Criminology, 10*, 399–428.

Goldstein, J. (1960). Police discretion not to invoke the criminal process: Low visibility decisions in the administration of justice. *Yale Law Journal, 69*, 543–594 (March).

Haberman, C.P., Groff, E.R., Ratcliffe, J.H., & Sorg, E.T. (2016). Satisfaction with police in violent crime hot spots: Using community surveys as a guide for selecting hot spots policing tactics. *Crime & Delinquency, 62*(4), 525–557.

Harris, C.J. (2012). The residual career patterns of police misconduct. *Journal of Criminal Justice, 40*, 323–332.

Harris, C. (2014). The onset of police misconduct. *Policing: An International Journal of Police Strategies & Management, 37*(2), 285–304.

Harris, C.J. (2016). Towards a career view of police misconduct. *Aggression and Violent Behavior, 31*, 219–228.

Hedberg, E.C., Katz, C.M., & Choate, D.E. (2016). Body-worn cameras and citizen interactions with police officers: Estimating plausible effects given varying compliance levels. *Justice Quarterly, 34*(4), 627–651.

Hunter, R.D., Barker, T., & de Guzman, M.C. (2018). *Police–community relations and the administration of justice* (9th ed.). New York: Pearson.

Hyland, S., Langton, L., & Davis, E. (2015). *Police use of nonfatal force, 2002–11*. Washington, DC: Bureau of Justice Statistics.

Jennings, W.G., Fridell, L.A., & Lynch, M.D. (2014). Cops and cameras: Officer perceptions of the use of body-worn cameras in law enforcement. *Journal of Criminal Justice, 42*, 549–556.

Johnson, T.C., & Hunter, R.D. (2017). Changes in homeland security activities since 9/11: An examination of state and local law enforcement agencies' practices. *Police Practice and Research, 18*(2), 160–173.

Kelling, G. (1999). *Broken windows and police discretion*. Washington, DC: National Institute of Justice.

Kleck, G., & Barnes, J.C. (2014). Do more police lead to more crime deterrence? *Crime & Delinquency, 60*(5), 716–738.

Lee, H., Vaughn, M.S., & Lim, H. (2014). The impact of neighborhood crime levels on police use of force: An examination at micro and meso levels. *Journal of Criminal Justice, 42*, 491–499.

Lee, J., & Gibbs, J. (2015). Race and attitudes toward the police: The mediating effect of social distance. *Policing: An International Journal of Police Strategies & Management, 38*(2), 314–332.

Lee, J., Zhang, Y., & Hoover, L.T. (2013). Police response to domestic violence: Multilevel factors of arrest decision. *Policing: An International Journal of Police Strategies & Management, 36*(1), 157–174.

Leo, R. (2001). Questioning the relevance of *Miranda* in the twenty-first century. *University of Michigan Law Review, 99*, 1000–1029.

Leo, R., & Thomas, G. (Eds.). (1998). *The* Miranda *debate: Law, justice, and policing*. Boston: Northeastern University Press.

Levanon, L. (2016). The law of police entrapment: Critical evaluation and policy analysis. *Criminal Law Forum, 27*, 35–73.

Ley, B., Jankowski, N., & Brewer, P. (2012). Investigating CSI: Portrayals of DNA testing on a forensic crime show and their potential effects. *Public Understanding of Science, 21*, 51–67.

Lyman, M.D. (2017). *Drugs in society: Causes, concepts, and control* (8th ed.). New York: Routledge.

Lynch, T. (2002). *Policy analysis: Breaking the vicious cycle: Preserving our liberties while fighting terrorism*. Washington, DC: CATO Institute.

Maher, T. (2003). Police sexual misconduct: Officers' perceptions of its extent and causality. *Criminal Justice Review, 28*, 355–381.

Manning, P.K. (1978). The police: Mandate, strategies and appearances. In Manning, P.K., & Van Maanen, J. (Eds.), *Policing: A view from the street* (pp. 7–31). Santa Monica, CA: Goodyear.

McNamara, R.H., Crawford, C., & Burns, R. (2013). Policing the homeless: Policy, practice, and perceptions. *Policing: An International Journal of Police Strategies & Management, 36*(2), 357–374.

McNeal. G. (2014). *Drones and aerial surveillance: Considerations for legislators.* Malibu, CA: Center for Technology Innovation at Brookings.

McNeeley, S., & Grothoff, G. (2016). A multilevel examination of the relationship between racial tension and attitudes toward the police. *American Journal of Criminal Justice, 41,* 383–401.

Mears, D.P., Craig, M.O., Stewart, E.A., & Warren, P.Y. (2017). Thinking fast, not slow: How cognitive biases may contribute to racial disparities in the use of force in police–citizen encounters. *Journal of Criminal Justice, 53,* 12–24.

Murphy, K., & Tyler, T.R. (2017). Experimenting with procedural justice policing. *Journal of Experimental Criminology, 13,* 287–292.

Newman, G.R. (2007). *Problem-oriented guides for police. Response guides series no. 6: Sting operations.* Washington, DC: U.S. Department of Justice Office of Community Oriented Policing Services.

New York Civil Liberties Union (2018). *Stop and frisk data.* Retrieved from: www.nyclu.org/en/stop-and-frisk-data

Nunn, S., Quinet, K., Rowe, K., & Christ, D. (2006). Interdiction day: Cover surveillance operations, drugs, and serious crime in an inner-city neighborhood. *Police Quarterly, 9*(1), 73–99.

Palmiotto, M. (Ed.). (1984). *Critical issues in criminal investigation.* Cincinnati, OH: Anderson.

Peterson, J., Sommers, I., Baskin, D., & Johnson, D. (2010). *The role and impact of forensic evidence in the criminal justice process* (Report No. 231977). Los Angeles: California State University.

Phillips, S., & Sobol, J. (2010). Twenty years of mandatory arrest: Police decision making in the face of legal requirements. *Criminal Justice Policy Review, 21,* 98–118.

Prenzler, T., Porter, L., & Alpert, G.P. (2013). Reducing police use of force: Case studies and prospects. *Aggression and Violent Behavior, 18,* 343–356.

Punch, M. (2009). *Police corruption: Deviance, reform, and accountability in policing.* Portland, OR: Willan.

Rabe-Hemp, C., & Braithwaite, J. (2012). An exploration of recidivism and the officer shuffle in police sexual violence. *Police Quarterly, 16*(2), 127–147.

Reaves, B.A. (2015). *Local police departments, 2013: Equipment and technology.* Washington, DC: Bureau of Justice Statistics.

Reaves, B.A. (2017). *Police response to domestic violence, 2006–2015.* Washington, DC: Bureau of Justice Statistics.

Reitzel, J., & Piquero, A. (2006). Does it exist? Studying citizens' attitudes of racial profiling. *Police Quarterly, 9*(3), 161–183.

Rhineberger-Dunn, G., Briggs, S.J., & Rader, N. (2016). Clearing crime in prime-time: The disjuncture between fiction and reality. *American Journal of Criminal Justice, 41,* 255–278.

Roberts, A. (2008). The influence of incident and contextual characteristics on crime clearance of nonlethal violence: A multilevel event history analysis. *Journal of Criminal Justice, 36*(1), 61–71.

Roberts, A., & Roberts, J.M., Jr. (2016). Crime clearance and temporal variation in police investigative workload: Evidence from National Incident-Based Reporting System (NIBRS) data. *Journal of Quantitative Criminology, 32,* 651–674.

Rojek, J., Alpert, G., & Smith, H. (2012). Examining officer and citizen accounts of police use-of-force incidents. *Crime & Delinquency, 58,* 301–327.

Rosenbaum, D., Schuck, A., Costello, S., Hawkins, D., & Ring, M. (2005). Attitudes toward the police: The effects of direct and vicarious experience. *Police Quarterly, 8*(3), 343–365.

Ross, D., & Myers, J. (2010). Recent legal developments: *Maryland v. Shatzer*: Revisiting reinterviewing after invoking *Miranda*'s right to counsel. *Criminal Justice Review, 35*(3), 371–385.

Sacks, H. (1978). Notes on police assessment of moral character. In Manning, P.K., & Van Maanen, J. (Eds.), *Policing: A view from the street* (pp. 187–202). Santa Monica, CA: Goodyear.

Schafer, J., & Mastrofski, S. (2005). Police leniency in traffic enforcement encounters: Exploratory findings from observations and interviews. *Journal of Criminal Justice, 33*(3), 225–238.

Schroeder, D., & White, M. (2009). Exploring the use of DNA evidence in homicide investigations: Implications for detective work and case clearance. *Police Quarterly, 12*(3), 319–342.

Schuck, A.M., Rosenbaum, D.P., & Hawkins, D.F. (2008). The influence of race/ethnicity, social class, and neighborhood context on residents' attitudes toward the police. *Police Quarterly, 11*(4), 496–519.

Shane, J.M., Lawton, B., & Swenson, Z. (2017). The prevalence of fatal police shootings by U.S. police, 2015–2016: Patterns and answers from a new data set. *Journal of Criminal Justice, 52,* 101–111.

Sherman, L.W. (1990). Police crackdowns. In Klockars, C., & Mastrofski, S. (Eds.), *Thinking about police: Contemporary readings* (pp. 188–211). New York: McGraw Hill.

Shjarback, J.A. (2015). Emerging early intervention systems: An agency-specific pre-post comparison of formal citizen complaints of use of force. *Policing, 9*(4), 314–325.

Smith, M., Makarios, M., & Alpert, G. (2006). Differential suspicion: Theory specification and gender effects in the traffic stop context. *Justice Quarterly, 23*(2), 271–295.

Smykla, J.O., Crow, M.S., Crichlow, V.J., & Snyder, J.A. (2016). Police body-worn cameras: Perceptions of law enforcement leadership. *American Journal of Criminal Justice, 41,* 424–443.

Sousa, W.H., Coldren, J.R., Jr., Rodriguez, D., & Braga, A.A. (2016). Research on body-worn cameras: Meeting the challenges of police operations, program implementation, and randomized controlled trial designs. *Police Quarterly, 19*(3), 363–384.

Spohn, C., White, C., & Tellis, K. (2014). Unfounding sexual assault: Examining the decisions to unfound and identifying false reports. *Law & Society Review, 48*(1), 161–192.

Stinson, P.M., Sr., Liederbach, J., Brewer, S.L., Jr., Schmalzried, H.D., Mathna, B.E., & Long, K.L. (2013). A study of drug-related police corruption arrests. *Policing: An International Journal of Police Strategies & Management, 36*(3), 491–511.

Stroshine, M., Alpert, G., & Dunham, R. (2008). The influence of "working rules" on police suspicion and discretionary decision making. *Police Quarterly, 11*(3), 315–337.

Sun, I., Payne, B., & Wu, Y. (2008). The impact of situational factors, officer characteristics, and neighborhood context on police behavior: A multilevel analysis. *Journal of Criminal Justice, 36*(1), 22–31.

Sunshine, J., & Tyler, T. (2003). The role of procedural justice and legitimacy in shaping public support for policing. *Law & Society Review, 37*, 103–135.

Telep, C., & Weisburd, D. (2012). What is known about the effectiveness of police practices in reducing crime and disorder? *Police Quarterly, 15*(4), 331–357.

Terrill, W. (2005). Police use of force: A transactional approach. *Justice Quarterly, 22*(1), 107–138.

Terrill, W., & Paoline, E.A., III. (2007). Nonarrest decision making in police–citizen encounters. *Police Quarterly, 10*, 308–331.

Terrill, W., & Paoline, E.A., III. (2013). Examining less lethal force policy and the force continuum: Results from a national use-of-force study. *Police Quarterly, 16*, 38–65.

The Washington Post (2018). *Fatal force.* Retrieved from: www.washingtonpost.com/graph ics/2018/national/police-shootings-2018

Time, V., & Payne, B. (2002). Police chiefs' perceptions about *Miranda*: An analysis of survey data. *Journal of Criminal Justice, 30*(1), 77–86.

Tyler, T. (2005). Policing in black and white: Ethnic group differences in trust and confidence in the police. *Police Quarterly, 8*(3), 322–342.

U.S. Customs and Border Patrol (2018). *CBP releases updated border search of electronic device directive and FY17 statistics.* Retrieved from www.cbp.gov/newsroom/national-media-release/cbp-releases-updated-border-search-electronic-device-directive-and

Waddell, K. (2017, April 12). The steady rise of digital border searches. *The Atlantic.* Retrieved from: www.theatlantic.com/technology/archive/2017/04/the-steady-rise-of-digital-border-searches/522723/

Walker, S. (2015). *Sense and nonsense about crime, drugs, and communities.* Stamford, CT: Cengage.

Warren, P. (2011). Perceptions of police disrespect during vehicle stops: A race-based analysis. *Crime & Delinquency, 57*(3), 356–376.

Wentz, E.A., & Schlimgen, K.A. (2012). Citizens' perceptions of police service and police response to community concerns. *Journal of Crime and Justice, 35*(1), 114–133.

Wilson, J.Q. (1988). *The exclusionary rule: Crime file study guide.* Washington, DC: U.S. Department of Justice.

Withrow, B., & Dailey, J. (2004). A model of circumstantial corruptibility. *Police Quarterly, 7*(2), 159–178.

Worden, R.E., Harris, C., & McLean, S.J. (2014). Risk assessment and risk management in policing. *Policing: An International Journal of Police Strategies & Management, 37*(2), 239–258.

Worden, R.E., McLean, S.J., Paoline, E., & Krupa, J. (2015). *Features of contemporary early intervention systems: The state of the art.* Albany, NY: The John F. Finn Institute for Public Safety.

Wu, Y., Sun, I.Y., & Triplett, R.A. (2009). Race, class or neighborhood context: Which matters more in measuring satisfaction with police? *Justice Quarterly, 26*(1), 125–156.

Zalman, M., & Smith, B. (2007). The attitudes of police executives toward *Miranda* and interrogation policies. *Journal of Criminal Law & Criminology, 97*(3), 873–942.

IMPORTANT CASES

Arizona v. Evans, 115 S. Ct. 1185 (1995).

Arizona v. Gant, 556 U.S. 332 (2009).

Brewer v. Williams, 430 U.S. 387 (1977).

Chambers v. Maroney, 453 U.S. 42 (1970).

Chimel v. California, 395 U.S. 752 (1969).

Coolidge v. New Hampshire, 403 U.S. 443 (1981).

Dickerson v. United States, 530 U.S. 428 (2000).

Fernandez v. California, 134 S. Ct. 1126 (2014).

Florida v. Riley, 488 U.S. 455 (1989).

Floyd v. City of New York, 959 F.Supp 2d 540 (2nd Cir. 2013).

Georgia v. Randolph, 547 U.S. 103 (2006).

Illinois v. Gates, 462 U.S. 213 (1983).

Katz v. United States, 389 U.S. 347 (1967).

Kentucky v. King, 131 S. Ct. 1849 (2011).

Mapp v. Ohio, 367 U.S. 643 (1961).

Maryland v. Shatzer, 559 U.S. 98 (2010).

Maryland v. Wilson, 519 U.S. 408 (1997).

Miranda v. Arizona, 384 U.S. 436 (1966).

New York v. Class, 475 U.S. 106 (1986).

Terry v. Ohio, 392 U.S. 1 (1968).

Thornton v. United States, 541 U.S. 615 (2004).

United States v. Jones, 565 U.S. 400 (2012).

United States v. Leon, 484 U.S. 897 (1984).

United States v. Matlock, 415 U.S. 164 (1974).

United States v. Ross, 456 U.S. 798 (1982).

United States v. Wade, 388 U.S. 218 (1967).

Utah v. Strieff, 136 S. Ct. 2056 (2016).

Warden v. Hayden, 387 U.S. 294 (1967).

Weeks v. United States, 232 U.S. 383 (1914).

CHAPTER 7

The Criminal Courts

There are several thousand courts in operation in the United States. Most of these are at the local level in cities and counties across the country, and they employ more than 27,000 justices, judges, magistrates, or other judicial officers (Malega & Cohen, 2013). As with police agencies, the large number of individual courts in the United States reflects our belief in local autonomy. Justice agents filed more than 18 million criminal cases (excluding traffic cases) in American state courts in 2015 (Schauffler et al., 2016). Another 46 million traffic cases also arose that year. In addition, courts process cases involving domestic relations (e.g., divorce) and civil disputes each year. Thus, crime is a part-time function for our courts. Figure 7.1 describes the workload of America's courts.

The basic purpose of the courts is to resolve disputes (Neely, 1983; Stumpf, 1988). As our society becomes increasingly complex, we not only encounter more disputes, but informal mechanisms of dispute resolution become less effective. Added to this is the fact that the large and increasing number of attorneys in our society makes it easier for people to obtain legal counsel and to use the courts. Thus, more people are bringing their disputes to the courts each year. Each of these individuals is seeking justice. In discussing the purpose of the courts, Rubin (1984:4) noted:

> If we try to describe the purpose of the courts, someone will usually first suggest that their purpose is to "do justice," to provide individualized justice in individual cases. This is true, but whether justice is done depends typically upon the interests or viewpoints of the affected or interested parties. We are confident that some guilty people have been found innocent in our courts, and that innocent persons have been found guilty. These trials may have been conducted fairly, but was justice done?

The point here is that justice is elusive of definition. Indeed, the courts are like jugglers attempting to keep many divergent interests in motion without

IMPORTANT TERMS

appellate courts

arraignment

bail

bench trial

charging

community courts

community prosecution

count

double jeopardy

dual system

exculpatory evidence

formal charges

general jurisdiction

grand jury

hung jury

indictment

jurisdiction

jury trial

limited jurisdiction

nolle prosequi

nolo contendere

preliminary hearing

preventive detention

release on recognizance (ROR)

restorative justice

special jurisdiction

tort

trial courts

unified court system

writ of certiorari

TYPES OF CASES FILED IN STATE COURTS, 2015

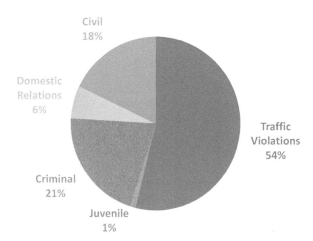

FIGURE 7.1
Types of Cases Filed in State Courts, 2015.

Source: Schauffler et al. (2016).

dropping anything. The courts must strike a balance between the rights of disputing parties. In the criminal courts, the principal balance is between the rights of an individual (and by extension, all individuals) and the rights of the state. The resolution of criminal cases more often entails compromise than competition. Perhaps the best example of this "juggling act" quality of the criminal courts is seen in the practice of plea bargaining.

In plea bargaining, the state and the individual defendant compromise, with neither side getting all that it would hope to achieve. The judge serves as the juggler, balancing the interests of the state (in securing a conviction and punishing a criminal) against the interests of the individual (in protecting his or her liberty and constitutional rights). The alternative to plea bargaining is trial. Trial is the epitome of competition filled with costs and uncertainties for both parties involved in the dispute. Either side may "win" the trial, but both sides will have to expend time and money, experience aggravation, and risk losing to the competition. In plea bargaining, the accused is convicted but avoids the full measure of punishment for the offense of which he or she is accused. Neither side is completely satisfied with the outcome, but both can accept it.

We must understand these compromises, which characterize the workings of the criminal courts, within the organizational context of the courts. They involve not only a defendant and a prosecutor as two competing parties but also a defense attorney (or defender's office), a prosecutor's office, a judge, witnesses, other court staff, the police, possibly jurors, and others. Each actor or set of actors in every criminal case affects the outcome (Walker, 2015). However complicated the issues involved in any case are, the complexity of the court process matches or exceeds the complexity of the issues (Mays & Taggart, 1986).

THE ORGANIZATION OF AMERICAN COURTS

The term "courts" covers a wide range of decision-making bodies, ranging from part-time justice of the peace (or mayor's) courts, in which the "judge" often is not trained in the law, to the nine-member U.S. Supreme Court. There are many different types of courts organized on two basic levels: federal courts and state courts. The existence of two sets of courts in the United States, one federal and the other state, has led observers to speak of a dual system of courts. Box 7.1 provides a diagram of the American court process.

Court systems contain two basic types of courts: trial courts and appellate courts. Trial courts are fact-finding bodies whose job it is to determine the facts of a case (i.e., did the defendant commit the crime?). Appellate courts are law-interpreting bodies whose job it is to determine if the laws were correctly applied and followed (i.e., should the defendant have been provided defense counsel at trial?).

Within trial courts, distinctions exist based on jurisdiction, which is the definition of a court's authority. Determination of a court's jurisdiction typically involves both geography (a county court hears cases arising in its county) and type of case. A court of general jurisdiction can hear civil and criminal cases of all sorts. A court of limited jurisdiction typically is constrained to hearing only minor cases, or conducting the early parts of more serious cases. Special jurisdiction courts are also created to deal with specific types of cases, such as family matters (court of domestic relations) or wills and estates (probate court).

Federal Courts

The federal judicial system comprises the U.S. Supreme Court, 12 U.S. courts of appeals, 94 district courts, more than 400 magistrates assigned to the district courts, and a number of special courts for tax, patent, customs, and contract cases. In comparison to most state court systems, the federal courts are relatively simply organized. U.S. magistrates, appointed by district court judges, are empowered to issue warrants, hear petty cases, and conduct the preliminary stages of more serious criminal cases.

District courts exist in each of the 50 states, the District of Columbia, and the federal territories. District courts are trial courts of the federal system and, combined with magistrates, they comprise the lower courts in the federal system. They hear both civil and criminal cases involving the federal government or violations of federal laws. The workload of the federal district courts is described in Figure 7.2.

BOX 7.1 THE STRUCTURE OF AMERICAN COURTS

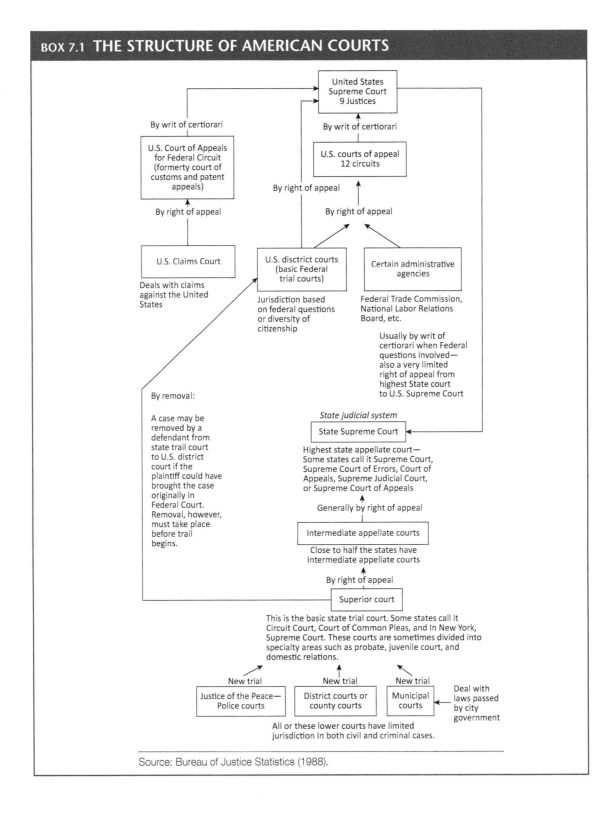

Source: Bureau of Justice Statistics (1988).

U.S. DISTRICT COURTS: NATURE OF OFFENSE, 2016

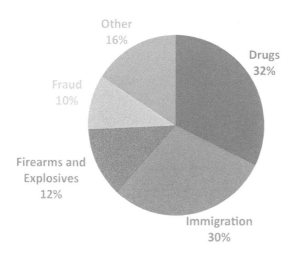

FIGURE 7.2
U.S. District Courts: Nature of Offense, 2016.

Source: Federal Court Management Statistics (2016).

U.S. courts of appeals represent the 50 states, the District of Columbia, and the federal territories. Numbers identify 11 of these courts; the twelfth is the Court of Appeals for the District of Columbia. These courts receive appeals from the decisions of the district courts and decide appeals from the decisions of many federal administrative agencies. Their decisions become binding on all federal district courts under their jurisdiction.

The U.S. Supreme Court is the nation's highest court (see Figure 7.3). Its nine justices hear appeals from the U.S. courts of appeals as well as those from state courts of last resort (usually state supreme courts) that involve questions of federal law or the U.S. Constitution. In most cases, the Supreme Court is not obligated to decide a case and only elects to do so by granting a **writ of certiorari**. This writ is an order to the lower court to send its records of the case so that the Supreme Court can review them. The Court receives thousands of applications for certiorari each year, but agrees to hear only around 80 cases per term (Moffett, Maltzman, Miranda, & Shipan, 2016).

State Courts

In terms of general organization, state court systems mirror the federal courts. They are divided into tiers of limited or special jurisdiction, general jurisdiction courts, and a court of last resort. Boxes 7.2 and 7.3 illustrate two state court systems that differ in terms of their complexity. From state to state, the courts vary greatly in name, number, administration, and power. General jurisdiction trial courts are commonly referred to as: circuit, district, superior, chancery, common pleas, supreme, county, or simply trial courts. Several court systems

FIGURE 7.3
The U.S. Supreme Court is the nation's highest court. It is located in Washington, DC.

Ellen S. Boyne

have multiple court names, such as circuit courts and district courts. Courts at this level employ more than 10,000 judges (Malega & Cohen, 2013).

State court systems take one of two basic organizational structures. A traditional court structure involves separate general jurisdiction and limited jurisdiction courts. A **unified court system** combines the two types of courts into one.

The numbers and names of state courts and their organization are diverse. Rubin (1984:11) observed

> At the state level, courts are created by state constitutions and legislative enactments, and by municipal and county-level legislation. In too many jurisdictions the proliferation of courts has left the citizen unsure of what court to go to for a particular cause of action (divorce, contract dispute, crime, etc.), and for that matter attorneys are not always sure of where to file a particular suit. Some actions can be filed in two or three different courts, and appeals can be taken to several forums.

BOX 7.2 TEXAS COURT STRUCTURE

SUPREME COURT
9 justices sit en banc

CSP case types:
- Discretionary jurisdiction in civil, administrative agency, juvenile certified questions from federal courts, original proceeding cases.

COURT OF CRIMINAL APPEALS
9 justices sit en banc

CSP case types:
- Mandatory jurisdiction in capital criminal, criminal, original proceeding cases.
- Discretionary jurisdiction is certified questions from federal court.

Court of last resort

COURTS OF APPEALS (14 courts)

80 justices sit in panels

CSP case types:
- Mandatory jurisdiction in civil noncapital criminal, administrative agency, juvenile, original proceeding, interlocutory decision cases.
- No discretionary jurisdiction.

Intermediate appellate court

DISTRICT COURTS (424 courts) 424 judges

DISTRICT COURT (414 courts)   A

414 judges

CSP case types:
- Tort, contract, real property rights ($200/no maximum), estate, miscellaneous civil, Exclusive administrative agency appeals jurisdiction.
- Domestic relations.
- Felony, misdemeanor, DWI/DUI miscellaneous criminal.
- Juvenile.

Jury trials.

CRIMINAL DISTRICT COURT
(10 courts)

10 judges

CSP case types:
- Felony, misdemeanor, DWI/DUI miscellaneous criminal cases.

Jury trials.

Court of general jurisdiction

COUNTY-LEVEL COURTS (482 courts) 482 judges

CONSTITUTIONAL COUNTY COURT
(254 courts)
254 judges

CSP case types:
- Tort, contract, real property rights, miscellaneous civil ($200/$5,000), estate, mental health, civil trial court appeals.
- Misdemeanor, DWI/DUI, criminal appeals.
- Moving traffic, miscellaneous traffic.
- Juvenile.
Jury trials.

PROBATE COURT
(17 courts)

17 judges

CSP case types:
- Eslate.
- Mental health.

Jury trials.

COUNTY COURT AT LAW (211 courts)
211 judges
CSP case types:
- Tort, contract, real property rights, miscellaneous civil ($200/$100,000), estate, mental health, civil trail court appeals.
- Misdemeanor, DWI/DUI, criminal appeals.
- Moving traffic, miscellaneous traffic.
- Juvenile.
Jury trials.

Court of limited jurisdiction

MUNICIPAL COURT* (894 courts)
1,345 judges
CSP case types:
- Misdemeanor.
- Moving traffic. parking, miscellaneous traffic. Exclusive ordinance violation jurisdiction.

Jury trials.

JUSTICE OF THE PEACE COURT* (827 courts)
827 judges
CSP case types:
- Tort, contract, real property rights ($0/$5,000), small claims ($5,000), mental health.
- Misdemeanor.
- Moving traffic, parking, miscellaneous traffic.
Jury trials.

*Some municipal and justice of the peace courts may appeal to the district court.

CSP = Court Statistics Project

Source: Rottman and Strickland (2006:312).

BOX 7.3 INDIANA COURT STRUCTURE

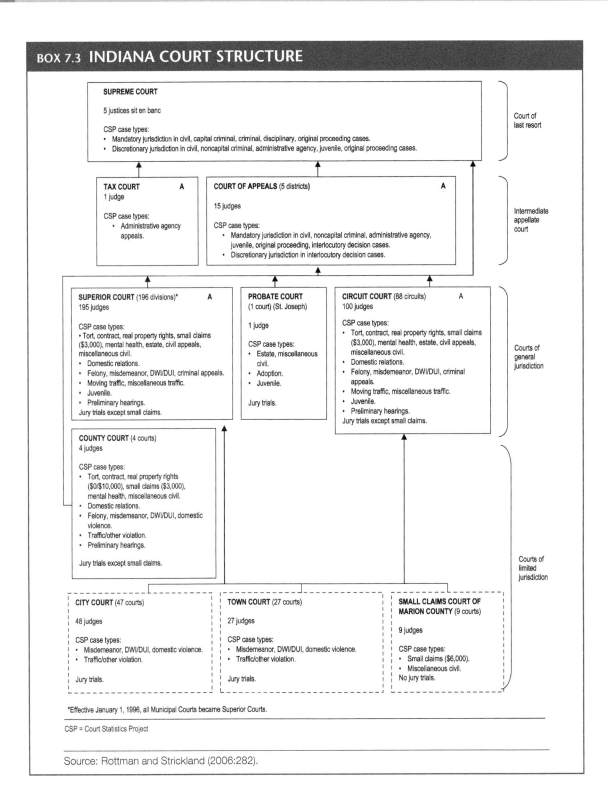

SUPREME COURT

5 justices sit en banc

CSP case types:
- Mandatory jurisdiction in civil, capital criminal, criminal, disciplinary, original proceeding cases.
- Discretionary jurisdiction in civil, noncapital criminal, administrative agency, juvenile, original proceeding cases.

Court of last resort

TAX COURT   A
1 judge

CSP case types:
- Administrative agency appeals.

COURT OF APPEALS (5 districts)   A

15 judges

CSP case types:
- Mandatory jurisdiction in civil, noncapital criminal, administrative agency, juvenile, original proceeding, interlocutory decision cases.
- Discretionary jurisdiction in interlocutory decision cases.

Intermediate appellate court

SUPERIOR COURT (196 divisions)*   A
195 judges

CSP case types:
- Tort, contract, real property rights, small claims ($3,000), mental health, estate, civil appeals, miscellaneous civil.
- Domestic relations.
- Felony, misdemeanor, DWI/DUI, criminal appeals.
- Moving traffic, miscellaneous traffic.
- Juvenile.
- Preliminary hearings.
Jury trials except small claims.

PROBATE COURT
(1 court) (St. Joseph)

1 judge

CSP case types:
- Estate, miscellaneous civil.
- Adoption.
- Juvenile.

Jury trials.

CIRCUIT COURT (88 circuits)   A
100 judges

CSP case types:
- Tort, contract, real property rights, small claims ($3,000), mental health, estate, civil appeals, miscellaneous civil.
- Domestic relations.
- Felony, misdemeanor, DWI/DUI, criminal appeals.
- Moving traffic, miscellaneous traffic.
- Juvenile.
- Preliminary hearings.
Jury trials except small claims.

Courts of general jurisdiction

COUNTY COURT (4 courts)
4 judges

CSP case types:
- Tort, contract, real property rights ($0/$10,000), small claims ($3,000), mental health, miscellaneous civil.
- Domestic relations.
- Felony, misdemeanor, DWI/DUI, domestic violence.
- Traffic/other violation.
- Preliminary hearings.

Jury trials except small claims.

Courts of limited jurisdiction

CITY COURT (47 courts)

48 judges

CSP case types:
- Misdemeanor, DWI/DUI, domestic violence.
- Traffic/other violation.

Jury trials.

TOWN COURT (27 courts)

27 judges

CSP case types:
- Misdemeanor, DWI/DUI, domestic violence.
- Traffic/other violation.

Jury trials.

SMALL CLAIMS COURT OF MARION COUNTY (9 courts)

9 judges

CSP case types:
- Small claims ($6,000).
- Miscellaneous civil.
No jury trials.

*Effective January 1, 1996, all Municipal Courts became Superior Courts.

CSP = Court Statistics Project

Source: Rottman and Strickland (2006:282).

Indeed, very often the greatest service an attorney can provide a client is to ensure that the proper papers are filed with the appropriate court. Jackson (1974) described the structure of American courts as one that would "make a chart-maker collapse in despair." He described the organizational chart of courts in the United States as "a bewildering maze of parallel, perpendicular, crisscrossing, and overlapping lines." It is frequently difficult to determine whether, how, and where to enter the courts with any given case.

Suppose that you have a disagreement with a neighbor about parking spaces on the street. For the moment (because your car is already parked in front of the neighbor's house), you have won. The next morning, you awaken to discover four flat tires. What should you do? The options are many. First, you may "grin and bear it." Not knowing for sure who did you wrong, you may opt to repair the tires and try to forget the incident (although you may spend several nights monitoring your car, or you may be more selective in deciding where you park). Second, you may confront your neighbor because you surmise that he is the guilty party. (If relations between you and your neighbor were better, you would have had an informal mechanism to avoid this incident in the first place.) Third, you may seek revenge by slashing your neighbor's tires that evening. Fourth, you may decide to call the police and report the vandalism. Calling the police could lead to a court appearance. There may be other options, but these four appear to be the most likely.

If you find the first three options unacceptable, in all probability you would call the police. This allows you to collect insurance payments for damages, and it transfers the decision of what to do to the responding officer. The dispute then may reach the courts as a criminal complaint, but deciding whether, where, and how to enter the courts would be someone else's problem. Unless you have extraordinarily expensive tires, you could take your case to small claims court and sue your neighbor for damages. You could also secure counsel by hiring an attorney or by contacting a legal aid office.

This hypothetical case resulted from a dispute among neighbors about parking privileges. It could go to a number of different courts as a **tort** (a dispute between private parties over a wrongful injury suffered; that is, damaged tires), a crime (willful destruction of property by your neighbor), or both. Given this complexity, how can the average citizen know how and where to file a court case?

The general structure of state courts includes courts of limited jurisdiction that are empowered to hear petty cases such as traffic violations, and to conduct the preliminary stages of more serious cases (much like federal magistrates). These limited jurisdiction courts include misdemeanor courts that hear cases involving less serious crimes. Commentators have frequently criticized misdemeanor courts, often including traffic courts, for their practices.

Feeley (1979) wrote a book on the misdemeanor courts arguing, "the process is the punishment." Feeley explained that sentences in these courts tend to be very lenient because the inconvenience and disruption defendants experience in the court process is sufficient penalty for most of the minor crimes tried in misdemeanor courts. Feeley described practices in the New Haven, Connecticut, misdemeanor courts 40 years ago that are remarkably similar to those of the current day. Geller (2016) described the legal and financial burdens faced by those arrested by New York City's stop and frisk policy from 2009 to 2012. She described a process in which conditions encouraged defendants to waive rights, decline trials, and move to complete the process as quickly as possible. The fear is that defendants in these situations might plead guilty to a crime that they otherwise would not have been found guilty of, leaving them with a criminal record which will impact their lives well into the future.

At a second level of courts are those of general jurisdiction, which conduct trials for civil and criminal cases. Most states also have a juvenile or family court. This court is a special jurisdiction trial court empowered to hear juvenile delinquency cases, but is not otherwise involved with criminal matters. Such courts also hear divorce, child custody, and related domestic relations cases. Finally, there are frequently other special-purpose courts, such as probate and surrogate's courts.

In most states, a third level of appellate courts exists, at which the courts operate like the U.S. courts of appeals (i.e., they hear appeals from the lower courts). Some states have more than one intermediate appeal level, so that a case may go through two or more appeals before reaching the third level of courts. Each state has a court of last resort or a supreme court, which receives appeals from all lower courts in the state and from the intermediate appellate courts. The decisions of these courts are binding on all other courts in the state.

Appellate courts do not decide issues of fact; rather, appellate courts serve to interpret the law. Appellate courts accept the findings of fact from the lower courts, and then decide whether the lower-court judges interpreted and applied the law correctly. They resolve questions of law raised by those who lost in the lower courts. Decisions of higher courts on the interpretation and application of the law are binding on all the courts beneath them. Thus, when the U.S. Supreme Court decides on an interpretation of the U.S. Constitution, that interpretation is binding on all the courts in the nation.

Problem-Solving Courts

In the past three decades, a growing emphasis has been placed on specialized courts aimed at solving specific types of problems. These courts range from drug courts that deal exclusively with offenders who have substance abuse problems to domestic violence courts and reentry courts that specialize in returning prisoners to productive lives in the community. In combination, these have been called "problem-solving courts" (Casey & Rottman, 2003) because the role of the court has been redefined as that of responding to specific problems such as drug abuse or domestic violence. Marlowe, Hardin, and Fox (2016) found over 4,300 problem-solving courts operating in 2014 (Table 7.1). Drug courts represent approximately three-quarters of all problem-solving courts, and are available in over half of all U.S. counties. Wolf (2007) outlined six principles that guide the development of problem-solving courts:

TABLE 7.1 Types of Problem-Solving Courts in the United States, 2014	
Court Type	Number Reported
Drug	3,057
Adult mental health	392
Truancy	314
Domestic violence	210
Child support	62
Juvenile mental health	37
Reentry	30
All other	266

Source: Marlowe, Hardin, and Fox (2016).

Enhanced information: Problem-solving courts are equipped with better staff training and specialized knowledge regarding the particular problem. This allows judges to make more informed decisions about the treatment needs and risks of an individual defendant.

Community engagement: Courts can engage the community to discuss topics of interest such as local "hot spots" of crime. Courts have used a variety of methods to engage the community, ranging from mailed surveys to community gatherings.

Collaboration: Courts can improve interagency communication by bringing together multiple parties such as judges, prosecutors, probation officers, social service partners, victims groups, and so on. This increased communication can foster new sentencing options to address problems.

Individualized justice: Courts can sentence offenders to individually tailored community-based services that can reduce recidivism, improve community safety, and enhance confidence in the system.

Accountability: The courts can improve the accountability of offenders by sending the message that all criminal behavior, even low-level crime, has an impact on community safety and has consequences.

Outcomes: The courts can analyze the outcomes, costs, and benefits of the court's operations. By doing so, the courts become more accountable to the public.

At the same time, some courts have broadened the problem-solving approach to community problems. "Community courts" have developed and spread to deal with a range of community problems through the application of the criminal law (and less formal adjudication options). The courts connect to community resources and seek to solve problems rather than simply find facts. Thus, the court tries not only to resolve a conflict but also to understand its underlying causes and secure necessary community services to prevent the conflict from happening again (Rottman & Casey, 1999). Some community courts are actually diversion programs or mediation centers that focus on less serious disputes. In some places, however, actual community courts exist. They tend to practice **restorative justice**, a form of negotiation and discussion between the parties involved in a case designed to create or improve understanding. Kuo, Longmire, and Cuvelier (2010:327) stated, "Restorative justice aims to heal victims and offender through dialog, which is the essence of the restorative justice process." After discussing the crime, its causes, and effects, the parties agree to what they believe is a fair settlement of their differences.

Rottman (1996) described the movement to community courts. His description of the development and spread of community courts mirrors similar accounts of the development of community policing. Initially, courts represented the community and served as forums for dispute resolution in which the resolution process reflected community life. As the courts became more professional and centralized, courts became divorced from the communities they served, in part to reduce the impact of "politics" on court processing. The community court movement seeks to reestablish the link between courts and communities. The community courts generally have a greater range of sentencing options than traditional courts, including community service, substance abuse treatment, and requirements to attend general educational development (GED) classes. In addition, community courts often provide defendants with referrals for job training and housing assistance (Frazer, 2006). A key component of community courts is citizen involvement (Terry, 2000). A part of this general movement involves **community prosecution**, in which the prosecutor is assigned to the case from initial appearance through disposition and works with the police, community, and other agencies not just to secure conviction, but also to solve the problems that led to the criminal behavior (Jansen & Dague, 2006).

Although not yet well evaluated, community courts hold promise for improving justice for less serious disputes and crimes that often find their way into the lower courts of the criminal justice system. Henry and Kralstein (2011) reported that the limited evaluations to date indicate that community courts (1) rely on community service sentencing, (2) are viewed favorably by both the public and defendant, (3) decrease jail sentences, (4) increase compliance with court conditions, and (5) conclude cases more quickly than traditional criminal courts. However, studies are less conclusive regarding other goals of community courts. For example, Grommon, Hipple, and Ray (2017) found no evidence that defendants processed through a community court had a reduced rate of recidivism compared to defendants processed through traditional courts.

THE DEVELOPMENT OF AMERICAN COURTS

During the colonial period, most political power was in the hands of the colonial governor (Neubauer, 1984:36). The colonists generally adopted the existing English system of courts and government but modified the system to fit the less complicated nature of colonial life. They did not fully transplant the many specialized courts that had evolved in Britain. Rather, they established a county court as the basic forum for adjudication. Appeals from rulings of the county court went directly to the governor. Although it was possible to appeal a gubernatorial decision to the English courts, it was seldom done (Glick, 1983:35).

As the colonies grew and commerce developed, the county courts became increasingly less able to handle the number and intricacy of cases that arose. Each colony created special courts to expedite the handling of cases. They developed new courts of general jurisdiction to reduce the need for litigants to travel great distances to have their cases heard in county courts. The addition and creation of new courts took place on a haphazard basis. New courts were created to resolve specific problems, not with any grand scheme of court structure in mind. Furthermore, court development occurred independently in each colony (Stumpf, 1988:70–73). One result of this method of development is that courts performing the same functions often have different names.

After the American Revolution, the greatest issues facing the courts were those surrounding the balance between federal and state powers. The nature of the U.S. Constitution, with its limitations on federal powers and reservation of certain governmental powers to the states, ensured the continuation of state judicial systems. The Federal Judiciary Act of 1789, designed to create a court system for the resolution of cases arising from federal laws, was a compromise between the federalists (who favored a strong central government) and the antifederalists (who championed states' rights and feared the results of a strong central government). The compromise was that they created federal courts, but that these courts would not cross state lines. To this day, no U.S. district court has more than one state within its jurisdiction.

State judges were required to swear an oath of allegiance to the U.S. Constitution. The Judiciary Act of 1789 gave the Supreme Court of the United States authority to review state decisions involving questions of a constitutional or federal nature. Although this fact gave the appearance that state courts were subservient to the federal courts (Jacob, 1984:162), state courts retained tremendous powers. The federal courts did not gain significantly in authority until social changes in the United States caused state courts to become ineffective.

Throughout the nineteenth century, the development of industry, the completion of the transcontinental railroad, and the growth in interstate commerce combined to create conditions in which state courts were not capable of resolving a large number of cases. In time, citizens increasingly called on federal courts to resolve disputes. As Jacob (1984:163) noted

> Just as state governments in general became less important to national policymaking, so state courts, to a somewhat lesser degree, lost their pre-eminent position in the judicial system and increasingly concentrated on private law cases, which enforce existing norms and affect only the immediate parties to a case.

THE FUNCTIONS OF COURTS

In describing the role of the courts, one of the first goals of the process is likely to be "justice" (Rubin, 1984). That is, the purpose of the courts is to do justice. However, we are not able to agree on exactly what is "just." Alan Dershowitz (1982:xvi) suggested that, in fact, "nobody wants justice." Rather, parties to a suit are only interested in winning.

It is not possible to understand the operations of the criminal courts without having an appreciation for the complexity of the job performed by the courts. Addressing this issue, Neely (1983:16) noted

> As long as we are talking only about criminal courts, the questions are relatively simple. As soon, however, as civil courts enter the picture, all bets are off. For while improved funding of the criminal courts would return economic dividends to the public, improved funding of the civil courts has mixed income effects.

Neely cited New York City increasing funds to courts of general jurisdiction to help alleviate caseload backlogs as an example where the creation of more courts and the appointment of more judges opened more opportunities for people to press civil lawsuits. In many of these accelerated cases, the defendant in the civil suit was the city of New York. In effect, the citizens of New York City had paid more money to create more courts that could then be used by plaintiffs to sue the city for still more money in damages.

Because most courts empowered to hear criminal cases also hear civil cases (as is true in New York City), it is not possible to enhance court capacities selectively. Creating new judgeships and courts to accelerate criminal trials, for example, will necessarily also serve to speed civil trials. In fact, the civil docket may be so crowded that increasing the number of courts or judges will not substantially affect the outcome of criminal cases. Whereas our focus is on the criminal courts for the remainder of this chapter and the next, we must remember that the civil caseload lurks in the background and represents the largest portion of court workload (with the exception of traffic cases).

The functions of the criminal courts are twofold: the repression of crime and the protection of the rights of individuals accused of crimes. Courts serve these goals within an environmental context that includes many factors. These factors include the characteristics of the various actors (defendants, witnesses, juries, judges, prosecutors, defense attorneys, etc.), the organizational goals of involved agencies, the political climate, present social forces, and the like. The law is only one of several factors that impinge on judicial outcomes. As Glick (1983:18) stated, "Most cases also are settled through informal negotiation, not trials. Personal decision making and compromise are the keys to understanding how disputes are settled."

THE CRIMINAL COURT PROCESS

The basic dispute addressed in criminal courts is that between the need to control criminal behavior and the desire to protect individual rights and liberties. This dispute is at the core of each decision in the court segment of the justice system. After arrest, the suspect proceeds to the court stage of the criminal justice process and moves along a series of hearings and decisions that result in either conviction and punishment, or release from custody. The principal decision points in the courts are initial appearance (bail determination), formal charging, preliminary hearing, arraignment on charges, and trial. Sentencing, or punishment, is the subject of Chapter 9.

Initial Appearance

Shortly after arrest, the suspect comes before a magistrate for the setting of bail. **Bail** is a security posted by the defendant to ensure appearance at later court proceedings. The Eighth Amendment to the U.S. Constitution provides that "excessive bail shall not be required." There is some controversy as to whether this provision creates a right to bail for criminal defendants, or whether it merely protects them from facing an excessive bail. The leading U.S. Supreme Court decision on bail was rendered in the case of *Stack v. Boyle* (1951). Under federal law, defendants in noncapital cases are entitled to bail. The issue in this case was whether bail set at $50,000 was excessive, given the defendant's inability to post that amount. The Court decided that the purpose of bail is to ensure appearance at later proceedings and, absent evidence to support an exception, bail that is not reasonably calculated—or that is higher than that normally fixed for a similar offense—is excessive.

Approximately 62% of state-level felony defendants are released before trial (Reaves, 2013). The average amount of bail set for defendants is largely influenced by the type of crime that the defendant is accused of committing (see Table 7.2). As a result, defendants who are facing serious charges are least likely to obtain pretrial release (see Table 7.3). The ability to obtain bail has been shown to have significant effects to the outcome of a defendant's case, as research has shown that defendants who do not secure release prior to trial are convicted at higher

TABLE 7.2 Mean Bail Amount, by Most Serious Charge

Most Serious Charge	Bail Amount ($)
Violent offense	115,000
Property offense	33,200
Drug offense	34,200
Public-order offense	33,800

Source: Reaves (2013).

TABLE 7.3 Pretrial Release Status of Felony Defendants and Percent Convicted, 2009

Case Characteristic	Released (%)	Held on Bail (%)	Denied Bail (%)
Total of all cases	62	34	4
Violent offense	55	39	6
Property offense	63	33	3
Drug offense	65	31	3
Public-order offense	63	32	6

Source: Reaves (2013).

rates and are punished more severely (Oleson, Lowenkamp, Wooldredge, Van-Nostrand, & Cadigan, 2017; Reaves, 2013; Sacks & Ackerman, 2014).

Several procedures for providing bail are used throughout the United States (Box 7.4). Full cash bonds require the defendant to pay the entire bail amount to the court in exchange for their release. As can be imagined, many defendants are unable to pay the full amount. Thus, a number of alternatives have been developed to assist in securing pretrial release. First, the defendant can use any real estate assets which they own as collateral for the bail amount. In this situation, the court may place a lien on the defendant's property until the conclusion of the case. However, it is common for criminal defendants to not have access to either the cash or property needed to obtain bail. These defendants must rely on surety bond with the assistance of a bail agent.

BOX 7.4 PRETRIAL RELEASE PROCEDURE

Financial Conditions	Non-financial Conditions
Full cash bond	Personal recognizance
Property bond	Conditional release
Surety bond	Unsecured bond
Deposit bond	

Cohen and Reaves (2007).

Bail bond agents are often small business owners who provide bail for criminal defendants for a fee. Usually, the bond agent charges 10% of the bail amount, which is not refundable. Thus, if you face a bail of $2,500, the bond agent will charge you $250 and post the full bond to secure your release. If you appear at the later stages of your trial, the court refunds the bail to the bond agent, but you are out the $250. Should you fail to appear at later hearings, the bond agent will seek you out and return you to court to protect his or her investment (Burns, Kinkade, & Leone, 2005). Some observers of the bail process have expressed concern with several factors: Profit as the motive behind pretrial release; the lack of regulation of bail agents; the errors made by bond agents in apprehending those who "skipped" bail; and irregularities in the posting of bond (using the same assets to secure release for several defendants) (Bradford, 2012; Goldfarb, 1965; Goldkamp, 1980; Johnson & Stevens, 2013). See Figure 7.4.

In response to these concerns, several innovations have occurred across the country. One of the more popular of these innovations involves **release on recognizance (ROR)**, whereby a defendant with ties to the community (a job, family, stable residence, etc.) is released on his or her own recognizance

FIGURE 7.4
A bail bond office near a jail and courthouse. Bail bond agents often seek to attract clients by conspicuously locating their offices close to county and municipal jails.

Ellen S. Boyne

without posting bail if there was reason to believe that he or she would not flee the jurisdiction (Maxwell, 1999). In other places, courts allowed defendants to post only 10% of the full amount directly to the court, because 10% was the percentage risked by a defendant when a bond agent posts bail. Another innovation has been the use of conditional release, in which the defendant may be required to report to a pretrial supervision officer or be subject to electronic monitoring. Finally, some courts provide unsecured bonds, where the court will set a bail amount but not require any upfront payment for release. Rather, the defendant simply has to pay the full bail amount if he/she does not appear in court.

Reaves (2013) reported that 16% of felony defendants in large jurisdictions were rearrested, and 29% engaged in some sort of misconduct prior to the resolution of their case. Therefore, a central question in bail determinations is to what extent the setting of a bail amount should reflect concern for protecting public safety. Historically, the only recognized purpose of bail was to ensure appearance by the defendant at later court proceedings (Baughman, 2018). A common concern of police, prosecutors, and magistrates (not to mention the public) is to keep suspected offenders off the streets until they can be convicted and punished. To accomplish this goal, some persons advocate **preventive detention**, in which the court denies pretrial release for defendants suspected of being dangerous (Walker, 2015). The federal Bail Reform Act of

1984 authorizes the pretrial detention of categories of offenses or offenders. In *United States v. Salerno* (1987), the court held that when using adversarial procedures in open court to determine that a defendant poses a risk of pretrial crime, it is permissible under the Eight Amendment to deny pretrial release.

Many counties use bail schedules which essentially set predetermined money bail amounts based on the defendant's most serious charge (Carlson, 2011). These are particularly common in misdemeanor courts which are tasked with processing dozens or hundreds of cases each day. Bail schedules allow court officials to quickly assign bail to defendants in a consistent manner while managing their large caseload. However, bail schedules that do not take into account the defendant's ability to pay and specific information about the defendant's risk have been criticized for being unfair, ineffective, and unconstitutional (Hurley, 2016; Wiseman, 2016). Additionally, the cost of housing large numbers of pretrial detainees can be substantial (Baughman, 2017).

As a result of the growing pressure, many states are beginning to look at possible reforms to their bail procedures (Box 7.5). These states often use risk assessment tools to determine whether a defendant is suitable for release. Those with prior criminal records, who have no visible means of support, who are accused of serious offenses, or are otherwise thought to be dangerous usually do not qualify for ROR or low-amount bail. Although the use of these revised bail procedures have been shown to reduce the jail population and enable more defendants to keep their jobs and social ties within the community while awaiting trial, critics have argued that the risk assessment tools used might have a disproportionately negative impact on those from lower socioeconomic classes (Knox & Keifer, 2017).

BOX 7.5 POLICY DILEMMA: MONEY BAIL

The United States is one of only two countries that require defendants to pay money in exchange for being released while awaiting their trial (Baughman, 2018). In recent years, this has come under attack from critics who argue that cash bail discriminates against those who do not have the means to secure release. As will be shown in Chapter 8, failure to receive bail can have a significant negative impact on a defendant's court outcome. In 2017, New Jersey replaced cash bail by using an evidence-based risk assessment tool to help judges make release decisions. Based on the score of the risk assessment, defendants were either released on their own recognizance or placed on a monitoring plan reflecting their level of risk. Using this system, only the most at-risk defendants are denied release pending trial. In its first year of operation, New Jersey courts released over 94% of defendants and effectively reduced the county jail population (Grant, 2018). Policymakers in other states are now considering potential changes to their bail process.

With the exception of preventive detention hearings, the bail decision is usually not concerned with assessing evidence. For defendants arrested without a warrant, the U.S. Supreme Court has held that there must be a "prompt" judicial determination of probable cause. In those cases, the magistrate or judge must assess the adequacy of the evidence on which the arrest was based. This assessment occurs at the initial appearance for defendants arrested without warrants. In *County of Riverside v. McLaughlin* (1991), the Supreme Court ruled that such a review must occur within 48 hours of the arrest.

Preliminary Hearing

Whether the defendant makes bail or secures other pretrial release, the next decision point in the court process is the **preliminary hearing**. At this stage, the case against the accused is reviewed by a neutral magistrate to determine whether the evidence is sufficient to justify binding the defendant over for trial (Chen, 1991). The preliminary hearing is not a full trial of the case but is an open court process in which the strength of the state's case against the defendant is tested. The prosecution must establish that there is sufficient evidence to show probable cause that the defendant committed a crime. The defense may cross-examine witnesses, testify, and call witnesses on his or her behalf. The "information" or presentment method of charging occurs at this point.

If the magistrate decides that there is sufficient evidence to justify further action, the defendant may be "bound over" for trial. In many states, this action means that the case will go to the grand jury for a decision about formal criminal charges. In other states, the preliminary hearing serves as a charging process. The purpose of the hearing, however, is to ensure that the state is justified in continuing to proceed against the defendant.

Formal Charging

At some point in the pretrial segment of the court process, the state must file **formal charges** against a defendant. These charges are allegations of the specific crimes for which the defendant will stand trial. They are termed "formal charges" to distinguish them from the arrest charge, which may not actually reflect the offense for which the state will try the accused. In **charging**, the prosecutor (or state's attorney or district attorney) applies the criminal law to the facts of the case and identifies which provisions of the criminal code the offender violated.

For example, the police may arrest someone who is holding a screwdriver while standing on the porch of a home and charge him or her (in the arrest report) with attempted burglary. Upon reviewing the case and the law, the prosecutor may decide that the evidence will not support so serious a charge and opt instead to charge criminal trespassing and possession of burglar's tools. Similarly, the police may arrest someone for first-degree murder, but use the formal charge of manslaughter.

The evidentiary standard for charging is the same as that for arrest: probable cause. However, there is a subtle difference in interpretation of probable cause among police and prosecutors. Police tend to be "backward looking" in attempting to justify an arrest, whereas the prosecutor is "forward looking" in attempting to predict the likelihood of successful prosecution (Newman,

1978). That is, the police require probable cause to believe the suspect has committed a crime, whereas the prosecutor requires probable cause to believe it is possible to convict the suspect (now defendant) of the crime (Adams, 1983; Boland & Forst, 1985).

Formal charges can be filed by either of two methods. As mentioned above, one way is through the use of a preliminary hearing. The other is through an **indictment** by the grand jury. The **grand jury** is a panel of citizens (often sitting for a month or longer) that reviews evidence in criminal cases to determine whether sufficient evidence exists to justify trial of an individual (Acker & Brody, 2004). Grand juries are usually larger than trial juries, with sizes ranging from 5 to 23. In addition to this charging function, grand juries also have investigative powers and occasionally investigate suspected criminality and can issue formal charges based on the results of that investigation (Morril, 2011; Rottman & Strickland, 2006).

In *United States v. R. Enterprises* (1991:292, 299), the Supreme Court had this to say about the grand jury: "A grand jury may compel the production of evidence or the testimony of witnesses as it considers appropriate, and its operation is unrestrained by the technical procedural and evidentiary rules governing the conduct of criminal trials." This virtually unrestrained investigatory power is open to abuse both by the grand jury itself and by the prosecutor. Some critics of the grand jury argue that police and prosecutors use it to obtain evidence by subpoena that they could not secure by search warrant (Dillard, Johnson, & Lynch, 2003; Morril; 2011).

Some have criticized the grand jury as a "rubber stamp" of the prosecutor. The prosecutor in secret proceedings without a magistrate or judge to instruct the jury presents cases to the grand jury. Dillard et al. (2003:3) stated, "As a practical matter, the prosecutor calls the shots and dominates the entire grand jury process." In about 95% of cases, the grand jury issues a "true bill" or indictment, a formal document that lists the specific violation of the criminal code of which the defendant is accused. In cases in which the defendant is accused of more than one crime, each violation of the law identified in the indictment is a **count**. In only about 5% of cases does the grand jury go against the wishes of the prosecutor and issue a "no bill," or fail to indict. Thus, it appears that the grand jury does not perform its function of checking the discretion of the prosecutor. The problem with this type of analysis, however, is that it assumes that the prosecutor wants an indictment in every case presented to the grand jury, and that a 5% rejection rate is evidence of inefficiency. Critics suggest that prosecutors probably have insufficient evidence for charging in more than the 5% of cases that grand juries reject.

In contrast, the preliminary hearing occurs in open court and the defendant and his or her attorney are present and allowed to examine witnesses. The proceedings of the preliminary hearing become part of an official record, and the hearing takes place before a magistrate or judge who ensures that rules of evidence are followed. The result of both processes is the same: Either the defendant is held for trial on formal charges or the evidence is found to be insufficient to support the charges and the case is dismissed. The choice of charging process often depends on jurisdictional tradition and laws and is at the discretion of the prosecutor. In some places, it is the prosecutor's choice regarding which process to follow.

The grand jury is more under the control of the prosecutor, but it does not allow a test of evidence sufficiency at trial (i.e., how well witnesses will "hold up" under cross-examination), and it yields no evidence that can be entered into the trial record directly. The U.S. Supreme Court has consistently held that the grand jury process is not subject to the same due process requirements as other decision points. In *United States v. Williams* (1992), the court held that the prosecutor is not required to disclose exculpatory evidence to the grand jury. Exculpatory evidence is evidence that tends to establish the innocence of the accused or defendant. Normally at trial, prosecutors are bound to disclose such evidence to the defense. The Supreme Court reasoned that such issues are dealt with at trial, and requiring strict due process protections at the grand jury stage would only delay proceedings and not add to trial fairness.

The grand jury also has the advantage of not disclosing the nature of the case or evidence to the defense attorney. The preliminary hearing, however, allows the state to test the strength of its case and, because the defendant faces his or her accusers and can cross-examine witnesses, testimony presented at this hearing can be used at trial if necessary. The disadvantage to the prosecution is that it must follow rules of evidence more strictly and, therefore, the defense gets a good indication of the nature and strength of the case against the defendant. For the defendant, the information process is preferred because of the stricter rules of evidence and the chance to preview the prosecution's case.

Traditionally, the prosecutor has had broad, almost uncontrolled discretion in charging decisions. Prosecutors have the ability to "nol pros" cases, meaning they can decide not to press formal charges regardless of the available evidence. The term comes from the Latin *nolle prosequi*, which means "I do not prosecute." Many observers—especially proponents of more serious criminal treatment of domestic violence—have criticized the broad charging discretion of prosecutors. In response, some prosecutors' offices have adopted "mandatory filing" policies by which the prosecutor files charges against every suspect arrested for domestic violence. Evaluations of this policy report no substantial benefit from mandatory filing over discretionary practices (Davis, O'Sullivan, Farole, & Rempel, 2008; Peterson & Dixon, 2005). Among other things, it appears that prosecutors exercise their charging discretion so as to reduce the number of charges filed in cases that are unlikely to result in conviction.

The broad discretion of the prosecutor in deciding criminal charges raises questions about fairness. The criminal code represents "the arsenal of the prosecutor."

The prosecutor can use the criminal law as a weapon to reward, punish, and motivate the defendant. A bigoted prosecutor can level charges that are more serious against minority group members, or refuse to reduce charges in plea agreements. The prosecutor can withdraw charges or recommend sentencing leniency for defendants who cooperate with an investigation or testify for the state.

The charging discretion of the prosecutor is not only broad, but it is also almost completely free from review. The lack of controls on prosecutor charging raises important questions about the fairness of charging. Spohn and Fornango (2009) found differences between federal prosecutors regarding their use of certain procedures. Whether a federal prosecutor recommended leniency for a defendant who assisted the case depended, in large measure, on the prosecutor. Frederick and Stemen (2012) found that whereas strength of evidence was the most important factor, characteristics of both the victims and offender also impacted the prosecutor's decision making. More recently, O'Neal, Tellis, and Spohn (2015) also found that both legal and extralegal factors contribute to the prosecutors' decision to bring charges among a sample of intimate partner sexual assault cases. The grand jury or the magistrates at the initial appearance have responsibility for controlling prosecutorial charging discretion.

The popular media generally depicts the resolution of a criminal case as being the result of a jury trial characterized by calculated strategy and dignified, formal courtroom demeanor and drama. In reality, most criminal cases end at the next stage of the court process, the arraignment. This is the point at which courts ask the defendant to plead to the charges.

Arraignment

Once formally charged with a crime, the defendant is called into court to be notified of the charges against him or her, and is asked to plead to the charges. In most criminal cases, the defendant will enter a plea of guilty at this point, and avoid a trial of the case on the facts. This high percentage of guilty pleas clearly illustrates the "compromise" nature of the court process. It is most often the result of what is known as plea bargaining.

At **arraignment**, the formal charges are read and the defendant is asked, "How do you plead?" The defendant (usually it is the defense attorney who speaks at this point) can answer in one of five ways: (1) not guilty, (2) not guilty by reason of some defense (i.e., a special, affirmative defense such as insanity or self-defense), (3) guilty, (4) *nolo contendere*, or (5) the defendant can stand silent. If the defendant is silent, the judge will enter a plea of "not guilty" for the defendant.

The only plea a judge must accept at this point is "not guilty." The plea of *nolo contendere* means that the defendant does not contest the charge and will be convicted of the offense. The difference is that the conviction cannot be used against the defendant in other proceedings, especially civil actions. This can be very important, depending on the nature of the offense. It is a relatively common plea in white-collar offenses.

Suppose you drank too much at a party, and while driving home, struck and injured a pedestrian. The responding officer who investigates the accident will probably discover that you are drunk, and charge you with driving under the influence. At arraignment, you plead *nolo contendere* and stand convicted, with all that the conviction entails (such as loss of driving privileges, mandatory jail term, a fine, and the like). The pedestrian you injured files a tort suit against you for pain and suffering, alleging negligence on your part. Conviction of drunken driving would establish your blame for the civil case in itself. By pleading *nolo contendere*, the pedestrian must establish your negligence at the civil trial, without reference to the outcome of your criminal trial.

Certain affirmative defenses involve special pleadings in which the defense states the reason for asserting innocence (e.g., "not guilty by reason of insanity"). In these cases, the defense generally is not contesting the facts of the case regarding the criminal act (*actus reus*), but rather is asserting that the requisite mental state (*mens rea*, or intention to commit a crime) does not exist for the act to be a crime. This special plea gives notice to the state that the defense will use an affirmative defense, and shifts the burden of proof from the state to the defense. As we will see, the state generally is required to prove all elements of a criminal offense beyond a reasonable doubt. Failure to do so results in an acquittal. With affirmative defenses, the defense is required to raise a question about the case; for example, the defense may give evidence that the defendant might have been legally "insane" at the time of the crime (Klofas & Weisheit, 1987). Box 7.6 summarizes the common defenses available to defendants.

Conceivably, the defense could remain silent throughout an entire trial and, after the state enters all its evidence, move for a directed verdict (dismissal) because the state failed to meet the burden of proof beyond a reasonable doubt. If the state's case is indeed not strong enough, the court dismisses the charges, acquitting the defendant. With most affirmative defenses, the

---

BOX 7.6 COMMON DEFENSES TO CRIMINAL LIABILITY

Justification: Defendants admit they were responsible for their acts but argue that what they did was right under the circumstances.

Examples:

- Self-defense
- Choice of evils
- Consent.

Excuses: Defendants admit what they did was wrong but argue that they were not responsible for their acts.

Examples:

- Insanity
- Diminished capacity
- Age
- Duress
- Intoxication
- Entrapment.

Source: Samaha (2017).

defendant admits the act but denies some element crucial to the mental aspect of the crime, such as intent. The burden then falls on the defense to raise a reasonable doubt about that element, which the state must then prove (beyond a reasonable doubt) actually existed at the time of the offense.

The "not guilty" plea results in setting a trial date and subsequent moving for trial. This stage includes jury selection and the filing of pretrial motions for disclosure of evidence, suppression of evidence, and the like, in preparation for the trial itself. Although this is a typical outcome of arraignment on television and in the movies, it is very rare in the actual operations of the criminal courts. Cohen and Kyckelhahn (2010) reported that 95% of felons convicted in large jurisdictions were convicted by a guilty plea. Ostrom, Kauder, and LaFountain (2004:61) reported that in 2001, only about 3% of all criminal cases (including misdemeanors and Driving While Intoxicated [DWI] offenses) went to trial.

The "guilty" plea is an admission of the offense and removes the need for a trial. It is left to the judge's discretion whether to accept a plea of guilty, for it entails a waiver of the right to trial. In many states, it is also a waiver of the right to appeal rulings on the admissibility of the evidence and other controversies. The Federal Rules of Criminal Procedure (see Box 7.7) instruct the judge to investigate the factual nature of the plea, the voluntariness of the plea, and the defendant's awareness of the effects of the plea before accepting a guilty plea.

BOX 7.7 FEDERAL RULES OF CRIMINAL PROCEDURE, RULE 11—PLEAS

A defendant may plead not guilty, guilty, or, with the consent of the court, *nolo contendere*. The court may refuse to accept a plea of guilty, and shall not accept such plea or a plea of *nolo contendere* without first addressing the defendant personally and determining that the plea is made voluntarily with understanding of the nature of the charge and the consequences of the plea. If a defendant refuses to plead or fails to appear, the court shall enter a plea of not guilty. The court shall not enter a judgment upon a plea of guilty unless it is satisfied that there is a factual basis for the plea.

If the defendant does not plead guilty, the next stage of the court process is trial. This is the point at which the court decides about conviction. Here, the state presents the case against the defendant and the defense attempts to discredit or otherwise cast doubt on the case presented by the prosecutor. The act of pleading guilty means that the state's case will go uncontested. Although contested trials are the exception in criminal law (if not all law), it is at trial that the full strength of the value placed on individual liberty is evident.

Trial

Although not often used, the jury trial is the "balance wheel" (Neubauer, 1984:284) of the court process. It is the possibility of the jury trial that serves to ensure "justice" in the more common event of plea bargaining. As Jacob (1984:207) explained

Although critics focus on plea bargains, bench and jury trials continue to constitute an essential part of the criminal justice process. It is true that only a very small proportion of cases go to trial, and an especially small proportion go to trial by jury. But the possibility of going to trial constrains the plea bargaining process. No one has to accept a bargain that is worse than the decision that could be obtained at a jury trial. Doubtful cases can be brought to a jury or bench trial, even when the prosecutor would rather close the case with a lenient bargain.

Trials occur in less than 10% of felony cases in the United States. These trials are of two types: bench trials and jury trials. A common form of criminal trial is the **bench trial**, held before a judge sitting alone, with no jury. These trials are routine for petty offenses for which the maximum penalty does not exceed incarceration for 6 months. Defendants in more serious cases often accept a bench trial in lieu of the more costly and time-consuming jury trials. The **jury trial** involves a panel of citizens who have the task of determining the facts of the case. In these trials, the judge rules on questions of law and presides over the trial, but the jury is responsible for questions of fact, how much weight to give the testimony of a witness, and the final decision about guilt or innocence. Considerable state-by-state variation exists in the rate of jury trials relative to bench trials (Frampton, 2012). Some states use jury trials almost exclusively to determine the guilt or innocence of defendants charged with serious crimes, while other states encourage defendants to opt for a bench trial.

The right to trial by jury is rooted in our legal tradition and firmly established in the U.S. Constitution. The U.S. Supreme Court ruled that any criminal defendant facing a punishment of incarceration of more than 6 months had the right to trial by jury in the case of *Baldwin v. New York* (1970). The Court later clarified that a defendant facing several petty charges which, when combined, created the possibility for imprisonment for more than 6 months was not entitled to a jury trial (*Lewis v. United States*, 1996). Individual states may grant the right to trial by jury for those facing less serious charges if they desire, but the U.S. Constitution does not require them to do so.

Most of us envision the jury as a body of 12 members who come to a unanimous decision about whether the defendant is guilty. To some, the phrase "jury of one's peers" connotes that the members of the jury should be representative of the defendant in terms of age, sex, education, place of residence, race, and other factors. "Peers," however, has been determined to refer to fellow citizens, and thus the jury does not need to reflect the characteristics of the defendant. Moreover, the jury does not have to consist of 12 members, and does not have to reach a unanimous verdict in all states. Box 7.8 shows state requirements of jury size for felony and misdemeanor cases.

The U.S. Supreme Court has addressed the unanimity of jury verdicts and the size of the jury. Although some questions remain, the Court decided that it

BOX 7.8 STATE REQUIREMENTS FOR JURY SIZE, FELONY, AND MISDEMEANOR TRIALS

All states with capital punishment require a 12-member jury in capital cases

Nine states allow fewer than 12 members in at least some felony cases

Fewer than 12 members

Arizona	Louisiana
Arkansas	Massachusetts
Connecticut	Pennsylvania
Florida	Utah
Indiana	

Thirty-seven states authorize juries of fewer than 12 members for at least some misdemeanor cases

Alaska	Iowa	New Mexico
Arizona	Kansas	New York
Arkansas	Kentucky	North Dakota
California	Louisiana	Ohio
Colorado	Massachusetts	Oklahoma
Connecticut	Michigan	Oregon
Florida	Minnesota	Pennsylvania
Georgia	Montana	South Carolina
Hawaii	Nebraska	Texas
Idaho	Nevada	Utah
Illinois	New Hampshire	Virginia
Indiana	New Jersey	West Virginia
		Wyoming

Source: Rottman and Strickland (2006:233–237).

was permissible to use a jury of six members in a robbery trial in the case of *Williams v. Florida* (1970). The Court further decided that nonunanimous verdicts are permissible by upholding convictions based on juror votes of 9–3 (*Johnson v. Louisiana*, 1972) and 10–2 (*Apodaca v. Oregon*, 1972). In 2013, the Supreme Court declined to hear a legal challenge to nonunanimous juries. Even so, some critics argue that allowing a nonunanimous guilty verdict undermines the idea of "beyond reasonable doubt" (Riordan, 2012). The Court has ruled that cases heard by juries of fewer than six members are unconstitutional (*Ballew v. Georgia*, 1978), and that in serious criminal cases, a jury of six must render a unanimous verdict (*Burch v. Louisiana*, 1979). Because most states and the federal system require unanimous verdicts, the failure of the jury to reach consensus means there is no decision. A jury that cannot reach consensus about the verdict is a **hung jury**. In these situations, the prosecution must decide whether to attempt to retry the case in front of a new jury at a later date.

At trial, the burden of proving guilt rests with the state, and the rights of the offender are strictly protected. In the case of *In re Winship* (1970), the U.S. Supreme Court ruled that the prosecution must prove every element of the offense "beyond a reasonable doubt." In *Gideon v. Wainwright* (1963), the Court held that the state must provide an accused with defense counsel at state expense if he or she is unable to provide for his or her own defense. This case involved a convicted felon who had asked for the assistance of a defense attorney at trial and was denied. The Court ruled that the right to counsel at trial applied to felony cases. In a later case, *Argersinger v. Hamlin* (1972), the Court extended the right to any criminal prosecution in which the possible penalty included incarceration. Thus, not only does the defendant have a right to contest the state at trial, but if he or she is unable to retain an attorney, the defendant has a right to counsel provided by the state.

Although relatively rare, the full jury trial is the center of the court process. As such, the trial is what we most often think of when we consider the courts. The jury trial epitomizes our adversary system of justice, in which we expect truth to emerge from the arguments of two sides of a question. If acquitted, the defendant is set free and generally cannot be tried again for the offense.

The U.S. Constitution protects criminal defendants from double jeopardy. **Double jeopardy** occurs when a defendant faces trial or punishment more than once for a single offense. The case of Byron de la Beckwith, who was convicted of the murder of civil rights leader Medgar Evers, illustrates the point. Although some believe that the double jeopardy protection should support a defendant's right to achieve finality—to put the case behind him or her (Hickey 1995), double jeopardy applies only if the original jeopardy terminates. In the de la Beckwith case, the defendant was tried twice in 1964, with both trials resulting in hung juries. In 1969, the prosecutor entered a *nolle prosequi* order for this case but did not terminate the case. In 1990, de la Beckwith was indicted, and his subsequent trial resulted in a conviction. In reviewing the double jeopardy issues, the Mississippi Supreme Court ruled (*Beckwith v. State*, 1992) that de la Beckwith was not entitled to double jeopardy protection because his jeopardy (risk of conviction) had never been terminated.

If convicted, the case proceeds to the next decision point of the justice system: sentencing. Sentencing and punishment are the focus of Chapter 9.

THEORY AND THE COURTS

How can we explain the workings of the criminal courts? Why do some cases result in trials whereas others result in dismissing criminal charges? How is it that prosecutors pursue some cases but decline to prosecute others? What explains differences in rates or outcomes of plea bargaining? Which defendants are likely to secure pretrial release?

In this chapter (and the next), we see that the law controls or explains a great deal of court practices. The seriousness of criminal charges and the risk the defendant poses influence pretrial release, bail, charging, and conviction decisions. At the same time, the characteristics of the defendant and those of the justice system officials frequently explain at least some outcomes. At a minimum, we have two sets of theories that seem to apply. First, a legal theory suggests that changing the law will change court outcomes. Second, an "interaction" theory suggests that changing the people involved (judge, prosecutor, defense attorney, and defendant) can change outcomes. Of course, changing the structure or organization of courts will also produce change.

In Chapter 8, we examine issues in the courts. As we review these issues, we can see how they relate to "explanations" or theories of the courts. Commentators actually present theories of the courts each time they suggest causes of these issues. We need to pay particular attention to how explanations for courts compare to explanations for police or corrections. If we can find explanations that apply to all three subsystems, we can identify theories of criminal justice.

DUE PROCESS, CRIME CONTROL, AND THE COURTS

If, as we suggest, the primary purpose of the courts is to resolve conflicts, it should not be surprising that the conflict between due process and crime control comes into high relief in the criminal courts. As the core issue in the criminal justice system, the conflicting interests in liberty and order are a common theme in court procedures. Our entire system of adversary trial illustrates a "combat" between the individual and the society. The procedures and decisions of the courts reflect the difficulty of resolving this conflict. Referring to the problems in bail decisions, Maxwell summarized this conflict between due process and crime control. She wrote (1999:127), "The difficulty arises in balancing the rights of defendants on the one hand, who are presumed innocent before conviction and should not be unnecessarily detained, and ensuring future court appearances and community safety, on the other."

Those who champion individual rights over public safety tend to support practices that increase the numbers of defendants who secure pretrial release through bail or other programs. They endorse practices such as the preliminary hearing and grand jury process, in which a neutral magistrate or panel of citizens reviews the government's justification for proceeding against a citizen. The provision of defense counsel, the requirement of high evidentiary standards for conviction, and the use of juries to decide criminal matters all promise to protect the rights of individuals. On the other hand, those who wish to protect public safety and control crime tend to support preventive detention of dangerous offenders. They encourage speedy resolution of criminal cases through guilty pleas, less-than-unanimous jury verdicts, and a generally streamlined decision process in the courts.

Community courts and other specialized courts (e.g., drug courts) aim to improve the workings of the judicial component of the justice system. Dissatisfaction with current procedures underlies these reform movements. Drug courts seek to use the justice system and its coercive control to ensure that drug offenders receive the treatment necessary to prevent future crime. Community courts try to restore harmony in the community by both resolving minor disputes and attacking the causes of community conflicts. In both of these cases, we use the criminal justice system to solve individual or community problems that manifest themselves in crime, but that are not limited to criminal behavior. It remains to be seen if these efforts represent an expansion of state power over individuals (crime control), or a focusing of state power for the benefit of individuals (due process). At present, the hope is to achieve a balance between the two that best serves the interests of the individual and of the community.

Each of the criminal justice system decision points discussed in this chapter revolves around the conflict between due process and crime control. Specific practices in pretrial and trial proceedings represent efforts to achieve a balance between concerns for individual liberties and social needs for order. In later stages of the justice process (sentencing and corrections), the guilt of the defendant has been established. Although concerns about the rights and interests of offenders continue, the fact of guilt renders crime control interests more salient.

See Box 7.9 for selected court cases on criminal procedure.

BOX 7.9 SELECTED COURT CASES ON CRIMINAL PROCEDURE

Stack v. Boyle 342 U.S. 1 (1951)	The right to bail established in noncapital cases. Noted the purpose of bail is to secure appearance at later court proceedings
Gideon v. Wainwright 372 U.S. 335 (1963)	An accused must be provided with defense counsel at state expense if he or she is unable to provide for his or her own defense
Baldwin v. New York 399 U.S. 66 (1970)	Any criminal defendant facing a punishment of more than 6 months of incarceration has the right to trial by jury
In re Winship 397 U.S. 358 (1970)	In deciding the burden of proof required for a juvenile adjudication, the Court held that "the Due Process Clause protects the accused against conviction except upon proof beyond a reasonable doubt of every fact necessary to constitute the crime ..."
Williams v. Florida 399 U.S. 78 (1970)	In a robbery case it is permissible to use a jury of six members
Apodaca v. Oregon 406 U.S. 404 (1972)	Nonunanimous verdicts are permissible. Court upheld a jury's vote of 10–2
Argersinger v. Hamlin 407 U.S. 25 (1972)	A defendant has a right to counsel at trial (either hired or appointed) if the possible penalty for the offense includes incarceration
Johnson v. Louisiana 406 U.S. 356 (1972)	Nonunanimous verdicts are permissible. Court upheld a jury's vote of 9–3
United States v. Salerno 481 U.S. 739 (1987)	Defendants can be detained prior to trial without bail if the government is able to establish that the defendant is dangerous to the community

REVIEW QUESTIONS

1. Briefly describe how the courts in the United States are organized. Explain how this particular court structure developed.

2. Distinguish between lower and higher courts.

3. Identify the principal decision points in the criminal court process.

4. What is bail, and what is its constitutional purpose?

5. What is meant by preventive detention?

6. Describe two ways in which formal criminal charges are brought against defendants.

7. What are community courts?

8. Differentiate between a bench trial and a jury trial.

REFERENCES

Acker, J., & Brody, D. (2004). *Criminal procedure: A contemporary perspective* (2nd ed.). Sudbury, MA: Jones & Bartlett.

Adams, K. (1983). The effect of evidentiary factors on charge reduction. *Journal of Criminal Justice, 11*(6), 525–538.

Baughman, S.B. (2017). Costs of pretrial detention. *Boston University Law Review, 97*(1), 1–29.

Baughman, S.B. (2018). *The bail book: A comprehensive look at bail in America's criminal justice system.* New York: Cambridge University Press.

Boland, B., & Forst, B. (1985). Prosecutors don't always aim to pleas. *Federal Probation, 54*(2), 10–15.

Bradford, S. (2012). *For better or for profit: How the bail bonding industry stands in the way of fair and effective pretrial justice.* Washington, DC: Justice Policy Institute.

Bureau of Justice Statistics (1988). *Report to the nation on crime and justice* (2nd ed.). Washington, DC: U.S. Department of Justice.

Burns, R., Kinkade, P., & Leone, M. (2005). Bounty hunters: A look behind the hype. *Policing: An International Journal of Police Strategies & Management, 28*(1), 118–138.

Carlson, L. (2011). Bail schedules: A violation of judicial discretion? *Criminal Justice, 26*(1), 12–17.

Casey, P., & Rottman, D. (2003). *Problem-solving courts: Models and trends.* Washington, DC: National Center for State Courts.

Chen, H. (1991). Dropping in and dropping out: Judicial decision making in the disposition of felony arrests. *Journal of Criminal Justice, 19*(1), 1–17.

Cohen, T., & Reaves, B. (2007). *Pretrial release of felony defendants in state courts.* Washington, DC: Bureau of Justice Statistics.

Cohen, T., & Kyckelhahn, T. (2010). *Felony defendants in large urban counties, 2006.* Washington, DC: Bureau of Justice Statistics.

Davis, R., O'Sullivan, C., Farole, D., & Rempel, M. (2008). A comparison of two prosecution policies in cases of intimate partner violence: Mandatory case filing versus following the victim's lead. *Criminology and Public Policy*, *7*(4), 633–662.

Dershowitz, A.M. (1982). *The best defense*. New York: Vintage Books.

Dillard, W., Johnson, S., & Lynch, T. (2003). A grand facade: How the grand jury was captured by government. In *Policy analysis*. Washington, DC: CATO Institute (May 13).

Federal Court Management Statistics (2016). U.S. *district courts national judicial caseload profile*. Retrieved from www.uscourts.gov/statistics-reports/federal-court-management-statistics-december-2016

Feeley, M. (1979). *The process is the punishment*. New York: Russell Sage Foundation.

Frampton, T.W. (2012). The uneven bulwark: How (and why) criminal jury trial rates vary by state. *California Law Review*, *100*(1), 183–222.

Frazer, M.S. (2006). *The impact of the community court model on defendant perceptions of fairness: A case study at the Red Hook Community Justice Center*. New York: Center for Court Innovation.

Frederick, B., & Stemen, D. (2012). *The anatomy of discretion: An analysis of prosecutorial decision making*. New York: Vera Institute of Justice.

Geller, A. (2016). The process is still the punishment: Low-level arrests in the broken windows era. *Cardozo Law Review*, *37*(3), 1026–1058.

Glick, H.R. (1983). *Courts, politics and justice*. New York: McGraw-Hill.

Goldfarb, R. (1965). *Ransom: A critique of the American bail system*. New York: Harper & Row.

Goldkamp, J. (1980). *Two classes of accused*. Lexington, MA: Lexington Books.

Grant, G.A. (2018). *Criminal justice reform report to the governor and the legislature for calendar year 2017*. Trenton, NJ: New Jersey Courts.

Grommon, E., Hipple, N.K., & Ray, B. (2017). An outcome evaluation of the Indianapolis community court. *Criminal Justice Policy Review*, *28*(3), 220–237.

Henry, K., & Kralstein, D. (2011). *Community courts: The research literature*. New York: Center for Court Innovation.

Hickey, T. (1995). A double jeopardy analysis of the Medgar Evers murder case. *Journal of Criminal Justice*, *23*(1), 41–51.

Hurley, G. (2016). *The constitutionality of bond schedules*. Williamsburg, VA: National Center for State Courts.

Jackson, D.D. (1974). *Judges*. New York: Atheneum.

Jacob, H. (1984). *Justice in America*. Boston: Little, Brown.

Jansen, S., & Dague, E. (2006). Working with a neighborhood community prosecutor. *The Police Chief*, *73*(7), 40–44. Retrieved July 16, 2007, from http://policechiefmagazine.org

Johnson, B., & Stevens, R. (2013). The regulation and control of bail recovery agents: An exploratory study. *Criminal Justice Review*, *38*(2), 190–206.

Klofas, J., & Weisheit, R. (1987). Guilty but mentally ill: Reform of the insanity defense in Illinois. *Justice Quarterly*, *4*(1), 39–50.

Knox, P. & Keifer, P. (2017). *The risks and rewards of risk assessments*. Maricopa, AZ: National Center for State Courts.

Kuo, S.Y., Longmire, D., & Cuvelier, S. (2010). An empirical assessment of the process of restorative justice. *Journal of Criminal Justice, 38*(3), 318–328.

Malega, R., & Cohen, T.H. (2013). *State court organization, 2011*. Washington, DC: Bureau of Justice Statistics.

Marlowe, D.B., Hardin, C.D., & Fox, C.L. (2016). *A national report on drug courts and other problem-solving courts in the United States*. Alexandria, VA: National Drug Court Institute.

Maxwell, S. (1999). Examining the congruence between predictors of ROR and failures to appear. *Journal of Criminal Justice, 27*(2), 127–141.

Mays, G.L., & Taggart, W.A. (1986). Court clerks, court administrators and judges: Conflict in managing the courts. *Journal of Criminal Justice, 14*(1), 1–8.

Moffett, K.W., Maltzman, F., Miranda, K., & Shipan, C.R. (2016). Strategic behavior and variation in the Supreme Court's caseload over time. *Justice System Journal, 37*(1), 20–38.

Morril, G.D. (2011). Prosecutorial investigations using grand jury reports: Due process and political accountability concerns. *Columbia Journal of Law and Social Problems, 44*(4), 483–512.

Neely, R. (1983). *Why courts don't work*. New York: McGraw-Hill.

Neubauer, D.W. (1984). *America's courts and the criminal justice system*. Monterey, CA: Brooks/Cole.

Newman, D.J. (1978). *Introduction to criminal justice* (2nd ed.). Philadelphia: J.B. Lippincott.

Oleson, J.C., Lowenkamp, C.T., Wooldredge, J., VanNostrand, M., & Cadigan, T.P. (2017). The sentencing consequences of federal pretrial supervision. *Crime & Delinquency, 63*(3), 313–333.

O'Neal, E.N., Tellis, K., & Spohn, C. (2015). Prosecuting intimate partner sexual assault: Legal and extra-legal factors that influence charging decisions. *Violence Against Women, 21*(10), 1237–1258.

Ostrom, B., Kauder, N., & LaFountain, N. (2004). *Examining the work of state courts, 2003*. Williamsburg, VA: National Center for State Courts.

Peterson, R., & Dixon, J. (2005). Court oversight and conviction under mandatory and non-mandatory domestic violence case filing policies. *Criminology & Public Policy, 4*(3), 535–558.

Reaves, B.A. (2013). *Felony defendants in large urban counties, 2009 – Statistical tables*. Washington, DC: Bureau of Justice Statistics.

Riordan, K. (2012). Ten angry men: Unanimous jury verdicts in criminal trials and incorporation after *McDonald*. *Journal of Criminal Law and Criminology, 101*(4), 1403–1433.

Rottman, D. (1996). Community courts: Prospects and limits. *National Institute of Justice Journal, 231*, 46–51 (August).

Rottman, D., & Casey, P. (1999). A new role for courts? *National Institute of Justice Journal, 240*, 12–19 (August).

Rottman, D., & Strickland, S. (2006). *State court organization 2004*. Washington, DC: Bureau of Justice Statistics.

Rubin, H.T. (1984). *The courts: Fulcrum of the justice system* (2nd ed.). New York: Random House.

Sacks, M., & Ackerman, A. (2014). Bail and sentencing: Does pretrial detention lead to harsher punishments? *Criminal Justice Policy Review, 25,* 59–77.

Samaha, J. (2017). *Criminal law.* Boston: Cengage.

Schauffler, R.Y., LaFountain, R.C., Strickland, S.M., Holt, K.A., Genthon, K.J., & Allred, A.K. (2016). *Examining the work of state courts: An overview of 2015 state court caseloads.* Washington, DC: National Center for State Courts.

Spohn, C., & Fornango, R. (2009). U.S. attorneys and substantial assistance departures: Testing for interprosecutor disparity. *Criminology, 47*(3), 813–846.

Stumpf, H.P. (1988). *American judicial politics.* New York: Harcourt, Brace, Jovanovich.

Terry, W. (2000). *Opening the courts to the community: Volunteers in Wisconsin's courts.* Washington, DC: Bureau of Justice Assistance Bulletin.

Walker, S. (2015). *Sense and nonsense about crime, drugs, and communities* (8th ed.). Belmont, CA: Wadsworth.

Wiseman, S.R. (2016). Fixing bail. *George Washington Law Review, 84*(2), 417–479.

Wolf, R. (2007). *Principles of problem-solving justice,* New York: Center for Court Innovation.

IMPORTANT CASES

Apodaca v. Oregon, 406 U.S. 404 (1972).

Argersinger v. Hamlin, 407 U.S. 25 (1972).

Baldwin v. New York, 399 U.S. 66 (1970).

Ballew v. Georgia, 435 U.S. 223 (1978).

Beckwith v. State, 615 So. 2d. 1134 (1992).

Burch v. Louisiana, 441 U.S. 130 (1979).

County of Riverside v. McLaughlin, 500 U.S. 44 (1991).

Gideon v. Wainwright, 372 U.S. 335 (1963).

In re Winship, 397 U.S. 358 (1970).

Johnson v. Louisiana, 406 U.S. 356 (1972).

Lewis v. United States, 518 U.S. 322 (1996).

Stack v. Boyle, 342 U.S. 1 (1951).

United States v. R. Enterprises, 498 U.S. 292 (1991).

United States v. Salerno, 481 U.S. 739 (1987).

United States v. Williams, 504 U.S. 36 (1992).

Williams v. Florida, 399 U.S. 78 (1970).

CHAPTER 8

People and Problems in the Courts

Four sets of people play major roles in the court segment of the justice system: (1) defense attorneys, (2) prosecutors, (3) judges, and (4) jurors. These people have a direct effect on decisions in the criminal courts. With the exception of jurors, who are usually typical citizens, all are trained in the law and therefore share a common culture (Glick, 1983:2–3; Holten & Lamar, 1991:115–119). In many ways, prosecutors and defense attorneys are similar, except for the fact that they are adversaries at trial. Very often, judges are recruited from the ranks of prosecutors and (to a lesser extent) defense attorneys. As a result, judges not only share a common training and educational experience, they often share common career paths.

DEFENSE ATTORNEYS

In the field of law, criminal law is not a particularly well-respected specialty (Neubauer, 1984:156). Wice (1978) studied defense attorneys across the country and reported that defense counsel generally were solo practitioners or attorneys working in small offices (with two or three associates). They usually began their private practices late in life, and many obtained their criminal law experience and training in prosecutors' offices. Most were not graduates of the nation's best law schools, and their salaries were generally on the lower end of the earnings scale for the legal profession. As Neubauer and Fradella (2014:184) observe, "Most lawyers view criminal cases as unsavory." Of the hundreds of thousands of attorneys in the United States, relatively few will take criminal cases. The number of criminal law specialists is quite low (Mays, 2012; Neubauer & Fradella, 2014).

Of those attorneys who routinely take criminal cases, many do so on a part-time basis to supplement their earnings from a general law practice. Criminal defense services typically are provided by young, inexperienced attorneys, or by older, somewhat less successful attorneys. As Holten and Lamar (1991:124) observed, "The field (criminal defense law) does seem to attract more than its share of marginal practitioners." Among several reasons given to explain the attraction of marginal practitioners is that much of criminal defense work is mundane and uncomplicated. By requiring that defense counsel be appointed for **indigent** (i.e., poor) defendants, the burden of paying for defense is placed on the government when the defendant is poor (Box 8.1).

BOX 8.1 FACTORS USED TO DETERMINE CLIENT INDIGENT STATUS

Financial ability to pay (income, assets, employment status, etc.).

Poverty guidelines

Nature of charge

Cost of private counsel

Age

Financial ability of family members

Ability to post bond.

Source: Strong (2016).

Traditional defense services were available to a defendant only if he or she could afford to retain the services of an attorney. **Privately retained counsel** is still an option for criminal defendants today, and involves the defendant hiring his or her own attorney, who for a fee then represents the defendant at trial and in all other stages of the court process. For those unable to pay an attorney, however, the effect of the Supreme Court's decisions on the right to counsel has been the development of alternative systems for the provision of defense services. The most common are the creation of a public defender's office and the use of assigned counsel. About half of all criminal jurisdictions use a combination of indigent defense services (Strong, 2016).

Privately retained counsel, public defenders, and assigned counsel comprise the three major methods of securing representation for criminal defendants. A fourth method, **contract systems** (those in which lawyers will submit bids to serve as counsel on an annual basis), is used in relatively few counties (Strong, 2016; Wood, Goyette, & Burkhart, 2016). In terms of the volume of cases handled, the most common of these forms is the public defender's office. Larger jurisdictions (those having the heaviest caseloads) are likely to operate a public defender's office that mirrors the prosecutor's office. In many jurisdictions, the **public defender** is elected and authorized to employ a number of assistant public defenders to serve all indigent defendants. In these systems, the defense attorneys work for the municipality on salary. Young attorneys often seek these positions to develop trial experience prior to beginning their own private practices. Some older attorneys, weary of the rigors of private practice, also take positions with the public defender's office.

Langton and Farole (2009) reported on 957 public defender offices operated in 49 states and the District of Columbia in 2007. Only Maine did not report any public defender offices. Twenty-two states operated state defender offices, with the other 27 having county-level organizations. Of these systems, states and counties funded 13 jointly. In 14 states, counties alone paid for public defender offices (see Figure 8.1).

The majority of criminal jurisdictions use an **assigned counsel** system (often in conjunction with some other form of defense, e.g., a public defender), but this system serves fewer than half of all indigent criminal defendants. In jurisdictions in which criminal caseloads usually are not large enough to justify the expense of developing and operating public defenders' offices, assigned counsel are used. In this system, attorneys voluntarily enter their names to be considered for criminal defense work or, in some places, all members of the local bar are enrolled on the defenders' list. As indigents come before the court, the judge appoints an attorney from the list, moving down the list of attorney names as needed. In this system, the assigned counsel generally receives a fee based on an hourly rate, but with a maximum. The fee paid generally is less than an attorney would charge a private client (especially when all members of the bar appear on the list). Again, because of the relatively low fees available under this system, only younger, less established attorneys tend to seek out assigned counsel appointments. In the federal system, assigned counsel are referred to as **panel attorneys** because defense counsel are assigned from a list or panel of approved lawyers.

The federal criminal courts also provide legal representation for defendants who are unable to afford private counsel. The framework for indigent defense within

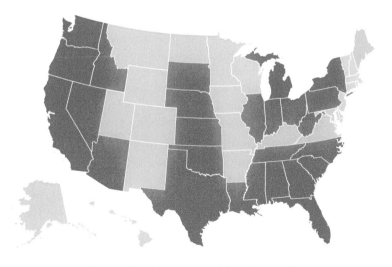

Organizational structure of public defender offices

▬ County-based offices

▬ State-based program

▬ No public defenders

FIGURE 8.1
Counties or Local Jurisdictions Funded and Administered Public Defender Offices in 27 States and the District of Columbia in 2007.

Source: Farole and Langton (2010:1).

the federal system was created with the passage of the Criminal Justice Act of 1964. This law originally relied on a form of assigned counsel (panel attorneys) to represent all indigent defendants charged with a federal crime. However, the law was amended in 1970 to create a system of federal public defender offices throughout the country. These federal public defender offices represent approximately 60% of federal defendants, with panel attorneys representing the remaining 40% (U.S. Courts, 2017). A key difference between the two systems is the flexibility to investigate cases. Unlike the federal public defenders' offices, panel attorneys must obtain permission from the trial judge for expenses related to hiring outside investigators or expert witnesses (Patton, 2017).

There has been some debate regarding which of these systems provides the best defense services to the accused. It is not unusual for each attorney employed by a county or state public defender office to represent several hundred cases per year (Benner, 2011). These large caseloads have led some scholars to suggest that public defenders are overworked and cannot provide any meaningful defense for their clients (Fairfax, 2013; Wood, Goyette, & Burkhart, 2016). However, analyses comparing privately retained counsel with publicly appointed attorneys have failed to find any significant differences in the quality of service (Hartley, Miller, & Spohn, 2010; Rattner, Turjeman, & Fishman, 2008).

Cohen (2014) found that although public defenders perform as well as privately retained attorneys, defendants represented by assigned counsel were convicted and sentenced to prison at higher rates than those with other forms of representation. Anderson and Heaton (2012) examined murder case outcomes, finding that those represented by a public defender experienced better court outcomes compared to assigned counsel. Finally, Williams (2017) found that defendants who were assigned a public defender received lower bail amounts and more nonfinancial release options than defendants who obtained privately obtained counsel. Keep in mind that most cases never go to trial, but instead most defendants plead guilty in exchange for a reduced sentence. Public defenders have regular working relationships with the prosecutors and the judge, which might allow them to negotiate more favorable sentences than privately retained attorneys.

A final option for defendants in criminal cases is to defend themselves. However, this is not an option commonly used for defendants facing serious charges (Harlow, 2000). Given the adage that "a lawyer who represents himself has a fool for a client," do defendants have the right to make fools of themselves? The U.S. Supreme Court has decided that they do. In *People v. Faretta* (1975), the defendant, Faretta, sought self-representation, but the trial judge denied his request, insisting that Faretta accept the assistance of appointed defense counsel. Faretta appealed, arguing that it was his right to represent himself at trial. The Supreme Court agreed with Faretta that self-representation (*pro se* defense) is a constitutional right of the defendant. In *McKaskle v. Wiggins* (1984), the Supreme Court ruled that states are allowed to appoint a standby counsel over the objections of a defendant who is representing themselves. The Court further clarified the right to self-representation in *Indiana v. Edwards* (2008). In this case, the defendant suffered from a mental illness and was determined to be competent to stand trial but not competent to represent himself. Upon appeal, the Supreme Court held that the defendant could not carry out the basic tasks needed to present his own defense. Therefore, the Court permitted the state to deny the defendant's request for self-representation.

Although relatively rare, *pro se* defenses do occur. Reasons that a defendant would insist on self-representation include a desire to communicate his or her motives to the jury, disagreement with his or her counsel's strategy, or the belief that he or she could convey a more genuine description of the events leading to their arrest than the appointed counsel or public defender (Cabell, 2012). Conversely, the defendant might have too much faith in the legal system and believe that his or her innocence will be clear to the jury (Homiak, 1976). Judges presiding over cases in which a defendant is representing himself must take an active role in ensuring a fair trial, including conducting competency hearings and making sure that the prosecution does not abuse the system (Willis, 2010).

PROSECUTORS

Unlike defense services, there is no privately retained prosecutor. Rather, prosecutors, usually elected officials, have the duty to provide legal counsel for the state in criminal trials. Sklansky (2017) described prosecutors as the bridge between the police and the courts, as they must simultaneously work with law enforcement professionals and judges to maintain an efficiently operating court system. Depending on the size of the jurisdiction and the size of its criminal caseload, prosecutors may be either full time or part time. Nationally, most chief prosecutors are full time, and often hire a large staff of assistant prosecutors, supervisory attorneys, victim advocates, investigators, and support staff (Perry & Banks, 2011). Depending on the size of the prosecutor's office, the prosecutor (frequently called the county attorney, state's attorney, or district attorney) may actually serve as a manager, rarely engaging in trial preparation or courtroom appearances. Table 8.1 describes the distribution of staff in prosecutors' offices in the United States.

TABLE 8.1 Personnel Categories in Prosecutor's Offices

Position	Percent of Personnel
Chief prosecutor	3
Assistant prosecutor	32
Support staff	33
Staff investigators	9
Managers/supervisors	7
Legal services (clerks/paralegals)	5
Civil attorneys	2
Victim advocates	6
Other	2
Total	100

Estimated personnel: 77,927

Source: Perry and Banks (2011).

Prosecutors are usually locally elected for a specified term, a system which was designed to keep district attorneys accountable to their communities (Ellis, 2012). In the largest jurisdictions, prosecutors work full time and hire assistants who provide the bulk of legal services for the office. These assistants are salaried employees of the municipality. The overwhelming majority of assistant prosecutor positions are full-time appointments. These positions are most attractive to young attorneys seeking trial experience and to older attorneys who do not wish to engage in private practice or who have not been able to secure a position with a larger law firm.

At the federal level, U.S. attorneys provide prosecutorial services. U.S. attorneys are presidential appointees who are empowered to employ assistants to handle federal criminal cases. In addition, the federal government (as well as some state and municipal governments) sometimes creates an Office of the Special Prosecutor, where an attorney is employed to investigate and prosecute cases arising out of some special circumstance, such as the Watergate investigation during the Nixon administration or the Russian election meddling investigation during the Trump administration. Jurisdictions hire a **special prosecutor** when there is concern about possible conflicts of interest, or when a case is so complex that the need to perform the ordinary duties of the prosecutor's office would render it impossible for that office to pursue the case adequately. In the Watergate and Russian meddling examples, because the president appoints U.S. attorneys, it might be difficult for one of them to investigate criminal allegations against the president.

Although similar in many respects, one major difference between the prosecutor's office and the public defender's office is that the prosecutor's office traditionally has been a stepping-stone for political careers. Partly because of its elective nature, and partly because the prosecutor can garner considerable media coverage by tackling certain cases (sex crimes, white-collar crime, child abuse, etc.), the office of prosecutor enables incumbents to prepare for political advancement. An attorney who is interested in a political career as a judge or other elected official often begins in the prosecutor's office (Miller, 2015). Unlike the prosecutor, the public defender generally does not gain voter support, mostly because he or she earns a living by defending the accused (who may be sex offenders, child abusers, or other unsavory characters).

JUDGES

There are approximately 30,000 state judgeships and 1,700 federal judgeships in the United States, including appellate courts and courts of limited jurisdiction (Institute for the Advancement of the American Legal System, 2017). To many attorneys, becoming a judge represents the pinnacle of a legal career (Neely, 1983). Compared to the general practice of law, being a judge is more prestigious, more secure, and less stressful. The overwhelming majority of the nation's judges achieve their offices through election. Unfortunately, judicial elections—at least as measured by voter turnout—generally do not provoke much citizen interest. Yet, the selection of judges is important because it is the judge's decisions on matters such as the admissibility of evidence and objections from the prosecutor and defense attorneys during a trial that can ultimately influence the nature of evidence presented to the jury. The judge holds the responsibility of ensuring that both the prosecutor and defense attorneys follow proper procedures to determine guilt or innocence.

Two basic methods are used to select judges in the United States: election and appointment. In the case of elections, states determine whether judicial races will be partisan or whether they must be nonpartisan in nature. In elections, judges run for office in much the same way as mayors or state legislators. In states having appointment processes, the governor or state legislature is empowered to appoint judges to office. For many years, there has been debate about how best to ensure that those selected for judicial positions have the requisite qualifications for the job (Bannon, 2016; Reddick, 2010). Generally, the debate has focused either on the election of judges or on the Missouri Plan.

In the **Missouri Plan** (so named because it originated in the state of Missouri), a judicial nominating commission rates the qualifications of candidates and identifies qualified candidates or recommends appointment to the governor (Reddick, 2002). The governor appoints someone from the list of qualified/recommended candidates. The appointed judge must then periodically face an uncontested "retention election," in which the sole question before the voters is whether the judge should be retained in office. Should the voters decide to oust the judge, the governor appoints another from a list of qualified candidates, and the process repeats. Supporters of this system of judicial selection argue that it provides the best of all worlds; that is, the system includes direct voter input in the selection of judges through the retention election, yet it ensures that only qualified candidates become judges and insulates those judges, at least somewhat, from political pressure and from partisan politics. Reddick (2010) reported that judges selected on merit were less likely to receive judicial discipline than those elected to office. Figure 8.2 shows the methods by which judges are selected in the United States.

The federal judicial system operates differently with federal judges appointed for life terms by the president of the United States. The president generally accepts nominations from the senators of the state in which the district court judge will serve, but is not required to do so. The president then nominates the candidate, whose final appointment is subject to the advice and consent of the Senate. Controversy about the Senate's role in the confirmation process has persisted for decades. Some senators tend to look closely at the nominee's ideological views to determine whether the potential judge can distinguish between law and politics

FIGURE 8.2
Methods of Judicial
Selection for State Judges.

Source: Bannon (2016).

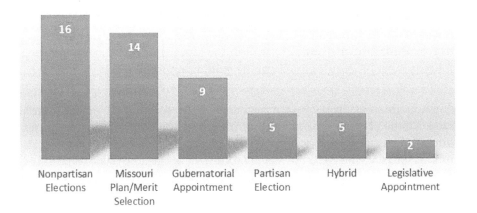

Judicial Selection

or whether the judge's political perspectives align with their own. Others take a much more deferential role, only examining whether the judge's ideologies are outside of the mainstream (Lambert, 2012). Since the 1980s, the confirmation process has seemingly become much more politically polarizing. The appointment of Supreme Court justices during the Reagan administration illustrated the hazards of Senate confirmation faced by federal judicial nominees.

In the autumn of 1987, President Reagan nominated three candidates before he was able to obtain the consent of the Senate. Seeking conservative justices who would be strict constructionists in their interpretation of constitutional issues, the president nominated Judge Robert Bork, followed by Judge Douglas Ginsburg and Judge Anthony M. Kennedy. An active senatorial committee dominated by Democrats closely investigated all three candidates. Judge Bork was found unacceptable because of his record on equal rights issues. Judge Ginsburg removed his name from consideration after acknowledging that he had experimented with marijuana in his youth (*Criminal Justice Newsletter*, 1987). The third nominee, Judge Kennedy, won Senate approval. The controversy surrounding the confirmation hearings on the appointment of Justice Clarence Thomas (President George H.W. Bush's nominee) also illustrates the confirmation process. The Senate eventually confirmed Justice Thomas, despite great controversy regarding charges of sexual harassment. President Barack Obama's nomination of Elena Kagan to the U.S. Supreme Court in 2010 led to several days of rancorous Senate hearings.

Traditionally, the Senate rules have required a 60 vote threshold to confirm federal judicial nominees. This threshold ensured that judges were not confirmed without either one political party holding a large majority in the Senate or at least some degree of bipartisan support for the nominee. In 2013, Senate Democrats voted to change the 60 vote threshold to a simple majority vote for most judicial nominees. The so-called "nuclear option" was viewed as necessary at the time to overcome opposition of then minority Senate Republicans. The exception to the procedural change was Supreme Court nominations, which were still subject to the 60 vote threshold. In 2017, Senate Republicans, who were then the majority party, voted to use the nuclear option for Supreme Court nominations to clear the way for the confirmation of Neil Gorsuch.

The principal difference between the federal and state judiciaries is the tenure of judicial office in each. Most state judges, no matter how selected, undergo periodic review and face possible removal without impeachment. Federal judges receive lifetime tenure upon appointment to the bench. Lifetime tenure insulates the judge from improper political or other influence, but it also means that it is difficult to remove incompetent or corrupt judges from office. For example, removal of federal judges requires impeachment in the U.S. Senate.

JURORS

The last important set of decision makers in the criminal courts is the jury. The **jury** is a panel of citizens, selected through the process of *voir dire* (literally, "to speak the truth"), who are charged with hearing cases and determining guilt or innocence. Although rarely used, the criminal trial jury is symbolically important as evidence that the citizenry reigns. The U.S. Constitution guarantees the right to a jury trial. Regardless of the testimony and other evidence presented at trial, the verdict of the jury is almost totally within its discretion, and it is binding on the court. Only in very rare instances can the defense attorney succeed in convincing the judge to order a directed verdict of acquittal. This occurs when, at the conclusion of presenting its case, the prosecution has failed to provide enough evidence about some important element of the offense. In such cases, the defense counsel requests the court to dismiss charges because the prosecution has failed to establish its case. Juries also play a role in sentencing. In *Apprendi v. New Jersey* (2000), the Supreme Court ruled that any aggravating factor that is used to justify a sentence above the statutory maximum must be presented to a jury and proven beyond a reasonable doubt. Additionally, most states which allow the death penalty require that the jury must agree on the death penalty before the judge can order execution.

Because of the tremendous discretionary power of the jury, prosecutors and defense attorneys are very careful in the jury selection (*voir dire*) process. Each side receives a number of **peremptory challenges**, which allow them to remove otherwise qualified jurors from service. In addition, each side may challenge any number of prospective jurors for cause. Thus, during the selection process, prospective jurors may be asked how they feel about the defendant, whether they know about the case, and other such questions. Any answer may be cause for a challenge. Attorneys use peremptory challenges when they cannot establish a just cause for keeping someone off the jury, yet they believe that the individual will not be receptive to their case.

Both the prosecution and the defense seek to seat jurors whom they believe will support their respective presentations of the case. In many ways, jury selection is the most important part of the trial process (Fried, Kaplan, & Klein, 1975). Just as judicial decisions reflect the characteristics of the judge, jury decisions reflect the characteristics of the jury (Devine, 2012; Simon, 1980). A juror's background, experience, personality type, and values all serve as a filter through which evidence and legal arguments are viewed (Frederick, 2012).

Prospective jurors usually are selected randomly from the population within a court's jurisdiction. These prospective jurors form a panel (called the **venire**) from which the attorneys select (through the *voir dire* process) members who will serve on the jury. The representativeness of the panel of jurors is a thorny issue.

To ensure that juries reflect the characteristics of the community, courts select prospective juror names from voter registration lists, driver's license records, public utility customer records, city/county directories, tax rolls, and other lists that are likely to include all potential jurors (Rottman & Strickland, 2006). The court then randomly selects members of the venire from the list. Still, this process frequently fails to produce a representative sample (Fukurai, Butler, & Krooth, 1991). Not only do juries frequently lack proportionate representation of minorities, those minorities selected for jury duty tend to come from higher social class standings than might be desired.

Jury selection seems to discriminate against minorities in general, and lower-class minorities in particular (Caprathe, Hannaford-Agor, Loquvam, & Diamond, 2016; Joy & McMunigal, 2016). Taylor, Ratcliffe, Dote, and Lawton (2007) reported that neighborhood characteristics are linked to the likelihood that those summoned for jury duty actually show up at the courthouse. People in poorer, less stable neighborhoods and those from minority neighborhoods are less likely to respond to a jury summons. The relationships are complex, but part of the explanation may be that people in these neighborhoods do not receive the summonses.

Some persons are exempt from jury duty (Rottman & Strickland, 2006). The list of exemptions varies from state to state but typically includes certain professions (attorneys, physicians, etc.), public safety and active military members, and those with health problems. Half the states do not identify any exemptions from jury duty, but the judge can choose to dismiss a prospective juror. Most states require that jurors speak English and disqualify those who do not speak the language.

During the *voir dire* process, both prosecution and defense seek to identify possible biases in the potential jurors through their responses to questions. At best, this is an imprecise process. As Judge Peter O'Connell (1988:183) stated, the process of asking a potential juror whether he or she can be impartial "is a little like asking a practicing alcoholic if he has his drinking under control; we are asking the person who has the prejudice to determine if the prejudice will affect his decision." Thus, lawyers often must be creative in their questioning of potential jurors to uncover any potential biases (Mulvaney & Little, 2015). Many lawyers have begun conducting internet searches, including the examination of social media accounts, during the *voir dire* process to research the background of potential jurors (Burke, 2014; Lundberg, 2012).

The selection of jury members is important in part because the jury is empowered to decide the ultimate fate of a case. The tradition of English and American law is that the jury can nullify a law by voting to acquit even if the evidence supports conviction. **Jury nullification** occurs when jurors refuse to convict a guilty defendant because of their belief that the law or the government's use of the law in that case is unjust (Fissell, 2013). Controversy about jury nullification has escalated in recent years, especially as crime control policies such as the "war on drugs" are seen to have disproportionate effects on minority group members. Some observers cite the highly publicized murder trial of retired football star O.J. Simpson as a case in which the jury ignored the facts and thereby nullified the law (Zatz, 2000). In fact, several jurors believed that the Los Angeles police planted the evidence against Simpson. In their view, the police "framed" Simpson and the evidence was not believable.

Although evidence suggests that jury nullification does occur, its existence represents a contradiction in American law. The Supreme Court has not endorsed the practice and has instead ruled that juries should apply the law as it is presented in each case. At the same time, the judicial process allows juries to meet in private and provides practically no legal recourse against juries which decide to nullify the law (Duvall, 2012). Some scholars have debated the merits of formally recognizing the power of the jury to nullify the law. One way to achieve this would be allowing defense attorneys to inform the jury of its power during the trial. Royer (2017) even suggested that defendants could be allowed to enter a "nullification" plea, which would essentially admit that the defendant has violated the law but argue that the law is being unjustly applied and the defendant should be spared from punishment for their actions.

ISSUES IN THE CRIMINAL COURTS

As with every stage of the justice system, the court process is plagued with problems and issues that defy easy resolution. We touched on some issues earlier in this chapter, such as the best mechanism for the provision of defense counsel, the effects of judicial selection, and the process of jury selection. We will assess five additional issues in this section: (1) speedy trial, (2) the courtroom work group, (3) the role of the press in the trial process, (4) victim services, and (5) plea bargaining. With severe overcrowding in the nation's jails and hundreds of persons accused of terrorist connections or plans, the right to a speedy trial is again an important practical and constitutional issue. Court delay contributes to jail crowding as defendants wait for hearings and trials. There is also a debate about whether suspects arrested after the September 11, 2001, terrorist attacks and as a result of ongoing antiterror investigations should be tried in criminal or military courts, and how long the government can hold these persons without trial (McCarthy, 2017). The role of the press in trials is also still unresolved. This issue, which was particularly important during the political trials of the student antiwar and civil rights activists in the early 1970s, has resurfaced and again gained prominence with the increased use of social media. Additional issues involve the impact of technology and political influence on court operations.

Speedy Trial

We most often consider the issue embodied in the phrase "speedy trial" as court delay. Speedy trial is the solution to the problem of what some perceive as unacceptable delays in the processing of criminal cases from arrest to disposition. Supporters of speedy trial argue one of three positions: (1) speedy trials protect the rights of the defendant; (2) speedy trials are fairer to crime victims; or (3) speedy trials assist crime control (Bureau of Justice Statistics, 1986; Thompson, 2011).

The Sixth Amendment to the U.S. Constitution guarantees defendants the right to a speedy and public trial. Lengthy delays between arrest and trial (or between charging and trial) can be to the disadvantage of criminal defendants. Especially in those cases in which the defendant is unable to make bail, denial of a speedy trial amounts to incarceration prior to conviction. By virtue of their inability to obtain pretrial release, defendants awaiting trial for long periods of time are likely to suffer further losses, such as loss of income (if not loss of job), possible loss of residence, separation from family and friends, and, upon conviction, higher rates of incarceration.

The determination of when the right to a speedy trial has been violated is rarely a matter of simple objective fact. In *Barker v. Wingo* (1972), the Supreme Court spelled out four factors for courts to weigh in determining if a defendant's constitutional right to a speedy trial has been denied. The length of the delay is the most important consideration, but it must be judged in light of the reasons for the delay. Deliberate attempts to delay by the government weigh heavily in favor of the defendant. Certain reasons, such as the absence of a key witness, are considered valid. The court must also determine if the defendant asserted his or her rights to a speedy trial and if the delay prejudiced the case against the defendant. Federal criminal proceedings are guided by the **Speedy Trial Act** of 1974. This federal law outlined specific time limits for which the indictment and trial must begin after an arrest. Charges are dismissed if the time limits are exceeded, unless an approved delay is requested in compliance with the provisions of the Speedy Trial Act. These delays are common, which has limited the actual effects of the Speedy Trial Act in terms of processing criminal actions (Hopwood, 2014). Many states have also enacted similar laws which provide procedural guidance ensuring defendants the right to a speedy trial.

Delays in the prosecution of defendants arrested as terror suspects have raised a number of issues about the right to a speedy trial. An example is the case of José Padilla, an American citizen arrested in Chicago and charged with terrorism. Police arrested Padilla as an "enemy combatant" and held him for more than 2 years without formal criminal charges. In 2004, the U.S. Supreme Court avoided ruling in the case when it noted that Padilla's *habeas corpus* appeal against Defense Secretary Donald Rumsfeld was mistaken and needed to be filed against the commander of the military prison in which he was housed (*Rumsfeld v. Padilla*, 2004). The courts have generally ruled that a period of military detention does not start the clock on a defendant's right to a speedy trial (Kuhn, 2012).

The Obama administration decided to proceed with criminal trials of the detainees from Guantanamo Bay. They planned to hold trials in U.S. District Court in New York City. In the first case to come forward, Ahmed Ghailani was acquitted on all but one charge and received a life imprisonment sentence for the single count for which he was convicted. In *United States v. Ghailani* (2010), the Second Circuit Court of Appeals denied a claim by the defendant that his lengthy detention prior to trial violated the Sixth Amendment. The court held that his detention as an enemy combatant was not a criminal punishment and that the detention did not justify a dismissal of criminal charges. Following this case, the government decided to use military commissions to try future terror-related cases. As of 2018, most of the detainees at Guantanamo Bay had either been released to other countries or are still awaiting criminal processing. The extended delays for terror-related defendants have drawn the ire of many civil liberties advocates. For example, O'Rourke (2014) expressed fear that the *Ghailani* decision could set a dangerous precedent in which the government decides to indefinitely postpone trials of defendants who are known to possess sensitive information (e.g., Edward Snowden) to prevent the public release of that information.

Many people believe that the slow disposition of criminal cases negatively impacts crime victims. Lengthy court proceedings, especially those in which the victim is involved as a witness, result in lost work days, mental anguish while awaiting resolution of the case, delays in recouping stolen property, and other costs to the victim. Although speedy trials act to the advantage of the accused (especially the innocent accused), it has been suggested that the crime victim has a stake in the early resolution of a criminal case (Walker, 2015). Speedy trial, some argue, provides quick closure to crime victims and minimizes the inconvenience of cooperating with the prosecution or simply watching for the outcome of the criminal process.

Yet another reason for speedy trials is the desire to control crime. Critics blame pretrial delays for high rates of plea bargaining. They blame plea bargaining, in turn, for the lack of deterrent effect of the criminal law. A more quantifi-able effect of case-processing delays on crime, however, may be the amount of crime committed by criminal defendants who are free on bail and at large in the community. Reaves (2013:21) reported that police rearrested 16% of felons granted pretrial release while awaiting trial. The longer a defendant is at lib-erty on bail, the greater the likelihood that he or she will commit a new crime while awaiting trial. Speedier trials and convictions (especially those resulting in incarceration) would reduce at least this one aspect of crime (Walker, 2015). Finally, the longer the period is between when a crime occurs and trial, the more difficult it is to secure a conviction. Witnesses sometimes move or die, memories fail, outrage at the offense lessens, and general interest in the case wanes. Indeed, defense attorneys often use delay strategies to win cases; that is, they seek repeated continuances in the hope that witnesses eventually will fail to appear.

The National Center for State Courts (Van Duizend, Steelman, & Suskin, 2011) released a model time frame for criminal prosecutions which was proposed as

TABLE 8.2 Time from Arrest to Adjudication for Felony Defendants in the Largest 75 Counties

NCSC Model Standard	75% within 3 months	90% within 6 months	98% within 1 year
Actual Percent of Cases Adjudicated within—			
Murder	6%	11%	33%
Rape	17	36	72
Robbery	64	57	74
Assault	39	63	81
Burglary	47	69	82
Larceny/theft	48	69	84
Motor vehicle theft	57	78	88
Drug trafficking	41	64	79
Weapons	43	66	81

Source: Reaves (2013); Van Duizend, Steelman, and Suskin (2011).

realistic goals for states to work toward achieving (Table 8.2). In spite of the relatively widespread support for speedy trial, it is not easily realized. Speedy trials are neither in the best interest of the guilty defendant, nor always in the best interest of the prosecutor. Especially when the defendant is incarcerated and awaiting trial, the length of time between arrest and trial serves as an inducement to the defendant to engage in plea bargaining, thereby assuring conviction.

The most promising development in reducing delay involves changes in **prosecutorial case management** (Jacoby, Ratledge, & Gramckow, 1992). This involves policies and procedures to expedite case processing, including moves toward early screening of cases to plan for later processing; assignment of cases to tracks that anticipate trial, plea, diversion, and so on; and continuous monitoring of case progress (Cooper, 1994). Jurisdictions implementing such case management systems have experienced dramatic reductions in case-processing time. In general, these changes reflect recognition of "typical" cases so that prosecutors have better guidance concerning how to proceed and what to expect in a given case. Terry Baumer (2007) reported on an evaluation of an expedited case screening process that provided case reviews "24/7" and resulted in significant reductions in detention time and quicker releases of cases being discharged. Steelman, Goerdt, and McMillan (2000) described **differentiated case management** programs in operation in the United States. In these programs, prosecutors select certain types of cases that can proceed quickly through the court system. These simple cases are distinguished from those cases that have contested facts or involve complex legal issues. Rather than treating each case as unique, screening classifies cases into types that typically receive different treatments (e.g., diversion, plea bargaining, trial). Based on this, prosecutors can know what to expect in each case and better plan and manage the total caseload.

Many prosecutors' offices operate diversion programs or deferred prosecution for first offenders. Diversion programs quickly identify and refer these offenders to community-based social services, job training, or community service. Diversion programs are particularly useful for offenders suffering from mental illness or drug addiction (Gill & Murphy, 2017; Sung, 2006). The goal is for the treatment to reduce future criminal behavior, while saving valuable time and money within the traditional criminal justice system. Most often, offenders are either not charged with a crime or the crime is expunged upon successful completion of the treatment.

Prosecutors have also adopted **vertical prosecution** for specific types of cases. In large prosecutors' offices, the tradition was to assign different assistant prosecutors to different stages of the court process. Thus, one assistant would handle bail hearings, another charging, yet another arraignment, and so forth. No single prosecutor followed a case from the initial appearance up through sentencing. Vertical prosecution, on the other hand, refers to the practice of assigning the responsibility of a case to a single prosecutor who then follows that same case throughout the entire court process. Some offices use this management technique for specific types of crimes such as sexual assault, drug offenses, homicide, and child abuse. One potential benefit of vertical prosecution is that with specialization, prosecutors develop greater consistency in their handling of cases so that defendants and victims receive fairer treatment (Kingsnorth, Macintosh, & Wentworth, 1999).

The Courtroom Work Group

Eisenstein and Jacob (1977) and other observers of the criminal courts have identified what has come to be called the **courtroom work group**. This designation refers to the "regular" players in U.S. criminal courts: judges, prosecutors, and defense attorneys. This group, through close working relationships that develop over time, comes to a shared understanding of what are appropriate and inappropriate court procedures and outcomes. The shared norms and definitions among this informal group have real consequences for case outcomes and attempts at court reform. The work group often determines the "going rate" for crimes. For example, the group may define a suitable penalty for breaking and entering as a 6-month jail sentence. Having this shared understanding, all parties are likely to agree to a negotiated plea in which a defendant accused of breaking and entering receives a 6-month jail term (e.g., criminal trespass).

Because all members of the group are responsible for the orderly processing of cases, the development of shared understandings is useful to each of them. Rather than treat every case as unique, it becomes possible to classify cases as typical (i.e., deserving the going rate) or unusual (i.e., requiring alternative processing). Once they develop this informal understanding, it is exceptionally difficult to change. Myers and Reid (1995) studied efforts to change sentencing outcomes in three Florida county courts. They discovered that sentences imposed on defendants differed across counties, but were quite similar within each county. Thus, they concluded that prosecutors and defense attorneys, as well as judges, were important to efforts to change sentencing. Their research confirms that the courtroom work group is alive and important in those courts.

Harris and Jesilow (2000) studied the implementation of the three-strikes sentencing law in California. The law passed as a public initiative, with more than 80% of state voters supporting the law. It requires prosecutors to charge defendants who have prior convictions, and it enhances sentences for those who have prior convictions, so that each prior conviction (strike) results in a longer prison term. It also means that what would be misdemeanor offenses for persons with no prior convictions are sometimes elevated to felonies if the defendant has prior "strikes." In general, Harris and Jesilow found that the effect of the three-strikes law was to destabilize the courtroom work group. The law greatly enhances the importance of the prosecutor's charging decisions (to allege prior strikes), but its popularity with voters prevents prosecutors from avoiding the law in many cases. Harris and Jesilow found some evidence that the work group had already adjusted to reduce the impact of the law, but that new norms for plea negotiations had not yet emerged.

The concept of the courtroom work group assumes that every member of the group is working toward the same goal of processing the cases as efficiently as possible. It is important to recognize that this does not always occur. Metcalfe (2016) reported on the court processing of one large county in Florida which had experienced some conflict between the prosecutor and public defenders' offices. In this setting, increased familiarity between the defense attorney and the other members of the courtroom group actually interfered with the plea process and resulted in an increased likelihood of a case going to trial. One judge from that county reflected on the impact this has on the court by saying "if they're not able to come to a negotiation on any of their cases, all of sudden we have a bunch of cases that have to go to trial and frankly the system can't handle everything being tried" (Metcalfe, 2016:663).

Recent courtroom innovations might be changing the power structures within the work group as well as the process of how the "going rate" is determined. Rudes and Portillo (2012) reported on the work group interactions among the growing number of problem-solving courts that have formed across the country. They found that probation officers maintained a great deal of power in these settings due to their close knowledge of the offender's daily living situation and specialized knowledge of the available treatment options. Rudes and Portillo concluded that probation officers may, in fact, possess the greatest influence of all the members of the work group in this setting. Finally, the development of risk assessment tools might be fundamentally altering the way in which the "going rate" is calculated. Examining juvenile detention decisions in five counties, Maloney and Miller (2015) found that risk assessment tools had decreased the reliance of informal decision making and closed the decision disparities across different court systems in the same state.

Free Press and Fair Trial

The circumstance of media coverage compromising the trial process, although relatively rare, is an important concern (Duncan, 2008; Loges & Bruschke, 2005). This deals with a balance of rights—the right of the defendant to a fair and impartial hearing, and the right of the press to report on operations of the government (or, as might be stated by members of the news media, the public's "right to know"). Press coverage of criminal trials may affect court processes in two significant ways. First, the effects of pretrial publicity may make it difficult, if not impossible, to select a jury not exposed to prejudicial coverage of the case (Meringolo, 2010). This rationale applied in two famous trials within the past decade: the trial of Jerry Sandusky for the sexual molestation of minors while Sandusky was the defensive coordinator for the Pennsylvania State University

FIGURE 8.3
Orlando police officers keep members of the media and spectators away from the entrance to a restaurant across the street from the Orange County Courthouse, while members of Casey Anthony's defense team and staff were inside.

AP Photo/Phelan M. Ebenhack

football team and the trial of Casey Anthony, the individual accused and ultimately acquitted of killing her 2-year-old daughter in Florida. Second, in cases in which members of the press are allowed into the courtroom, it is possible that the entire trial will take on a circuslike atmosphere. Not only will the behavior of reporters disrupt the court proceedings, the very presence of reporters may alter the behavior of the judge, prosecutor, defendant, jury, and others (Giglio, 1982; Hannaford-Agor, 2008). See Figure 8.3.

The 2013 trial of George Zimmerman for the shooting of Trayvon Martin demonstrates how the media can have substantial effects on the criminal justice process. The case garnered national attention and created intense pressure on the police and prosecutors to make an arrest. After weeks of nonstop media attention, a resignation from the city's police chief, and the appointment of a special prosecutor, Zimmerman was arrested and charged for the killing of Martin. By that time, Zimmerman had been portrayed as a racist, paranoid vigilante who was the aggressor against an unarmed innocent teenager who had been the victim of racial profiling (Aizenman, 2012). Though Zimmerman was ultimately acquitted, the pretrial media attention had created many potential issues for Zimmerman's defense team during the trial. Further, Zimmerman's subsequent encounters with the police have received far more media attention than the same crimes would have received if committed by someone less notorious.

Aside from specific cases, the attention of the media does appear to have effects on the operation of courts. Surette (1999) examined the effect of a well-publicized case on the overall work of courts. Comparing cases processed before and after a highly publicized case of child molestation in a daycare facility, he found that court practices changed in response to the attention. The basic change was that more molestation cases went to court. It appeared that one effect of media exposure was to make law enforcement and prosecutors more sensitive to those cases. Prosecutors were less likely to dismiss charges. Of course, this meant that "weaker" cases came to trial, resulting in acquittal of those defendants. Other than increasing the number of such cases brought to court, the effects of the publicized case seemed to cancel each other out (more cases, but more acquittals). Loges and Bruschke (2005) reported that publicity seemed to make no difference in terms of trial outcomes (i.e., convictions), but high-profile cases seem to result in stiffer sentences. What is important is that the media seems to influence prosecutors (and other justice system actors).

Because this issue involves two constitutional guarantees, the right to a fair trial and the right to a free press, the courts generally have adopted a "totality of circumstances" test, wherein they attempt to balance the rights of the media to cover trials against the rights of the state and the accused to proceed in an orderly fashion. The issue becomes important only in those cases that receive the lion's share of media attention—those cases that involve particularly grue-some offenses or are otherwise newsworthy (involving well-known victims or defendants). Surette (2015) concluded that most "media trials" focus on spe-cific types of themes that are typical of entertainment media. These types of trials dominate news coverage of the courts and, Surette argued, distort public understanding of the courtroom. He noted (2015:128), "in these productions, personalities, personal relationships, physical appearances, and idiosyncrasies are commented on regardless of legal relevance." These extralegal factors are most often excluded from entering the trial and thus should not be known by the jury. At best, this creates a situation where the public might have a consider-ably different opinion of the eventual verdict based on their analysis of factors that the jury did not even consider in their deliberations. At worst, these factors may become known to the jury through the use of the media and used to help determine the defendant's guilt.

It is instructive to compare the American method of handling news media cov-erage of criminal trials with that of England (Brandonwood, 2000). English law forbids the press from reporting on pending criminal cases, under penalty of contempt of court. Once a criminal case has been opened, the British press is banned from commenting on it. Violation of this ban can (and usually does) result in jailing the offending reporter for contempt of court. In the United States, such a ban on reporting would violate the First Amendment's prohibi-tion against "prior restraint" (censoring free speech before it occurs). Rather,

U.S. courts have relied on posttrial remedies, such as declaring convictions invalid and ordering new trials, as in *Sheppard v. Maxwell* (1966) and *Estes v. Texas* (1965).

The strong American value placed on individual rights, such as free speech, and the nation's commitment to the sovereignty of the citizen prevent us from banning press coverage of trials. The U.S. Supreme Court has ruled that criminal trials must remain public (*Richmond Newspapers, Inc. v. Virginia*, 1980). The other side of the dilemma is the commitment to fairness and recognition that media coverage of criminal cases can bias juries and result in unfair verdicts. The American solution of posttrial remedies and the British use of prior restraint reflect the different values of the two societies. As Giglio (1982:349) stated, "The British emphasis, therefore, is on justice being done, rather than witnessed." Conversely, the American emphasis is on justice being witnessed, to ensure that justice is done.

The U.S. Supreme Court has ruled that cameras in the courtroom do not inherently violate a defendant's right to a fair trial (*Chandler v. Florida*, 1981). However, this ruling does not permit unlimited media access to the court (Campbell, Green, Hance, & Larson, 2017). Instead, the trial judge often has discretion to decide in which parts of the court proceeding, if any, that cameras or audio recorders are allowed to be present. Cameras are generally not allowed to show the jurors or be present at any closed session that is not open to the general public. Further, cameras are usually not permitted at federal trial courts and have never been allowed to record the U.S. Supreme Court, though the Court recently began to allow audio recording of the oral arguments (Papandrea, 2012).

Just as many courts became comfortable with cameras in the courtroom, the emergence of social media created a new set of dilemmas (see Box 8.2). The most serious concern involves the jurors' potential access to information which may not be admissible and presented as evidence in court. Indeed, even jurors who actively try to avoid receiving additional details about the case might not be able to control what they are exposed to when they log onto their social media accounts. Likewise, the long-standing (but relatively rarely used) practice

---

BOX 8.2 POLICY DILEMMA: TWEETS FROM THE COURTROOM?

Journalists are increasingly using social media formats such as Twitter to provide the public with live updates from highly publicized legal cases. These can be especially informative in cases where cameras are not allowed in the courtroom. However, courts have been divided regarding whether to allow these live social media updates. Winnick (2014) describes several concerns surrounding the use of Twitter in the courtroom. Among these is the fear that jurors may become prejudiced by unintentionally viewing these posts. Another possible concern is that witnesses who have been barred from the courtroom until their testimony may be exposed to information about other witnesses' testimony. How much real-time information should the public be exposed to during a criminal trial?

of sequestering the jury does not ensure that jurors can avoid being exposed to this information. Simpler (2012) suggested that courts should repeatedly instruct the jurors throughout the process about the expectation to not communicate with anyone about the case, explain to jurors the legal reasoning behind these rules, and remind them of the possible punishments associated with violating these rules.

The question of fair trial versus free press is complex because the coverage of criminal cases by the media, although sometimes misleading (e.g., by presenting the jury trial as the norm), contributes to better understanding of criminal justice processes for the citizenry. Media coverage of criminal cases supports the ultimate accountability of justice system officers to the public. On the other hand, we cannot ignore the possible biasing effects of media commentary about criminal cases. In comparing American and British practices, Giglio concluded that one result of freer press coverage of criminal cases in the United States is that American citizens have a better understanding of their system of justice than do citizens of Britain. The inherent difficulty in balancing the good against the bad of media coverage of criminal cases ensures that this issue will remain relevant for some time to come.

Victim Services

For at least the past 100 years, the role of average citizens in the court process has been shrinking. Citizen involvement in the courts has been limited to jury service and service as witnesses. Citizens have often tried to avoid jury duty, and the crime victim had a role only as a witness. In recent decades, however, there has been an effort to expand the role of citizens in the courts and to increase the voice of victims in court proceedings. The reasons for increasing citizen involvement in the courts are both practical and philosophical.

On the practical side, lack of citizen cooperation hinders the orderly process of court cases. When victims and other witnesses refuse to testify, it is impossible for agents of the justice system to gain convictions of guilty offenders. A perception of increased witness and victim intimidation, especially in relation to crimes associated with gang drug activity (McDonough, 2013) has spurred efforts to improve witness and juror protection. This can prove to be difficult in the digital media age that we now live in where social media accounts are dedicated to identifying and retaliating against witnesses who testify in criminal trials (Browning, 2014). Responding to this problem, courts developed a number of solutions to victim and witness intimidation. Courts have implemented emergency relocation and support, requests for higher bail or denial of release, increased pretrial and courtroom security, protective custody for victims and witnesses, banning of cell phones and other electronic devices from the courtroom, and community outreach programs seeking to better educate the public and gain the cooperation of community service organizations in assisting victims and witnesses.

The voice of the victim has been amplified in criminal court proceedings throughout the past several decades. Prosecutors often consult with crime victims before accepting a negotiated plea (Dawson, Smith, & DeFrances, 1993), and most jurisdictions have provisions for victim input into the sentencing decision. Judges, too, will usually thank citizens who serve on juries and recognize their contribution to the workings of the court. One reason for this concern with citizens' experiences in courts, of course, is that most prosecutors and judges are elected officials and thus need to serve the citizenry. A more important reason, however, is the growing recognition that victims, witnesses, and jurors deserve better treatment than they have traditionally received.

In response to this shift in our thinking about what kinds of consideration are owed to victims, witnesses, and jurors, courts have improved the jury selection process, streamlined jury service terms, and improved the physical facilities in which jurors convene. Similar changes have improved conditions for witnesses and victims. In the case of victims, legislation in most states requires the courts (generally the prosecutor's office) to provide services to crime victims. These services include notification about case progress; providing help with and information about restitution, compensation, and victim impact statements; and offering an orientation to the court process. Some jurisdictions require the prosecutor's office to provide escort services for victims and witnesses and to assist with referral to community services, the provision of counseling, and the return of property held as evidence. Box 8.3 describes the rights available to victims of federal crimes.

BOX 8.3 VICTIMS' RIGHTS IN FEDERAL CRIMINAL CASES

1. The right to be reasonably protected from the accused.

2. The right to reasonable, accurate, and timely notice of any public court proceeding or any parole proceeding involving the crime, or of any release or escape of the accused.

3. The right not to be excluded from any such public court proceeding, unless the court, after receiving clear and convincing evidence, determines that testimony by the victim would be materially altered if the victim heard other testimony at that proceeding.

4. The right to be reasonably heard at any public proceeding in the district court involving release, plea, sentencing, or any parole proceeding.

5. The reasonable right to confer with the attorney for the government in the case.

6. The right to full and timely restitution as provided in law.

7. The right to proceedings free from unreasonable delay.

8. The right to be treated with fairness and with respect for the victim's dignity and privacy.

Source: Office for Victims of Crime, www.ovc.gov/rights/legislation.html

Plea Bargaining

As shown in Table 8.3, the overwhelming majority of criminal convictions come as a result of a negotiated plea of guilty. The Bureau of Justice Statistics (Cohen & Kyckelhahn, 2010) reported that 95% of felony convictions resulted from guilty pleas. **Plea bargaining** refers to the process in which the state (through the prosecutor) and the defendant negotiate the terms under which the defendant will enter a plea of guilty. The importance of plea bargaining in the court process cannot be denied. There are divided opinions, however, as to the appropriateness of pleading guilty for considerations (Abrams, 2011; Bar-Gill & Ben-Shahar, 2009).

TABLE 8.3 Adjudication Outcomes for Felony Defendants

Outcome	Percentage
Plea guilty to a felony	53
Dismissed	25
Plea guilty for a misdemeanor	11
Convicted at trial	2
Acquitted at trial	1
Other outcome (diversion, deferred adjudication)	9

Source: Reaves (2013).

On the one hand, there are those who argue that plea bargaining violates the rights of the accused. Some believe the incentives to plead guilty are too enticing, especially in cases in which possible penalties are very severe, or in which the defendant is in jail awaiting trial. In most jurisdictions, by pleading guilty, the defendant has not only waived his or her right to a jury trial, but also to appeal questions of evidence admissibility. On the other hand, there are those who suggest that plea negotiations are unfair to the state and to the victim. These critics argue that plea bargains serve to let the guilty off without sufficient punishment. They contend that negotiated guilty pleas serve to lessen the deterrent effect of the law because offenders avoid their due penalties. Further, the idea that justice is open to negotiation is repugnant to the critics of plea bargaining.

Plea negotiations, however, do have some supporters. Some observers argue that the criminal justice system could not possibly provide the number of trials required if every criminal defendant were to demand one. The justice system would collapse under the burden of so many trials. The plea bargain results in a sure conviction of the offender, whereas a jury trial may not, and the defendant who pleads guilty is entitled to a sentence concession as a result of his or her honesty in admitting guilt, his or her demonstration of remorse in pleading guilty, or his or her cooperation with the state (Champion, 1987). Deciding the case of *Santobello v. New York* (1966), the U.S. Supreme Court declared that plea bargaining, when properly carried out, is something to be encouraged.

There are several competing views of the reality and purposes of plea bargaining. Some believe that plea bargaining is the effect of better representation of defendants. Feeley (1982) argued that plea bargaining is a product of increased adversariness. Prior to the twentieth century and the court reforms that added protections to defendants, trials were perfunctory. In an ideal world, plea bargaining occurs because the defendant now has an advocate, someone who can negotiate for him or her. In reality, defense attorneys are overworked and often fail to meaningfully negotiate for their client. The Supreme Court has ruled that plea bargaining is a critical point of the criminal justice process, and a point at which the defendant is entitled to effective counsel. In *Missouri v. Frye* (2012), the Court reversed the sentence of a man convicted of driving with a revoked license after his defense attorney failed to disclose the presence of the prosecution's plea bargain offer. The Court also reversed the sentencing of a man who had rejected a plea bargain offer after his defense attorney convinced him that the prosecution could not prove their case (*Lafler v. Cooper*, 2012). Even though the man was ultimately found guilty at the conclusion of a fair trial, the Court ruled that the trial itself was the result of ineffective counsel.

The National Association of Criminal Defense Lawyers (NACDL) (Boruchowitz, Brink, & Dimino, 2009) criticized the process of misdemeanor courts that encouraged high rates of guilty pleas. They reported that in many misdemeanor court systems up to 70% of cases were resolved at the initial appearance. Prosecutors offer limited-time plea deals to encourage defendants to waive their right to counsel and to plead guilty immediately. Even those cases that proceed beyond the initial appearance, the NACDL argues, most often result in guilty pleas. A lack of time and resources, coupled with low compensation to defense counsel, encourages plea bargaining.

Whatever else we may say about plea bargaining, it appears to be a functional practice. Again, the concept of a courtroom work group is important. Whether or not pleas of guilty result from formal negotiations between the prosecution and defense, the negotiations will continue, as will the pleas of guilty. Failure to recognize the informal structure of the courthouse—and the professional relationships that exist among the prosecutors, judges, and defense attorneys— leads reformers to believe that plea negotiations are a form of uncontrolled discretion. Instead, given that the purpose of the court is to resolve disputes, it is only natural that a certain amount of compromise develops among the actors that engage in dispute resolution daily. Indeed, the history of plea bargaining indicates that it is a useful and traditional practice (Sanborn, 1986). Plea bargaining is a good example of the effect of people on the justice system. In assessing the operations of the justice process, it is important to remember that people make up the system and the system operates through people (Champion, 1989).

Additional Issues in the Courts

The requirement that a defendant cannot be formally punished for a crime before they are found to be guilty beyond a reasonable doubt is thought of as a safeguard against levying punishment upon an innocent person. As we have described, every citizen is promised a fair process (whether they plea bargain or go to trial) to determine their guilt or innocence to a crime for which they have been charged. Impartiality is a fundamental tenet of the court system. In fact, the popular image of Lady Justice depicts a judicial system that applies the law fairly and equally. However, recent developments have caused some scholars to question whether the reality of the court process holds up to these promises.

The Supreme Court's decision in *Citizens United v. FEC* (2010) allowed corporations more flexibility in spending money to influence political campaigns. While most of the attention related to this decision has been dedicated to high-profile presidential and congressional races, the decision has also had an impact on the election of state supreme court seats. Bannon, Lisk, and Hardin (2017) reported that over 40% of spending in recent state supreme court elections has come from outside special interest groups, which frequently run advertisements criticizing a particular candidate's ruling in a past case or their stance on any number of other issues. Recent evidence (Kang & Shephard, 2015, 2016) suggests that party loyalty and/or campaign contributions might impact how these judges vote on key issues, renewing the attention given to find the most appropriate method to select judges that will reduce judicial partisanship.

We previously described that state judges are typically either elected by popular vote or appointed into their positions by governors, often with the assistance of a judicial nominating committee (i.e., Missouri Plan). Unfortunately, it seems that each of these selection methods has the potential for political influence. Lim and Snyder (2015) found that voters who select judges in partisan elections (in which the judge's party affiliation appears on the ballot) most often simply vote for the judge representing their favored political party regardless of the candidate's qualifications. Many states choose judges in elections without the candidate's political affiliation being placed on the ballot, but these elections attract the large levels of spending by special interest groups described above (Hall, 2016). Lim and Snyder point out that most voters simply do not take the time to follow judicial elections enough to make an informed decision about judicial qualifications. Voters are then left to choose a judge based on the candidate's political affiliation, information obtained by advertisements paid for by special interest groups, or perhaps simply which candidate is listed first alphabetically on the ballot.

Many states have responded to these criticisms by switching to a process of filling state supreme court seats by gubernatorial appointments. McLeod (2012) found that while governors typically will still select a judge who politically aligns with his or her judicial philosophy, this can be mitigated somewhat by the use of a judicial nominating committee or the requirement of a supermajority confirmation vote by the state's legislature. Still, the political influence of judicial nominations persists.

A second development affecting the operations of criminal courts consists of the availability of new technologies and the improvement of existing ones. These technological changes include the increasing use of forensic evidence such as DNA testing, videotape evidence, 9-1-1 emergency system audiotapes, polygraph evidence, and expert witness testimony. Popular dramas such as the *CSI* television shows create an expectation of sophisticated forensic investigations. The "CSI effect" is a term that summarizes the likely impact of these shows on citizen expectations (Kim, Barak, & Shelton, 2009).

These improvements in technology, especially DNA testing and information system technology, have raised new issues about the fairness and accuracy of our courts. Improved evidence and information have increased the recognition of **wrongful convictions**, cases in which courts convict innocent persons of criminal acts. It is unlikely that we will ever know the full extent to which wrongful convictions occur in the United States. When asked, justice officials estimate wrongful conviction occurs in as many as 3% of felony cases (Zalman, Smith, & Kiger, 2008). The number of convicted offenders later exonerated has increased dramatically in recent years. The National Registry of Exonerations reports that more than 2,000 exonerations have occurred since 1989 (National Registry, 2018).

Innocence projects are efforts to investigate claims of innocence maintained by persons convicted of criminal offenses (see www.innocenceproject.org). The growth of these projects and the rate of exonerations suggest that courts convict many thousands of innocent persons each year in the United States (Gross et al., 2004:11). Although most of these convictions are for less serious offenses, for which the convicted person serves a relatively short sentence, the likelihood that innocent defendants are convicted of crimes raises questions about the quality of justice. As Leo (2017:99) summed: "to be sure, there is no worse error that the criminal justice system itself causes than the wrongful conviction of the innocent."

Krieger (2011) surveyed the most common reasons that innocent persons are convicted. His analysis showed that the most common reasons for wrongful convictions are:

Inaccuracy of eyewitnesses: The most common cause of wrongful convictions. Eyewitness errors often take the form of live lineups, photos spreads, etc.

Perjured testimony: Knowingly false testimony by witnesses at trial.

Availability of DNA testimony: Although DNA has helped exonerate many wrongful conviction cases, the availability of DNA is relatively rare and is often not preserved for postconviction appeals.

Accuracy of DNA testing and scientific evidence: Many types of forensic science are very subjective, and methods of analyzing scientific evidence that were acceptable in the past are now considered inaccurate.

Prosecutorial misconduct: Prosecutors often do not assist in postconviction reviews of the evidence regardless of the merits of the appeal.

Ineffective defense representation: In general, defense attorneys have far fewer resources than prosecutors.

Ineffective capital representation: Defense representation in capital punishment cases have been the source of many legal challenges.

Police misconduct: A false confession obtained by the police is very strong evidence against the defendant.

Pretrial criminal procedure processes: Many procedural processes are aimed at efficiency but can hamper the defense's efforts to find witnesses, view evidence, etc.

Dervan and Edkins (2013) showed that the plea bargaining process might also play an important role in wrongful convictions, as innocent defendants might falsely admit guilt in return for a reduced sentence.

Each instance of a wrongful conviction threatens public confidence in the fairness and accuracy of the courts. On the other hand, that we eventually recognize innocence and punish official misbehavior is a positive. Although not downplaying the seriousness of wrongful conviction, it remains true that exonerations have occurred in only a small percentage of cases. Even counting the thousands of cases in which wrongful convictions might occur, these are in relation to the millions of convictions each year. As with perhaps no other issue, the problem of wrongful convictions highlights the tension between due process and crime control. How do we ensure that we convict and punish guilty persons and prevent the conviction of an innocent person?

A survey conducted by Zalman, Larson, and Smith (2012) found most respondents felt that wrongful convictions occurred frequently enough to justify reforms to the criminal justice system. Many states have implemented policy safeguards aimed at reducing the likelihood of wrongful convictions. These safeguards include increased oversight of crime laboratories, requiring the preservation of DNA evidence, allowing inmates access to postconviction DNA testing, reform to eyewitness procedures, and regulations involving the use of informants during the trial. Kent and Carmichael (2015) found that a state's likelihood of adopting such reforms was influenced by each state's political climate along with the presence of advocacy groups that actively lobby the legislature for the passage of such protections.

Those who have been wrongfully convicted experience many difficulties readjusting to society. Among these are the psychological trauma that is often associated with being incarcerated for a crime that they did not commit, the societal changes that occur during the incarceration, and ongoing stigma from the community which may persist even after an exoneration (Clow, Leach, & Ricciardelli, 2012). Despite public support for providing wrongfully convicted individuals financial compensation for their experiences (Clow, Blandisi, Ricciardelli, & Schuller, 2011), many states do not offer such compensation. In fact, many states do not even expunge the criminal records for offenders who were wrongfully convicted. Shlosberg, Mandery, West, and Callaghan (2014) found that expungement was a key factor related to the individual's ability to successfully reintegrate into society.

THE CRIMINAL COURTS IN THE WHOLE SYSTEM

Like the police, the criminal courts are a subsystem of the criminal justice system, which itself exists within the whole system of American society. This context can be understood when we consider the fact that those courts that hear criminal cases do so on a part-time basis. Most cases are civil in nature, and criminal cases are a part-time responsibility for the courts. Thus, we cannot easily resolve delay in criminal case processing by merely hiring more court personnel and building more courtrooms, because these changes will have the effect of accelerating both civil and criminal cases.

Black's (1976) theory of the sociology of law suggests that as a society becomes more diverse and larger, the use of the law will expand. In the criminal courts, we find the majority of cases and court personnel in a few very large jurisdictions. Moreover, the practice of plea bargaining represents, at some level, a less formal means of social control than trial and conviction. Even as we see proposals to reduce or eliminate plea bargaining and make formal trials more common, we also see a movement to develop less formal, problem-solving courts. We should not be surprised that the majority of cases are disposed of by plea bargaining. Trials seem to be reserved for those cases in which less formal resolutions are unworkable.

As our thinking about the role of citizens in the courts has changed, the courts themselves have become more sensitive to the needs and desires of victims, witnesses, and jurors. As communications technology improves, issues emerge surrounding media coverage, especially television and social media coverage, of courts. The courts themselves adopt, and adapt to, new technologies to improve operations and reduce the likelihood of wrongful convictions.

The courts reflect changes in the larger society and, in turn, the courts affect the larger society (Calvi & Coleman, 1989). Many of the conflicts and controversies existing in the courts represent efforts to manage the relationship between the courts and social forces. The role of juries and the selection of judges and prosecutors are evidence of the commitment of the United States to a legal system that is sensitive to our perceptions of what is right and what is wrong. As perceptions change (e.g., marijuana offenses are seen as less serious than in the past), there is pressure for the courts to change. Yet, we want the courts to be predictable and fair as well as free from improper outside influence. Thus, we want to know what the courts are doing, but we do not want court decisions to be biased by preconceived notions. As with achieving a balance between due process and crime control, the courts must seek a balance between responsiveness to social change and the commitment to impartiality.

REVIEW QUESTIONS

1. Identify three ways in which defendants in criminal cases may obtain defense counsel.

2. Compare criminal defense attorneys with prosecutors. In what ways are their jobs similar?

3. Decide which method of judicial selection you think is most likely to yield competent judges who are responsive to the public. Explain how the system you select will accomplish this goal.

4. Describe the *voir dire* process and explain how it allows attorneys to protect the interests of their clients.

5. How can the right to a speedy trial be expected to reduce strains in other areas of the justice system?

6. The conflict between a free press and a fair trial reflects one of the ideological dilemmas of American criminal justice. Explain this conflict and propose a solution.

7. Take a position on the issue of plea bargaining (i.e., either support or oppose the practice) and give arguments against those who take the opposing position to yours.

8. How do the courts reflect changes in the larger American society?

REFERENCES

Abrams, D. (2011). Is pleading really a bargain? *Journal of Legal Studies, 8,* 200–221.

Aizenman, H. (2012). Pretrial publicity in a post-Trayvon Martin world. *Criminal Justice, 27*(3), 12–17.

Anderson, J.M., & Heaton, P. (2012). How much difference does the lawyer make? The effect of defense counsel on murder case outcomes. *Yale Law Journal, 122*(1), 154–187.

Bannon, A. (2016). *Rethinking judicial selection in state courts.* New York: Brennan Center for Justice.

Bannon, A., Lisk, C., & Hardin, P. (2017). *Who pays for judicial races?* New York: Brennan Center for Justice.

Bar-Gill, O., & Ben-Shahar, O. (2009). The prisoners' (plea bargain) dilemma. *Journal of Legal Analysis, 1*(2), 737–773.

Baumer, T. (2007). Reducing lockup crowding with expedited initial processing of minor offenders. *Journal of Criminal Justice, 35*(3), 273–281.

Benner, L.A. (2011). Eliminating excessive public defender workloads. *Criminal Justice, 26*(2), 24–33.

Black, D. (1976). *The behavior of law*. New York: Academic Press.

Boruchowitz, R., Brink, M., & Dimino, M. (2009). *Minor crimes, massive waste: The terrible toll of America's broken misdemeanor courts*. Washington, DC: National Association of Criminal Defense Lawyers.

Brandonwood, J. (2000). You say "fair trial" and I say "free press": British and American approaches to protecting defendants' rights in high profile trials. *New York University Law Review, 75*, 1412–1451.

Browning, J. (2014). #Snitches get stitches: Witness intimidation in the age of Facebook and Twitter. *Pace Law Review, 35*(1), 192–214.

Bureau of Justice Statistics (1986). *Felony case-processing time*. Washington, DC: U.S. Department of Justice.

Burke, M. (2014). Online research in preparation for *voir dire. Nevada Lawyer, 22*(6), 12–15.

Cabell, K. (2012). Calculating an alternative route: The difference between a blindfolded ride and a road map in *pro se* criminal defense. *Law & Psychology Review, 36*, 259–274.

Calvi, J., & Coleman, S. (1989). *American law and legal systems*. Englewood Cliffs, NJ: Prentice Hall.

Campbell, S.A., Green, T.M., Hance, B.S., & Larson, J.G. (2017). The impact of courtroom cameras on the judicial process. *Journal of Media Critiques, 3*(10), 101–113.

Caprathe, W., Hannaford-Agor, P., Loquvam, S.M., & Diamond, S.S. (2016). Assessing and achieving jury pool representativeness. *The Judges' Journal, 55*(2), 16–20.

Champion, D. (1987). Felony offenders, plea bargaining, and probation: A case of extra-legal exigencies in sentencing practices. *Justice Professional, 2*(2), 1–18.

Champion, D. (1989). Private counsels and public defenders: A look at weak cases, prior records, and leniency in plea bargaining. *Journal of Criminal Justice, 17*(4), 253–263.

Clow, K.A., Leach, A.M., & Ricciardelli, R. (2012). Life after wrongful conviction. In Cutler, B.L. (Ed.), *Conviction of the innocent: Lessons from psychological research* (pp. 327–341). Washington, DC: American Psychological Association.

Clow, K.A., Blandisi, I.M., Ricciardelli, R., & Schuller, R.A. (2011). Public perception of wrongful conviction: Support for compensation and apologies. *Albany Law Review, 75*(3), 1415–1438.

Cohen, T. (2014). Who is better at defending criminals? Does type of defense attorney matter in terms of producing favorable case outcomes? *Criminal Justice Policy Review, 25*, 29–58.

Cohen, T., & Kyckelhahn, T. (2010). *Felony defendants in large urban counties, 2006*. Washington, DC: Bureau of Justice Statistics.

Cooper, C. (1994). *Expedited drug case management*. Washington, DC: Bureau of Justice Assistance.

Criminal Justice Newsletter (1987). President cites crime issue in choosing Kennedy for high court. *Criminal Justice Newsletter, 18*(22), 4.

Dawson, J., Smith, S., & DeFrances, C. (1993). *Prosecutors in state courts, 1992*. Washington, DC: Bureau of Justice Statistics.

Dervan, L., & Edkins, V. (2013). The innocent defendant's dilemma: An innovative empirical study of plea bargaining's innocence problem. *Journal of Criminal Law & Criminology, 103*, 1–48.

Devine, D.J. (2012). *Jury decision making: The state of the science (psychology and crime)*. New York: New York University Press.

Duncan, S. (2008). Pretrial publicity in high profile trials: An integrated approach to protecting the right to a fair trial and the right to privacy. *Ohio Northern University Law Review, 34*, 755–795.

Duvall, K. (2012). The contradictory stance of jury nullification. *North Dakota Law Review, 88*(2), 409–451.

Eisenstein, J., & Jacob, H. (1977). *Felony justice*. Boston: Little, Brown.

Ellis, M.J. (2012). The origins of the elected prosecutor. *Yale Law Journal, 121*(6), 1528–1569.

Fairfax, R.A., Jr. (2013). Searching for solutions to the indigent defense crisis in the broader criminal justice reform agenda. *Yale Law Journal, 122*(8), 2316–2335.

Farole, D., & Langton, L. (2010). *County-based and local public defender offices, 2007*. Washington, DC: Bureau of Justice Statistics.

Feeley, M. (1982). Plea bargaining and the structure of the criminal process. *Justice System Journal, 7*(Winter), 338–355.

Fissell, B.M. (2013). Jury nullification and the rule of law. *Legal Theory, 19*, 217–241.

Frederick, J. (2012). *Mastering* voir dire *and jury selection: Gaining an edge in questioning and selecting your jury*. Chicago: ABA Publishing.

Fried, M., Kaplan, K., & Klein, K. (1975). Juror selection: An analysis of *voir dire*. In Simon, R. (Ed.), *The jury system in America*. Beverly Hills, CA: Sage.

Fukurai, H., Butler, E., & Krooth, R. (1991). Cross-sectional jury representation or systematic jury representation? Simple random and cluster sampling strategies in jury selection. *Journal of Criminal Justice, 19*(1), 31–48.

Giglio, E. (1982). Free press—Fair trial in Britain and America. *Journal of Criminal Justice, 20*(5), 341–358.

Gill, K.J., & Murphy, A.A. (2017). Jail diversion for persons with serious mental illness coordinated by a prosecutor's office. *BioMed Research International, 2017*, 1–7.

Glick, H. (1983). *Courts, politics and justice*. New York: McGraw-Hill.

Gross, S., Jacoby, K., Matheson, D., Montgomery, N., & Patil, S. (2004). *Exonerations in the United States: 1989 through 2003*. Chicago: American Judicature Society.

Hall, M.G. (2016). Partisanship, interest groups, and attack advertising in post-*White* era, or why nonpartisan judicial elections really do stink. *Journal of Law and Politics, 31*, 429–456.

Hannaford-Agor, P. (2008). When all eyes are watching: Trial characteristics and practices in notorious trials. *Judicature, 92*(4), 197–201.

Harlow, C. (2000). *Defense counsel in criminal cases*. Washington, DC: Bureau of Justice Statistics, *4*, 8.

Harris, J., & Jesilow, P. (2000). It's not the old ball game: Three strikes and the courtroom workgroup. *Justice Quarterly, 27*(1), 186–203.

Hartley, R., Miller, H., & Spohn, C. (2010). Do you get what you pay for? Type of counsel and its effect on criminal court outcomes. *Journal of Criminal Justice, 38*, 1063–1070.

Holten, N.G., & Lamar, L. (1991). *The criminal courts: Structures, personnel, and processes*. New York: McGraw-Hill.

Homiak, R. (1976). *Faretta v. California* and the *pro se* defense: The constitutional right of self-representation. *American University Law Review, 25*, 897–937.

Hopwood, S. (2014). The not so speedy trial act. *Washington Law Review, 89*, 709–745.

Institute for the Advancement of the American Legal System (2017). *Judges in the United States.* Denver, CO: University of Denver.

Jacoby, J., Ratledge, E., & Gramckow, H. (1992). *Expedited drug case management programs: Issues for program development.* Washington, DC: U.S. Department of Justice.

Joy, P.A., & McMunigal, K.C. (2016). Racial discrimination and jury selection. *Criminal Justice, 31*(2), 43–45.

Kang, M.S., & Shepherd, J.M. (2015). Partisanship in state supreme courts: The empirical relationship between party campaign contributions and judicial decision making. *Journal of Legal Studies, 44*(1), 161–185.

Kang, M.S., & Shepherd, J.M. (2016). The long shadow of *Bush v. Gore*: Judicial partisanship in election cases. *Stanford Law Review, 68*, 1411–1452.

Kent, S.L., & Carmichael, J.T. (2015). Legislative responses to wrongful conviction: Do partisan principals and advocacy efforts influence state-level criminal justice policy? *Social Science Research, 52*, 147–160.

Kim, Y., Barak, G., & Shelton, D. (2009). Examining the "CSI-Effect" in the cases of circumstantial evidence and eyewitness testimony: Multivariate and path analyses. *Journal of Criminal Justice, 37*(5), 452–460.

Kingsnorth, R., Macintosh, R., & Wentworth, J. (1999). Sexual assault: The role of prior relationship and victim characteristics in case processing. *Justice Quarterly, 26*(2), 275–302.

Krieger, S. (2011). Why our justice system convicts innocent people, and the challenges faced by innocence projects trying to exonerate them. *New Criminal Law Review, 14*(3), 333–402.

Kuhn, W. (2012). The speedy trial rights of military detainees. *Syracuse Law Review, 62*, 209–253.

Lambert, W.G. (2012). The real debate over the Senate's role in the confirmation process. *Duke Law Journal, 61*(6), 1283–1327.

Langton, L., & Farole, D. (2009). *Public defender offices, 2007—Statistical tables.* Washington, DC: Bureau of Justice Statistics.

Leo, R.A. (2017). The criminology of wrongful conviction: A decade later. *Journal of Contemporary Criminal Justice, 33*(1), 82–106.

Lim, C.S.H., & Snyder, J.M., Jr. (2015). Is more information always better? Party cues and candidate quality in U.S. judicial elections. *Journal of Public Economics, 128*, 107–123.

Loges, W., & Bruschke, J. (2005). *Free press vs. fair trials: Examining publicity's role in trial outcomes.* Oxford: Routledge.

Lundberg, J.C. (2012). Googling jurors to conduct *voir dire*. *Washington Journal of Law, Technology & Arts, 8*(2), 123–136.

Maloney, C., & Miller, J. (2015). The impact of a risk assessment instrument on juvenile detention decision-making: A check on "perceptual shorthand" and "going rates"? *Justice Quarterly, 32*(5), 900–927.

Mays, G.L. (2012). *American courts and the judicial process*. New York: Oxford University Press.

McCarthy, B.J. (2017). Ethical issues in confronting terrorism. In Braswell, M.C., McCarthy, B.R., & McCarthy, B.J. (Eds.), *Justice, crime and ethics* (pp. 451–479). New York: Routledge.

McDonough, K.M. (2013). Combating gang-perpetrated witness intimidation with forfeiture by wrongdoing. *Seton Hall Law Review, 43*(4), 1282–1323.

McLeod, A. (2012). The party on the bench: Partisanship, judicial selection commissions, and state high-court appointments. *Justice System Journal, 33*(3), 262–274.

Meringolo, J.C. (2010). The media, the jury, and the high-profile defendant: A defense perspective on the media circus. *New York Law School Law Review, 55*, 981–1012.

Metcalfe, C. (2016). The role of courtroom workgroups in felony case dispositions: An analysis of workgroup familiarity and similarity. *Law & Society Review, 50*(3), 637–673.

Miller, M.C. (2015). *Judicial politics in the United States*. Boulder, CO: Westview Press.

Mulvaney, M.D., & Little, J.A., Jr. (2015). The importance of *voir dire*: Essential techniques for choosing finders of fact. *American Journal of Trial Advocacy, 39*, 313–338.

Myers, L., & Reid, S. (1995). The importance of county context in the measurement of sentence disparity: The search for routinization. *Journal of Criminal Justice, 23*(3), 223–241.

National Registry of Exonerations (2018). *Exonerations in 2017*. Retrieved from: www.law.umich.edu/special/exoneration/Documents/ExonerationsIn2017.pdf

Neely, R. (1983). *Why courts don't work*. New York: McGraw-Hill.

Neubauer, D. (1984). *America's courts and the criminal justice system*. Monterey, CA: Brooks/Cole.

Neubauer, D., & Fradella, H. (2014). *America's courts and the criminal justice system* (11th ed.). Belmont, CA: Wadsworth Cengage Learning.

O'Connell, P. (1988). Pretrial publicity, change of venue, public opinion polls—A theory of procedural justice. *University of Detroit Law Review, 65*(2), 169–197.

O'Rourke, A. (2014). The speedy trial right and national security detentions. *Journal of International Criminal Justice, 12*(4), 871–885.

Papandrea, M.R. (2012). Moving beyond cameras in the courtroom: Technology, the media, and the Supreme Court. *Brigham Young University Law Review, 2012*(7), 1901–1952.

Patton, D.E. (2017). The structure of federal public defense: A call for independence. *Cornell Law Review, 102*(2), 335–411.

Perry, S., & Banks, D. (2011). *Prosecutors in state courts, 2007—Statistical tables*. Washington, DC: Bureau of Justice Statistics.

Rattner, A., Turjeman, H., & Fishman, G. (2008). Public versus private defense: Can money buy justice? *Journal of Criminal Justice, 36*(1), 43–49.

Reaves, B.A. (2013). *Felony defendants in large urban counties, 2009*. Washington, DC: Bureau of Justice Statistics.

Reddick, M. (2002). Merit selection: A review of the social scientific literature. *Dickinson Law Review, 106*, 729–744.

Reddick, M. (2010). *Judging the quality of judicial selection methods: Merit selection, elections, and judicial discipline*. Chicago: American Judicature Society. Retrieved July 18, 2010, from www.acjs.org

Rottman, D., & Strickland, S. (2006). *State court organization, 2004*. Washington, DC: Bureau of Justice Statistics.

Royer, C.E. (2017). The disobedient jury: Why lawmakers should codify the jury nullification. *Cornell Law Review, 102*(5), 1401–1430.

Rudes, D.S., & Portillo, S. (2012). Roles and power within federal problem solving courtroom workgroups. *Law & Policy, 34*(4), 402–427.

Sanborn, J.B. (1986). A historical sketch of plea bargaining. *Justice Quarterly, 3*(2), 111–138.

Shlosberg, A., Mandery, E.J., West, V., & Callaghan, B. (2014). Expungement and post-exoneration offending. *Journal of Criminal Law and Criminology, 104*(2), 353–388.

Simon, R. (1980). *The jury: Its role in American society*. Lexington, MA: Lexington Books.

Simpler, M.F., III. (2012). The unjust "web" we weave: The evolution of social media and its psychological impact on juror impartiality and fair trials. *Law & Psychology Review, 36*, 275–296.

Sklansky, D.A. (2017). The nature and function of prosecutorial power. *Journal of Criminal Law & Criminology, 106*(3), 473–520.

Steelman, D., Goerdt, J., & McMillan, J. (2000). *Caseflow management: The heart of court management in the new millennium*. Williamsburg, VA: National Center for State Courts.

Strong, S.M. (2016). *State-administered indigent defense systems, 2013*. Washington, DC: Bureau of Justice Statistics.

Sung, H. (2006). From diversion experiment to policy movement: A case study of prosecutorial innovation. *Journal of Contemporary Criminal Justice, 22*(3), 220–240.

Surette, R. (1999). Media echoes: Systemic effects of news coverage. *Justice Quarterly, 26*(3), 601–631.

Surette, R. (2015). *Media, crime, and criminal justice: Images, realities, and policies*. Stamford, CT: Cengage.

Taylor, R., Ratcliffe, J., Dote, L., & Lawton, B. (2007). Roles of neighborhood race and status in the middle stages of juror selection. *Journal of Criminal Justice, 35*(4), 391–404.

Thompson, C. (2011). A shooting suspect's release revives the right to a speedy trial in Missouri. *Missouri Law Review, 76*, 971–998.

U.S. Courts (2017). *Defender services*. Retrieved from: www.uscourts.gov/services-forms/defender-services

Van Duizend, R., Steelman, D.C., & Suskin, L. (2011). *Model time standards for state trial courts*. Williamsburg, VA: National Center for State Courts.

Walker, S. (2015). *Sense and nonsense about crime, drugs, and communities*. Stamford, CT: Cengage.

Wice, P. (1978). *Criminal lawyers: An endangered species*. Beverly Hills, CA: Sage.

Williams, M.R. (2017). The effect of attorney type on bail decisions. *Criminal Justice Policy Review, 28*(1), 3–17.

Willis, R. (2010). A fool for a client: Competency standards in *pro se* cases. *Brigham Young University Law Review, 2010*(1), 321–334.

Winnick, J.K. (2014). A tweet is(n't) worth a thousand words: The dangers of journalists' use of Twitter to send news updates from the courtroom. *Syracuse Law Review, 64*(2), 335–355.

Wood, L.C., Goyette, D.T., & Burkhart, G.T. (2016). Meet-and-plead: The inevitable consequence of crushing defender workloads. *Litigation, 42*(2), 20–26.

Zalman, M., Smith, B., & Kiger, A. (2008). Officials' estimates of the frequency of "actual innocence" convictions. *Justice Quarterly, 25*(1), 72–100.

Zalman, M., Larson, M.J., & Smith, B. (2012). Citizens' attitudes toward wrongful convictions. *Criminal Justice Review, 37*(1), 51–69.

Zatz, M. (2000). The convergence of race, ethnicity, gender, and class on court decision making: Looking toward the 21st century. In Horney, J. (Ed.), *Criminal Justice 2000: Policies, processes, and decisions of the criminal justice system: Vol. 3.* (pp. 503–552). Washington, DC: National Institute of Justice.

IMPORTANT CASES

Apprendi v. New Jersey, 530 U.S. 466 (2000).

Barker v. Wingo, 407 U.S. 514 (1972).

Chandler v. Florida, 449 U.S. 560 (1981).

Citizens United v. FEC, 558 U.S. 310 (2010).

Estes v. Texas, 381 U.S. 532 (1965).

Indiana v. Edwards, 554 U.S. 164 (2008).

Lafler v. Cooper, 132 S.Ct. 1376 (2012).

McKaskle v. Wiggins, 465 U.S. 168 (1984).

Missouri v. Frye, 132 S.Ct. 1399 (2012).

People v. Faretta, 422 U.S. 806 (1975).

Richmond Newspapers, Inc. v. Virginia, 448 U.S. 555 (1980).

Rumsfeld v. Padilla, 542 U.S. 426 (2004).

Santobello v. New York, 404 U.S. 257 (1966).

Sheppard v. Maxwell, 384 U.S. 333 (1966).

United States v. Ghailani, S10 98 Crim. 1023 (2010).

CHAPTER 9

Sentencing: The Goals and Process of Punishment

After conviction, the next major decision point in the criminal justice system is sentencing, the decision about punishment. In many ways, the sentencing decision represents the crux of the justice system, for it is at this point that we determine what to do to, for, with, or about the criminal offender.

Criminal sentences may involve the imposition of fines, community supervision, or incarceration. In some cases, a criminal sentence includes a combination of all of these. Excluding fines and short-term incarceration in a local jail, millions of people each year receive criminal sentences of probation or imprisonment. In 2015, approximately 1 in 37 adults in the United States was under some form of correctional supervision. However, less than one-third of these individuals were incarcerated in either a jail or prison (Kaeble & Glaze, 2016).

To understand how the distinction is drawn between who is incarcerated and who is allowed to remain in the community under supervision, we need to understand the purposes of punishment and the motivations of the people who make the punishment decision. This chapter examines these factors.

THE PURPOSES OF PUNISHMENT

Traditionally, four purposes or justifications for criminal penalties have been advanced: (1) deterrence, (2) incapacitation, (3) treatment (or rehabilitation), and (4) just deserts (or retribution). These theories of punishment answer the question "Why punish at all?" The average citizen does not often hear this question in the context of criminal law. Yet, most of us have heard the expression that "two wrongs do not make a right." Applied to the criminal law, this suggests that imposing a punishment on someone who has broken the law is a "second wrong," which does not make it right. If I steal your laptop and am sent to prison for a year because of it, have things turned out right? You are deprived of your laptop, I am deprived of a year of my life, and all of us pay the costs of trial and imprisonment.

IMPORTANT TERMS

boot camp

concurrent term

consecutive term

determinate sentencing

deterrence

disparity

false negative

false positive

focal concerns

general deterrence

good time

habitual offender statutes

incapacitation

indeterminate sentencing

intermediate sanctions

just deserts

mandatory minimum sentences

parole eligibility

presentence investigation (PSI)

presumptive
sentencing

rehabilitation

retribution

sentencing

sentencing
commissions

specific deterrence

split sentences

three-strikes laws

treatment

truth in sentencing

From this point of view, the punishment of crime seems useless, if not wasteful. However, we usually do not think of it this way. Rather, it seems almost automatic that we punish someone who breaks the law (von Hirsch, 1976). What can justify a system of penalties that we can argue involves nothing but costs to everyone affected?

Deterrence is a purpose of punishment based on the idea that punishment of the individual offender produces benefits for the future by making the crime less attractive. Deterrence has two parts: **specific deterrence**, in which the object of the deterrent effect is the specific offender, and **general deterrence**, in which the object of the deterrent effect is a wider audience (the general public). According to deterrence theory, punishment is an example of what awaits law violators and serves to educate would-be offenders, so that they will weigh the costs of crime against its benefits (Paternoster, 1987).

With specific deterrence, after release from prison, I will think twice before I steal another laptop because I now know that I will face a year in prison if the police catch me stealing again. With general deterrence, after seeing what happened to me, you will rethink your plan to replace your missing laptop by stealing your neighbor's tablet. In both cases, the punishment serves to prevent future crimes. If society does not punish, there may be no reason for us to obey the law.

Deterrence has a certain intuitive appeal. On the surface, it makes sense that deterrence will work. Unfortunately, the evidence of its effectiveness is mixed (Chalfin & McCrary, 2017). Few (if any) criminals believe they will be caught and punished (indeed, if they expected to be caught, they most probably would not commit the offense). Although we have used the term "educate," deterrence really operates on fear. Would-be offenders must fear the penalty, and thus we are left with a society in which "proper" behavior is based on fear—a basis that most of us do not favor. Finally, when someone has thought about it and decided not to commit a crime, it is unclear whether the person made the decision based on fear of punishment, fear of public ridicule, fear of eternal damnation, or some other factor entirely.

The theory of deterrence assumes that humans are rational beings guided by a pleasure principle. That is, humans do things that please them and avoid things that hurt them. Further, it holds that we are able to assess the likely effects of our behavior and guide ourselves according to these assessments. As rational beings, we will avoid bad behaviors, such as committing crimes, when the behaviors will produce unpleasant results (punishments). Rationality, of course, is difficult to establish. Indeed, many crimes, especially violent offenses among friends and family, are more emotional than rational.

For a punishment to deter a would-be offender, it must meet two conditions. First, the penalty must be severe enough that the pain of the punishment will outweigh the pleasure of the criminal act. Using an economic example, if the penalty for theft were a $100 fine, it may be severe enough to stop someone from stealing $50. Given the opportunity to steal $500, however, the rational person may take the money and pay the fine, keeping a $400 profit. Thus, deterrence depends in part on the severity of the punishment.

The second condition is that the offender must believe we will impose the penalty. Not only must the punishment be severe enough to outweigh the gain realized by crime, but also the likelihood of punishment must be high enough that the offender takes the threat seriously. If the punishment for theft were 10 years in prison, but the chance of punishment were only one in a million, the threat of punishment probably would not deter. Deterrence theory suggests that the lower the risk of punishment, the higher the likelihood that a crime will be committed.

Research indicates that, of these two conditions, certainty of punishment is the more important aspect of deterrence. Given a rational offender, uncertainty of punishment makes deterrence even more troublesome. The offender will assess the likelihood of capture, and then take steps to reduce the chance of detection before committing an offense. Deterrence may best serve to make offenders more cautious rather than less criminal. In a review of the deterrence literature, Paternoster (2010) suggested that criminals do appear to be rational actors, but that because of the delays in administering punishment, the criminal justice system is not able to exploit this rationality.

Supporters of deterrence often engage in "penalty escalation" because it is easier to alter the severity of punishment than its certainty (Newman, 1983). In this case, the penalty for a given offense—for example, 5 years of imprisonment for theft—is increased (perhaps to 10 years). It is easier to increase penalty severity than to become more efficient at catching and punishing thieves. Furthermore, deterrence rests on an argument that the good of the penalty (in terms of crimes prevented) must outweigh its harm in terms of injury to the punished. It is possible, then, to impose a deterrent penalty on an innocent person, as long as the social good outweighs the individual harm.

In a study of the specific deterrent effects of punishment for DWI offenses, Bouffard, Niebuhr, and Exum (2017) found that the perception of an arrest as being a likely outcome to drinking and driving was more influential to one's decision to drive drunk than exposure to the actual punishment. In other words, would-be offenders must perceive that the crime is likely to result in arrest and punishment (Schoepfer, Carmichael, & Piquero, 2007). As Bruce Jacobs (2010) put it, it may be impossible to deter some offenders because they do not consider the consequences of their actions. Finally, some researchers have found that the experience of being incarcerated actually increases the likelihood of future crime in some offenders (Bales & Piquero, 2012; Wood, 2007). Whereas this might seem counterintuitive, one must realize that the incarcerated offender is isolated from conventional society, placed in an environment where they are in constant contact with other offenders, and face significant stigma along with difficulty finding housing upon release.

Incapacitation, like deterrence, suggests that punishment serves to prevent future crime, but not by education or fear. An incapacitative punishment prevents future crimes by removing opportunities for crime. One good reason to imprison a criminal is to ensure that he or she does not have the chance to commit a crime in society again. I cannot steal your laptop if I am in prison. Probably the most effective incapacitative penalty is capital punishment. Not only do the dead tell no tales but they also commit no crimes. See Figure 9.1.

Incapacitation has two major drawbacks. First, it is expensive to incapacitate most offenders. The costs associated with imprisonment have forced many states to look at alternatives to incarceration. Second, it is difficult to predict who is likely to commit an offense in the future. While the tools used by the court to identify those persons likely to pose a threat of serious crime in the future have improved dramatically in recent years, they are nonetheless imperfect tools. The prediction problem is simply too complex.

In any attempt to predict "dangerousness" among a population of criminal offenders (not to mention the general population), one runs a risk of making two types of errors: false positives and false negatives. A **false positive** refers to someone who is predicted to be dangerous (positive on danger), but who turns out not to be a threat (false). A **false negative** refers to someone who is predicted to be safe (negative on danger), but who turns out to be dangerous (false). We incapacitate false positives as if they were dangerous, wasting

FIGURE 9.1
The traditional cage-like appearance of the prison or jail cell, shown here at the Old Salem Jail, is symbolic of incapacitation as a purpose of sentencing. The Old Salem Jail in Salem, Massachusetts, was operational from 1813 until 1991.

Ellen S. Boyne

prison resources and needlessly infringing on individual freedoms. We fail to incapacitate false negatives leaving them free to victimize others. Every crime prediction scheme available makes both types of errors. Given these limitations, policymakers must decide how much error they are willing to tolerate (Smith & Smith, 1998), balancing concerns about due process with concerns about crime control. Upon reviewing the available evidence about incapacitation, Cullen, Jonson, and Nagin (2011:60S) concluded, "[W]e need to take a giant collective step backward and understand that imprisonment is not a panacea for the crime problem."

We can argue that incapacitative penalties have the effect of punishing the innocent. The determination of how severe a penalty should be for incapacitation does not rest on the seriousness of the crime committed but on a prediction that the offender will commit another crime. Like the queen in *Alice in Wonderland*, proponents of incapacitation invoke the penalty first, and the crime comes later, after the penalty. If there in fact is no crime, so much the better.

Treatment (also called rehabilitation) is another rationale based on a reduction of future crime. Unlike deterrence or incapacitation, treatment is concerned with the offender as an individual (Cullen & Gilbert, 2013). Here, the punishment imposed is one that fits the individual and is most likely to result in a change in the individual's desire to commit crime. Treatment suggests that individuals commit crimes for a variety of reasons and that the solution to the problem of crime will be achieved through changing individuals so that they will not wish to engage in crime, and by having other options available to them that will allow them to avoid criminality.

Treatment penalties assume an "identity of interest" between the state and the offender. The state wants what is best for the offender, that is, to improve his or her chances in life and to reduce the offender's desire to commit crimes. The offender either wants the same thing or would want it if he or she were competent. Thus, the imposition of a punishment is for the offender's own good. A sanction that seems disproportionate to the seriousness of the offense is not a problem, as long as it is "good" for the offender.

What appears from the research on treatment effectiveness is that programs work best when they follow the principles of the risk–need–responsivity (RNR) model. This model first suggests that treatment should be focused on high-risk offenders rather than "waste" it on those who do not need it. It may also be useful to vary the duration of treatment based on the offenders' risk (Sperber, Latessa, & Makarios, 2013). The second principle of the RNR model is that treatment should target the factors that influence offenders to violate the law. Andrews and Bonta (2010) identify several areas in which offenders often need help to improve, including antisocial personalities, antisocial associates, unstable family and marital attachments, deficiencies in work or school attainment, availability of leisure activities, and substance abuse. To be most effective, a rehabilitation program will match the proper treatment to the needs of each specific offender. For example, offering drug abuse counseling to an offender who has no prior indication of drug usage is not only a waste of resources but also not likely to have an impact on that offender's recidivism risk.

The responsivity principle of the RNR model acknowledges that not every offender will react identically to the treatment offered. Rather, everyone has their own cognitive style, cultural experiences, social support, and level of motivation to change (Andrews & Bonta, 2010). In general, high-risk drug offenders often benefit from intensive cognitive behavioral therapy to learn relapse prevention techniques. However, each treatment might be most effective on a different type of offender (Spiropoulos, Salisbury, & Van Voorhis, 2014; Van Voorhis, Spiropoulos, Ritchie, Seabrook, & Spruance, 2013; Wright, Pratt, Lowenkamp, & Latessa, 2011). Perhaps a 20-year-old marijuana user with several convictions as a juvenile and a breakdown in his/her family would benefit from a different type of treatment compared to an older offender who had begun using heroin relatively recently and has a significant amount of family and social support. Hamilton (2011) found that substance abusers can be effectively matched to the most effective types of treatments by using responsivity techniques.

Just deserts (also called **retribution** or desert) is the only justification for criminal punishment that is not "forward looking." It does not offer a reduction in future crime as the principal justification for the imposition of a punishment. Rather, the idea of "just deserts" operates from the belief that whoever breaks the law "deserves" to be punished. Breaking the law in itself is justification for punishment, whether that punishment reduces, increases, or has no effect on future levels of crime. The criminal law is a promise wherein we (the state) promise to punish anyone who violates the law, and in the desert rationale, we must keep our promise.

Just deserts places limits on the degree to which we can punish someone. Punishment is expected to be commensurate with (or proportionate to) the severity of the crime committed. For example, public torture and execution of parking violators would deter most of us from parking illegally (if not from driving altogether), and these methods would be acceptable in a purely deterrent system of punishment. In contrast, these extreme forms of punishment for parking violations would be unthinkable in a desert scheme. Although a purely deterrent system of punishment does not require that the person punished be convicted first, the just deserts rationale holds that only convicted offenders should be punished.

People often state the just deserts justification for punishment in biblical terms as "an eye for an eye." It fits rather well with our beliefs that, regardless of the reason, those who break the law should be punished (Johnson & Sigler, 1995). However, retribution is also difficult to achieve in practice. There tends to be widespread disagreement about how much punishment those convicted of different kinds of crimes deserve (Kubiak & Allen, 2011; Ramirez, 2013). Nearly everyone has an opinion regarding how punitive the justice system is or should be, regardless of their actual experiences with the system. These opinions can have important influences on the implementation of sentencing laws. For example, those who are afraid of crime or feel that prison conditions are too pleasant for their inmates seem to support more punitive punishments (Baker, Falco, Metcalfe, Berenblum, Aviv, & Gertz, 2015; Wozniak, 2016). A societal dominance of punitive attitudes, in turn, can create a "bidding war" where politicians attempt to position themselves as tougher on crime than their counterparts (Krisberg, 2016:138). Thus, it is easy to see how punishment can be imposed simply for the sake of punishment.

Practical considerations also make punishments difficult to administer consistently. As Wonders (1996) observed, sentencing is located in the middle of the criminal justice system. Attempts to make sentencing "fair" must contend with both earlier (investigation, arrest, charging) and later (correctional placements, parole release) decisions. Finally, statewide legislation is implemented in different localities, and each of these has its own courtroom work group and definition of appropriate penalties (Myers & Reid, 1995). These factors make it difficult to determine an appropriate level of punishment for each offender as well as ensure that the punishment is administered fairly across different localities.

These four rationales for the imposition of criminal punishment—deterrence, incapacitation, treatment, and just deserts—have been presented separately and analyzed as if they were required to stand alone. In practice, however, sentencing systems support many purposes. The sentencing judge may "throw the book" at a defendant for deterrent purposes, whereas correctional authorities seek to "treat" the offender's problems and rehabilitate him or her, only to have a paroling authority refuse to release the offender because of a fear that the offender will commit a new crime and, therefore, must be incapacitated.

Not only do most sentencing systems expressly serve all four of these functions, the decision makers in those sentencing systems (judges, parole authority members, correctional workers, and administrators) favor different justifications for punishment. To complicate matters even more, these different actors favor different rationales for punishment of different offenders at different points in time. Furthermore, each criminal jurisdiction has its own structure for the determination and implementation of criminal sentences.

SENTENCING STRUCTURES IN THE UNITED STATES

Criminal sentencing has been the topic of a great deal of interest and debate (Jefferson-Bullock, 2016; Tonry, 2016; Travis & Lytle, 2017). Since 1975, many states have adopted changes in the structures used for criminal sentencing. The result has been the addition of a number of innovative formats for sentencing. If anything, criminal sentencing in the United States has become even more complicated in the past four decades. Tonry (2016:1) observed, "Each state and the federal system has a crazy quilt of incompatible and conflicting laws, policies, and practices." Almost anything that can be said about the nation's sentencing policies would be inaccurate for at least some jurisdictions. Nonetheless, the traditional distinction made between determinate and indeterminate sentencing is useful.

Determinate sentencing structures are those in which at the time of sentencing the offender knows the exact length and nature of his or her punishment. Indeterminate sentencing structures are those in which the precise length of the penalty is unknown until some time has passed since the imposition of the penalty. Within each of these classes are several types of sentencing structures that represent a balance of power among the legislative, judicial, and executive branches of government. The structure of sentencing reflects varying degrees of emphasis on each of the goals of criminal sentencing, and it affects how sentencing is conducted.

Box 9.1 depicts sentencing structures found across the United States. The majority of sentencing authority rests with the legislature, which establishes criminal penalties and defines crimes. There has traditionally been a certain level of sentencing power vested in the executive branch in the form of clemency. The governor of a state or the president of the United States can grant pardons, commutations, reprieves, and other forms of mercy. Legislatures usually delegate sentencing power to parole authorities and sentencing judges.

BOX 9.1 BASIC SENTENCING STRUCTURES IN THE UNITED STATES

Indeterminate sentencing: The judge specifies minimum and maximum sentence lengths. These set upper and lower bounds on the time to be served. The actual release date (and therefore the time actually served) is determined later by parole authorities within those limits.

Partially indeterminate sentencing: A variation of indeterminate sentencing in which the judge specifies only the maximum sentence length. An associated minimum sentence automatically is implied but is not within the judge's discretion. The implied minimum may be a fixed time (such as 1 year) for all sentences or a fixed proportion of the maximum. In some states the implied minimum is zero; thus, the parole board is empowered to release the prisoner at any time.

Determinate sentencing: The judge specifies a fixed term of incarceration, which must be served in full (less any "good time" earned in prison). There is no discretionary parole release.

Mandatory minimum sentencing: A minimum sentence that is specified by statute for all offenders convicted of a particular crime with special circumstances (e.g., robbery with a firearm). Mandatory minimum sentences can be used in both determinate and indeterminate sentencing structures.

Presumptive sentencing guidelines: The appropriate sentence for an offender in a specific case is presumed to fall within a range authorized by sentencing guidelines that are adopted by a sentencing body such as a sentencing commission. Judges are expected to sentence within this range or provide written justification for departure. These guidelines often provide for some appellate review for a departure.

Voluntary sentencing guidelines: Recommended sentencing policies that are not required by law and are based on past sentencing practices.

Source: Bureau of Justice Assistance (1998:1–2); Bureau of Justice Statistics (1988:91).

In indeterminate sentencing systems, the legislature establishes a range of penalties (minimum to maximum), and the sentencing judge is then authorized to impose a sentence that is not less than the minimum and not more than the maximum. In some cases, the legislature establishes a minimum term and then allows the judge to set a maximum within some absolute outer limit. In other cases, the legislature sets a minimum term that is some fraction of the maximum term decreed by the judge. In any indeterminate sentencing system, a paroling authority has power to grant release sometime between the end of the minimum term and the end of the maximum term. Thus, the judge has powers delegated by the legislature, and the paroling authority has power that is limited both by the legislature and by the decision of the judge.

In determinate sentencing systems, the legislature generally reserves most of the sentencing power for itself. In some cases, the legislature will actually determine the sentence for persons convicted of a specific offense. Here, the judge and parole authority have no sentencing power. More common is the model in which the legislature describes the expected or normal penalty for an offense, and then allows the sentencing judge to modify it (with reasons given). This is presumptive sentencing. The law establishes what the legislature "presumes" will be the sentence, and then requires the judge to explain any case in which the judge does not impose the presumed penalty. In determinate sentencing systems, the parole authority typically has no sentencing power.

In most cases, the legislature grants the sentencing judge the ability to choose between a sentence of incarceration and one of community supervision. This is considerable sentencing power, even in a system in which the judge's power to determine the length of a prison sentence has been strictly curtailed. The decision as to which decision maker receives how much sentencing power reflects a different emphasis on each of the four purposes of criminal sentencing.

Griset (2002) examined the development of sentencing policy in Florida over 20 years, as the general sentencing structure in the state moved from essentially indeterminate to determinate sentencing. She observed that sentencing policy is the product of many different decision makers, including the legislature, judiciary, and various executive branch officials. With regard to Florida, Griset (2002:299) observed, "Like a dysfunctional family, the three branches of government in different places and at different times, have employed a variety of pathological adaptations in exercising their punishment powers, often resulting in muddled or destructive punishment policy." Whereas one branch might be trying to lengthen terms, another might be seeking to reduce them. One set of decision makers might be trying to increase the use of imprisonment, and another seeking to increase diversion from prison. In short, the shared nature of sentencing power, coupled with the political importance of crime and punishment, often led to inconsistent and contradictory sentencing actions.

Although presumptive and mandatory sentencing schemes entered the criminal justice system as part of America's "get tough" movement away from rehabilitation, the actual impact of these sentencing reforms on prison admissions is unclear. Harmon (2013, 2016) found that presumptive sentencing laws generally increased the imprisonment rates over time. However, other researchers (e.g., Rengifo & Stemen, 2015; Stemen & Rengifo, 2011) have found determinate sentencing schemes to lower imprisonment rates. The decision to grant parole has become increasingly politicized (Hatheway, 2017), which resulted in less parole releases during the get tough era. Alternatively, increased financial pressure during economic recessions might encourage the increased use of parole. These types of public pressures impact determinate sentencing systems most on the front end (judges), while indeterminate sentencing options are influenced by these public pressures at the back end (paroling authority).

Deterrent sentences are best defined and imposed by the legislature, so that it is clear to everyone beforehand that a specific punishment will follow a certain criminal conviction. This is the rationale behind **mandatory minimum sentences**. In these sentences, the legislature decrees that anyone convicted of a particular offense (say, use of a handgun in the commission of a crime or drunk driving) receives a minimum number of days or years of incarceration. The prison sentence is mandatory and legislators expect it will deter offenders. Most states and the federal government have mandatory minimum prison sentences for offenders convicted of committing at least some felonies. However, as states continue to face challenges in funding the large prison population, many have begun to relax their mandatory minimum sentences on certain offenses (Porter, 2016).

Just deserts, or retributive, sentencing is perhaps best accomplished by granting limited sentencing power to the judge. The legislature sets an expectation or limit on how severe a penalty may be imposed for a particular criminal act, but the judge is expected to "fine-tune" the sentence so that the severity of the sentence matches the severity of the offense.

Rehabilitative (treatment) and incapacitative sentences work best by granting substantial control in sentencing to the paroling authority. This authority can then determine when the offender is safe for release (that is, cured of criminal tendencies) and can adjust the sentence accordingly. The legislature and sentencing judge are ill-suited to this task because either they do not deal directly with the offender or their roles occur too early in the process. Both of these rationales require that someone continually monitor the offender for progress.

Alternatively, for the purpose of incapacitation, the legislature can develop **habitual offender statutes** that allow for increased penalties for repeat offenders. One such law is the California "three strikes and you're out" law. This law was enacted as a result of a voter initiative (referendum) and provides for a doubling of the sentence upon a second felony conviction "strike," and a tripling of the sentence (or 25-year term, whichever is longer) on the third

FIGURE 9.2
A group of children display photos of their father, who was jailed for a term of 25 years to life, during a protest against California's three-strikes law.

AP Photo/Damian Dovarganes

"strike" (Harris & Jesilow, 2000). The problem with such laws is that they seek to imprison all offenders meeting the criteria (e.g., a third felony conviction) and risk relatively high rates of false positives (Saint-Germain & Calamia, 1996; Turner, Sundt, Applegate, & Cullen, 1995) (see Figure 9.2). In practice, the prosecutors and judges often have discretionary powers to divert three-strikes eligible offenders to less severe sentences (Chen, 2014). Most contemporary **three-strikes laws** have been written so they are applied only to the most violent repeat offenders (Chen, 2008).

Datta (2017) assessed the long-term deterrence and incapacitation impacts of California's three-strikes law. She found that, although the law's incapacitation effect was minimal, there did seem to be a deterrent effect for both property and violent crime. Interestingly, she found that the deterrent effect was strongest for the second instead of the third strike. The policy question, of course, is whether the increased costs in terms of incarceration are justified by the modest reductions in future crime.

SENTENCING IN THE JUSTICE SYSTEM

Criminal sentencing is the final decision point in the court segment of the criminal justice system (with the possible exception of probation revocation, which we discuss in Chapter 12). It represents a transitional decision point at which judicial and correctional officials jointly determine the fate of the offender. After taking into account the statutory sentencing guidelines, testimony presented to the court by the offender and/or victim, and information presented to the court in the presentence investigation, it is usually the judge who makes the final sentencing decisions. In capital cases, the jury will often either make the sentencing decision or recommend a sentence to the judge.

The decision about sentence has two stages. First, the sentencing judge decides whether to incarcerate the convicted offender. Next, the conditions of sentence are determined. These conditions range from the restrictions and obligations placed on those who receive probation to the length of term for those incarcerated.

Historically, sentencing judges had a greater voice in setting the conditions of prison confinement for sentenced offenders. The judge used to be able to specify the institution to which an offender would be sent, or to require that the sentence be served "at hard labor." Today, most jurisdictions restrict the judge to setting the term of confinement, and leave to the correctional authorities the discretionary power to establish the place and conditions of confinement.

Especially in indeterminate sentencing systems based on a rehabilitative rationale, the sentencing decision is expected to be based on the results of a **presentence investigation (PSI)** conducted by a probation officer. The PSI report describes the offense and offender and, in many jurisdictions, includes a recommendation of sentence from the reporting officer (Neubauer & Fradella, 2019). PSIs are common in most felony cases across the nation, thereby involving probation officers in the sentencing decision.

Presentence Investigation

After conviction, the judge generally sets a date for sentencing that is delayed long enough to allow the probation department to conclude a PSI. This investigation includes a detailed assessment of the offense and the offender's criminal and social background. The probation department prepares and submits the PSI report to the sentencing judge for use in arriving at an appropriate sentence. In addition to its usefulness at the sentencing stage, the PSI serves many other purposes. It is a basic information source for the correctional programming used in developing probation plans and institutional classifications; it provides background data used by parole boards in their decisions; and it serves as a basic resource document for research on correctional authorities, offenders, and corrections (Ellis, 2014; Stinchcomb & Hippensteel, 2001).

The typical PSI report includes a face sheet, which contains basic offense and offender demographic information. The textual part of the report generally covers (1) the offense giving the official (police) version and the offender's version of the facts (occasionally the victim's version is also added); (2) the social history of the defendant (describing his or her childhood) and current family, employment, economic, and educational situations; and (3) the prior criminal record of the offender. The probation officer obtains much of this information through interviews with the offender, members of the offender's family, and others who know the offender. The material in the report need not meet the strict evidentiary standards of trial. Carman and Harutunian (2004) assessed the accuracy of federal presentence reports and found that the reports often contained disputed or inaccurate information that was based on hearsay evidence or information otherwise not permitted to be used during the trial. They urged that we improve the process by which disputed information is resolved to ensure that the decisions made based using the reports are fair.

Traditionally, the court need not disclose the PSI report to the defendant. In 1949, the U.S. Supreme Court decided the case of *Williams v. New York*, which dealt with the issue of disclosure of the PSI report. Williams was convicted of capital murder, but the jury had recommended leniency. After reviewing the PSI report, the judge imposed the death penalty. Williams's attorney was not allowed to see the report and verify its accuracy. The Court ruled that there was no constitutional right to review the report. However, in 1977, the Court ruled in *Gardner v. Florida* that when the death penalty is imposed because of information contained in a PSI report, the defense has a right to review that report. It is now commonplace for the presentence investigation to be disclosed to the defense attorney as well as the prosecutor prior to sentencing (Fruchtman & Sigler, 1999).

In jurisdictions where probation officers make recommendations to the sentencing judge, sentencing judges concur with these recommendations in the great majority of cases (Freiburger & Hilinski, 2011; Leifker & Sample, 2011). This is one reason why some argue for disclosure to the defendant. Disclosure of the PSI report, however, is opposed by others, who argue that letting the defendant see the report will cause those interviewed to withhold information, or that disclosure might jeopardize the rehabilitation of the offender, should he or she learn things from the report that would be detrimental to rehabilitative programming.

The Sentencing Hearing

There is no constitutional right to a separate hearing for sentencing, but generally, a sentencing hearing is held for felony defendants. Many states have statutory provisions for sentencing hearings (Rottman, Flango, Cantrell, Hansen, & LaFountain, 2000). In jurisdictions or cases in which there is no PSI, sentencing typically follows conviction. Although there is not a separate hearing, the defendant is afforded the opportunity to speak on his or her own behalf (usually through counsel), and the state (prosecutor) is asked to comment on sentencing. Sometimes the victim is present and allowed to speak or submit a written statement for consideration at sentencing.

At the point of sentencing, whether there is a separate hearing for that purpose, the defendant has the right to counsel. In 1967, in *Mempa v. Rhay*, the U.S. Supreme Court ruled that sentencing was a "critical stage" of the justice system, at which the defendant stands to lose protected rights and interests. The Court held that the offender had the right to representation by counsel at sentencing.

The Parole Hearing

In cases where the offender receives a sentence to incarceration in prison under an indeterminate sentence model, a parole authority usually makes the final sentencing decision. In these cases, the judge imposes a sentence that has a minimum and maximum term, but the actual duration of the penalty is unknown. The paroling authority has the power to grant release (i.e., limit the duration of confinement) at some point between the end of the minimum term and the end of the maximum term. The entire board or a panel of board members usually makes the decision after a hearing with the inmate. Some parole boards advise the governor, but the governor holds the authority to grant parole. Most states continue to use parole as a means of release for prison inmates (Kaeble, 2018). In these states, the parole release decision is an important part of the sentencing process.

Parole eligibility defines which types of inmates the parole board can release at what points in their sentence. The legislature determines parole eligibility. For example, the legislature might require that the offender serves a minimum term before release is authorized. Similarly, the legislature can define certain offenses (or offenders) as ineligible for parole. In cases in which states use the "life without parole" sentence for murder, the legislature has defined murderers as "ineligible" for parole. Given eligibility, the parole authority can grant release to a prison inmate.

At the parole hearing, the paroling authority reviews the criminal and social history of the offender, assesses his or her adjustment to prison, and evaluates the offender's potential for success under parole supervision. This is often accomplished with the assistance of a risk assessment tool, though the parole board can depart from its recommendation due to other factors presented at the hearing or simply the board's determination that the defendant has not served sufficient time in prison (Rhine, Petersilia, & Reitz, 2017). Morgan and Smith (2008) reported that institutional factors such as prison crowding and fiscal pressures influenced overall rates of parole. The inmate can speak at this hearing and present whatever evidence he or she thinks is relevant. In recent years, the victim or victim's family has also been granted increased access to attend or speak at parole hearings.

Many have criticized discretionary release on parole for failing to protect the community by releasing dangerous offenders early as well as for violating individual rights by limiting the defendants' procedural protections. McGarraugh (2013) argued that the use of risk assessment tools is problematic at the parole stage because the defendant has no right to counsel at the hearing and therefore is unlikely to successfully challenge the reliability of the instrument. For the purposes of an offender's right to counsel, the Court has not recognized a parole hearing as a critical stage. From the offender's viewpoint, however, the parole hearing is obviously an important procedure and officials should have the most accurate information available to make a decision.

CONDITIONS OF SENTENCE

Whether the judge chooses probation or incarceration, the second decision in sentencing relates to the conditions of sentence. In cases of probation, the judge retains broad discretionary power to set the conditions of supervision. With incarceration sentences, the legislature shares sentencing power with correctional authorities.

Probation

Probation, as a punishment for criminal behavior, is conditional liberty. The convicted offender can remain in the community under the supervision of a probation officer, provided he or she abides by certain conditions of conduct established by the judge. Should the probationer violate the conditions of release, he or she may be taken before the court for a hearing, at which time probation may be revoked and the probationer sentenced to prison.

Probation may be imposed in a variety of ways. The most common way in which probation is imposed is when a pronounced prison sentence is ordered suspended; that is, when the execution of a sentence (the taking of an offender to prison) is suspended. Another very common practice is suspended sentencing, in which the offender is placed on probation before a sentence (prison term) is actually pronounced. A third method entails a direct sentence to probation, in which the sentence is a probation term. Other, less formal mechanisms for probation may also be used, but these more closely mirror "diversion" programs than criminal sentences. In diversion cases, the court does not enter or record the conviction if the defendant successfully completes the probation term.

A probation officer who works for the court supervises the offender on probation, but the probationer is responsible to the judge. The judge establishes the conditions of release, including such things as "punish lessons," curfews, partial incarceration, and community service (Parisi, 1980; Umbreit, 1981). Because of the large number of judges, and their individual discretion, probation conditions include a wide variety of restrictions and prescriptions (Siegel, Schmalleger, & Worrall, 2011). We discuss these more fully in Chapter 12.

Incarceration

Although we examine incarceration in detail in Chapter 10, the conditions of a sentence of incarceration involve two dimensions: the type of facility and the length of term. In any jurisdiction with more than one correctional facility, there will be differences in the experience of being imprisoned depending on where the offender is incarcerated. Prisons differ in terms of population (hardened, dangerous criminals, or first offenders), type of programs offered (educational/vocational or industrial), and level of custody (maximum, medium, or minimum security). As an example of the variance in prison experiences, serving 3 years by working outside on a prison farm for half of the day and attending school for the other half is qualitatively different from spending the same amount of time locked inside a huge maximum-security prison making automobile tags all day.

Sometimes statute controls the decision as to where an offender will serve his or her sentence, so that offenders older than a certain age or those convicted of a specified crime must be incarcerated in a maximum-security prison. What the offender will do while incarcerated, and where he or she will be incarcerated, however, are questions that often are left to the discretion of correctional authorities.

The length of sentence is controlled by statutory provisions that define parole eligibility (minimum term) and by the **good time** (reductions in length of prison sentence for good behavior while incarcerated) policies and laws of the jurisdiction. Most states shorten an offender's prison term for good behavior while incarcerated (Rottman et al., 2000; Steiner & Cain, 2017). Some states only credit good time against the maximum term, which serves to advance the date of mandatory release. In others, it counts against the minimum term, advancing the date of parole eligibility. Still others apply good time credit against both the minimum and maximum terms. Thus, the legislature, judge, and parole board all have some power regarding sentence length, but correctional administrators who award and revoke good time also share in this decision.

Most states with provisions for good time award reductions of sentence on a "per days served" basis. Good time is to be an incentive to encourage inmates to obey prison rules. Emshoff and Davidson (1987) recognized that sentence reductions may affect inmate behavior, but contended that other factors are at least as important. States vary widely on how they administer good time (National Conference of State Legislatures, 2016). However, a model might award 1 day of good time for every 2 served, so that after serving 20 days, officials add 10 days to complete the month. This model reduces a 6-year sentence to 4 years (one-third reduction). Some states award good time on a sliding scale so the longer the sentence, or the longer the term served, the more good time days received. An inmate could earn 10 days for every 30 for the first 3 years, and 15 days for every 30 for the next 3 years, and so on, "sliding" up the scale of good time award. To this sentence may be added "meritorious" good time, earned by the inmate for special accomplishments such as exceptional industry, donating blood, and the like. Reductions for good behavior, then, can be substantial.

Whatever the reductions for good behavior, other factors complicate time calculations. For an offender unable to secure pretrial release on bail, what part of his or her pretrial time counts as time served? The time spent in jail after conviction but before going to prison generally counts toward a prison sentence. The decision whether to count the time spent in jail prior to conviction is discretionary with the judge or correctional authorities. Thus, an offender who is in jail for 6 months prior to conviction and 3 months after conviction, prior to transport to the prison, may be granted 9 months off his or her sentence, and generally must be granted 3 months.

ISSUES IN SENTENCING

Like the rest of the justice system, sentencing is fraught with questions and unresolved issues. The sentencing process is constantly emerging and evolving, sometimes returning to earlier practices and procedures previously abandoned as inappropriate. Today sentencing is undergoing considerable scrutiny and reform. In the latter part of the nineteenth century, we began to abandon corporal and capital penalties and long, harsh prison terms in favor of rehabilitative strategies that included more flexible prison sentences. In the last quarter of the twentieth century, we abandoned rehabilitative, indeterminate sentences in favor of more retributive fixed terms. Today, we are moving away from retributive sentences toward an unclear future.

We will examine four core issues that allow us to understand better the complex problems of criminal sentencing. Disparity is a central issue today and is the concern at the base of many efforts to change criminal sentencing. Capital punishment has reemerged (perhaps continued) as an issue in sentencing. The calculation of prison time and "truth in sentencing" are still points of contention, and, finally, we continue to grapple with the development of intermediate or alternative sanctions.

Disparity

Disparity refers to the unequal treatment of similar offenders at sentencing (Gottfredson, 1979). Most of us would agree that offenders with similar criminal histories convicted of the same offense ought to receive similar penalties. When differences in sentences appear among similar offenders, the differences generally are termed "disparity." Yet, disparity actually refers to *unwarranted* differences (Gottfredson, Cosgrove, Wilkins, Wallerstein, & Rauh, 1978).

For example, if a jurisdiction follows an incapacitation sentencing rationale and has two first-offender burglars, it would not be disparity if one offender who was determined to pose a great risk of further crime receives a prison sentence, when the other, thought to pose little risk, receives probation. Under the concerns of incapacitation, these offenders are not similar. The current debate over disparity hinges, in large measure, on the definition of "similar" (Vining, 1983).

Since 1976, several states and the federal government have enacted legislation that alters their sentencing structures. In large measure, these changes seek to increase equity (reducing disparity) in criminal sentences (Anspach & Monsen,

1989; Ulmer, 1997). Approximately half of the states and the federal government have created **sentencing commissions**, formal bodies assigned to assess and oversee criminal sentencing and recommend reforms (Watts, 2016.). In coordination with these commissions, several states have implemented presumptive or determinate sentencing to ensure that similarly situated offenders receive similar sanctions. The results of these reforms are still unclear and evaluations continue (Grunwald, 2015; Wooldredge, 2009).

One type of determinate sentencing that states and the federal government have adopted in hopes of controlling disparity and ensuring more certainty in punishment is the use of sentencing guidelines. Sentencing guidelines identify the factors that judges should consider in sentencing and give an indication of what would be an acceptable penalty for each of many different types of cases. By 2017, 15 states and the federal government had developed sentencing guidelines (Mitchell, 2017). As shown in Figure 9.3, in 5 jurisdictions, the guidelines are mandatory; in 10, they are advisory. Alabama has both advisory and mandatory sentencing guidelines, depending on which type of felony is committed. Mitchell (2017) reports that, in practice, each state allows judges to deviate from the sentencing guidelines.

In the spring of 2004, the U.S. Supreme Court decided the case of *Blakely v. Washington*. The sentencing guidelines of the state of Washington allowed the judge to increase a prison term if the judge found that the offense or offender

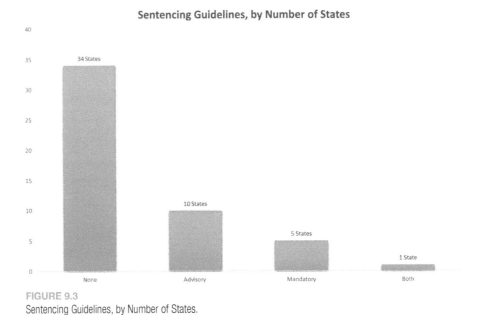

FIGURE 9.3

Sentencing Guidelines, by Number of States.

Source: Mitchell (2017).

represented an aggravated or more serious case than normal. The Supreme Court ruled that relying only on a judicial determination of facts in this instance violates a constitutional right to have facts determined by a jury. It is still unclear what might be the full effects of this decision. Although some commentators feared the decision spelled the end to sentencing guidelines, a more cautious and restrained assessment of the decision suggests that sentencing guidelines, especially in most state systems, can survive constitutional challenge (Frase, 2007; Skove, 2004). The next year the Court decided *United States v. Booker* (2005), in which it ruled that even though federal judges must consider the federal sentencing guidelines, they can alter sentences for other factors. Following the *Booker* decision, the judge's impact on a defendant's sentence length has dramatically increased (Scott, 2010; Yang, 2014).

Defendant race and sex seem to influence sentencing. A large body of research reports that the age and sex of the defendant explain at least some variation in criminal sentences (Goulette, Wooldredge, Frank, & Travis, 2015; Kutateladze, Andiloro, Johnson, & Spohn, 2014; Ulmer, 2012). Women generally receive less severe sentences than men do. This difference not only reflects differences in the types of offenses and prior records of men and women but also probably reflects the justice system's recognition of the sex differences that exist in society. There are similar persistent differences in the sentencing of African Americans as compared to Caucasians. African-American defendants generally receive more severe sentences than Caucasians do, but again the picture is complicated by differences in rates of conviction for certain types of offenses and in prior criminal records of Caucasians and African Americans. It may be that these differences in offenses and criminal records explain some of the differences in sentencing, yet it is also likely that criminal sentencing reflects the disadvantaged status of minority group members in American society (Schlesinger, 2011).

BOX 9.2 POLICY DILEMMA: RACIAL IMPACT STATEMENTS

Research has shown that even seemingly "colorblind" crime policies often have disproportionate impacts on racial minorities. For example, this can occur when a state increases the punishment for actions that are most likely committed by minorities, or when risk assessment tools increase the punishment based on traits that minorities are most likely to have (Schlesinger, 2011). Some states have responded to this reality by requiring racial impact statements to accompany any proposed law change involving the criminal justice system. These statements are similar to fiscal and environmental impact statements that are common in many states. Though racial impact statements are currently only required in a handful of states, several states have debated the merits of this idea. Many scholars (e.g., Smith, 2017) advocate for this as a way to force legislators to at least consider the racial outcomes of proposed legislation. However, Erickson (2014) points out that the presence of a racial impact statement does not require legislators to act on this information. She proposes that states should go an extra step and require legislators to take some sort of action (e.g., public comment, comparing alternatives, etc.) if an impact statement indicates a probable disparate impact on racial minorities. Would you support a requirement that a racial impact statement be considered prior to your state adopting a new crime-related law?

Commentators identify the culprit for sentencing disparity as the individual whims of justice system officials (judges, parole board members, etc.) that come out in the exercise of discretionary authority. Steffensmeier, Ulmer, and Kramer (1998) identified the "focal concerns" of judges at sentencing. These are public protection, blameworthiness, and practical considerations. Judges base sentences on how "bad" and how dangerous the defendant is, as well as the effect of decisions on families, communities, and the justice system. Cano and Spohn (2012) reported that "sympathetic" or "salvageable" offenders received lower sentences in federal courts. For example, offenders who were employed, educated, female, and had dependent children often received sentences below the federal mandatory minimum sentence.

Several other explanations have been proposed to examine the racial and gender differences in sentencing outcomes. Some support exists for the idea that females might benefit from some "chivalrous" treatment in sentencing (Embry & Lyons, 2012; Griffin & Wooldredge, 2006). The racial threat hypothesis assumes that punishment of minorities will increase in locations that minorities represent greater proportions of the population. However, evaluations of the racial threat hypothesis have been mixed (Feldmeyer & Ulmer, 2011; Feldmeyer, Warren, Siennick, & Neptune, 2015; Wang & Mears, 2010). Doerner and Demuth (2010) observed that the race and sex of defendants interact and combine with earlier decisions (arrest, charging, etc.) to complicate an understanding of criminal sentences. Indeed, the impact of a "cumulative disadvantage" faced by minorities in the justice system has received more attention in recent years (Kutateladze et al., 2014).

Capital Punishment

In America, from colonial times through the early part of the nineteenth century, most punishments for serious offenses involved physical pain, such as branding, maiming, flogging, and death. In recent history, we have abandoned those forms of punishment in favor of the more humane alternative of incarceration. Yet, the execution of those convicted of heinous murders has continued. In 2017, 23 inmates were executed and more than 2,800 inmates were on death row. Between 1977 and 2017, there were more than 1,400 executions in 35 states and the federal system (Death Penalty Information Center, 2018). However, the number of death penalty sentences have significantly declined since the peak in the mid-1990s. Figure 9.4 shows the trend in death penalty executions.

Many factors have contributed to the decline in death sentences, including the increased presence of anti-death penalty activism and the impact of several court rulings limiting the scope of the punishment. The U.S. Supreme Court has not ruled the death penalty to be cruel and unusual punishment or to be otherwise unconstitutional. Moreover, in the 1976 case of *Gregg v. Georgia*, the Court held that the death penalty, per se, is not unconstitutional. Earlier, in

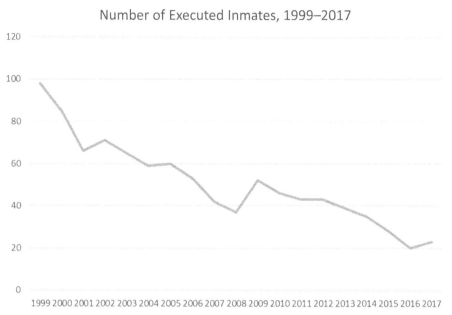

Number of Executed Inmates, 1999–2017

FIGURE 9.4
Number of Inmates
Executed, by Year of
Sentencing.

Source: Davis and Snell
(2018); Snell (2014,
2017).

Furman v. Georgia (1972), the Court focused on the procedures by which states imposed the death penalty. Having found the existing procedures too vague and unstructured, the Court did not address the issue of the death penalty itself. The *Gregg* decision was the first of several in which the Court reviewed revised procedures for imposing the death penalty. Given an acceptable procedure (jury recommendation based on a presentation of aggravating and mitigating factors identified in the statute), the Court also indicated that the death penalty was constitutional.

More recently, the Court has heard several challenges related to the constitutionality of the death penalty for certain types of offenders. First, the Court ruled in *Coker v. Georgia* (1977) that the death penalty was disproportionate punishment for a man convicted of raping an adult. In 2002, the Court decided that the death penalty was unconstitutional for those suffering from certain intellectual disabilities (*Atkins v. Virginia*, 2002). Three years later, the Court extended this ban to offenders who were juveniles at the time of the crime (*Roper v. Simmons*, 2005). Finally, the Court disallowed the death penalty as a punishment for a defendant convicted of child rape (*Kennedy v. Louisiana*, 2008).

The most contemporary challenge to the death penalty relates to the method of execution. Virtually all modern executions are carried out by lethal injection. Traditionally, states have used a three drug combination which would effectively render the inmate unconscious, paralyze their muscles, and then induce a heart attack. In 2008, the Supreme Court upheld this lethal injection process (*Baze v. Rees*, 2008). However, death penalty opponents successfully mounted a public pressure campaign against most of the domestic and international drug manufacturers which supplied the drugs needed for states to carry out their planned executions. As the struggle to obtain the drugs intensified, states began to pass new laws changing the execution protocol and/or not publicly disclosing the protocol or the identity of the state's supplier of the required drugs. In turn, this has prompted many (e.g., Crider, 2014; Mennemeier, 2017) to question the constitutionality of these state secrecy laws. Others argue that, ironically, the efforts of anti-death penalty advocates to slow down the pace of executions has had the unintended consequence of making the lethal injection process less safe (Fan, 2015). In any event, the shortage of lethal injection drugs and subsequent legal challenges have dramatically decreased the number of executions carried out in recent years.

Some support the death penalty as an incapacitative sanction, saying that murderers pose too great a risk to society. Vito and Wilson (1988) addressed this issue and found that most prisoners sentenced to death in their sample were not particularly dangerous. Fewer than 25% of death row inmates who were released committed new crimes, and none committed a new homicide. Others suggest that an alternative to capital punishment is to impose life sentences without possibility of parole (Whitehead, Braswell, & Burgason, 2017).

The debate over capital punishment continues. The public, it appears, is in the midst of a shift regarding its view of capital punishment. An annual Gallup public opinion survey revealed that in 2017, 55% of Americans favored the death penalty for an offender who had been convicted of murder. However, this was the lowest level of support for capital punishment since 1972. Not surprisingly, political affiliation is a strong predictor of one's views of the death penalty. Whereas 72% of Republicans currently favor capital punishment, only 39% of Democrats support the punishment. It is important to note that support among all political parties has declined in recent years (Jones, 2017).

Using Time as a Penalty: Truth in Sentencing

We already have seen how provisions for good time affect sentencing decisions by adding correctional administrators to the list of officials who exercise authority in sentencing decisions. What is not as clear is the complexity of using time as a penalty. Although most states allow the death penalty for heinous murders, the basic limitation on criminal penalties in the United States is time. Although conditions of confinement vary, for the individual offender the length of term is critical. How long an offender serves is dependent on several factors, each of which has importance for the severity of punishment and the operation of correctional institutions.

An important consideration is the ability of the judge to impose terms for multiple convictions to run either consecutively or concurrently. A **consecutive** term is one in which each sentence must be served in order, one following the other. A concurrent term is one in which all the sentences run at the same time. Two 5-year terms imposed consecutively total 10 years; the same terms imposed concurrently total 5 years.

The unresolved problem in the calculation of prison time is that the imposed sentence generally is not the same as the actual term served. von Hirsch and Hanrahan (1979) suggested that one of the major difficulties faced in sentencing reform is that we do not sentence offenders in "real time." The public sees the sentence imposed as being the maximum term ordered by the judge, without reference to good time or to the fact that judges frequently impose concurrent terms. If an offender sentenced to three concurrent 5-year terms is released in 2 years, the public may begin to believe that the law does not mean what it says. Felons sentenced to prison in state courts typically serve only about two-thirds of their sentences before release (Durose & Langan, 2007).

Concern about the gap between the length of sentence imposed by the court and that actually served by convicted offenders spurred a call for "truth in sentencing." In 1995, Congress passed legislation calling for states to develop truth in sentencing so that those convicted of violent offenses will serve at least 85% of the term they receive. Federal aid was made available to states that adopt such truth in sentencing laws, and many states changed their sentencing laws to achieve sentences that are more truthful (see Box 9.3). Thus far, the solutions sought by the states involve reserving prison space for those convicted

---

BOX 9.3 THE "TRUTH IN SENTENCING" REFORM

Highlights

Three decades of sentencing reform—1970s through 1990s

- *Indeterminate sentencing*: Common in the early 1970s, parole boards have the authority to release offenders from prison.

- *Determinate sentencing*: States introduced fixed prison terms that could be reduced by good time or earned time credits.

- *Mandatory minimum sentences*: States added statutes requiring offenders to be sentenced to a specified amount of prison time.

- *Truth in sentencing (TIS)*: First enacted in 1984, TIS laws require offenders to serve a substantial portion of their prison sentence. Parole eligibility and good time credits are restricted or eliminated.

- Violent offenders released from prison in 1996 were sentenced to serve an average of 85 months in prison. Prior to release they served about half of their prison sentence or 45 months.

- Under TIS laws requiring 85% of the sentence, violent offenders would serve an average of 88 months in prison based on the average sentence for violent offenders admitted to prison in 1996.

- Nearly 7 in 10 state prison admissions for a violent offense in 1997 were in states requiring offenders to serve at least 85% of their sentence.

- By 1998, states and the District of Columbia met the federal Truth in Sentencing Incentive Grant Program eligibility criteria. Eleven states adopted TIS laws in 1995, a year after the 1994 Crime Act.

Source: Ditton and Wilson (1999:1).

of violent crimes by reducing the rate of imprisonment and length of terms for nonviolent offenders. Of course, one solution to the problem would be the imposition of prison sentences in real time. However, most legislators are unwilling to admit to the public that violent offenders, on average, will serve a little more than 4 years instead of the 10 years to which they are sentenced.

The impact of truth in sentencing is likely to be the diversion of repeat, non-violent offenders to community sanctions such as probation. The effects of these changes on both community corrections and prison populations are difficult to estimate. It is clear, however, that truth in sentencing affects both prison and community corrections populations. Turner, Greenwood, Fain, and Chiesa (2006) reviewed the impact of truth in sentencing laws, noting that whereas many states did in fact pass legislation to ensure that violent offenders would serve at least 85% of their sentence, most of the states would have done so without the federal aid. Instead, public concern over crime was a key factor influencing legislation. In addition, many states complied with the 85% threshold by reducing the average sentence length, making it easier for offenders to serve the required 85% of their sentence without significantly impacting actual time served. This finding was confirmed by Durose and Langan (2007), who reported that violent felons now receive shorter sentences but serve a greater percentage of their sentence.

Owens (2011) reported that truth in sentencing laws might have a broader impact than first envisioned. Specifically, offenders arrested for violent felonies but convicted on lesser charges are treated more punitively in states that have adopted truth in sentencing laws. Judges in these cases seem to honor the spirit of the truth in sentencing laws, showing yet again how a punishment targeting a specific group of defendants can have impact on the system as a whole.

Intermediate Sanctions

Throughout the past two decades, there has been increasing interest in the development and implementation of intermediate sanctions for crime or punishments that fall somewhere between imprisonment and traditional probation. Historically, sentencing judges chose between community supervision and confinement in prison or jail as the primary sanctions for crime. Especially as the numbers of people convicted of crime have increased, and prison and jail populations have exceeded the capacity of institutions, these choices have been seen as inadequate. DuPont (1985) called for the development of a meaningful continuum of sanctions ranging from no restriction to maximum-security incarceration. Others echoed the call (Klein, 1997; Morris & Tonry, 1990; Petersilia, 1987).

In response to the perceived need for a wider variety of sanctions, a number of innovative practices emerged. These include various combinations of incarceration and community supervision dispositions. We discuss specific forms of these sanctions in more detail in Chapters 12 and 13, but a brief listing is in order. Sanctions combining incarceration with supervision have emerged, including shock probation, shock parole, placement in community residential facilities, home incarceration, and **split sentences** (sentences combining a period of incarceration with a period of probation supervision) (Petersilia, Lurigio, & Byrne, 1992). So, too, penalties have been developed that make the experience of community supervision more severe. These include day reporting, intensive supervision, electronic monitoring, and community service orders.

The growth of these penalties blurs the traditional distinction between incarceration and community supervision. Prison **boot camps** involve a short, intense incarceration. The offender goes to prison, but for a much shorter period. Observers note that these sanctions allow gradations in punishment along a continuum (Holsinger & Latessa, 1999; McCarthy, 1987; Schwartz & Travis, 1996). Thus, penalties can escalate from traditional probation supervision through intensive supervision to day reporting (offenders report to the probation office daily). If we desire more restrictive punishments, the offender could be placed under house arrest (restricted to his or her home), sent to a residential facility (halfway house), or sent to prison. As Table 9.1 shows, judges often impose multiple penalties on convicted offenders.

See Table 9.2 for a table of selected court cases on sentencing.

Recalling our discussion of sentencing disparity, the development of a range of punishments is potentially troublesome. If we suspect sentencing decisions of reflecting prejudice when only a few choices were available, the provision of more choices increases the chances of disparate treatment. The more things

TABLE 9.1 Conditions of Probation Sentence Received Most Often by Convicted Defendants in the 75 Largest Counties, by Most Serious Conviction Offense, 2009				
	Percent whose sentence to probation included—			
Most Serious Conviction Offense	Fine	Treatment	Community Service	Restitution
All offenses	28%	27%	23%	20%
All felonies	28	29	22	21
Violent offenses	26	29	27	22
Property offenses	19	18	23	37
Drug offenses	36	45	21	7
Public-order offenses	27	16	22	26
Misdemeanors	27	13	26	14

Source: Reaves (2013:32).

TABLE 9.2 Selected Court Cases on Sentencing

Williams v. New York 377 U.S. 241 (1949)	There is no constitutional right to review a presentence investigation (PSI) report. See later controlling case, *Gardner v. Florida* (1977)
Mempa v. Rhay 389 U.S. 128 (1967)	An offender has the right to representation by counsel at the point of sentencing
Furman v. Georgia 408 U.S. 238 (1972)	Current procedures for imposing the death penalty violate the equal protection clause of the Fourteenth Amendment and the prohibition against cruel and unusual punishment
Gregg v. Georgia 428 U.S. 153 (1976)	Death penalty statutes that contain sufficient safeguards against arbitrary and capricious imposition are constitutional
Gardner v. Florida 430 U.S. 349 (1977)	When the death penalty is imposed as a result of information contained in a presentence investigation (PSI) report, the defense has a right to review that report
Apprendi v. New Jersey, 530 U.S. 466 (2000)	Other than the presence of a prior conviction, any factor that increases the penalty beyond the statutory maximum must be proven beyond a reasonable doubt by a jury
Ring v. Arizona, 536 U.S. 584 (2002)	The presence of any aggravating factors that would increase punishment from life imprisonment to death must be determined by a jury
United States v. Booker, 543 U.S. 220 (2005)	Federal sentencing guidelines are advisory rather than mandatory

Source: Portions of this table were adapted from del Carmen, Ritter, and Witt (2008).

the judge can do to punish an offender, the greater the judicial discretion. Conversely, intermediate sanctions may reduce the relative degree of disparity. Rather than a simple "either/or" issue of incarceration, disparate sentences may be closer in degree of restrictiveness. The difference between intensive supervision and day reporting is not as great as that between probation and prison. The Sentencing Project reports that, faced with financial emergencies, many states have changed sentences and developed alternatives to incarceration (Porter, 2010). O'Hara (2005) questioned the ability of alternative sanctions or processes to ensure uniformity in sentencing. Specifically regarding restorative justice, O'Hara argued that if we seek equivalent sentencing outcomes, we probably need to reduce, rather than increase, sentencing choices.

THEORY AND SENTENCING

What explains criminal punishments? How is it that persons convicted of the same offense are often given different penalties? If we want to change criminal sentencing, how can we best accomplish it? Do changes in the law or changes in the characteristics of sentencing officials lead to changes in sentencing?

As with other aspects of the justice system, our answers to these questions reveal our theoretical understanding of the issue. Our review of sentencing identified several factors that can, and do, influence sentencing. The serious-ness of the crime, the risk posed by the offender, and the characteristics of both defendants and officials influence the sentencing outcome. Research indicates that the characteristics of the court, its organization and culture, also affect sentencing decisions. It is important to recall that sentencing decisions also reflect the directional flow of decisions in the process. Earlier decisions about investigation, arrest, bail, and charging exert influence on sentence outcomes.

Although there are some specific sentencing theories such as focal concerns, racial threat, and the chivalry hypothesis, theory has not guided much of the research in this area. The same kinds of factors that appear to explain police decisions and earlier court outcomes also relate to sentencing decisions. It is useful to remember these explanations as we proceed. In the end, we may be able to identify some broad categories of criminal justice theory.

DUE PROCESS, CRIME CONTROL, AND SENTENCING IN THE WHOLE SYSTEM

The sentencing decision links the court subsystem to the correctional sub-system of the justice process. Sentencing decisions reflect the demands and stresses placed on corrections, and indicate how the justice process reacts to social change. For example, with the emergence of drug abuse as a serious crim-inal problem, the sentencing of those convicted of drug-related offenses has changed. Many more drug offenders now receive prison sentences than in the past, and their sentences to prison are longer than before. As society came to see drug offenders as more serious criminals, the increased seriousness was reflected in more severe sentences. These sentences have changed the compo-sition of correctional (especially prison) populations (Barnes & Kingsnorth, 1996; Beck, 1997; Carson, 2018).

Perhaps the most contemporary wave of social change that has influenced crim-inal justice sentencing pertains to sex offenders. Partly as a result of increased media attention, sex offenders are currently among the country's most hated criminals. As a result, the public has demanded the implementation and enforcement of specialized punishments for those convicted of sex offenses (Koon-Magnin, 2015). These laws often include increased prison time, sex offender registrations, housing restrictions, and electronic monitoring. The interplay between the public's perceptions, the legislator's actions, and the cor-rectional officials who are tasked with enforcing these specialized punishments illustrates the systematic nature of the justice process.

Remember that one of the primary motivations for reform of criminal sen-tencing and the development of alternative sanctions is a desire for fairness. Not only do we seek punitive and incapacitative alternatives to prison for crime control reasons, we seek equitable sanctions for diverse individuals. The sentencing decision is the point at which the full weight of state power bears on the individual. The challenge for sentencing is to achieve crime control goals while respecting the rights of the individual offender. Yet, available evidence suggests that criminal sentences reflect differences in sex, race, and age. If we seek equal treatment of persons at sentencing, it is likely that we will first have to develop equal treatment of persons in the broader society.

Perhaps the major obstacle to reforming criminal sentencing in the United States lies not in sentencing's considerable complexity of operation and calculation, but more in the lack of effort among our citizens to achieve understanding of the sentencing process. To achieve a rational and workable system of sentencing, it may first be necessary to be more open and honest about the meaning of criminal sentences—and more consistent in our purposes served by criminal sanctions.

REVIEW QUESTIONS

1. Identify the four traditional purposes of criminal penalties.
2. What types of errors are involved in the prediction of dangerousness at sentencing?
3. Distinguish between determinate and indeterminate criminal sentencing structures.
4. What is the presentence investigation, and what purposes does it serve?
5. What is the sentencing hearing, and what takes place at one?
6. Identify two principal types of sanctions imposed on serious criminal offenders.
7. What is meant by "parole eligibility"?
8. What is a sentencing disparity?
9. Besides sentencing disparity, identify two other current issues in criminal sentencing.

REFERENCES

Andrews, D.A., & Bonta, J. (2010). *The psychology of criminal conduct*. New Providence, NJ: Matthew Bender.

Anspach, D., & Monsen, S. (1989). Determinate sentencing, formal rationality, and khadi justice in Maine: An application of Weber's typology. *Journal of Criminal Justice, 17*(6), 471–485.

Baker, T., Falco Metcalfe, C., Berenblum, T., Aviv, G., & Gertz, M. (2015). Examining public preferences for the allocation of resources to rehabilitative versus punitive crime policies. *Criminal Justice Policy Review, 26*(5), 448–462.

Bales, W.D., & Piquero, A.R. (2012). Assessing the impact of imprisonment on recidivism. *Journal of Experimental Criminology, 8*(1), 71–101.

Barnes, C., & Kingsnorth, R. (1996). Race, drug, and criminal sentencing: Hidden effects of the criminal law. *Journal of Criminal Justice, 24*(1), 39–55.

Beck, A. (1997). Growth, change, and stability in the U.S. prison population, 1980–1995. *Corrections Management Quarterly, 1*(2), 1–14.

Bouffard, J.A., Niebuhr, N., & Exum, M.L. (2017). Examining specific deterrence effects on DWI among serious offenders. *Crime & Delinquency, 63*(14), 1923–1945.

Bureau of Justice Assistance (1998). *1996 national survey of state sentencing structures.* Washington, DC: Bureau of Justice Assistance.

Bureau of Justice Statistics (1988). *Report to the nation on crime and justice* (2nd ed.). Washington, DC: U.S. Department of Justice.

Cano, M., & Spohn, C. (2012). Circumventing the penalty for offenders facing mandatory minimums: Revisiting the dynamics of "sympathetic" and "salvageable" offenders. *Criminal Justice and Behavior, 39*(3), 308–332.

Carman, G.W., & Harutunian, T. (2004). Fairness at the time of sentencing: The accuracy of the presentence report. *St. John's Law Review, 78*(1), 1–14.

Carson, E.A. (2018). *Prisoners in 2016.* Washington, DC: Bureau of Justice Statistics.

Chalfin, A., & McCrary, J. (2017). Criminal deterrence: A review of the literature. *Journal of Economic Literature, 55*(1), 5–48.

Chen, E.Y. (2008). Impacts of "three strikes and you're out" on crime trends in California and throughout the United States. *Journal of Contemporary Criminal Justice, 24*(4), 345–370.

Chen, E.Y. (2014). In the furtherance of justice, injustice, or both? A multilevel analysis of courtroom context and the implementation of three strikes. *Justice Quarterly, 31*(2), 257–286.

Crider, N.A.W. (2014). What you don't know will kill you: A First Amendment challenge to lethal injection secrecy. *Columbia Journal of Law and Social Problems, 48,* 1–55.

Cullen, F.T., & Gilbert, K.E. (2013). *Reaffirming rehabilitation.* Boston: Elsevier.

Cullen, F.T., Jonson, C.L., & Nagin, D.S. (2011). Prisons do not reduce recidivism: The high cost of ignoring science. *The Prison Journal, 9*(3), 48S–65S.

Datta, A. (2017). California's three-strikes law revisited: Assessing the long-term effects of the law. *Atlanta Economic Journal, 45*(2), 225–249.

Davis, E., & Snell, T.L. (2018). *Capital punishment, 2016.* Washington, DC: Bureau of Justice Statistics.

Death Penalty Information Center (2018). *Facts about the death penalty.* Retrieved March 16, 2018, from https://deathpenaltyinfo.org/documents/FactSheet.pdf

del Carmen, R.V., Ritter, S.E., & Witt, B.A. (2008). *Briefs of leading cases in corrections* (5th ed.). Newark, NJ: LexisNexis Matthew Bender (Anderson Publishing).

Ditton, P., & Wilson, D. (1999). *Truth in sentencing in state prisons.* Washington, DC: Bureau of Justice Statistics.

Doerner, J., & Demuth, S. (2010). The independent and joint effects of race/ethnicity, gender, and age on sentencing outcomes in U.S. federal courts. *Justice Quarterly, 27*(1), 1–27.

DuPont, P. (1985). *Expanding sentencing options: A governor's perspective.* Washington, DC: U.S. Department of Justice.

Durose, M., & Langan, P. (2007). *Felony sentences in state courts, 2004*. Washington, DC: Bureau of Justice Statistics.

Ellis, A. (2014). Federal presentence investigation report. *Criminal Justice, 29*(3), 49–50.

Embry, R., & Lyons, P.M., Jr. (2012). Sex-based sentencing: Sentencing discrepancies between male and female sex offenders. *Feminist Criminology, 7*(2), 146–162.

Emshoff, J.G., & Davidson, W.S. (1987). The effect of "good time" credit on inmate behavior: A quasi-experiment. *Criminal Justice and Behavior, 14*(3), 335–351.

Erickson, J. (2014). Racial impact statements: Considering the consequences of racial dispro-portionalities in the criminal justice system. *Washington Law Review, 89*, 1425–1465.

Fan, M.D. (2015). The supply-side attack on lethal injection and the rise of execution secrecy. *Boston University Law Review, 95*, 427–460.

Feldmeyer, B., & Ulmer, J.T. (2011). Racial/ethnic threat and federal sentencing. *Journal of Research in Crime and Delinquency, 48*(2), 238–270.

Feldmeyer, B., Warren, P.Y., Siennick, S.E., & Neptune, M. (2015). Racial, ethnic, and immi-grant threat: Is there a new criminal threat on state sentencing? *Journal of Research in Crime and Delinquency, 52*(1), 62–92.

Frase, R. (2007). The Apprendi-Blakely cases: Sentencing reform counter-revolutions? *Crimi-nology & Public Policy, 6*(3), 403–432.

Freiburger, T.L., & Hilinski, C.M. (2011). Probation officers' recommendations and final sen-tencing outcomes. *Journal of Crime and Justice, 34*(1), 45–61.

Fruchtman, D.A., & Sigler, R.T. (1999). Private pre-sentence investigation. *Journal of Offender Rehabilitation, 29*(3/4), 157–170.

Gottfredson, D.M., Cosgrove, C.A., Wilkins, L.T., Wallerstein, J., & Rauh, C. (1978). *Classifi-cation for parole decision policy*. Washington, DC: U.S. Government Printing Office.

Gottfredson, M.R. (1979). Parole guidelines and the reduction of sentencing disparity: A pre-liminary study. *Journal of Research in Crime and Delinquency, 16*(2), 218–231.

Goulette, N., Wooldredge, J., Frank, J., & Travis, L., III. (2015). From initial appearance to sentencing: Do female defendants experience disparate treatment? *Journal of Criminal Justice, 43*, 406–417.

Griffin, T., & Wooldredge, J. (2006). Sex-based disparities in felony dispositions before and after sentencing reform in Ohio. *Criminology, 44*(4), 893–923.

Griset, P. (2002). New sentencing laws follow old patterns: A Florida case study. *Journal of Criminal Justice, 30*(4), 287–301.

Grunwald, B. (2015). Questioning Blackmun's thesis: Does uniformity in sentencing entail unfairness? *Law & Society Review, 49*(2), 499–534.

Hamilton, Z.K. (2011). *Treatment matching for substance-abusing offenders*. El Paso, TX: LFB Scholarly Publishing.

Harmon, M.G. (2013). "Fixed" sentencing; The effect on imprisonment rates over time. *Jour-nal of Quantitative Criminology, 29*(3), 369–397.

Harmon, M.G. (2016). Opening the door for more: Assessing the impact of sentencing reforms on commitments to prison over time. *American Journal of Criminal Justice, 41*(2), 296–320.

Harris, J., & Jesilow, P. (2000). It's not the old ball game: Three strikes and the courtroom workgroup. *Justice Quarterly, 17*(1), 185–203.

Hatheway, K. (2017). Creating a meaningful opportunity for review: Challenging the politicization of parole for life-sentenced prisoners. *American Criminal Law Review, 54*(2), 601–625.

Holsinger, A., & Latessa, E. (1999). An empirical evaluation of a sanction continuum: Pathways through the juvenile justice system. *Journal of Criminal Justice, 27*(2), 155–172.

Jacobs, B. (2010). Deterrence and deterrability. *Criminology, 48*(2), 417–442.

Jefferson-Bullock, J. (2016). How much punishment is enough?: Embracing uncertainty in modern sentencing reform. *Journal of Law & Policy, 24*(2), 345–409.

Johnson, I., & Sigler, R. (1995). Community attitudes: A study of definitions and punishment of spouse abusers and child abusers. *Journal of Criminal Justice, 23*(5), 477–487.

Jones, J.M. (2017). *U.S. death penalty support lowest since 1972.* Washington, DC: Gallup.

Kaeble, D. (2018). *Probation and parole in the United States, 2016.* Washington, DC: Bureau of Justice Statistics.

Kaeble, D., & Glaze, L. (2016). *Correctional populations in the United States, 2015.* Washington, DC: Bureau of Justice Statistics.

Klein, A.R. (1997). *Alternative sentencing, intermediate sanctions and probation* (2nd ed.). Cincinnati, OH: Anderson.

Koon-Magnin, S. (2015). Perceptions of and support for sex offender policies: Testing Levenson, Brannon, Fortney, and Baker's findings. *Journal of Criminal Justice, 43*, 80–88.

Krisberg, B. (2016). How do you eat an elephant? Reducing mass incarceration in California one small bite at a time. *Annals of the American Academy of Political and Social Science, 664*, 136–154.

Kubiak, S.P., & Allen, T. (2011). Public opinion regarding juvenile life without parole in consecutive statewide surveys. *Crime & Delinquency, 57*(4), 495–515.

Kutateladze, B.L., Andiloro, N.R., Johnson, B.D., & Spohn, C.C. (2014). Cumulative disadvantage: Examining racial and ethnic disparity in prosecution and sentencing. *Criminology, 52*(3), 514–551.

Leifker, D., & Sample, L. (2011). Probation recommendations and sentences received: The association between the two and the factors that affect recommendations. *Criminal Justice Policy Review, 22*(4), 494–517.

McCarthy, B. (Ed.). (1987). *Intermediate punishments.* Monsey, NY: Criminal Justice Press.

McGarraugh, P. (2013). Up or out: Why "sufficiently reliable" statistical risk assessment is appropriate at sentencing and inappropriate at parole. *Minnesota Law Review, 97*(3), 1079–1113.

Mennemeier, K.A. (2017). A right to know how you'll die: A First Amendment challenge to state secrecy statutes regarding lethal injection drugs. *Journal of Criminal Law & Criminology, 107*(3), 443–492.

Mitchell, K.L. (2017). State sentencing guidelines: A garden full of variety. *Federal Probation, 81*(2), 28–36.

Morgan, K., & Smith, B. (2008). The impart of race on parole decision-making. *Justice Quarterly, 25*(2), 411–435.

Morris, N., & Tonry, M. (1990). *Between prison and probation: Intermediate punishments in a rational sentencing system.* Oxford: Oxford University Press.

Myers, L., & Reid, S. (1995). The importance of county context in the measurement of sentence disparity: The search for routinization. *Journal of Criminal Justice, 23*(3), 223–241.

National Conference of State Legislatures (2016). *Good time and earned time policies for state prison inmates (as established by law)*. Retrieved from https://docs.legis.wisconsin.gov/misc/lc/study/2016/1495/030_august_31_2016_meeting_10_00_a_m_room_412_east_state_capitol/memono4g

Neubauer, D.W., & Fradella, H.F. (2019). *America's courts and the criminal justice system* (13th ed.). Boston: Cengage.

Newman, G.R. (1983). *Just and painful*. New York: Macmillan.

O'Hara, M. (2005). Is restorative justice compatible with sentencing uniformity? *Marquette Law Review, 89*, 305–325.

Owens, E. (2011). Truthiness in punishment: The far reach of truth-in-sentencing laws in state courts. *Journal of Empirical Legal Studies, 8*, 239–261.

Parisi, N. (1980). Combining incarceration and probation. *Federal Probation, 44*(2), 3–12.

Paternoster, R. (1987). The deterrent effect of the perceived certainty and severity of punishment: A review of the evidence and issues. *Justice Quarterly, 4*(2), 173–217.

Paternoster, R. (2010). How much do we really know about criminal deterrence? *Journal of Criminal Law & Criminology, 100*(3), 765–823.

Petersilia, J. (1987). *Expanding options for criminal sentencing*. Santa Monica, CA: RAND.

Petersilia, J., Lurigio, A., & Byrne, J. (1992). Introduction: The emergence of intermediate sanctions. In Byrne, J., Lurigio, A., & Petersilia, J. (Eds.), *Smart sentencing: The emergence of intermediate sanctions* (pp. ix–xv). Beverly Hills, CA: Sage.

Porter, N. (2010). *The state of sentencing 2009: Developments in policy and practice*. Washington, DC: The Sentencing Project.

Porter, N. (2016). *The state of sentencing 2015: Developments in policy and practice*. Washington, DC: The Sentencing Project.

Ramirez, M.D. (2013). American's changing views on crime and punishment. *Public Opinion Quarterly, 77*(4), 1006–1031.

Reaves, B.A. (2013). *Felony defendants in large urban counties, 2009*. Washington, DC: Bureau of Justice Statistics.

Rengifo, A.F., & Stemen, D. (2015). The unintended effects of penal reform: African American presence, incarceration, and the abolition of discretionary parole in the United States. *Crime & Delinquency, 61*(5), 719–741.

Rhine, E.E., Petersilia, J., & Reitz, K.R. (2017). The future of parole release. *Crime and Justice, 46*(1), 279–338.

Rottman, D., Flango, C., Cantrell, M., Hansen, R., & LaFountain, N. (2000). *State court organization, 1998*. Washington, DC: Bureau of Justice Statistics.

Saint-Germain, M., & Calamia, R. (1996). Three strikes and you're in: A streams and windows model of incremental policy change. *Journal of Criminal Justice, 24*(1), 57–70.

Schlesinger, T. (2011). The failure of race neutral policies: How mandatory terms and sentencing enhancements contribute to mass racialized incarceration. *Crime & Delinquency, 57*(1), 56–81.

Schoepfer, A., Carmichael, S., & Piquero, N. (2007). Do perceptions of punishment vary between white-collar and street crimes? *Journal of Criminal Justice, 35*(2), 151–163.

Schwartz, M., & Travis, L. (1996). *Corrections: An issues approach* (4th ed.). Cincinnati, OH: Anderson.

Scott, R.W. (2010). Inter-judge sentencing disparity after *Booker*: A first look. *Stanford Law Review*, *63*, 1–66.

Siegel, L.J., Schmalleger, F., & Worrall, J.L. (2011). *Courts and criminal justice in America*. Upper Saddle River, NJ: Prentice Hall.

Skove, A. (2004). Blakely v. Washington*: Implications for state courts*. Williamsburg, VA: National Center for State Courts.

Smith, J.M. (2017). Racial impact statements, knowledge-based criminology, and resisting color blindness. *Race and Justice*, *7*(4), 374–397.

Smith, W., & Smith, D. (1998). The consequences of error: Recidivism prediction and civil-libertarian ratios. *Journal of Criminal Justice*, *26*(6), 481–502.

Snell, T.L. (2014). *Capital punishment, 2013*. Washington, DC: Bureau of Justice Statistics.

Snell, T.L. (2017). *Capital punishment, 2014–2015*. Washington, DC: Bureau of Justice Statistics.

Sperber, K.G., Latessa, E.J., & Makarios, M.D. (2013). Examining the interaction between level of risk and dosage of treatment. *Criminal Justice and Behavior*, *40*(3), 338–348.

Spiropoulos, G.V., Salisbury, E.J., & Van Voorhis, P. (2014). Moderators of correctional treatment success: An exploratory study of racial differences. *International Journal of Offender Therapy and Comparative Criminology*, *58*(7), 835–860.

Steffensmeier, D., Ulmer, J., & Kramer, J. (1998). The interaction of race, gender, and age in criminal sentencing: The punishment cost of being young, black, and male. *Criminology*, *36*(4), 763–797.

Steiner, B., & Cain, C.M. (2017). Punishment within prison: An examination of the influences of prison officials' decisions to remove sentencing credits. *Law & Society Review*, *51*(1), 70–98.

Stemen, D., & Rengifo, A.F. (2011). Policies and imprisonment: The impact of structured sentencing and determinate sentencing on state incarceration rates, 1978–2004. *Justice Quarterly*, *28*(1), 174–201.

Stinchcomb, J.B., & Hippensteel, D. (2001). Presentence investigation reports: A relevant justice model tool or a medical model relic? *Criminal Justice Policy Review*, *12*(2), 164–177.

Tonry, M. (2016). *Sentencing fragments: Penal reform in America, 1975–2025*. New York: Oxford University Press.

Travis, L.F., & Lytle, D.J. (2017). Balancing harms: The ethics and purposes of criminal sentencing. In Braswell, M.C., McCarthy, B.R., & McCarthy, B.J. (Eds.), *Justice, crime, and ethics* (pp. 170–187). New York: Routledge.

Turner, M., Sundt, J., Applegate, B., & Cullen, F. (1995). "Three strikes and you're out" legislation: A national assessment. *Federal Probation*, *53*(3), 16–35.

Turner, S., Greenwood, P., Fain, T., & Chiesa, J. (2006). An evaluation of the federal government's violent offender incarceration and truth-in-sentencing incentive grants. *The Prison Journal*, *86*(3), 364–385.

Ulmer, J.T. (1997). *Social worlds of sentencing: Court communities under sentencing guidelines*. Albany, NY: SUNY Press.

Ulmer, J.T. (2012). Recent developments and new directions in sentencing research. *Justice Quarterly*, *29*(1), 1–40.

Umbreit, M.S. (1981). Community service sentencing: Jail alternative or added sanction? *Federal Probation*, 45(3), 3–14.

Van Voorhis, P., Spiropoulos, G., Ritchie, P.N., Seabrook, R., & Spruance, L. (2013). Identifying areas of specific responsivity in cognitive–behavioral treatment outcomes. *Criminal Justice and Behavior*, 40(11), 1250–1279.

Vining, A.R. (1983). Developing aggregate measures of disparity. *Criminology*, 21(2), 233–252.

Vito, G.F., & Wilson, D.G. (1988). Back from the dead: Tracking the progress of Kentucky's Furman commuted death row population. *Justice Quarterly*, 5(1), 101–111.

von Hirsch, A. (1976). *Doing justice*. New York: Hill & Wang.

von Hirsch, A., & Hanrahan, K.J. (1979). *The question of parole*. Cambridge, MA: Ballinger.

Wang, X., & Mears, D.P. (2010). Examining the direct and interactive effects of changes in racial and ethnic threat on sentencing decisions. *Journal of Research in Crime and Delinquency*, 47(4), 522–557.

Watts, A.L. (2016, Aug. 5). *The composition of sentencing commissions*. Retrieved from: https://sentencing.umn.edu/content/composition-sentencing-commissions

Whitehead, J.T., Braswell, M.C., & Burgason, K.A. (2017). To die or not to die: Morality, ethics, and the death penalty. In Braswell, M.C., McCarthy, B.R., & McCarthy, B.J. (Eds.), *Justice, crime, and ethics* (pp. 208–231). New York: Routledge.

Wonders, N. (1996). Determinate sentencing: A feminist and postmodern story. *Justice Quarterly*, 13(4), 611–648.

Wood, P. (2007). Exploring the positive punishment effect among incarcerated adult offenders. *American Journal of Criminal Justice*, 31, 8–22.

Wooldredge, J. (2009). Short- versus long-term effects of Ohio's switch to more structured sentencing on extralegal disparities in prison sentences in an urban court. *Criminology and Public Policy*, 8(2), 285–312.

Wozniak, K.H. (2016). Perceptions of prison and punitive attitudes: A test of the penal escalation hypothesis. *Criminal Justice Review*, 41(3), 352–371.

Wright, K., Pratt, T.C., Lowenkamp, C.T., & Latessa, E.J. (2011). The importance of ecological context for correctional rehabilitation programs: Understanding the micro- and macro-level dimensions of successful offender treatment. *Justice Quarterly*, 29(6), 775–798.

Yang, C.S. (2014). Have interjudge sentencing disparities increased in an advisory guidelines regime? Evidence from *Booker*. *New York University Law Review*, 89(4), 1268–1342.

IMPORTANT CASES

Atkins v. Virginia, 536 U.S. 304 (2002).

Apprendi v. New Jersey, 530 U.S. 466 (2000).

Baze v. Rees, 553 U.S. 35 (2008).

Blakely v. Washington, 542 U.S.296 (2004).

Coker v. Georgia, 433 U.S. 584 (1977).

Furman v. Georgia, 408 U.S. 238 (1972).

Gardner v. Florida, 430 U.S. 349 (1977).

Gregg v. Georgia, 428 U.S. 153 (1976).

Kennedy v. Louisiana, 554 U.S. 407 (2008).

Mempa v. Rhay, 389 U.S. 128 (1967).

Ring v. Arizona, 536 U.S. 584 (2002).

Roper v. Simmons, 543 U.S. 551 (2005).

United States v. Booker, 543 U.S. 220 (2005).

Williams v. New York, 377 U.S. 241 (1949).

CHAPTER 10

Incarceration

On any given day more than 2.2 million people are incarcerated in thousands of jails and hundreds of prisons across the United States (Carson, 2018; Zeng, 2018). Approximately one in every 114 adult residents of the United States are either in prison or jail (Kaeble & Glaze, 2016). The overwhelming majority of those incarcerated is male, and racial minorities make up a disproportionate number of those behind bars. A recent survey of 45 state prison systems found that the average cost per inmate was over $33,000 per year (Mai & Subramanian, 2017). For a variety of reasons, the prison traditionally is our response to crime.

There is good reason to believe that prisons are central to American corrections. By virtue of their size, history, and cost, prisons receive the lion's share of attention from correctional administrators (and, until recently, from persons studying corrections as well). Furthermore, the threat of imprisonment is considered necessary to make less severe sanctions workable (Connolly, 1975). The argument is that without the threat of imprisonment, offenders would not take seriously such sanctions as fines or probation.

Incarceration of criminals (and those accused of criminal behavior) takes place in prisons and jails. These are two distinct types of institutions. **Prisons** generally are state or federal facilities, segregated according to the sex of inmates, that house persons convicted of felonies who are serving sentences of 1 year or more. **Jails** are usually local municipal institutions and house a variety of people convicted of misdemeanors or at various stages of criminal case processing. As might be expected, there are far more jails in the country than prisons, but most of the jails are small and experience a greater turnover in population than prisons (Stephan, 2008; Stephan & Walsh, 2011).

This chapter examines incarceration in the American criminal justice system. It describes the history, organization, and practice of incarceration in both prisons and jails. In the following chapter, we address several contemporary issues in the operation of prisons and jails (see Box 10.1).

IMPORTANT TERMS

clear and present danger

compelling state interest

congregate system

"hands-off" doctrine

institutionalization

jailhouse lawyers

jails

least restrictive alternative

multijurisdictional jails

pains of imprisonment

penitentiary

Prison Litigation Reform Act

prisonization

prisons

reformatory

role stress

segregate system

smug hack

total institution.

BOX 10.1 POLICY DILEMMA: FAMILY IMMIGRANT DETENTION

The U.S. Immigration and Customs Enforcement (ICE) detains over 30,000 immigrants on a daily basis, using a combination of their own detention facilities and contracts with private or state prison systems (Kerwin & Alulema, 2015). Often, these detainees have been caught while trying to illegally cross the U.S. border and are awaiting deportation proceedings. The most controversial issue related to immigration-related detention involves families or minor children who are unaccompanied by a parent while crossing the border. Beginning in 2014, the United States has experienced a surge in women and children seeking asylum from violence in their home countries (e.g., El Salvador, Honduras, Guatemala). The United States was largely unprepared for such an influx of asylum requests, as most of the existing detention facilities were not certified for housing minor children. As a result, many children are separated from their parents after arrival and sent to separate facilities or to sponsor families while they await their case's processing. Many scholars have expressed concern about the welfare of these individuals and the appropriateness of detaining families seeking asylum in the United States (Musalo & Lee, 2017; Schriro, 2017). However, others warn that readily accepting these families into the United States would pose a national security risk and would encourage others to seek refuge in a similar manner. How you do feel is the best way to process families or unaccompanied children who seek asylum in the United States?

THE ORIGINS OF AMERICAN INCARCERATION

The jail was established in England during the reign of Alfred the Great. Its purpose was to serve as a detention facility for those accused of seriously breaching the peace. In addition to tax collecting and other duties, the shire reeve (sheriff) was responsible for maintaining the jail. Then, as now, the jail operated on a local level in holding prisoners for a centralized authority.

The American colonists brought the jail with them to the New World but generally did not need to use it within their small, close-knit communities. As had happened earlier in England, however, towns in the colonies grew larger and jails began to receive more use. Those incarcerated in jails were arrested persons not yet convicted, or were debtors or people who failed to pay fines. We did not yet use incarceration as a punishment for crime (Moynahan & Steward, 1980).

Incarceration as a response to criminal behavior developed as part of a larger "discovery" of the asylum in American society in the early 1800s (Rothman, 1971). The American Revolution changed the nature of American society. No longer was it common for people to know all of their neighbors and to maintain a sense of small community in their dealings with others. Mobility (caused by the war), the beginnings of industrialism and immigration, and the growth of commerce and cities led to a more impersonal, less intimate social climate. Problems of poverty, dependency, and crime became increasingly common and seemed to require a more centralized solution than had been characteristic of the colonial response. Americans found the solution to social problems of poverty, insanity, and crime in the asylum or institution. Poorhouses, insane asylums, and jails became more common around the nation in the latter part of the 1700s. The first prison in the United States opened in Newgate,

Connecticut, before the Revolution in 1773. The general use of imprisonment did not spread until afterward (Durham, 1990).

The **penitentiary**, a place of punishment and repentance, as a response to crime was particularly attractive. The harsh criminal code of England had been transported to the colonies with the result that most offenses were punished with what Langbein (1976) termed "blood punishments." It was common practice to torture, mutilate, or execute offenders. These barbaric penalties violated the assumptions of the Enlightenment, which underlay the New Republic. When Patrick Henry addressed the Virginia House of Burgesses and declared, "Give me liberty or give me death!" he unwittingly identified the perfect penalty. Incarceration gives the convicted offender neither liberty nor death.

Pragmatically, incarceration solved a pressing problem of administration for penal codes that provided for severe punishments. When the penalties for crimes were perceived as being too harsh (e.g., lashes or branding for petty theft, death for repeat offenses), juries dealt with the dilemma by failing to convict the offenders. Some observers saw this as a major obstacle to meeting the deterrent functions of the law. Furthermore, in a rational penalty system, it was difficult to grade penalties to crimes when physical pain was the standard. Incarceration seemed more humane and likely to result in higher conviction rates. It also made an easier task of matching lengths of term to seriousness of offense.

The problem of harsh penalties was very troublesome. If the penalty for stealing a pig was death, what should be the penalty for stealing a cow? What would deter a burglar who faced hanging from killing the homeowner? Jurors faced tough decisions as well. Voting to convict a hungry offender who stole a loaf of bread could ensure that the offender suffered branding, mutilation, or death. By establishing time as a punishment, and describing the prison as a harsh but humane environment, it was possible to better match penalties to offenses, such as 1 year for stealing a pig or 3 years for stealing a cow. The availability of incarceration as a sentencing option also led to a higher rate of conviction, as juries were more willing to see a hungry thief spend a few months in prison (where food and clothing would be available).

Jeremy Bentham established the centrality of the prison as an appropriate sanction for deterrent purposes, although later historians have overlooked Bentham's focus on deterrence. Sullivan (1996) reviewed Bentham's original work, concluding that, to Bentham, the prison provided a technology that allowed precise alteration of the intensity and duration of punishment. According to Bentham, this precision would allow punishers to achieve, via scientific application, optimal deterrent value from criminal punishment.

Ideologically, incarceration appeared well suited to the needs of offender reform. Without the opportunity to transport prisoners to penal colonies, penitentiaries provided internal penal colonies to which offenders could be sentenced. The penitentiary removed the offender from the evil environment of the city, allowed the person to reflect on the error of his or her ways, and taught good work habits. Indeed, in the early days of incarceration, some viewed the penitentiary as a Utopia. Most proponents of the penitentiary believed crime was the result of an evil environment, and believed the penitentiary would insulate the offender from further criminal influences. The French philosopher Foucault (1979) is perhaps the best known proponent of the view that the prison emerged as a means of disciplining or controlling people. The prison emerged as a primary means of criminal punishment for a variety of reasons (Garland, 1990).

The Congregate/Segregate System Debate

During the 1820s, two systems of penitentiary discipline developed. They are compared in Box 10.2. The first emerged in Pennsylvania, known as the **segregate system** or the Pennsylvania system. Here, inmates were housed separately in individual cells, took their meals in their cells, exercised in separate yards, and never interacted with other offenders. Any industry conducted was cottage industry, in which inmates completed the entire product in their cell.

BOX 10.2 SEGREGATE VERSUS CONGREGATE SYSTEMS

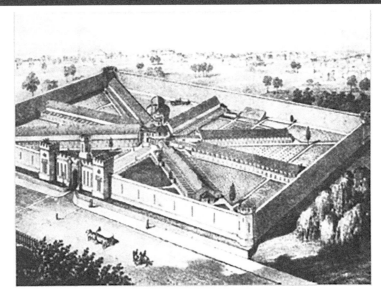

The Eastern State Penitentiary, designed by John Haviland and completed in 1829, became the model and primary exponent of the Pennsylvania "separate" system. The prison had seven original cell blocks radiating from the hublike center, a rotunda with an observatory tower, and an alarm bell.

A corridor ran down the center of each block, with the cells at right angles to the corridor. Each cell had a back door to a small, uncovered exercise yard and double front doors, the outer one made of wood, and the other of grated iron with a trap so that meals could be passed to prisoners.

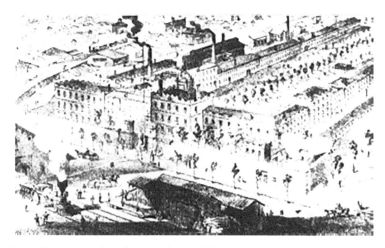

In 1816, New York began the construction of a new prison at Auburn. It was patterned after other early American prisons with a few solitary cells to conform to the law of solitary confinement to be used for punishment, and with sizable night rooms to accommodate most prisoners.

To test the efficiency of the Pennsylvania system, an experiment was tried in 1821 with a group of inmates who were confined to their cells without labor. Many of these inmates became insane and sick. The experiment was abandoned as a failure in 1823, and most of the inmates studied were pardoned.

A new plan was adopted whereby all inmates were locked in separate cells at night, but worked and ate together in congregate settings in silence under penalty of punishment.

Source: American Correctional Association (1983).

This system had the advantage of ensuring that offenders were protected from the corrupting influence of other offenders. The major disadvantages were that offenders suffered psychologically from isolation, and the prison was expensive to operate and was not always able to produce a profit from its industry. Nonetheless, the Pennsylvania system had its supporters because it was true to an ideal of penance and isolation.

In contrast, the second penitentiary in the United States developed in Auburn, New York, and was known as the **congregate system** or the Auburn system of prison discipline. Here, although housed in separate cells, inmates ate, worked, and exercised in groups. The prison maintained isolation by a strict rule of silence that prohibited inmates from conversing with each other. This organization reduced costs because of mass movement and feeding of prisoners. Further, the ability of offenders to work in groups allowed assembly-line methods of production and a wider variety of prison-made products.

The strengths of this system were the more humane mingling of prisoners (even if they had to remain silent, inmates at least were able to see each other) and the cost-effectiveness of the system. The disadvantages were that the congregate system required closer surveillance of inmates to enforce silence, and that prison administrators were unable to keep inmates isolated. In addition, in the congregate system, there was greater potential for riots and fights among inmates.

For several decades, a debate raged about which of these systems was the better method for handling inmates. In the end, the cost-effectiveness of the congregate system emerged the victor of the debate, at least in the United States. Later generations of American penitentiaries most often operated under the Auburn system of congregate feeding, work, and exercise.

The experience of the Virginia Penitentiary illustrates this point. In an attempt to benefit from both the congregate and segregate systems, Virginia law required that inmates, housed in larger dormitories under a "silent system" at first, spend their last 3 months in solitary confinement. Keve (1986:41) cited the 1832 report of a legislative study committee, which found the solitary confinement requirement of Virginia prison terms counterproductive:

> Upon the subject of the three month's solitary confinement required by law to be inflicted upon convicts immediately preceding their discharge, the committee have had much reflection; and they have come to the conclusion that this portion of the close confinement ought to be abolished. They believe that it is productive of no substantial benefit, but is on the contrary, decidedly injurious. It obliterates the habits of industry previously acquired. Upon the score of more interest to the state it is inexpedient because it abstracts from the institution the most valuable portion of the time and labor of the convicts. It exceeds the requisitions of stern justice. … But above all it ruins the health of the victim and indirectly takes away human life. … In any aspect which it may be viewed experience proves its inexpediency, if not its absolute inhumanity and injustice.

The Changing Purposes of Prisons

Initially, people believed that the experience of incarceration alone would lead to improved behavior on the part of offenders. The learning of good work habits, the removal from contaminating influences, and other benefits of incarceration were supposed to result in better citizens. However, over time, it became clear that the reform (prison) was itself in need of reform. By the middle of the nineteenth century, penitentiaries were replaced with reformatories in which offenders were trained to be law-abiding citizens and released from incarceration as soon as it was clear that they had been reformed.

Created as a humanitarian and practical alternative to the corporal and capital punishments of the eighteenth century and earlier, the penitentiary soon came to be recognized as flawed. The early developers of the penitentiary supported isolation of prisoners because they sincerely believed that criminality was caused when the environment tempted people. The term "penitentiary" was applied because it was designed to be a place where the offender could do penance. Designers thought lack of adequate discipline and training from the family, combined with the lure of taverns and brothels, led otherwise law-abiding citizens into a life of crime. Isolation, solitary contemplation on right and wrong, and the discipline of hard work in a sanitized environment free from criminal temptations would work to improve offenders and prepare them for lawful living.

By the Civil War period, however, observers had come to believe that simple incarceration was not enough. The penitentiary lacked an incentive to offenders to make them want to change. Furthermore, the imposition of a definite term of confinement seemed counterproductive. If a prisoner with a 5-year term had reformed in 1 year, these observers saw no reason to continue his or her confinement for an additional 4 years. It was important that prisons take active steps to reform criminals. These steps would include instilling discipline and industry as habits in the inmates, and offering the possibility of early release from incarceration as an incentive to reform.

In the **reformatory**, inmates were raw materials to be shaped into law-abiding citizens (that is, reformed). Upon entering the reformatory, the prisoner lost his or her civilian identity, contact with the outside world, and nearly all rights. The reformatory was a factory producing useful citizens. The process through which this production was to be accomplished was training. Prisoners were conditioned to industry and discipline in the belief that once the habits were established in the prison, they would not be broken after release. As an ex-convict, author Braly (1976:202) noted that the habits acquired in the penitentiary did endure, at least for a while. Describing his first few days of freedom after release from San Quentin, Braly wrote, "That first morning and every morning for several weeks I woke exactly at six-thirty when the big bell had begun to pound in the blocks. Rise and shine. It's daylight in the swamps."

Within 50 years of the creation of reformatories, a new correctional ideology developed (O'Leary & Duffee, 1971). The mere training of offenders was not sufficient. Rather, offenders were regarded as ill and in need of treatment to cure them of their proclivities toward crime. The rehabilitative correctional institution had arrived, with a treatment staff in addition to the custody and industry staffs.

Several problems had plagued the reformatory ideal. First, the habits established in prison did not last long for many prisoners after they were released. Second, over time, prison administrators increasingly used early release as a prison management tool rather than recognition of inmate reformation. If the prison became crowded, they released inmates. If a particular inmate was needed (e.g., if he or she was the only barber in the institution), he or she might not be released, no matter how reformed he or she became. Finally, the view of the cause of criminality also changed.

The growth of the social and behavioral sciences, and the development of service professions related to these disciplines, fostered a reexamination of institutional corrections. Spalding (1897), secretary of the Massachusetts Prison Association, remarked (1897:47), "The State is not an avenger, with a mission to right the wrong which the criminal has done, but is to try to right the criminal, that he may cease to do wrong." The emphasis on the individual offender, and on the motives and causes of the individual's criminality, grew in the early twentieth century (Fogel, 1979:50–61). This was the Progressive Era, when there was a general trust in the ability of the state to do good for individuals, and when reformers wanted prisons that met the needs of individual criminals rather than the general needs of society (Rothman, 1980).

In the mid-1960s, yet another shift occurred in correctional ideology. Observers now viewed prisons themselves as part of the problem of crime. The best solution to reforming criminal offenders was thought to rest in keeping the criminal in the community where he or she could learn how to live a law-abiding life. Prison populations fell, and those kept in prison increasingly were able to take advantage of furlough programs and to enjoy increased contact with the outside world. Observers of the nation's prisons came to believe that attempts at individualizing treatment, and the focus on the cause of crime as being inside the offender, were ineffective. Rather, to succeed in changing offenders into law-abiding citizens, it would be necessary to deal with the criminal in society.

Another prison writer, Torok (1974:91), succinctly summarized the arguments in favor of the reintegration philosophy of corrections, which flourished for a decade from the mid-1960s:

> It costs the taxpayer up to six thousand dollars each year to keep one convict locked up uselessly in prison. On the other hand, it would only cost about three hundred dollars a year to keep the same man in the community, under close supervision, on parole or probation, where he will pay his own way, earn a salary, keep his family off welfare and live a law-abiding life. Both society and the offender would benefit from this approach but so many people are unwilling to examine the facts and figures objectively. They continue to pour millions of dollars of tax money into an archaic prison system which does not correct, does not reform, simply does not, in any sense work.

In the mid-1970s, correctional ideology again shifted. The purpose of imprisonment came to be defined principally as punishment. Inmates served time as punishment for criminal offenses. Whatever programs and industries were available to inmates were there for voluntary use. Release from incarceration was based on service of sentence, not on evidence of reform or rehabilitation. Further, more offenders were expected to serve prison sentences as punishment, although the lengths of terms were reduced for most offenders (Twentieth Century Fund, 1976).

One argument in favor of this newest purpose of prisons was that, in the end, punishment had been the only purpose that the tradition of the prison had shown incarceration could serve. Torok (1974:88) wrote, "In actual practice, prisons do little more than punish." Similarly, Braly (1976:362) wrote of the new prisons, "The old timers scorned these new prisons and dismissed them as Holiday Motels. We couldn't be conned by departmental [Department of Corrections] window dressing. We were still under the Man, and the Man still had a gun locked away somewhere nearby. ... Essentially, it was only Folsom with Muzak."

A number of commentators support the notion of the prison as a place of punishment. Logan and Gaes (1993) suggested that prisons follow a confinement model in which the purpose is the secure, safe, humane custody of inmates. This notion of the purpose of the prison fits well with a resurgence of retributive punishment. If the purpose of punishment is to return harm for harm, then the prison should impose pain safely and humanely (Clear, 1994; Cullen, 1995). When combined with tremendous crowding in prisons, the retributive ideal led to what some called a "new penology" (Feeley & Simon, 1992). In this new penology, the primary purpose of the prison was the efficient management of a large population (Holcomb & Williams, 2003; Rutherford, 1993).

FIGURE 10.1
Eastern State Penitentiary in Philadelphia became the model and primary exponent of the Pennsylvania "separate" system. It is now a museum open to visitors.

Ellen S. Boyne

At the turn of the century, the purpose of prisons was again the subject of intense debate (Clear, 1997). Several observers noted that the move to more punitive policies concerning the use of prisons and jails is at least partly responsible for the tremendous growth in America's inmate population during the past 45 years (Cook, 2009). Critics blamed punitive conditions in prisons and jails for failing to prepare inmates for life after incarceration, and for making the management of correctional institutions more difficult (Rhine, 1992; Seiter, 1997; Wright, 2000). In response, a growing number of commentators called for the reinstitution of rehabilitative programming in correctional institutions (Listwan, Jonson, Cullen, & Latessa, 2008; Pollock, Hogan, Lambert, Ross, & Sundt, 2012).

The American prison developed as a humane alternative to the harsh punishments of colonial justice. In a history of slightly less than 200 years, the purpose

of prisons changed several times in accord with changing public and social attitudes regarding crime and human behavior. Through it all, the prison has survived. It is instructive to recognize that one of the first prisons ever erected, Auburn Penitentiary, is still in operation today. It has undergone several renovations and name changes, yet it still houses inmates. Pennsylvania opened the Eastern State Penitentiary in Philadelphia in 1829 (see Figure 10.1). In 1970, the state closed the penitentiary, but the city of Philadelphia used it to house inmates from the county prison after a riot there. In 1971, the city finally closed the penitentiary as a prison. It is now a museum open to visitors (Eastern State Penitentiary, 2010).

The history of prisons in the United States reveals that these institutions are enduring (Schwartz & Travis, 1996). The prison represents the core of our views on how to respond to criminal behavior. It is likely to do so at least into the foreseeable future.

THE ORGANIZATION OF AMERICAN INCARCERATION

Given the distinct (although related) natures and functions of jails and prisons, it is necessary to give separate treatment to the organizations of the two types of institutions. Although both types of institutions hold convicted offenders, their widely divergent structures cannot be combined easily.

Jail Organization

Local control, multiple functions, and a transient, heterogeneous population have shaped the major organizational characteristics of jails. Box 10.3 provides a partial answer to the question, "Who is in jail?" Typically, jails are under the jurisdiction of county government. In most instances, the local area has neither the necessary tax base from which to finance a jail adequately nor sufficient size to justify even the most rudimentary correctional programs. In addition,

---

BOX 10.3 WHO IS IN JAIL?

- Persons pending court proceedings such as arraignment, trial, or sentencing, including both those unable to secure pretrial release and those denied release

- Probation, parole, and bail violators and absconders

- Juveniles pending transfer to juvenile authorities

- Persons with mental illnesses pending movement to mental health facilities

- Military personnel for protective custody, contempt, or as witnesses

- Persons awaiting transfer to state or federal correctional facilities

- Federal and state prisoners held because of prison crowding

- Convicted offenders sentenced to short terms (generally under 1 year)

- Material witnesses

- Prison inmates on furlough or other temporary release status.

Source: James (2004).

local control has inevitably meant involvement with local politics. Jails are in a paradoxical situation: Although clung to tenaciously by localities, they often have been unwilling or unable to meet even minimal accreditation standards.

County sheriffs operate most jails in the United States. The majority of jail staff consists of custodial officers, and because of the administration of jails by sheriffs and police, many custodial officers are sworn police officers who would rather not be serving in jails. As local institutions, jails must often compete for resources within a larger sheriff or police department and with other municipal services such as public works, sanitation, health, and education departments. Moreover, over half of American jails are designed to house fewer than 100 inmates. Yet, these small jails only house approximately 8% of the nation's jail inmates. The largest 5% of jails house over 43% of all jail population (Zeng, 2018). In some areas, the use of multijurisdictional jails has increased. These facilities serve more than one municipality and provide the economic benefits of scale.

On any given day, our nation's jails hold more than 740,000 inmates (Zeng, 2018). Annually, jails process approximately 12 million inmates with a weekly turnover rate of approximately 60%. Over half of those incarcerated in jails are awaiting trial and not yet convicted of a criminal offense. Including those who will receive pretrial release or a sentence to a penalty other than incarceration, the Bureau of Justice Statistics estimated the average length of stay for persons entering jails at 23 days (Minton, Ginder, Brumbaugh, Smiley-McDonald, & Rohloff, 2015). Figure 10.2 shows the average daily jail population from 2005 to 2016.

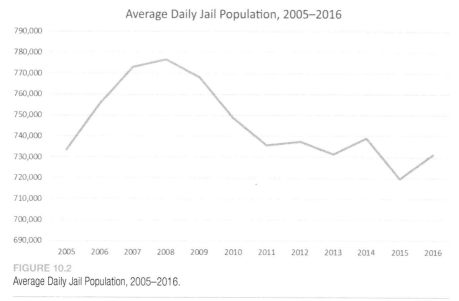

FIGURE 10.2
Average Daily Jail Population, 2005–2016.

Source: Zeng (2018).

TABLE 10.1 Persons under Jail Supervision by Confinement Status

Percent of Jail Inmates Confined

Inside jails	92.1
Outside jails	7.9
Status of Jail Inmates Not Confined in Jail (%)	
Weekender program	15.3
Electronic monitoring	22.4
Home detention	1.0
Day reporting	7.0
Community service	22.6
Other pretrial supervision	13.6
Other work programs	11.0
Treatment programs	3.3
Other	3.8

Source: Minton and Zeng (2015).

Because of short terms and small total population sizes, programs for jail inmates generally either do not exist or are inadequate. Most jails have programs for education and health care. Jail inmates spend much of their incarceration time in idleness, viewing television, or otherwise attempting to pass the hours. Jail inmates have high levels of medical needs and substance dependence and abuse, posing problems for jail treatment programs (Bronson, Stroop, Zimmer, & Berzofsky, 2017; Maruschak, 2006). It is an understatement to say that the jail population comprises offenders with high service needs.

Newer jails often incorporate a redesign of jail facilities. An increasing number of new jails have been built using alternatives to the traditional row of cells. Often called direct supervision designs, these jails contain housing units with cells arranged along the exterior walls with a common living/dining area in the center. This design is coupled with the delivery of most services to inmates within the housing unit. Rather than placing the correctional staff in a control center or instruct them to walk along the long corridors of traditional jails, the staff are instead housed inside the living areas with direct and frequent communication with the inmates. Researchers have reported that this design seems to reduce tension and inmate misbehavior (Morin, 2016; Wener, 2006). Unfortunately, in many places the construction of these new facilities has not meant the closing of older, traditional jails (Bikle, 2000). Instead, because of growing populations, many jurisdictions now operate both types of jails. In addition, nearly 8% of persons under jail supervision nationally are supervised outside the jail facility, as shown in Table 10.1.

Prison Organization

Prisons exist on either the state or federal level. Unlike jails, prisons suffer from problems associated with their large size. The National Advisory Commission on Criminal Justice Standards and Goals observed the detrimental consequences of overly large institutions (1973:355):

> The usual response to bigness has been regimentation and uniformity. Individuals become subjugated to the needs generated by the institution. Uniformity is translated into depersonalization. A human being ceases to be identified by the usual points of reference, such as his name, his job, or family role. He becomes a number, identified by the cellblock where he sleeps. Such practices reflect maladaptation resulting from size.

State governments operate the majority of prisons in the United States. Every state has a prison, or state confinement facility, and about one-third of states operate more than 20 (Stephan, 2008). Approximately 8% of prisoners under the jurisdiction of a state or federal government are housed in privately operated facilities, while another 5% are housed in local jails. The Bureau of Justice Statistics reported that three jurisdictions, California, Texas, and the federal prison system, housed roughly one-third of all inmates in 2016 (Carson, 2018). See Figure 10.3.

At the end of 1984, the Bureau of Justice Statistics (1985) reported more than 480,000 offenders held in American prisons. That population increased more than 17% by the end of 1986, when there were 546,659 prison inmates in the United States (Bureau of Justice Statistics, 1987:1). At the end of 1996, there were nearly 1,200,000 inmates in state and federal prisons, an increase of almost 250% (Mumola & Beck, 1997). By the end of 2008, there were more than 1.6 million people in prison. However, in 2009, prison releases exceeded admissions for the first time in 31 years. The prison population at year-end 2016 had fallen to approximately 1.5 million prisoners (Carson, 2018). Because prisons have a much lower turnover rate than jails, fewer than 1.3 million inmates were admitted and released in prisons each year. For the past several years, releases have outpaced admissions.

FIGURE 10.3

An aerial view of the Butner Federal Correctional Complex in Butner, North Carolina. The complex consists of a low-security facility, two medium-security facilities, and a medical center that houses inmates of all security levels with health issues.

AP Photo/Gerry Broome

Because of the longer terms of confinement and generally larger populations, prisons are able to provide a wide range of programs to inmates, including educational, vocational, recreational, social, and psychological counseling programs. Stephan (2008:5–6) reported that 85% of confinement facilities offered educational programs, 92% offered counseling programs, and most facilities provided work assignments to inmates. In fact, some states require inmates to either obtain education or work while incarcerated. Despite this, many inmates find themselves idle, having neither an academic nor a work assignment (Batchelder & Pippert, 2002).

Similar to jails, the majority of prison staff holds custodial positions (66%). There are, however, higher percentages of prison staff holding administrative (2%), education (3%), professional/technical (10%), maintenance/clerical (12%), and other (7%) job titles (Stephan, 2008:4). Unlike those working in jails, prisons hire personnel solely as correctional personnel. The organization and administration of prisons have changed little over the years and do not seem likely to change much in the future.

DOING TIME

Incarceration is the foundation of American corrections. With the exception of capital punishment, it is the most severe sanction available to the state. Incarceration is the "stick" that supports community corrections programs, such as halfway houses, probation, and parole (Reasons & Kaplan, 1975). Prisons are an American invention, and Americans rely on the use of incarceration as a response to criminal behavior.

The experience of incarceration differs depending on where an inmate serves the sentence; yet, in large measure, it is the same wherever and whenever it occurs. The inmate in either prison or jail is typically under control and is not a contributing member of any policymaking body. This fact leads to a similarity of experience for all inmates. The experience of "doing time" is painful. Gresham Sykes (1969) identified the **pains of imprisonment** as (1) deprivation of liberty, (2) deprivation of goods and services, (3) deprivation of heterosexual relations, (4) deprivation of autonomy, and (5) deprivation of security. Hassine (1996:18), an inmate in the Pennsylvania prison system, described these pains as follows: "At first, I missed the obvious, sex, love, family, and friends. But it wasn't long before I stopped missing these things and started focusing on the next wave of things I no longer have: privacy, quiet, and peace of mind, intangibles that I have never stopped missing to this day."

Many observers have identified what has been called **prisonization**. This refers to the apparent fact that prison inmates become socialized into a specific prison subculture. Some contend that the experience of deprivation creates a prison subculture that provides inmates with norms and rules for living in prison. In contrast, others suggest that offenders import the prison subculture from outside. That is, the type of people sent to prison already have a set of values, norms, and beliefs that is different from and counter to conventional society (Zaitzow, 1999). The research evidence on whether the prison subculture is a product of the characteristics of inmates coming to prison (importation model) or emerges from the experience of imprisonment (deprivation model) has shown that both models are important (Cihan, Davidson, & Sorensen, 2017; Tewksbury, Connor, & Denney, 2014). Marcum, Hilinski-Rosick, and Freiburger (2014) studied prison misconduct committed by both male and female prisoners in California. Their research supported a merger of the importation and deprivation models. Certain deprivation-related factors such as an inmate being housed in maximum security or being incarcerated for a longer period of time were associated with increased misconduct. However, the prisoners' age also significantly impacted their propensity toward violent misconduct. Hochstetler and DeLisi (2005) observed that different types of inmates are more likely to place themselves in circumstances in which they are exposed to opportunities for good or bad behavior in prison. They suggested that the reality of living in prison is quite complex and that it is a combination of inmate and institution characteristics that explains inmate behavior in prison. Whatever the source, the subculture provides inmates with a means of making sense of the prison and enduring the hardships of incarceration.

Deprivation of Liberty

By definition, those confined in correctional institutions do not have any liberty of movement. Even within the institution, the inmate is not at liberty to move around. As Sykes (1969:65) observed, "In short, the prisoner's loss of liberty is a double one: first, by confinement to the institution and second, by confinement within the institution." Figure 10.4 shows the Federal Correctional Complex, Terre Haute.

Prisoners move en masse from cellblock to activity, to dining facility, to activity, and back to cellblock. Few inmates get the privilege of moving about the institution without an escort. Inmates also are not free to choose to whom they may write or with whom they may otherwise interact. The net effect of this deprivation is isolation from the outside community. This loss of liberty is symbolic of a loss of status as a trusted member of society. As one inmate put it:

> Freedom is the only meaningful thing to a human. Without freedom things lose meaning. The whole system in prison is designed to degenerate a human being, to break him as a man. They take away all of his freedom, his freedom to express himself and his feelings. How can you be human if you can't express yourself? (Wright, 1973:146).

The Federal Correctional Complex, Terre Haute (FCC Terre Haute) is a U.S. federal prison complex for male inmates in Terre Haute, Indiana. It consists of two facilities: the Federal Correctional Institution, Terre Haute (FCI Terre Haute), a medium-security facility with a communication management unit, and the U.S. Penitentiary, Terre Haute (USP Terre Haute), a high-security facility that holds federal death row for men.

Ellen S. Boyne

Deprivation of Goods and Services

Although inmates have more options now than when Sykes made his observations, prisons still deprive inmates of access to, and ownership of, a wide variety of goods and services. Most prisons do not allow inmates to possess money, and most require standardization of clothing and other possessions.

Upon admission to a prison or jail, the "civilian" possessions of inmates (e.g., jewelry, money, and clothing) are confiscated and either stored until release or shipped to a destination chosen by the inmate (usually home). On admission to the prison, the inmate loses all of the amenities of free society by which we make a statement about who we are, such as the clothes we wear, the way we wear our hair, the car we drive, and the like. Instead, the inmate receives a uniform, a prison haircut, and generally cannot exercise personal taste in the selection and purchase of goods and services. Hassine (1996:19–21) described his first disciplinary infraction as the result of a desire to have a hamburger "his way." He wrote, "My first misconduct at Graterford resulted from missing one of life's simplest pleasures: a fresh-cooked burger." He bought contraband hamburger, buns, butter, onions, and a heating element. While preparing his hamburger, he was discovered and written up. He concluded this tale, writing, "For many years afterward, my prison handle became 'Burger King,' even though I had never gotten a chance to taste the object of my crime."

Most penal facilities now have commissaries, where inmates can purchase toiletries, candy bars, and other small items. Some prisons allow inmates to wear certain articles of civilian clothing, such as hats or T-shirts, but the range of options available to the inmate is very restricted. The effect of this deprivation is that the inmate feels impoverished. His or her self-worth is lessened by virtue of the reduced net worth.

> In prison the slightest distinction is cherished and enlarged. … Bob, since he had bad feet, had been allowed to keep his own shoes rather than wear the Santa Rosa hightops which were standard issue. He had polished these shoes until they glittered, and, as we spoke, he continued to rub one shoe and then the other against his pants leg. He also wore the watch Big John had given him, and he glanced at it frequently as if he had an important appointment and wasn't just standing around, as I was, killing time until lunch (Braly, 1976:156).

Deprivation of Heterosexual Relations

As noted earlier, most penal facilities in the United States are segregated by sex. The inmate is not only denied liberty and impoverished by incarceration but is also forced to endure involuntary celibacy. The lack of members of the opposite sex in the prison society leads to anxieties about sexual identity among inmates. Some observers have characterized the prison culture as being "ultra-masculine" (Lutze & Murphy, 1999).

The effect of sex segregation is not only physical but also psychological in that the inmate has lost one-half of the audience and comparison group from whom he or she receives a validation of sexual identity. That is, it is more difficult to be masculine or feminine in an all-male or all-female society (respectively) than in mixed company. One of the principal ways in which we know who we are sexually is by comparison with members of the opposite sex. In prison, these comparisons are largely absent (Nacci & Kane, 1984).

> Some men look feminine and looks are enough alone for a man behind these walls to try and get him. It is a hell of a thing to say, but here you are another man and you are behind these walls and before long another man begins to look like a woman to you (Lockwood, 1982:54).

Deprivation of Autonomy

The prison is a "total institution," as described by Goffman (1961). The **total institution** provides all of the necessities for the individual and makes all of the decisions for its residents. This fact leads to what has been called "institutionalization," which is the formation of individuals who are almost wholly dependent on the institution and incapable of caring for themselves in the free society.

The prison inmate is not allowed to decide when to eat, what to eat, when or how often to take a shower, when to go to sleep, when to awaken, what job to do, and how to make other seemingly trivial decisions. Rather, the inmate is subjected to a life in which all major, and most minor, decisions are made by others. The refusal of inmate requests is generally not accompanied by any explanation, thereby adding insult to injury.

The net effect of the loss of autonomy is a reduction in feelings of self-worth, as inmates come to recognize that they are no longer in control of their own destinies. The inmate is reduced to a state of childlike dependency on the parent (the state). As Sykes (1969:76) explained, "But for the adult who has escaped such helplessness with the passage of years, to be thrust back into childhood's helplessness is even more painful." In the words of an inmate at San Quentin:

> The worst thing here is the way your life is regulated, always regulated, day in and day out. They tell you what to do almost every moment of the day. You become a robot just following instructions. They do this, they say, so that you can learn to be free on the outside (Wright, 1973:146).

Deprivation of Security

The final pain of imprisonment mentioned by Sykes is paradoxical: the loss of security. In a prison, even one classified as maximum security, the prisoner experiences a real loss of personal security. Prisons are not safe places in which to reside. Consider living in a neighborhood where all of those around you are accused criminals, convicted felons, and other criminal offenders. How many people would voluntarily move into such a neighborhood?

At any time, the prisoner must be prepared to fight to protect his or her belongings and personal safety. Living with the constant threat of victimization is stressful, and this constant stress adds to the pain of imprisonment. Cell doors may serve not only to lock in the inmate but also to lock out others who may harm the prisoner. Many prison inmates request placement in segregation (solitary confinement) for reasons of personal safety. Existing in the general prison population is generally a frightening experience.

> It used to be a pastime of mine to watch the change in men, to observe the blackening of their hearts. It takes place before your eyes. They enter prison more bewildered than afraid. Every step after that, the fear creeps into them. … No one is prepared for it.
>
> Everyone is afraid. It is not an emotional, psychological fear. It is a practical matter. If you do not threaten someone, at the very least, someone will threaten you (Abbott, 1981:144).

CORRECTIONAL OFFICERS: THE OTHER INMATES

If imprisonment is frustrating and painful to inmates, it is not much less so for those whose job it is to work within the walls of penal facilities each day. The correctional officer serves a dual role: (1) manager of inmates and (2) line-level worker within the prison. As a line worker, the officer is subjected to frequent "shake-downs" (to control the possibility of officers smuggling contraband into the prison), supervision, and disciplinary action by superior officers, and other controls, which make the occupation of correctional officer similar to the role of prison inmate (Clear & Cole, 1986:306).

The daily responsibilities of custodial officers in prisons and jails can be very stressful (Dowden & Tellier, 2004; Griffin, 2006; Lambert, 2003; Lambert, Hogan, & Tucker, 2009). Research links this stress to physical illness, job dissatisfaction, and job turnover among correctional officers. What happens on the job, stress, and level of job satisfaction also spill over to influence general satisfaction with life. Research indicates that correctional officers who report being less satisfied at work are also likely to report lower levels of satisfaction with their lives in general (Armstrong, Atkin-Plunk, & Wells, 2015; Lambert et al., 2010). Although they may be able to leave the institution at the end of their shifts, most correctional officers, over the span of their careers, will spend more time in prison or jail than will the inmates. They too suffer several pains of imprisonment.

Research shows that concerns about safety and career development are important sources of stress for correctional officers (Gordon & Baker, 2017; Hartley, Davila, Marquart, & Mullings, 2013; Triplett, Mullings, & Scarborough, 1996). Much of the current research focuses on what many commentators have termed **role stress**. For example, an early survey of correctional officers in New Jersey (Cheek & Miller, 1983) identified 21 items that the officers thought were the most stressful aspects of the job. The most important of these fell into the categories of organizational or administrative problems. The officers reported being most troubled by a lack of clear job description, absence of support from superiors, and not being able to exercise personal judgment. Cheek and Miller (1983:119) concluded:

> The officer gets no respect from anyone. Not from the outside community, which sees him as the brute portrayed in the old James Cagney movies, not from the inmates who use him as a dumping ground for their hostility, not from prison administrators who expect him to play the tin soldier. … A stressful job indeed!

Recent research has confirmed the existence of role stress. It appears that role stress can be alleviated by creating a work environment that clarifies the roles and responsibilities of the officers, provides the officers an input into decision making, and gives officers greater control over their jobs (Dowden & Tellier, 2004; Garland, Hogan, & Lambert, 2013; Lambert et al., 2009; Steiner & Wooldredge, 2015). Finally, research shows that officers who believe the prison administration treats them fairly are less likely to experience stress or leave the job (Lambert, Hogan, & Griffin, 2007; Taxman & Gordon, 2009).

The Bureau of Labor Statistics (2017) reports that the average pay for correctional officers is approximately $47,000. However, the entry-level salary in some states is in the $25,000 range. Requirements for officers are also minimal and turnover in the custodial ranks is high. By virtue of the low entrance criteria, relatively low pay, and low status associated with their job, custodial officers suffer many of the same kinds of deprivations as do inmates (Pollock, 2004).

Although the correctional officer is the backbone of the correctional institution (Dial, 2010), the role of correctional officer has often been ignored. Investigations of the correctional officer's role have identified the importance of the officer to the operation of the prison (Lerman & Page, 2012; Rembert & Henderson, 2014). Guards have the greatest contact with prison inmates and are most directly responsible for the smooth operation of the prison (Johnson, 2001; Pollock, 2004). Correctional officers have been shown to engage in many types of misconduct, which can serve to undermine the safety and operation of the prison (see Box 10.4).

As with any occupational group, it is difficult to generalize about guards, because they differ widely among themselves in terms of how they perceive and perform their jobs. However, some research has examined the effect of diversity among correctional officers. Traditionally, women have faced obstacles when working in a correctional facility. Among these obstacles was the idea that women could not respond reliably to violent situations or that they were not able to handle the mental stress of working in a prison (Jurik, 1985). To an extent, these perceptions still exist today.

BOX 10.4 TYPES OF CORRECTIONAL OFFICER DEVIANCE

Deviance against the Institution

Improper use of agency equipment

Purposely failing to perform one's duties

Theft of correctional facility property

Abuse of sick time

Accepting gifts from inmates and contractors.

Deviance against Inmates

Abuse of authority

Mishandling/theft of inmate property

Discrimination toward inmates

Violence/excessive force

Sexual relations with or assault of inmates.

Deviance against Other Correctional Officers

Drinking on the job

General boundary violations (minimize professional distance between correctional staff and prisoners)

Discrimination

Sexual harassment of fellow correctional officer

Smuggling contraband.

Source: Ross (2013).

The research on the effects of diversity in the correctional officer field largely undermines these perceptions. For example, Jackson and Ammen (1996) investigated changes in the composition of the correctional officer force resulting from affirmative action regulations in Texas. They concluded that the recruitment of females and individuals from minority groups resulted in a wider range of officer perceptions of inmates and prison treatment. Camp, Saylor, and Wright (2001) found that diversity in the prison workforce was associated with lower job commitment among white, male officers, but did not seem to affect teamwork and job performance. Further, Tewksbury and Collins (2006) found that women respond similarly to instances of inmate challenges and misbehavior. Finally, evidence is mixed regarding gender's effect on work stress. Griffin (2006) reported no significant difference in the levels of stress experienced between male and female correctional officers. However, she also found that female officers report greater concerns regarding safety and perceptions of organizational support for equitable treatment of employees. More recently, Dial, Downey, and Goodlin (2010) found that female correctional officers experienced greater levels of work stress compared to males.

Toch and Klofas (1982) reported that perhaps one-quarter of all custodial officers fit the Hollywood image of the brutal, uncaring guard. They called this type of officer the "smug hack." However, this estimate means that 75% of the custodial force does not fit the stereotype. Johnson (2001) suggested that most guards are not "hacks," but play an important role in providing services to inmates. The guards do this, Johnson argued, at least partly in response to their need to make the job more challenging and important than either the public or the prison administration believes it to be.

Many officers, however, try to solve the problem of alienation by expanding their roles and making them more substantial and rewarding. These officers discover that in the process of helping inmates, and thereby giving them more autonomy, security, and emotional support, the officers gain the same benefits: more control over their environment, more security in their daily interactions with prisoners, and a sense of community—however, inchoate or ill defined— with at least some of the people under their care. In other words, in solving inmate adjustment problems, staff solve their own problems as well. Hassine (1996:117–118) commented on the symbiotic relationship between guards and inmates:

> [A]n unwritten agreement has been established between inmates and guards: inmates get what they want by being friendly and nonaggressive, while guards ensure their own safety by not strictly enforcing the rules. For the most part, inmates manipulate the guards' desire for safety, and guards exploit the inmates' need for autonomy.

Prison custodial officers then, like inmates and like police in the community, reflect the nature of their positions within the justice system. Traditionally, correctional officers are less supportive of treatment efforts than any other group in the prison (Young, Antonio, & Wingeard, 2009). In a classic experiment on the effects of incarceration, Zimbardo (1972) concluded that incarceration profoundly affected both the inmates and the guards. He concluded that the social situation (role definitions of guard or prisoner, presence or absence of power, etc.) determines how people will act. Riley (2000) explored the influence of social setting on guard behavior. He found that correctional officers developed "sense-making" tactics whereby they could define inmates as different, and less than human, thus justifying their exercise of control. Marquart (2005) suggested that relationships between inmates and officers are generally cooperative as both groups attempted to maintain order in the institution. Whether because of administrative rules, a lack of training, or some other reason, prison officers have a limited range of actions open to them. As part-time prison inmates, correctional officers also face deprivations of liberty, autonomy, and security. What these officers seem to want is not just more respect or support from the public and superiors, but more options. That is, officers seem to believe they will achieve greater security in the prison if they are granted the liberty to perform their jobs in a more autonomous fashion (Hepburn, 1987).

INCARCERATION IN THE CRIMINAL JUSTICE SYSTEM

For years, the operations of penal facilities were not open to public or court scrutiny. Rather, the prisons and jails were assumed to require no supervision by outsiders, and indeed, it was believed that outside interference would be more harmful than beneficial. In the 1970s, however, this condition changed with the emergence of prisoners' rights.

The unreviewed, nearly total power over inmates that traditionally rested with correctional authorities is now subject to judicial review upon the filing of lawsuits by the inmates themselves. In the 1941 case of *Ex parte Hull*, the U.S. Supreme Court ruled that inmates had the right to access the courts. Initially, this right of access was more theoretical than practical. For years, appellate courts adopted a **"hands-off" doctrine**, refusing to consider prisoner petitions and keeping the court's hands off correctional institutions when deciding cases concerning the rights of inmates (Vito & Kaci, 1982). In a series of appellate court decisions, though, courts identified the rights of inmates and placed due process controls on the exercise of discretion by prison authorities.

In 1964, the U.S. Supreme Court decided the case of *Cooper v. Pate*, ruling that prisoners in state and federal institutions were protected from arbitrary and capricious violations of their civil rights. Prison inmates were entitled to the protections of the Civil Rights Act of 1871. In deciding the case, the Court provided a vehicle by which inmates could challenge conditions of confinement in the nation's courts. A later decision, *Johnson v. Avery* (1969), further strengthened the position of inmates. In this case, the Court ruled that prison authorities must either allow inmates the use of jailhouse lawyers (inmates who assist others in the preparation of court documents) or provide an adequate alternative. In combination, these two rulings meant that inmates were not only entitled to certain rights but also that the state must provide them with the necessary resources to secure those rights in court. As Cohen (1972:862) noted, "To hold, for example, that a prisoner must be guaranteed reasonable access to the courts, that he must suffer no reprisals for his efforts, and that there is a right to some form of assistance, recognizes the prisoner as a jural entity."

Wallace (1992) observed that commentators often rely on the decision of the Virginia Supreme Court in *Ruffin v. Virginia* (1871) as support for the hands-off approach of the courts to prisoners' rights. In this decision, the court wrote that the prisoner was a "slave of the state," and thus had no protected rights. Wallace observed that indeed there has been a tradition of judicial oversight of prisons and concern with prisoners' issues. The problem for prisoners, however, was that courts were reluctant to intervene directly, except in the most extreme cases, and inmates were often unable to get the attention of either the courts or the public. Abandonment of a general hands-off attitude made court oversight of prisons more likely.

With access to the courts thereby ensured, the next stage in the development of prisoners' rights began. It was now time for prisoners to seek protections from the more onerous conditions of confinement. In a flurry of litigation, this is precisely what happened. The more important developments in prisoners' rights occurred in three areas, relating to the protections of the First, Eighth, and Fourteenth Amendments.

Unfortunately, some inmates also began to increasingly fight for reforms through violence. On September 9, 1971, inmates rioted at the New York State Correctional Facility at Attica, where they took control of a large part of the prison. On September 13, state police and prison authorities stormed the prison and took command. When the smoke cleared, 43 people were dead; most of them (30) were inmates killed during the attack. It was the bloodiest prison riot in American history. Nine years later, in 1980, inmates rioted for 36 hours at the New Mexico State Penitentiary. When the incident was over, officials found that the rioting inmates had killed 33 prisoners, and many of these deaths occurred after the victims had been barbarically tortured (Rolland, 1997; Serrill & Katel, 1980). Both of these deadly incidents were directly related to the poor conditions of the prison system and the inmates' perceptions that their concerns were not being taken seriously. Although these prison riots successfully got the public's attention, the reaction to this type of prison violence was not always productive to the inmates' cause. Instead, the riots resulted in a public backlash against efforts to reform prison conditions (Thompson, 2016). In recent years, the courts have retreated from their active reform of prison life to a more limited role.

In deciding questions of prisoners' rights, the courts generally have applied three tests to the reasonableness of prison conditions and regulations: (1) compelling state interest, (2) least restrictive alternative, and (3) clear and present danger. A **compelling state interest** is any concern of the state (prison administration) that is so important that it overrides the protections afforded in the Constitution. Such an interest, for example, would be evident if an inmate were to request the right to go on a pilgrimage for religious reasons; the state interest in custody is compelling and justifies the denial of the right to go on a pilgrimage. The **least restrictive alternative** refers to the desire to be no more oppressive than is necessary to meet the needs of the state. For example, a rule punishing an inmate for possession of lewd photographs would probably be too restrictive, given that a rule prohibiting display of the material would meet the state's interests in not arousing other inmates. Finally, **clear and present danger** refers to conditions or behaviors that pose an immediate threat to safety or order and relates to controls on the activities of inmates that pose a direct threat to the smooth operation of the facility. For example, courts will uphold rules prohibiting inmates from assembling and making inflammatory speeches, because such activities pose a clear and present danger of instigating riots.

PRISONERS' RIGHTS

First Amendment Protections

The First Amendment to the U.S. Constitution provides for the freedoms of religion, speech, the press, and assembly, as well as the right to petition the government for redress of grievances. Some of the more noteworthy prisoners' cases focused on the question of freedom of religion, although other First Amendment issues also emerged. The examples above illustrate some of the kinds of issues that have been decided in the courts. The U.S. Supreme Court upheld restrictions on the right of inmates to correspond or communicate with other prisoners, but struck down rules that prevented prisoners from marriage [*Turner v. Safley*, 482 U.S. 78 (1987)]. These issues relate to the First Amendment right to freedom of assembly.

The major decisions regarding freedom of religion dealt with the emergence and spread of Islam in the prisons. Black Muslims in prisons were initially (and perhaps are to this day) viewed with suspicion by prison administrators. The Muslims were seen more as a political group than as a religious sect. Correctional personnel saw requests for special diets, spiritual mentors (chaplains), access to the Koran, and the like as challenges to prison authority rather than as attempts to practice a religion. The Muslims were required to litigate almost every issue involved in the practice of their religion in prison.

Similar litigation was required to clarify other First Amendment issues, such as access to the media, censorship of mail, visitation rights, formation of prisoners' unions, and other activities. Employing the three reasonableness tests described above, the courts have decided hundreds of cases dealing with First Amendment rights of prisoners.

Eighth Amendment Protections

The Eighth Amendment to the U.S. Constitution prohibits cruel and unusual punishments. Prisoners sued under this amendment to protest perceived deficiencies in nearly every aspect of prison life, from food to medical treatment. With the recent surge in prison populations, these suits have become more controversial as inmates seek relief from overcrowded prisons and overburdened prison resources. In *Rhodes v. Chapman* (1981), the U.S. Supreme Court ruled that crowding (housing two inmates in cells designed for one) by itself does not constitute cruel and unusual punishment. However, in *Brown v. Plata* (2011), the Court required the California Department of Corrections to reduce its prison population by 137.5% to alleviate serious overcrowding issues. The Court determined that California could not provide adequate medical care to inmates owing to the excessive overcrowding problem within its prison system.

Eighth Amendment suits have been responsible for the cessation of corporal punishment and a general improvement in prison conditions. There is evidence, however, that the courts are reluctant to interfere in the operation of prisons. An important case in prison conditions was *Holt v. Sarver* (1970), the basis for the movie *Brubaker*. In this case, inmates in Arkansas protested a wide range of prison conditions, from poor sanitary facilities to the use of inmate trustees (including allegations of inadequate medical care, food, sleeping quarters, and almost everything else). The Court found that many of the specific allegations did not constitute cruel and unusual punishment, but when it considered the claims together (the totality of the circumstances), the Court decided that conditions at the prison combined to make life there cruel and unusual. Later cases similar to this have led to the appointment of "masters," who are charged by the Court with the responsibility of bringing a prison, or even an entire prison system, into compliance with the Constitution.

Fourteenth Amendment Protections

The third major area of development in prisoners' rights lies within the purview of the Fourteenth Amendment, which provides that no state may deprive any citizen of life, liberty, or property without due process of law. This amendment also ensures that the federal courts can apply the requirements of the U.S. Constitution to the states.

The U.S. Supreme Court decided what might be the most important due process case in the area of prisoners' rights in 1974. In *Wolff v. McDonnell*, the Court determined what due process rights applied to prison disciplinary hearings. Prior to this decision, prisoners often faced a presumption of guilt and were granted no constitutional protections in disciplinary hearings. The *Wolff* decision required that correctional authorities provide the following to any prisoner charged with a rules violation: (1) written notice of the charges, (2) a hearing within 72 hours of notice, (3) warnings of possible criminal proceedings that could result from the hearing, (4) a written statement of the findings and evidence, and (5) the right to appeal within 5 days.

The impact of these rights on the daily lives of prisoners is unclear. Although prisons are still not country clubs by any stretch of the imagination, one effect of the prisoners' rights movement has been to make prisons less oppressive than they had been. Even though the pains of imprisonment still do exist, inmates today have somewhat more freedom and autonomy within the institution than did inmates at the time Sykes identified the five deprivations associated with prison life. Yet, correctional officials have ways to circumvent the requirements of court rulings such as the procedural protections afforded in *Wolff* (Thomas, Mika, Blakemore, & Aylward, 1991). Further, a number of court decisions have limited the ability of inmates to bring suits alleging violations of their rights. Alexander (1993:115) contended that the courts have "slammed the door" on inmates: "Whether in personal liability, habeas corpus, or rights emanating from the amendments to the U.S. Constitution, the Supreme Court of the 1980s and early 1990s has made it much more difficult for inmates to prevail." Two decades after Alexander penned these comments, the rate of prisoner filings and the likelihood of success have continued to decline (Schlanger, 2015).

Table 10.2 provides a table of selected court cases on prisoners' rights.

TABLE 10.2 **Selected Court Cases on Prisoners' Rights**	
Ruffin v. Virginia, 62 VA 790 (1871)	Declared prison inmates to be "slaves of the state" while imprisoned
Cooper v. Pate, 378 U.S. 546 (1964)	Granted inmates protection from arbitrary and capricious violations of civil rights
Ex parte Hull, 312 U.S. 546 (1941)	Granted prison inmates the right of access to the courts
Johnson v. Avery, 393 U.S. 483 (1969)	Required prison administrators to provide legal assistance to inmates
Holt v. Sarver, 309 F. Supp. 362 (1970)	Used "totality of circumstances" test to determine that conditions which shocked the conscience of the court violated the Eighth Amendment
Wolff v. McDonnell, 418 U.S. 539 (1974)	Specified due process protections for inmates facing disciplinary charges
Rhodes v. Chapman, 452 U.S. 337 (1981)	Declared that prison crowding, by itself, does not violate the ban on cruel and unusual punishment
Porter v. Nussle, 534 U.S. 516 (2002)	Held that the exhaustion requirement of the Prison Litigation Reform Act (PLRA) of 1995 applies to all inmate suits about prison life, regardless of whether they involve systemic conditions or isolated acts of wrongdoing
Brown v. Plata, 563 U.S. 493 (2011)	Ordered the release of California prisoners to alleviate serious overcrowding which had led to inadequate medical care

DUE PROCESS, CRIME CONTROL, AND INCARCERATION

The issue of prisoners' rights demonstrates the conflict between due process and crime control in the incarceration process. Even after conviction of a felony offense and a sentence to prison, individuals retain rights and interests protected by our laws. Before the government can restrict the liberty of individuals (even incarcerated individuals), it must justify its decisions. Requirements for tests of state interest and the least drastic alternative represent limits on the authority of prison and jail staff. Viewing these concerns within a framework of the totality of circumstances, however, allows courts to consider the operational and crime control effects.

After conviction, and certainly after incarceration, the balance between due process and crime control concerns shifts to some degree. Alexander's (1993) observation that the federal courts are slamming the door on inmates is partly a reaction to this shift in balance. The courts are attempting to limit the disruption of crime control and operational efforts of correctional administrators by defining causes of action more narrowly. The U.S. Congress passed the **Prison Litigation Reform Act** of 1995 (PLRA), a law requiring inmates to exhaust all administrative remedies (appeals through the prison administration and department of corrections) before they could file a suit in federal court. In 2002, the U.S. Supreme Court decided *Porter v. Nussle,* ruling that the requirement of exhausting all administrative remedies was mandatory, no matter what the basis of the inmate's appeal, including allegations of excessive force.

This narrowing of definition, compared to the heyday of prisoners' rights, seems to be a victory for crime control supporters over those concerned with due process. In all likelihood, it is a natural attempt by the justice system to achieve equilibrium between these conflicting concerns. One thing that has not changed is the right of inmates to protection from unreasonable or malicious interventions by correctional staff. The creation of a prisoners' rights movement settled the question of the existence of due process protections for inmates. What remains is to determine how we can balance those protections against the crime control concerns of society.

REVIEW QUESTIONS

1. Why can we consider incarceration to be the cornerstone of American corrections?

2. Briefly relate the history of American prisons and jails. Distinguish between the Auburn and Pennsylvania systems of incarceration.

3. With reference to organizational and legal differences, distinguish between prisons and jails.

4. Identify the pains of imprisonment.

5. Who are the other inmates?

6. What is the hands-off doctrine?

7. What are the basic tests the courts use to assess prisoners' rights claims?

8. Identify three areas in which courts recognized prisoners' rights.

REFERENCES

Abbott, J.H. (1981). *In the belly of the beast: Letters from prison.* New York: Vintage.

Alexander, R. (1993). Slamming the federal courthouse door on inmates. *Journal of Criminal Justice, 21*(2), 103–116.

American Correctional Association (1983). *The American prison: From the beginning … A pictorial history.* College Park, MD: Author (pp. 39, 48).

Armstrong, G.S., Atkin-Plunk, C.A., & Wells, J. (2015). The relationship work–family conflict, correctional officer job stress, and job satisfaction. *Criminal Justice and Behavior, 42*(10), 1066–1082.

Batchelder, J.S., & Pippert, J.M. (2002). Hard time or idle time: Factors affecting inmate choices between participation in prison work and education programs. *The Prison Journal, 82*(2), 269–280.

Bikle, B. (2000). *Direct supervision jails: What have we learned?* Paper presented at the annual meeting of Midwestern Criminal Justice Association, October.

Braly, M. (1976). *False starts: A memoir of San Quentin and other prisons.* New York: Penguin Books.

Bronson, J., Stroop, J., Zimmer, S., & Berzofsky, M. (2017). *Drug use, dependence, and abuse among state prisoners and jail inmates, 2007–2009.* Washington, DC: Bureau of Justice Statistics.

Bureau of Justice Statistics (1985). *Jail inmates, 1983.* Washington, DC: U.S. Department of Justice.

Bureau of Justice Statistics (1987). *Prisoners in 1986.* Washington, DC: U.S. Department of Justice.

Bureau of Labor Statistics, U.S. Department of Labor (2017, May). *Occupational employment statistics.* Retrieved from www.bls.gov/oes/current/oes333012.htm

Camp, S., Saylor, W., & Wright, K. (2001). Racial diversity of correctional workers and inmates: Organizational commitment, teamwork, and worker's efficacy in prisons. *Justice Quarterly, 18*(2), 411–427.

Carson, E.A. (2018). *Prisoners in 2016.* Washington, DC: Bureau of Justice Statistics.

Cheek, F.E., & Miller, M.D. (1983). The experience of stress for correction officers: A double-bind theory of correctional stress. *Journal of Criminal Justice, 11*(2), 105–120.

Cihan, A., Davidson, M., & Sorensen, J. (2017). Analyzing the heterogeneous nature of inmate behavior: Trajectories of prison misconduct. *The Prison Journal, 97*(4), 431–450.

Clear, T.R. (1994). *Harm in American penology: Offenders, victims, and their communities.* Albany, NY: SUNY Press.

Clear, T.R. (1997). Ten unintended consequences of the growth in imprisonment. *Corrections Management Quarterly, 1*(2), 25–31.

Clear, T.R., & Cole, G.F. (1986). *American corrections.* Monterey, CA: Brooks/Cole.

Cohen, F. (1972). The discovery of prison reform. *Buffalo Law Review, 21*(3), 855–887.

Connolly, P.K. (1975). The possibility of a prison sentence is a necessity. *Crime & Delinquency, 21*(4), 356–359.

Cook, P. (2009). Explaining the imprisonment epidemic. *Criminology & Public Policy, 8*(1), 25–28.

Cullen, F. (1995). Assessing the penal harm movement. *Journal of Research in Crime and Delinquency, 32*(2), 338–358.

Dial, K.C. (2010). *Stress and the correctional officer.* El Paso, TX: LFB Scholarly Publishing.

Dial, K.C., Downey, R.A., & Goodlin, W.E. (2010). The job in the joint: The impact of generation and gender on work stress in prison. *Journal of Criminal Justice, 38*(4), 609–615.

Dowden, C., & Tellier, C. (2004). Predicting work-related stress in correctional officers: A meta-analysis. *Journal of Criminal Justice, 32*(1), 49–62.

Durham, A. (1990). Social control and imprisonment during the American revolution: Newgate of Connecticut. *Justice Quarterly, 7*(2), 293–323.

Eastern State Penitentiary (2010). *Timeline.* Retrieved July 26, 2010, from http://easternstate.org/history.

Feeley, M., & Simon, J. (1992). The new penology: Notes on the emerging strategy of corrections and its implications. *Criminology, 30*(3), 449–474.

Fogel, D.F. (1979). *We are the living proof … The justice model for corrections.* Cincinnati, OH: Anderson.

Foucault, M. (1979). *Discipline and punish: The birth of the prison.* New York: Vintage Books.

Garland, B., Hogan, N., & Lambert, E. (2013). Antecedents of role stress among correctional staff: A replication and expansion. *Criminal Justice Policy Review, 24*(5), 527–550.

Garland, D. (1990). *Punishment and modern society.* Chicago: University of Chicago Press.

Goffman, E. (1961). *Asylums: Essays on the social situations of mental patients and other inmates.* Chicago: Aldine.

Gordon, J., & Baker, T. (2017). Examining correctional officers' fear of victimization by inmates: The influence of fear facilitators and fear inhibitors. *Criminal Justice Policy Review, 28*(5), 462–487.

Griffin, M. (2006). Gender and stress: A comparative assessment of sources of stress among correctional officers. *Journal of Contemporary Criminal Justice, 22,* 4–25.

Hartley, D.J., Davila, M.A., Marquart, J.W., & Mullings, J.L. (2013). Fear is a disease: The impact of fear and exposure to infectious disease on correctional officer job stress and satisfaction. *American Journal of Criminal Justice, 38*(2), 323–340.

Hassine, V. (1996). *Life without parole: Living in prison today.* Los Angeles: Roxbury.

Hepburn, J. (1987). The prison control structure and its effects on work attitudes: The perceptions and attitudes of prison guards. *Journal of Criminal Justice, 15*(1), 49–64.

Hochstetler, A., & DeLisi, M. (2005). Importation, deprivation, and varieties of serving time: An integrated-lifestyle-exposure model of prison offending. *Journal of Criminal Justice*, *33*(3), 257–266.

Holcomb, J., & Williams, M. (2003). From the field: "Bad Time": The rise and fall of penal policy in Ohio. *Journal of Crime and Justice*, *26*(2), 153–175.

Jackson, J., & Ammen, S. (1996). Race and correctional officers' punitive attitudes toward treatment programs for inmates. *Journal of Criminal Justice*, *24*(2), 153–166.

James, D. (2004). *Profile of jail inmates 2002*. Washington, DC: Bureau of Justice Statistics.

Johnson, R. (2001). *Hard time: Understanding and reforming the prison* (3rd ed.). Monterey, CA: Brooks/Cole.

Jurik, N. (1985). An officer and a lady: Organizational barriers to women working as correctional officers in men's prisons. *Social Problems*, *32*(4), 375–388.

Kaeble, D., & Glaze, L. (2016). *Correctional populations in the United States, 2015*. Washington, DC: Bureau of Justice Statistics.

Kerwin, D., & Alulema, D. (2015). Piecing together the US immigrant detention puzzle one night at a time: An analysis of all persons in the DHS-ICE custody on September 22, 2012. *Journal on Migration and Human Security*, *3*(4), 330–376.

Keve, P.W. (1986). *The history of corrections in Virginia*. Charlottesville, VA: University Press of Virginia.

Lambert, E. (2003). The impact of organizational justice on correctional staff. *Journal of Criminal Justice*, *31*(2), 155–168.

Lambert, E., Hogan, N., & Griffin, M. (2007). The impact of distributive and procedural justice on correctional staff job stress, job satisfaction, and organizational commitment. *Journal of Criminal Justice*, *35*(6), 644–656.

Lambert, E., Hogan, N., & Tucker, K. (2009). Problems at work: Exploring the correlates of role stress among correctional staff. *The Prison Journal*, *89*(4), 460–481.

Lambert, E., Hogan, N., Jiang, S., Elechi, O., Benjamin, B., Morris, A., et al. (2010). The relationship among distributive and procedural justice and correctional life satisfaction, burnout, and turnover intent: An exploratory study. *Journal of Criminal Justice*, *38*(1), 7–16.

Langbein, J.H. (1976). The historical origins of the sanction of imprisonment for serious crime. *Journal of Legal Studies*, *5*(1), 35–60.

Lerman, A.E., & Page, J. (2012). The state of the job: An embedded work role perspective on prison officer attitudes. *Punishment & Society*, *14*(5), 503–529.

Listwan, S., Jonson, C., Cullen, F., & Latessa, E. (2008). Cracks in the penal harm movement: Evidence from the field. *Criminology & Public Policy*, *7*(3), 423–466.

Lockwood, D. (1982). The contribution of sexual harassment to stress and coping in confinement. In Parisi, N. (Ed.), *Coping with imprisonment* (pp. 45–64). Beverly Hills, CA: Sage.

Logan, C., & Gaes, G. (1993). Meta-analysis and the rehabilitation of punishment. *Justice Quarterly*, *10*(2), 245–262.

Lutze, F., & Murphy, D. (1999). Ultramasculine prison environments and inmates' adjustment: It's time to move beyond the "boys will be boys" paradigm. *Justice Quarterly*, *16*(4), 709–733.

Mai, C., & Subramanian, R. (2017). *The price of prisons: Examining state spending trends, 2010–2015*. New York: Vera Institute of Justice.

Marcum, C.D., Hilinski-Rosick, C.M., & Freiburger, T.L. (2014). Examining the correlates of male and female inmate misconduct. *Security Journal, 27*(3), 284–303.

Marquart, J. (2005). Understanding the power of social contexts on criminal justice institutions. *Journal of Criminal Justice Education, 16*(2), 213–225.

Maruschak, L. (2006). *Medical problems of jail inmates.* Washington, DC: Bureau of Justice Statistics.

Minton, T., & Zeng, Z. (2015). *Jail inmates at midyear 2014.* Washington, DC: Bureau of Justice Statistics.

Minton, T., Ginder, S., Brumbaugh, S.M., Smiley-McDonald, H., & Rohloff, H. (2015). *Census of jails: Population changes, 1999–2013.* Washington, DC: Bureau of Justice Statistics.

Morin, K.M. (2016). The late-modern American jail: Epistemologies of space and violence. *Geographic Journal, 182*(1), 38–48.

Moynahan, J.M., & Steward, E. (1980). *The American jail.* Chicago: Nelson-Hall.

Mumola, C., & Beck, A. (1997). *Prisoners in 1996.* Washington, DC: Bureau of Justice Statistics.

Musalo, K., & Lee, E. (2017). Seeking a rational approach to a regional refugee crisis: Lessons from the summer 2014 "surge" of Central American women and children at the US–Mexico border. *Journal on Migration and Human Security, 5*(1), 137–179.

Nacci, P.L., & Kane, T.R. (1984). Sex and sexual aggression in federal prisons: Inmate involvement and employee impart. *Federal Probation, 48*(1), 46–53.

National Advisory Commission on Criminal Justice Standards and Goals (1973). *Corrections.* Washington, DC: U.S. Government Printing Office.

O'Leary, V., & Duffee, D. (1971). Correctional policy: A classification of goals designed for change. *Crime & Delinquency, 17*(4), 373–386.

Pollock, J. (2004). *Prisons and prison life.* Los Angeles: Roxbury.

Pollock, J.M., Hogan, N.L., Lambert, E.G., Ross, J.I., & Sundt, J.L. (2012). A utopian prison: Contradiction in terms? *Journal of Contemporary Criminal Justice, 28*, 60–76.

Reasons, C.E., & Kaplan, R.L. (1975). Tear down the walls? Some functions of prisons. *Crime & Delinquency, 21*(4), 360–372.

Rembert, D.A., & Henderson, H. (2014). Correctional officer excessive use of force: Civil liability under Section 1983. *The Prison Journal, 94*(2), 198–219.

Rhine, E. (1992). Sentencing reform and correctional policy: Some unanswered questions. In Hartjen, C., & Rhine, E. (Eds.), *Correctional theory and practice* (pp. 271–287). Chicago: Nelson-Hall.

Riley, J. (2000). Sensemaking in prison: Inmate identity as a working understanding. *Justice Quarterly, 17*(2), 359–376.

Rolland, M. (1997). *Descent into madness: An inmate's experience of the New Mexico State Prison riot.* Cincinnati, OH: Anderson.

Ross, J. (2013). Deconstructing correctional officer deviance: Toward typologies of actions and controls. *Criminal Justice Review, 38*, 110–126.

Rothman, D. (1971). *The discovery of the asylum.* Boston: Little, Brown.

Rothman, D. (1980). *Conscience and convenience: The asylum and its alternatives in progressive America.* Boston: Little, Brown.

Rutherford, A. (1993). *Criminal justice and the pursuit of decency*. Oxford: Oxford University Press.

Schlanger, M. (2015). Trends in prisoner litigation as the PLRA enters adulthood. *University of California Irvine Law Review, 26*(10), 153–178.

Schriro, D. (2017). Weeping in the playtime of others: The Obama administration's failed reform of ICE family detention practices. *Journal on Migration and Human Security, 5*, 452–480.

Schwartz, M., & Travis, L. (1996). *Corrections: An issues approach* (4th ed.). Cincinnati, OH: Anderson.

Seiter, R. (1997). A view from the inside: Setting correctional policies. *Corrections Management Quarterly, 1*(2), 81–83.

Serrill, M., & Katel, P. (1980). New Mexico: The anatomy of a riot. *Corrections Magazine, 6*(2), 6–24.

Spalding, W.F. (1897). *Indeterminate sentences*. In Proceedings of the National Conference of Charities and Correction at the 24th annual session. Toronto, Canada: July (pp. 7–14, 46–51).

Steiner, B., & Wooldredge, J. (2015). Individual and environmental sources of work stress among prison officers. *Criminal Justice and Behavior, 42*(8), 800–818.

Stephan, J. (2008). *Census of state and federal correctional facilities, 2005*. Washington, DC: Bureau of Justice Statistics.

Stephan, J., & Walsh, G. (2011). *Census of jail facilities, 2006*. Washington, DC: Bureau of Justice Statistics.

Sullivan, R. (1996). The birth of the prison: Discipline or punish? *Journal of Criminal Justice, 24*(5), 449–458.

Sykes, G.M. (1969). *The society of captives*. Princeton, NJ: Princeton University Press.

Taxman, F., & Gordon, J. (2009). Do fairness and equity matter?: An examination of organizational justice among correctional officers in adult prisons. *Criminal Justice and Behavior, 36*(7), 695–711.

Tewksbury, R., & Collins, S.C. (2006). Aggression levels among correctional officers: Reassessing sex differences. *The Prison Journal, 86*(3), 327–343.

Tewksbury, R., Connor, D.P., & Denney, A.S. (2014). Disciplinary infractions behind bars: An exploration of importation and deprivation theories. *Criminal Justice Review, 39*(2), 201–218.

Thomas, J., Mika, H., Blakemore, J., & Aylward, A. (1991). Exacting control through disciplinary hearings: "Making do" with prison rules. *Justice Quarterly, 8*(1), 37–57.

Thompson, H.A. (2016). *Blood in the water: The Attica Prison uprising of 1971 and its legacy*. New York: Pantheon.

Toch, H., & Klofas, J. (1982). Alienation and desire for job enrichment among correctional officers. *Federal Probation, 36*, 35–47.

Torok, L. (1974). *Straight talk from prison: A convict reflects on youth, crime and society*. New York: Human Sciences Press.

Triplett, R., Mullings, J., & Scarborough, K. (1996). Work-related stress and coping among correctional officers: Implications from organizational literature. *Journal of Criminal Justice, 24*(4), 291–308.

Twentieth Century Fund (1976). *Fair and certain punishment*. New York: McGraw-Hill.

Vito, G.F., & Kaci, J.H. (1982). Hands on or hands off? The use of judicial intervention to establish prisoners' rights. In Parisi, N. (Ed.), *Coping with imprisonment* (pp. 79–100). Beverly Hills, CA: Sage.

Wallace, D. (1992). *Ruffin v. Virginia* and slaves of the state: A nonexistent baseline of prisoners' rights jurisprudence. *Journal of Criminal Justice, 20*(4), 333–342.

Wener, R. (2006). Effectiveness of the direct supervision system of correctional design and management: A review of the literature. *Criminal Justice and Behavior, 33*(3), 392–410.

Wright, E.O. (Ed.). (1973). *The politics of punishment: A critical analysis of prisons in America*. New York: Harper Torchbooks.

Wright, K. (2000). The evolution of decision making among prison executives, 1975–2000. In Horney, J. (Ed.), *Policies, processes, and decisions of the criminal justice system* (pp. 177–224). Washington, DC: National Institute of Justice.

Young, K., Antonio, M., & Wingeard, L. (2009). How staff attitude and support for inmate and rehabilitation differs by job category: An evaluation of findings from Pennsylvania's Department of Corrections' employee training curriculum "Reinforcing Positive Behavior." *Journal of Criminal Justice, 37*(5), 435–441.

Zaitzow, B. (1999). Doing time: A case study of a North Carolina youth institution. *Journal of Crime and Justice, 22*(2), 91–124.

Zeng, Z. (2018). *Jail inmates in 2016*. Washington, DC: Bureau of Justice Statistics.

Zimbardo, P.G. (1972). The pathology of imprisonment. *Society, 3*(6), 4–8.

IMPORTANT CASES

Brown v. Plata, 563 U.S. 493 (2011).

Cooper v. Pate, 378 U.S. 546 (1964).

Ex parte Hull, 312 U.S. 546 (1941).

Holt v. Sarver, 309 F. Supp. 362 (1970).

Johnson v. Avery, 393 U.S. 483 (1969).

Porter v. Nussle, 534 U.S. 516 (2002).

Rhodes v. Chapman, 452 U.S. 337 (1981).

Ruffin v. Virginia, 62 VA. 790 (1871).

Turner v. Safley, 482 U.S. 78 (1987).

Wolff v. McDonnell, 418 U.S. 539 (1974).

CHAPTER 11

Problems and Issues in Incarceration

It is clear that the United States incarcerates a large number of individuals, but we do not know enough about whom we incarcerate and how patterns of incarceration change over time. As described in Chapter 10, the dramatic annual increases in jail and prison populations that the United States experienced from the 1970s until 2009 appears to have stabilized and are declining from their peak levels. However, states vary greatly in the extent to which they have experienced prison declines. While many states have significantly reduced their prison population in recent years, other states have experienced an increase in the number of prisoners. Based on the current pace of prison decarceration, it will take 75 years to reduce the prison population to 1990s levels (Ghandnoosh, 2018).

Phelps and Pager (2016) suggest that many state-level factors have influenced the reductions in prison population. First, violent crime rates appear to have declined. States which have experienced a greater decline in violent crime were more likely to reduce their prison population. Second, economic struggles have prompted many states to rely less on incarceration as a criminal punishment. Each prison inmate costs over $33,000 annually (Mai & Subramanian, 2017), leading many states to examine possible reforms to their correctional systems as a way to save money. However, the extent that states look to alternate punishments largely depends on party politics within each state. Phelps and Pager found that Democratic-controlled states have reduced their prison population at greater levels than states led by Republicans. This should not be surprising, as the political parties have long disagreed on the causes and solutions of crime. Just as "get tough" policies were largely responsible for increasing the prison population, reducing the prison population will require the intentional scaling back of these policies. Finally, Phelps and Pager found that race and class inequalities have impacted the incarceration practices over time.

The concept of incarceration—especially as practiced in the United States—is itself controversial. Critics often note that, among all the nations in Western

IMPORTANT TERMS

classification

consent decrees

contract system

deprivation model

design capacity

importation model

lease system

net-widening

operational capacity

piece price system

prison industrial complex

Prison Industry Enhancement Certification Program (PIECP)

Prison Rape Elimination Act

privatization

protective custody

public account system

public works system

rated capacity

state-use system

therapeutic community

civilization, the United States still has the highest rate of imprisonment and imposes the longest prison terms on offenders (Austin & Irwin, 2012). Many have described the negative effects of imprisonment and questioned its impact on reducing crime and helping criminals (Gau, 2019; Zweig, Yahner, Visher, & Lattimore, 2015). Mears, Cochran, and Cullen (2015) point out that despite years of research, the effects of incarceration are still largely unknown. Public opinion also seems to have shifted to favor alternatives to incarceration (Benenson Strategy Group, 2017; Greenberg Quinlan Rosner Research, 2018).

Even with the apparent shift toward alternate punishments, some scholars defend the use of incarceration. Wright and DeLisi (2016) noted that incarceration is much more lenient than some of the physically painful sanctions that were once used to punish offenders. DeLisi (2015) also pointed out that most offenders are given multiple opportunities for alternate punishments before incarceration is finally used for those who continue to commit criminal acts. When no other criminal sanction has been effective, perhaps incarceration is the only option that will protect society. Indeed, the debate seems to be less whether incarceration should be a punishment option, but rather the degree to which incarceration is used and which offenders should be subject to this form of punishment.

While most of us understand that there must be consequences for committing crime, increasing attention has been given to the collateral consequences of incarceration. Collateral consequences are penalties that offenders face but which are not explicitly a part of the sentence. In reality, collateral consequences exist for any criminal conviction. A conviction may impact one's eligibility for public housing, voting rights, ability to obtain professional licenses, etc. For many offenders, the collateral consequences of their conviction might be more punitive than the punishment itself. However, the decision to incarcerate an offender comes with the potential for even greater societal consequences.

By its very nature, incarceration disrupts family and social life, which can impact entire neighborhoods. There might be benefits to removing dangerous and abusive individuals from their communities. However, these benefits must be weighed by the economic and social effects of breaking up families and, in some cases, removing a family's primary breadwinner. Research has consistently demonstrated that incarceration can negatively impact the children of those incarcerated. For example, these children seem to associate with more at-risk peer groups and become more involved in certain types of delinquent activities (Bryan, 2017; Cochran, Siennick, & Mears, 2018; Porter & King, 2015). These effects have also been shown to persist into adulthood (Burgess-Proctor, Huebner, & Durso, 2016).

The economic costs of incarceration also spread more widely than the direct spending dedicated to operating prisons and jails. In fact, one study estimated that the total societal costs of incarceration was 11 times greater than the amount that federal and state governments spend on corrections (McLaughlin, Pettus-Davis, Brown, Veeh, and Renn (2016)). According to this estimate, the cost of incarceration to families, children, and our communities exceeds $531 billion annually. This estimate includes, among other things, the costs associated with adverse health effects and lower average wages experienced by the families of incarcerated individuals.

It is difficult to choose the issues on which to focus because so many problems are found in our prisons and jails. The increased lengthening of prison terms imposed has created a new class of geriatric prisoners. Increasing numbers of elderly offenders are serving time in American prisons, and these inmates pose management and medical problems for prison operation (Maruschak & Beck, 2001; Williams, Stern, Mellow, Safer, & Greifinger, 2012). In addition, higher rates of infectious disease among intravenous drug users have produced a disproportionate number of infected inmates, partly because of the war on drugs. The health problems posed by these inmates are another contemporary problem in prisons and jails (Maruschak, 2004).

A renewed focus on prison treatment of offenders to reduce the risk of future criminality has produced a variety of new programs and rekindled interest in traditional programming, such as education (Pompoco, Wooldredge, Lugo, Sullivan, & Latessa, 2017), work (Duwe, 2015), and activities by religious groups (Schaefer, Sams, & Lux, 2016). Personnel issues, ranging from correctional officer recruitment and selection through staff training (Seiter, 2017), as well as officer stress discussed in Chapter 10, have received much attention. Continuing issues, such as the influence of gangs in prisons, the prisoner subculture, the civil liability of prison administrators and systems, and facility design and size remain.

In this chapter, we discuss five continuing issues in American incarceration: (1) privatization, (2) crowding, (3) prison industries, (4) prison violence, and (5) medical care. These problem areas represent continuing issues around incarceration that reveal the structural limitations on the ability of correctional institutions to change—and show the effects of broader social changes on criminal justice system operations.

PRIVATIZATION

For decades, correctional institutions have contracted for services with private corporations, ranging from facility design and construction to the provision of medical care or food service for the inmate population. There has been increasing **privatization**, that is, contracting with a private vendor for the entire operation of a prison or jail (The Sentencing Project, 2017). The private company will typically charge the government a daily rate per inmate that covers the operating costs of the prison in addition to the company's profit. Greenwood (1981) suggested that private enterprise could do a better job of running the nation's prisons, at less cost. In 1985, the delegate assembly of the American Correctional Association (ACA) passed a policy statement that was generally supportive of further privatization (*Criminal Justice Newsletter*, 1985:1–2).

Privatizing the operation of correctional facilities began with several detention facilities operated by private businesses (Krajick, 1984). Most of these institutions dealt with special offender populations, such as illegal immigrants or juvenile offenders. Shover and Einstadter (1988) suggested that privatization came to the fore in the 1980s because it was congruent with a national social view that the private sector and the marketplace are proper forums for the resolution of social problems. This is another example of how corrections reflects changes in social and political ideology.

The rapid growth of privately operated prisons and detention centers was part of what Dyer (2000) called the "prison industrial complex." The **prison industrial complex** refers to the relationships between governments, correctional authorities, and private corrections companies that combine to support increased use of prison. President Dwight Eisenhower identified a "military industrial complex" in his farewell speech in 1961. He warned that the combination of a large-scale weapons industry and the military produced pressure to maintain and increase military spending at the federal level. Pollock (2004) described the effects of this prison industrial complex on prison policy. Facing economic hardship, communities agree to the construction of private prisons. These prisons become so important to the local economy that voters and local officials exert pressure to ensure that the prisons remain open, even if prison populations were to decline. These economic incentives for keeping prisons, of course, also apply to public institutions.

At year-end 2016, prisons operated by private contractors housed more than 128,000 inmates, comprising more than 8% of the total prison population and representing a 47% increase since year-end 2000 (Carson, 2018; West, 2010). Five states (Hawaii, Montana, New Mexico, Oklahoma, and Tennessee) had at least a quarter of their prison population housed in private prisons at year-end 2016. More than 18% of federal inmates were in private facilities at that time (Figure 11.1).

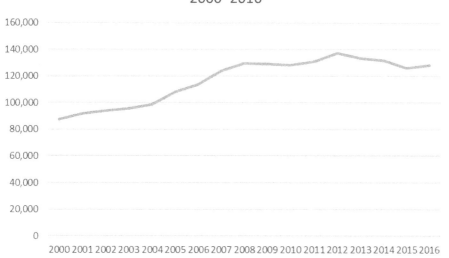

State and Federal Prisoners held in Private Facilities, 2000–2016

FIGURE 11.1
State and Federal Prisoners Held in Private Facilities, 2000–2016.

Sources: Carson (2014, 2015, 2018); Carson and Anderson (2016); Carson and Golinelli (2013); Carson and Sabol (2012); Guerino, Harrison, and Sabol (2011).

Ethical Issues of Privatization

A primary concern regarding the privatization of prisons is the unresolved ethical issues involving the privatization movement (Lindsey, Mears, & Cochran, 2016; Whitehead, Edwards, & Griffin, 2017). The basic question is whether a private company can be empowered to deprive citizens of their liberty. Many critics argue that private prison management runs counter to our moral and philosophical tradition concerning liberal–democratic constitutional government (Reisig & Pratt, 2000). The main motivation for privatization involves cost savings. Proponents argue that private companies can operate more efficiently than state or federal governments. However, these corporations must make a profit to stay in business. The most direct way for these corporations to become more cost-efficient than their public counterparts (and thus make a profit) is for them to cut spending on staff salaries and rehabilitation programming.

Another set of concerns deals with accountability for the operation of the institution. If a prison or a jail is contracted, the contracting governmental authority can ensure that certain services are provided through a carefully constructed contract (Whitehead, Edwards, & Griffin, 2017). The government's task changes from that of managing the facility to that of overseeing and managing the contract. In addition, other problems of accountability arise because a new layer of administration has been created.

Do we allow the manager of a private prison to cap the population of the facility? Can a private facility refuse to accept inmates suspected of being troublemakers? It is most likely that the governmental agency contracting for the private prison remains accountable for its operation, but now must negotiate with the contractor over most aspects of institutional operation (Cooper, 1993). What happens if the contracting government cannot supply enough inmates to make the private facility profitable? For-profit prison corporations spend part of their profit lobbying and making campaign contributions to those who they feel will help pass (or at least keep in place) "get tough" legislation (Ashton & Petteruti, 2011). Critics worry that these efforts might undermine evidence-based rehabilitation policies.

Even though private prisons house a substantial number of inmates, there has not been a rapid growth in their use. Private prisons and jails have emerged as a supplement to public facilities rather than a replacement (Burkhardt, 2017). One of the greatest users of private incarceration is the federal government. However, with many states facing severe economic times and large prison populations, interest in private prisons persists.

Effectiveness of Private Prisons

Even as many of the ethical issues surrounding privatization remain unsettled, the fact is that the privatization of prisons continues. Thus, it is important to examine the evidence regarding how effective private prisons are compared to their public counterparts. As mentioned, the primary motivation behind the privatization of prisons appears to be costs. With many services, contracting with private agencies appears to be more cost-efficient than governmental operation. Yet, the question of cost is less clear when discussing the possibility of an entire institution operated by a contractor. Any cost savings realized may come from lowering quality control. Might a private company hire custodial officers at minimum wage, thereby limiting the pool from which candidates might be drawn, and enhancing the likelihood of employee turnover? Assessments of private prison and jail operations must attend to both cost concerns and the impact of different types of facilities on future criminality by those incarcerated there (Pratt & Maahs, 1999). In a meta-analysis, Lundahl, Kunz, Brownell, Harris, and Van Vleet (2009) found mixed evidence regarding cost-effectiveness. Private prisons were more cost-effective in half of the studies included in the analysis. Of the remaining studies, private prisons were either as costly as or more costly than their publicly managed counterparts. Ultimately, it is very difficult to compare private and public prison costs. Comparisons of public and private prisons are sensitive to choices made by evaluators. Results will vary depending on what costs evaluators include and how they measure costs (Gaes, 2008). It is also difficult for researchers to compare prisons of different ages, design, and security levels, all of which will affect the cost of operating the prison. Culp (2011) pointed out that the number of private prison companies has decreased substantially since the mid-1990s, resulting in less competition and thus less cost savings.

Results are also mixed regarding future recidivism of inmates from private prisons. In one of the few published comparisons of private and public correctional facilities, Lanza-Kaduce, Parker, and Thomas (1999) concluded that privately operated juvenile facilities were more effective than their public counterparts. This study had some limitations, especially the fact that insufficient time had passed to be sure the youths housed in private facilities would continue to refrain from crime. Geis, Mobley, and Shichor (1999) criticized this evaluation because of links between its authors and private prison operations. They suggested that a potential conflict of interest might have influenced the results of the evaluation. A later comparison of public and private prisons in Florida revealed no difference in recidivism rates for inmates housed in those facilities (Bales, Bedard, Quinn, Ensley, & Holley, 2005). The two most recent evaluations of recidivism found inconclusive evidence that inmates serving their time in private prisons may have a slightly increased rate of recidivism than those serving their time in public prisons (Duwe & Clark, 2013; Powers, Kaukinen, & Jeanis, 2017).

Recent evidence has failed to find that public prisons offer superior conditions compared to private facilities (Burkhardt & Jones, 2016). Baćak and Ridgeway (2018) found that private prisons offered less substance abuse, psychological, and HIV/AIDS related programs. However, these differences were mostly related to the characteristics of public prisons compared to private facilities. They found that private prisons were more likely to house both genders, be lower security, have less inmate assaults, and have less crowding compared to public prisons. When these characteristics were controlled for, the differences in prison programming disappeared. Lindsey, Mears, and Cochran (2016) argued that there is still much that we do not know about the effectiveness of private prisons.

CROWDING

Perhaps the most pressing concern of correctional administrators in many jurisdictions is crowding. At midyear 1990, jails in the United States were operating at 104% capacity (Stephan & Jankowski, 1991). Between 1990 and 2016, the jail inmate population nearly doubled from just over 400,000 inmates to more than 740,000 inmates (Zeng, 2018). Clearly, this large increase to an already overcrowded jail system placed stress on many local jurisdictions to increase the amount of available jail space. Many of these local jurisdictions have responded over the last 3 decades by building new, larger facilities (Figure 11.2). From 1990 to 2016, jail capacity in the United States rose from

FIGURE 11.2
Rated Capacity of Local
Jails, 2000–2016.

Sources: Harrison and
Beck (2005); Zeng (2018).

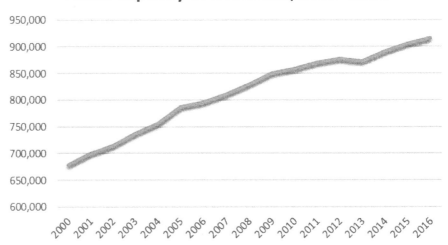

Rated Capacity of Local Jails, 2000–2016

approximately 295,000 to over 900,000. By 2016, jails were operating at an average of 80% capacity, reflecting the lowest level of jail crowding in decades. The reduction in crowding in jails is a positive development, but one that has come at a significant cost to local jurisdictions.

A similar situation exists with prisons. In the early 1980s, the majority of prison systems in the United States were crowded to the point that court intervention had occurred or inmates threatened lawsuits. In 1983, 8 state prison systems (Alabama, Florida, Michigan, Mississippi, Oklahoma, Rhode Island, Tennessee, and Texas) were either operating under court order to change or courts had declared their operations unconstitutional. In addition, the District of Columbia and 21 states had one or more institutions operating under court order, while 9 others had litigation pending and 2 states were operating under **consent decrees** (whereby the court and the state enter into a voluntary agreement about issues raised in court).

The Bureau of Justice Statistics (1986) reported that 32 jurisdictions were housing more inmates than their highest capacity. Using a different estimate of capacity, it was reported that as many as 41 jurisdictions could be classified as crowded. At the end of 2016, the federal system and 14 states were operating at or above their highest capacity (Carson, 2018). One of the difficulties in assessing prison crowding is the determination of prison population capacity. The state, federal, and District of Columbia jurisdictions apply a wide variety of capacity measures to reflect both the available space to house inmates and the ability to staff and operate an institution. Box 11.1 presents three definitions of prison capacity: (1) **rated capacity**, (2) **operational capacity**, and (3) **design capacity**.

BOX 11.1 DEFINITIONS OF PRISON CAPACITY

Rated capacity: The number of beds or inmates assigned by a rating official (such as a fire marshal) to institutions within the jurisdiction.

Operational capacity: The number of inmates that can be accommodated based on a facility's staff, existing programs, and services.

Design capacity: The number of inmates that planners or architects intended for the facility.

Source: Harrison and Beck (2006:7).

Crowding in correctional institutions poses several problems for inmates, correctional staff, and correctional administrators. For these reasons, it is simply not acceptable to jam more prisoners into existing space, at least not as a long-term response to crowding. Rather, we must develop alternatives that ensure adequate capacity to house the large numbers of offenders now populating the nation's prisons and jails.

For inmates, crowding places severe demands on available resources. Simple things that would generally be taken for granted (e.g., hot water for showers) become scarce resources. Privacy, always at a premium in penal facilities, becomes almost totally beyond the reach of most inmates. A crowded facility generally lacks the ability to provide activities or meaningful work experiences for inmates. They must share recreational, educational, vocational, and other resources with more inmates. It may even become difficult to find a seat from which to view television.

The Supreme Court has been reluctant to rule that prison overcrowding by itself is unconstitutional. However, overcrowding can lead to inhumane and dangerous conditions, which, at that point, violate the constitutional protections against cruel and unusual punishment. In a landmark decision (*Brown v. Plata*, 2011), the Supreme Court ruled that the California prison system had become so crowded that the state could not provide adequate supervision. In the majority opinion, Justice Kennedy illustrates an example of the conditions in California:

> A medical expert described living quarters in converted gymnasiums or dayrooms, where large numbers of prisoners may share just a few toilets and showers as "breeding grounds for disease." Cramped conditions promote unrest and violence, making it difficult for prison officials to monitor and control the prison population. … After one prisoner was assaulted in a crowded gymnasium, prison staff did not even learn of the injury until the prisoner had been dead for several hours.

At the time of the *Plata* ruling, California's prison population was at roughly 200% capacity. The ruling required the state to reduce the prison population to

137.5% of capacity. In response to the ruling, the state implemented a policy designed to reduce the prison population by targeted releases of inmates and incarcerating new nonviolent, nonsexual offenders under county jurisdiction in local jails.

Prison staff also feel the pressures of crowding. In addition to the increased tension in the institution and the perceived increase in the likelihood of attack by inmates, the workload of staff increases. Doubling the inmate population means doubling the caseloads of correctional counselors, increasing the class sizes of educational staff, doubling the demands on food service staff, and increasing activity for the custody staff. There are twice as many inmates to count, escort, search, counsel, and watch. The sheer numbers of inmates can overwhelm correctional officers. The crowded institution becomes more anonymous and impersonal, making it difficult for officers to recognize and know inmates or to provide adequate service and protection to them. As a result, correctional officers in crowded institutions are likely to be less attuned to the population and more distant from it.

Martin, Lichtstein, Jenkot, and Forde (2012) interviewed correctional officers to determine the effects of prison crowding. A prominent theme among the officers interviewed was the fear that being outnumbered would lead to increased levels of violence. As one correctional officer said, "[I]nmates allow us to be in this institution every day, and they allow us to leave. I think that anytime they want to take over, then they can" (Martin et al., 2012:97). Criminal justice system officials continue to identify prison and jail crowding as one of their most pressing problems (Pitzer, 2013).

Solutions to the crowding problem are of two basic types (see Schwartz & Travis, 1997:77–80): (1) capacity expansion and (2) demand reduction. We must either increase the amount of prison space available or reduce our demand for prison space. These two solutions involve a variety of strategies. The capacity expansion strategy relies on construction of more prisons, the conversion of facilities built for other purposes (such as mental hospitals or military bases), and renovations of existing facilities to increase the number of prisoners they can accommodate. The demand reduction solution is a bit more complex. As Clear and Austin (2009) point out, the size of America's prison population is a result of two factors: (1) the number of people sentenced to prison and (2) how long they stay there. One strategy would be to reduce the number of offenders sent to prison by increasing diversion programs and the use of community-based alternative sentences. However, evaluations of diversion programs often show a **net-widening** effect. In other words, instead of reducing the prison population, diversion programs are often focused on offenders who would otherwise not have been arrested or prosecuted at all (Walker, 2015). Another strategy is to reduce the amount of time inmates are kept in prison and, thereby, reduce prison populations. If release dates were accelerated so that offenders served only one-half as much time as they currently do, the present supply of prison space could accommodate twice as many inmates. This strategy could also offer false hope for reducing the prison population, though. For example, if a substantial number of inmates who are released early on parole are quickly rearrested or violate the terms of parole, those offenders will return to prison. The recent economic pressures have resulted in many state-level efforts to reduce prison populations. Clear and Schrantz (2011) describe a variety of the strategies that have been considered to help relieve prison overcrowding (see Box 11.2).

BOX 11.2 STRATEGIES FOR REDUCING THE PRISON POPULATION

Reduce Prison Intake Rate

- Strengthening probation
- Creating fiscal incentives for community corrections
- Reducing or eliminating mandatory penalties.

Reduce Length of Stay

- Effective parole release and supervision

- Special early release (e.g., increasing amount of good time an inmate can receive)
- Reform stringent truth in sentencing laws
- Offense-specific statutory changes
- Eliminate repeat offender sentencing enhancements.

Source: Clear and Schrantz (2011).

PRISON INDUSTRIES

In the early days of prisons in America, government officials expected inmates to pay for the costs of incarceration by working in prison industries. Indeed, the early proponents of incarceration hoped that prisons would not only be self-sufficient, but that they might even become profit centers for the state. After decades of prison industry, including contracting prison labor with private companies and state-run industries, prison factories have not been profitable.

"Self-sufficiency was a goal pursued by all states where prisons were instituted. Legislators persistently demanded that their penitentiaries pay their way, even though private industries lobbied against prison manufacturing," observed Keve (1986:28). Officials believed the prison industry had a number of beneficial effects. First, especially in the early years of the penitentiary, they hoped that convict labor could defray the expense of constructing and operating prisons and jails, that is, that inmates could pay for their own punishment (Durham, 1989). Second, they hoped the activity of work would occupy inmates and keep them from breaking prison rules. Finally, they expected that a regimen of work for inmates would serve to instill good habits in them. As Johnson (1987:26–27) summarized the approach:

> It was hoped that silent laboring days and solitary contemplative nights would encourage communion with God and effect a transformation of at least some of the wayward prisoners' souls. But simple conformity to the prison routine, a life of pure habits if not pure intentions, was enough to get a prisoner by in the congregate systems.

According to Fox (1983), prisons have organized their industry in many ways over time. Six of the more common of these are (1) piece price, (2) contract, (3) lease, (4) public account, (5) state use, and (6) public works. Each has its strengths and weaknesses. The **piece price system** was common in the early years of prisons. The manufacturer supplied the raw materials to the prison, and inmates constructed the finished product, which the manufacturer purchased at an established price. The **contract system** involved an entrepreneur contracting with the prison for labor and the use of prison shops for the production of goods. The highest bidder won the right to use prison labor and shops. The **lease system** was common in Southern prisons after the Civil War. The prison simply leased its convicts to a contractor who produced goods with convict labor. All of these methods involved having a private contractor.

The other common methods of organizing prison industry relied on prison-run operations. In a **public account system**, prisoners in correctional industries produce goods and sell them on the open market. There have traditionally been severe restrictions on what prisons may produce and to whom they may sell their products. In a **state-use system**, the prison produces almost any products, but the sale or distribution of the product is limited to governmental agencies. The state-use method of organization does not allow the prison industry to compete on the open market. The **public works system** uses inmate labor for public service projects, such as road maintenance, construction of parks, and other government services. Sing Sing Prison in the state of New York was an early public works prison project; it was constructed largely through the labors of inmates brought from the penitentiary at Auburn.

Opposition to the use of "slave" convict labor and the unfair advantage given to manufacturers employing convicts led to the passage of several laws limiting the sale of prison-made goods across state lines, and within states to non-governmental units (Cullen & Travis, 1984). The resulting decline in markets, as well as the lack of capital investment in prison industries, rendered most prison factories noncompetitive with free-world enterprises. More recently, however, there has been a resurgence of prison industries competing in free-world markets.

The new prison industries tend to be operated by either federal or state governments. These industries produce products such as furniture, signage, and garments to be sold to governmental agencies, public organizations, tax-supported entities, or markets in other countries (Chang & Thompkins, 2002). Box 11.3 describes some of the products produced by inmates housed in the federal bureau of prisons and sold to governmental agencies. An alternate model for prison industries is the **Prison Industry Enhancement Certification Program (PIECP)**. The PIECP was created by Congress in 1979 to encourage states to provide inmates with employment opportunities that are similar to private-sector work opportunities (Bureau of Justice Assistance, 2004). Under PIECP, governments create joint ventures with private industries. Though different models exist, inmates are often employed by, trained by, and even supervised by the private company (Sexton, 1995). Unlike the government-operated prison industry, PIECP programs pay inmates wages comparable to the private sector, and the products produced can be sold in the free market. Inmate participation in prison industries is generally voluntary. Inmates are required to pay taxes, child support, and other expenses. Prison officials consult local unions prior to the development of industries to ensure that the new prison jobs do not cause unemployment among law-abiding citizens (Auerbach, 1982).

BOX 11.3 UNICOR PRODUCTS

Apparel	License Plates
Awards and Plaques	Mattresses and Linens
Electronics	Office Furniture
Eyewear	Office Seating
Food Service	Print Products
Industrial Storage	Signage

Source: www.unicor.gov

The cycle of prison industry—progressing from the production of goods for the open market, through restricted public use, and back again to the free market—shows how difficult it is to resolve the issue of what to do with prison inmates. Some analyses have suggested that introducing free-market practices and pay scales into prison industry programs will benefit both the inmates (Schwarz, 1986) and the general economy of the state (Lonski, 1986:52). Still, there is opposition to the idea of prisoners manufacturing and selling goods. Finally, the old hopes for the prison industry have not changed. Those advocating the expansion of prison industry programs still suggest that the results of these efforts will achieve many goals. The Bureau of Justice Assistance (2004:3) identified the goals of the Prison Industry Enhancement Certification Program, to benefit:

- The corrections administrator, as a cost-effective way to occupy a portion of the ever-growing prison population
- The crime victim, by providing a means of partial repayment for harm sustained
- The inmate, through offering a chance to work, meet financial obligations, learn job skills, and increase chances of meaningful employment on release
- The private sector, by providing a stable and readily available workforce
- The public, through reducing costs with inmate worker contributions to room and board, family support, victim compensation, and taxes.

As is apparent from these goals, when compared to the history of prison industries, the only change has been that the supporters of industries no longer expect full self-sufficiency or complete rehabilitation to result from convict labor. Figure 11.3 presents the distribution of inmate wages under the PIECP. It shows that although inmates retained much of their earnings, almost one-half of gross wages went to inmate financial obligations (Moses & Smith, 2007). In addition, inmates used a portion of the net pay category (48% of total wages) for living expenses, including some health care costs, food, and toiletries. These payments by inmates further reduced state expenses.

FIGURE 11.3
Distribution of Wages Paid Inmates in PIECP.

Source: Moses and Smith (2007:35).

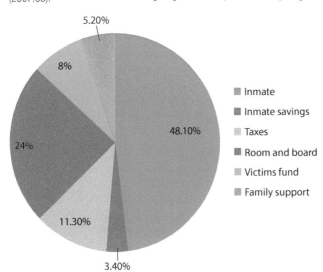

Industry programs for inmates have expanded to include jail populations. Miller, Sexton, and Jacobsen (1991) reported on jail industry programs in 15 local and 1 federal jail/detention center operating inmate work programs. The purposes of jail work programs are similar to those in prison programs and include development of inmate work habits and skills, reduction of costs, reduction of inmate idleness, and the meeting of community needs. Most jail programs compensate inmates with "good time" reductions in sentence length, but some pay wages. Some counties allow inmates to continue to work at their own jobs, reporting to the jail after work each day. In cases in which inmates earn wages, it is typical for the jail to bill for room and board or other fees.

Still, relatively few inmates participate in prison industries. Almost all prisons operate some sort of work or employment programs, most commonly using inmates to perform tasks required for the operation of the prison (see Figure 11.4). Not surprisingly, inmates rank prison jobs differently. Alarid (2003) found that although male and female inmates preferred somewhat different jobs, the best jobs according to inmates were those that provided some useful job skills, allowed a measure of independence, and were highest paid. No inmates indicated physical labor on prison maintenance under close supervision was attractive. Alarid also found no racial discrimination in inmate job assignments. The best jobs went to inmates having served the longest terms and those who were the lowest risks for escape or violence.

FIGURE 11.4
An inmate works inside a greenhouse at the Louisiana State Penitentiary in Angola, Louisiana. Inmates at the penitentiary are able to get state licenses in landscaping and horticulture.

AP Photo/Gerald Herbert

The resurgence of prison industries is a reflection of the current efforts to improve incarceration. Having learned from the past problems, correctional administrators are attempting to accentuate the positive aspects of past efforts while controlling the negative effects (Cullen & Travis, 1984). In a sense, the revitalization of prison industries represents a general return to incarceration as a preferred criminal sanction, which is an effect that has occurred since the early 1970s. The appeal of putting prisoners to work, both for their own good and for the good of the state, is strong. Inmate wage rates in most cases are substantially lower than wages in the free society. The question is whether it is possible to overcome objections to convict labor and to obtain the expected benefits of prison industries. Hopper (2013) reported that prisoners who participate in industry programs tend to have a lower recidivism rate upon release from prison. Cox (2016) found that those who had participated in the prison industry program are generally able to find a job after prison quicker than those who had not participated in the program. However, the vast majority of prisoners (whether or not they had participated in industry) were unable to maintain their job.

PRISON VIOLENCE

Violence has long been a part of the incarceration experience (Braswell, Montgomery, & Lombardo, 1994). The earliest prison riot in the United States occurred in 1774, when inmates at Newgate Prison of Connecticut revolted

(Fox, 1983:114). Newgate was an abandoned copper mine where offenders served terms in the old underground mine shafts. Riots and nonviolent strikes by inmates have occurred throughout the history of prisons and jails. In addition, violence on a smaller scale, between inmates and officers, and among inmates, has a long tradition. Prison violence occurs in many forms. We shall briefly examine interpersonal assaults, homicide, and sexual misconduct within the prison setting.

Interpersonal Assaults and Homicide

Conrad (1982) suggested that prison violence results from five factors: (1) violence-prone inmates; (2) the "lower-class value system," which emphasizes masculinity, toughness, and violence; (3) the use of violence by correctional administrators to control inmates; (4) the anonymity of large (especially crowded) prisons; and (5) the utility of violence in furthering inmate objectives. Given that prisons are places where aggressive, often violent, people live in proximity to each other, it is not surprising that so much violence occurs in prisons. Rather, it is surprising that so little violence occurs. Table 11.1 describes the types of offenses for which inmates are imprisoned in both state and federal prisons. More than half of state inmates have a current conviction for a violent crime. It is important to note that the federal justice system targets different types of criminals compared to the states. Most violent and property crimes fall under the jurisdiction of state courts. Thus, a larger percentage of federal inmates are incarcerated due to drug charges or weapons and immigrations (public order) charges than state inmates rather than violent crimes.

Whether a high level of violence exists in any particular institution, the potential for violence is always there, and the threat of harm influences the behavior of inmates and staff alike (Bowker, 1980). Zoukis (2017:346), a convict-writer who wrote a guide to surviving the federal prison system, wrote, "most long-time prisoners can attest to witnessing or being involved in physical confrontations that arose from simple misunderstandings. One prisoner cuts in front of another in the chow line. A chair is moved without permission. Gossip and other careless speech escalate." Zoukis suggested that life is most dangerous inside the higher security prisons which house those serving longer sentences. Wolff, Shi, and Siegel (2009) reported that the climate of the prison relates to levels of violence. "Climate" refers to how the inmates perceive both staff and other inmates treat them. In prisons where inmates are satisfied with their treatment, there is less violence.

TABLE 11.1 Percent of State and Federal Prison Inmates by Type of Crime Conviction

Type of Crime	State Prisons	Federal Prisons
Violent	54.5	7.7
Property	18.0	6.1
Drug	15.2	47.5
Public order	11.6	38.2
Other	0.7	0.5
Total	100.0	100.0

Source: Carson (2018).

Most of what we know about violence and misbehavior in prisons comes from research focused on male offenders (Steiner, Butler, & Ellison, 2014). Some recent research suggests that the predictors of violence and misbehavior may be similar among male and female inmates (Reidy, Cihan, & Sorensen, 2017; Thomson, Towl, & Centifanti, 2016). However, other research has found differences between violence in male and female prisons. Teasdale, Daigle, Hawk, and Daquin (2016) found male prisons to be more violent than female prisons. Their findings also showed that some factors which reduced the likelihood of violent victimization among men actually increased the likelihood of violent victimization among women.

It is difficult to determine exactly how much violence occurs within prisons. The Bureau of Justice Statistics (Noonan, 2016) reported 83 prison inmate homicides in 2014. Approximately 2% of inmate deaths were reported to be the result of homicides. Each year, there are also thousands of inmate assaults on other inmates. Teasdale et al. (2016) found that an estimated 13% of inmates reported being the victim of a physical assault at some point during their current incarceration. Previous research has generally found much higher levels of violence. For example, Wolff, Blitz, Shi, Siegel, and Bachman (2007) found that an estimated 20% of inmates reported being the victim of a physical assault either by staff or a fellow inmate during a 6-month period. In another self-report survey, 17% of inmates admitted to committing at least one inmate-on-inmate assault in the previous year (Lahm, 2008). Many factors contribute to the rates of reported physical violence in prisons, including the physical conditions of the prison in which the study is conducted and the way in which victimization is measured by researchers (Bierie, 2012; Wolff, Shi, & Bachman, 2008).

Because of the closed nature of the prison environment, most Americans do not give much thought to daily life within a prison setting. As described in Chapter 10, the only recourse most inmates have to improve their conditions is the court system. Occasionally, prisoners will choose to engage in organized violence (e.g., riot) as a way to bring attention to the conditions that they face. The paradoxical nature of prison violence, including riots, is that the violence may be the result of efforts to reform the prison and improve the quality of life for prison inmates. In effect, the prisoners' rights movement and the prevalence of "professional" prison administrators served to lessen the authority of correctional staff by humanizing the prison.

Because of their intensity and the numbers of inmates and staff who are at risk, as well as the possibility of danger to the facility itself, riots are a major concern to correctional administrators. The problem is that we do not know enough about the causes of prison riots to allow adequate planning to avoid them (Rynne, Harding, & Wortley, 2008). Marquart (2008) argued that correctional authorities must maintain control over the institution to prevent riots. The ability for the correctional institution to maintain control depends on factors such as the institution's security level, level of crowding, and characteristics of the inmate population (Griffin & Hepburn, 2013).

The dilemma we face is to solve the problem, if possible, without resorting to the oppressive conditions characteristic of prison life prior to the 1970s. The question is whether it is possible to grant legal rights and protections to prison inmates, as well as giving inmates some level of self-determination, while maintaining control over the operation of the prison. If we can answer yes to that question, there is a chance that we can reduce the current level of prison violence and retain what progress we have made in conditions of confinement. The most promising developments are in the area of inmate classification.

Classification involves the testing and assessment of inmates to determine inmate treatment needs and prison custody and security needs. Accurate, effective inmate classification can lead to safer, less violent prisons (Berk, Ladd, Graziano, & Baek, 2003). Wooldredge (2003) observed that effective classification can produce benefits beyond reduced violence. He argued that controlling prison violence is a positive goal to protect inmates and staff, but also a requirement for effective correctional treatment.

Sexual Assaults and Violence

Sexual assaults in prison are a special case of individual violence among inmates. Davis (1968) conducted the first reported study of sexual assaults among inmates. Davis studied this type of inmate violence in the Philadelphia prison system and in sheriffs' vans transporting prisoners. He included verbal assaults with actual physical attacks in his definition of sexual assault. As a result, Davis concluded that nearly 5% of the prison population had been victims of sexual assaults. The problem of sexual violence in prisons has entered common knowledge so that the popular media depict and discuss sexual assaults in prisons.

In 2003, the U.S. Congress enacted the Prison Rape Elimination Act. This law provides for the development of better information about the nature and incidence of rape and sexual assault in prisons, and provides funding to correctional authorities to reduce and control sexual violence in prisons. In passing the law, Congress (45 USC 15601, Sec. 2) made several findings about sexual violence in prisons. They found that young, first offenders are at greater risk of sexual assault than older offenders are; prison rapes generally go unreported and victims are untreated; and the high incidence of prison rape "increases the levels of violence, directed at inmates and at staff, within prisons." See Figure 11.5.

The academic literature and popular media have portrayed sexual violence in correctional institutions as widespread (Clifford & White, 2017). In fact, the actual incidence of sexual victimization is relatively low. The Bureau of Justice Statistics conducts annual surveys of prison and jail inmates, as well as youths confined in juvenile institutions (Beck & Stroop, 2017). Beck, Rantala, and Rexroat (2014) found the rate of alleged sexual victimization to be less than five per 1,000 inmates. Following an investigation by prison authorities, most of these allegations were found to be unsubstantiated. Correctional staff were reported to be the offenders in approximately half of the victimizations, and roughly three-quarters of the staff sexual misconduct was reported by the inmate as being consensual (Beck, Rantala, & Rexroat, 2014). Any sexual relations between staff and inmates, willing or not, is considered illegal and is reported as a sexual victimization.

True levels of sexual violence are very difficult to determine. Studies that rely on official records of sexual victimization are likely to underestimate the true prevalence of victimization in prisons (Jones & Pratt, 2008). For example, a survey of inmates at 10 prisons revealed that only 22% of the men and 34% of the women who had been victimized reported their incident to prison staff (Struckman-Johnson & Struckman-Johnson, 2006). Fowler, Blackburn, Marquart, and Mullings (2010) found that many of the same groups who are most at-risk for sexual victimization (e.g., homosexuals and those with prior victimizations) are the least likely to self-report the incident to prison officials. Victims might fear retaliation for reporting the offense, or they may simply feel that the prison staff will not seriously investigate the allegation. Due to the limitations of official records, it is possible that self-report data could provide a more accurate representation of sexual victimization. Results of one survey found that 4% of prison inmates reported some sort of sexual victimization within the past 6 months. The prisoners in this survey were more afraid of physical or property victimization than from being sexually victimized.

In terms of deterring inmate sexual assault, one useful result of fighting in prison is self-defense. Especially given the media portrayal of widespread rape and sexual victimization in prison, inmates may attempt to build a reputation in the institution to ensure they will not become targets of sexual aggressors.

FIGURE 11.5
Kendell Spruce, *right*, testifies at a public hearing before the National Prison Rape Elimination Commission. Spruce conveyed how he was assaulted by as many as 27 prisoners in an Arkansas prison. Listening at *left* is T.J. Parsell, who testified that as a 17-year-old he was gang-raped in an adult prison in Michigan. The National Prison Rape Elimination Commission was established by the 2003 Prison Rape Elimination Act to set new national policies to help prevent prison sexual assaults.

AP Photo/Eric Risberg

Thus, some inmates may assault others in nonsexual circumstances to avoid sexual attack. One inmate interviewed by Lockwood (1980:95–96) explained an assault on another in the dining hall as self-defense. The inmate Lockwood interviewed had been the victim of a sexual assault, and he attacked another inmate in the dining hall to prevent future sexual victimization. As the inmate explained his actions:

> They had a code in the prison on the chow line that said no man should cut ahead of you in the chow line. And this one man passed me a couple of times in the line, and he knew that I was aware of what he was doing. If I failed to do what I was supposed to do here, then I was lost again [would be a sexual victim again]. So the next day, when they come through the chow line, when this guy cut in front of me, I hit him in the head with a tray as hard as I could. And when he went to the ground, I hit him several more times before the guard could reach me. It is regrettable but it is the only way that you can handle it. And I didn't want to do it, but I did what I had to do to protect myself.

As Lockwood suggested, the importance of sexual violence in prisons may come more from the perception than the reality of sexual assaults. It appears that, in reality, there are relatively few cases of actual sexual assault in prisons. It also appears, however, that inmates believe the threat of sexual assault is both real and great. Many instances of individual violence in prisons, therefore, may result from the fear of sexual assault. In this way, it is possible that most prison violence is a form of sexual violence. We can infer a similar explanation for inmate violence from Hassine's (1996:23) observations on life in prison. He wrote,

> In the life of an inmate, if you catch someone stealing from you, you're compelled to deal with it physically. This is not because you want to or you think it's the right thing to do, but because you absolutely must. … If you choose to ignore the theft, the man will steal from you again and tell his friends, who in turn will also steal from you. Eventually, you will be challenged for more than just minor belongings.

From the perspective of the inmate, few choices are available for dealing with other inmates. Reliance on correctional authorities is a sign of weakness and is generally seen as ineffective. Not only are correctional officers often unable to protect victimized inmates, those inmates who report incidents to officers come to be labeled "snitches" and are then targets for increased violence. Often, inmates will find themselves being coerced into sexual behavior with a fellow inmate to gain protection from physical violence. Many inmates describe this behavior as consensual, although for others it is a form of sexual enslavement (Trammell, 2011). Another alternative is to seek protective custody. **Protective custody** is usually a housing unit kept separate from the general inmate population so that inmates in the unit are separated from attackers. Most protective custody units, however, offer little programming, and inmates are kept locked in their cells for most of the day. The lack of alternatives to confrontation adds to the sense that the prison is a violent place.

SPECIAL POPULATION TREATMENT

Many inmates incarcerated in America's correctional institutions have special needs that staff must care for. These inmates may include those with physical and developmental disabilities, terminally ill offenders, or sex offenders. This is not a new problem, as there have always been inmates with special needs (Bronson, Maruschak, & Berzofsky, 2015; Whitehead, Dodson, & Edwards, 2013). However, the recent trends toward longer sentences and the war on drugs have increased the number of these inmates. Proper care for these individuals requires compassion, resources, and properly trained staff. While a comprehensive discussion of these issues is beyond the scope of this text, we will briefly discuss the needs of inmates who require specialized treatment for chronic illnesses, mental health, and substance abuse.

Medical Treatment

Approximately half of state and federal prisoners report having a chronic condition such as cancer, high blood pressure, or diabetes. Another 21% report having an infectious disease such as hepatitis C, a sexually transmitted disease, or tuberculosis (Maruschak, Berzofsky, & Unangst, 2015). The frequency of medical problems among inmates has increased in recent years, a trend that is likely to continue as the prison population ages. Since the early 1990s, the number of prisoners age 55 or older has increased approximately 400% (Carson & Sabol, 2016). Many factors have contributed to this increase, including the aging of the general U.S. population (more arrests of older offenders), higher imprisonment rates of older arrestees compared to younger offenders, longer sentences for these offenders (largely due to criminal history and three-strikes laws), and the continued aging of existing inmates who are serving long sentences. These older inmates are more likely to have chronic medical conditions, which increases the costs of incarcerating older inmates (Maruschak et al., 2015).

Each state operates a health care system within its prisons. In fact, prisons are the only place where medical care is constitutionally required. Most states screen at least some of their inmates for the presence of infectious diseases, cardiovascular risk, and mental health as part of the admission procedures. Ongoing medical care, including specialty health services (e.g., dental, optometry, and mammography) is also typically available in prisons (Chari, Simon, DeFrances, & Maruschak, 2016). However, the quality of the medical treatment provided to inmates is often far from ideal.

The Supreme Court has ruled that deliberate indifference to those with serious medical problems is unconstitutional. The exact level of care that must be provided is unclear, and states vary greatly on how much they spend on inmate health care. For example, annual health care costs range from nearly $20,000 per inmate in California to just over $2,000 per inmate in Louisiana (The Pew Charitable Trusts, 2017). The amount spent on health care costs appears related to the number of health care-related staff employed by each state. Some states employ their own medical staff, whereas others contract with private clinicians to provide care for their inmates. Most states also require small copays from certain inmates who wish to visit the prison medical clinic for treatment of common illnesses. These copays are designed to prevent inmates who are only marginally sick or who might fake an illness from unnecessarily seeking treatment. However, some inmates indicate that these copays can serve as a deterrent from seeking needed medical care (Loeb & Steffensmeier, 2011).

Mental Health Treatment

The evolution of mental health care has significantly affected the growth of America's jails and prisons. Prior to the 1960s, the most common treatment given to those exhibiting symptoms of severe mental illness was institutionalization in mental hospitals. In the 1960s, these facilities began to receive criticism that they were overcrowded and failing to provide adequate treatment (Grob & Goldman, 2006). The facilities' poor conditions, combined with the increased availability of psychiatric medication, led the public to demand that the mentally ill be cared for in the community instead of inside mental hospitals. The result was a deinstitutionalization of the mentally ill and subsequently the dramatic reduction of available psychiatric hospital beds as many mental hospitals were shut down.

Unfortunately, the promise of providing quality community-based mental health care was never fulfilled. Nearly 60 years later, our communities still struggle to provide adequate mental health care to those in need. Because of this lack of support, those with mental illness often find themselves in contact with the justice system after either committing a criminal offense or often simply due to their unusual behavior. Many researchers suggest that the unintentional outcome of the 1960s deinstitutionalization movement has been to place those with serious mental health issues in penal institutions (Kim, 2016; Primeau, Bowers, Harrison, & Xu, 2013). Today, approximately 37% of prisoners and 44% of jail inmates report having a history of a mental health problem (see Figure 11.6).

The prevalence of mental illness poses a particularly difficult problem for correctional administrators. Simply put, jails and prisons are not the ideal places to house those with severe mental illness. Due to the inherent stressors of

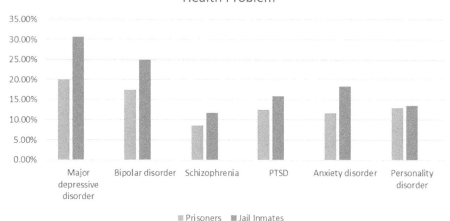

FIGURE 11.6
Percentage of Prisoners
and Jail Inmates Reporting
Mental Health Problems.

Source: Bronson and
Berzofsky (2017).

incarceration, mentally ill inmates have been found to be involved, as both attackers and victims, in a disproportionate amount of prison violence (Blitz, Wolff, & Shi, 2008; Fazel, Hayes, Bartellas, Clerici, & Trestman, 2016). To make matters worse, the most common institutional response to those who engage in prison misbehavior might further exacerbate the symptoms of mental illness (see Box 11.4).

BOX 11.4 POLICY DILEMMA: RESTRICTIVE HOUSING

On an average day, approximately 4% of state and federal inmates are housed in restrictive housing. Inmates can be held in restrictive housing for many reasons, including protection from potential attackers, as a sanction for violating prison rules, or while awaiting classification. Annually, nearly 20% of inmates will spend some amount of time in restrictive housing (Beck, 2015). The use of restrictive housing has come under increased scrutiny, especially as it is used on juveniles or those with mental illness. In 2016, President Obama banned the use of restrictive housing for juvenile offenders in the federal prison system. Although evidence is not conclusive (Astor, Fagan, & Shapiro, 2018), some critics argue that restrictive housing can have detrimental effects on an inmate's psychological well-being. Several states have also banned or placed significant limits on the use of restrictive housing with mentally ill inmates. However, a 2017 report showed that the federal prison system did not have limits on the amount of time that mentally ill inmates could spend in a restrictive housing setting (Office of the Inspector General, 2017). In some cases, mentally ill inmates have spent months or even years in this environment. One could argue that a ban on restrictive housing, including for protective purposes, could place vulnerable inmates in increased danger of physical harm. On the other hand, restrictive housing itself could potentially harm an inmate's mental health. Labrecque (2018) suggested that one solution might be to transform restrictive housing into a more therapeutic setting. Should we limit the use of restrictive housing for inmates with existing mental illness? If so, what remedies would a prison have to maintain security and keep all inmates safe?

A solution to these issues is not always clear. Correctional officers rarely have the specialized training needed to identify and respond to mental health symptoms, and may misinterpret an inmate's actions to be intentional misbehavior. The psychiatric professionals working in the prisons often report difficulties providing care in a setting that can negatively affect their client's mental health and limit the available treatment options (Ellis & Alexander, 2017). Some scholars have advocated for the increased use of diversion programs (e.g., mental health court) to ensure better access to therapeutic treatment outside of a prison setting (Atkin-Plunk & Sloas, 2018). However, we must also keep in mind that many of these offenders have committed serious crimes that are not eligible for diversion. The delicate balance between maintaining institutional safety and providing adequate care for those with mental illness will likely continue in the future.

Substance Abuse Treatment

Approximately 70% of offenders incarcerated in America's jails and prisons report using drugs on a regular basis (Table 11.2). In fact, state prisoners are 12 times more likely than the general population to meet the criteria for drug dependence or abuse. Unfortunately, less than a third of these inmates receive any type of drug treatment while incarcerated (Bronson, Stroop, Zimmer, & Berzofsky, 2017). This lack of treatment participation represents an opportunity, and perhaps a responsibility, to assist inmates in overcoming their dependency. Although prison-based drug treatment can be expensive, the positive long-term societal impact typically outweighs the treatment costs (National Institute of Drug Abuse, 2018).

Several types of drug treatment programs are available inside correctional facilities. Options include educational classes, self-help groups (e.g., Narcotics Anonymous), professional counseling, detoxification units, maintenance drugs (e.g., methadone), and residential treatment. Each of these treatment programs can be effective at different stages of an inmate's incarceration. Initially, the prison may use detoxification and maintenance drugs to manage the effects associated with stopping drug use (National Institute of Drug Abuse, 2018). During the final year of an inmate's sentence, residential treatment might be the best option.

The best-known residential drug treatment model is the **therapeutic community** (TC) which houses substance abusers together in specific areas of the prison where they receive a variety of individual

TABLE 11.2 Prisoners Who Report Regular Use of Drugs Prior to Incarceration

Drug Type	State Prisoners	Jail Inmates
Marijuana	62.7%	64.4%
Cocaine	34.2	38.5
Heroin	16.6	18.9
Depressants	18.3	20.8
Stimulants	23.4	23.9
Hallucinogens	21.7	22.5
Inhalants	6	5

Source: Bronson, Stroop, Zimmer, and Berzofsky (2017).

and group therapies. This setting allows for a highly structured plan where an inmate's entire day is controlled as part of the treatment. The idea is to create a strong sense of community where inmates are able to identify the underlying issues related to their own substance abuse and then to hold each other accountable for their actions. The final stage of this model often involves a work release program where offenders obtain employment and learn how to budget for expenses related to housing, utilities, etc. (Substance Abuse and Mental Health Services Administration, 2013). Following their release from prison, offenders are then encouraged, and sometimes required, to participate in community-based aftercare.

Olson and Lurigio (2014) evaluated the effects of a prison-based therapeutic community in Illinois. Upon release, offenders who had completed the treatment were 15% less likely to return to prison compared to offenders who had not participated in the treatment. However, the study found that completion of a community-based aftercare service was essential to the program's success. Offenders who had participated in both the prison-based therapeutic community and aftercare were 44% less likely to return to prison compared to those who had not participated in the therapeutic community program.

There are important limitations to the therapeutic community model. First, it is usually only offered toward the end of an inmate's sentence. As such, many inmates spend several years incarcerated before becoming eligible for participation. Van Wormer and Persson (2010) stressed the need for substance abuse education to be made available immediately upon entry to the prison setting. Kerrison (2018) also found that the therapeutic community model might not be effective for all racial groups. She pointed out that the model is based on the need to identify personal flaws that are responsible for the drug addiction so that these flaws can be addressed through the treatment. The assumption that the addiction is their fault might be more difficult to accept for a marginalized group whose exposure to poverty and institutional racism could have influenced their substance abuse habits. Indeed, interviews with therapeutic community graduates found that blacks were less likely to buy into the program's design compared to white participants.

As shown above, effective therapeutic community treatment begins during incarceration but is optimized when both institutional and community-based programs work together toward achieving a common goal. The same seems to be true with other types of drug treatment options. For example, several recent studies have shown that the availability of medication-based opioid treatment (e.g., methadone, naltrexone) in the jail and prison settings improves the continuation of such treatment once the offender is released (Brinkley-Rubinstein et al., 2018; Lincoln, Johnson, McCarthy, & Alexander, 2018; Rich et al., 2015).

This once again demonstrates the important linkages between the different parts of the criminal justice system and the community.

CONTINUING PRESSURES ON INCARCERATION

It does not seem likely that we will soon solve the current problems of incarceration in the criminal justice system. Although prison populations seem to have decreased from their peak level, it seems unlikely that the pace of this decline will substantially relieve the overcrowding issues facing many jurisdictions. As more states legalize marijuana and look toward alternative punishments for other drug offenses, the composition of jails and prisons might change. However, efforts to achieve truth in sentencing for violent offenders promise to increase the proportion of violent offenders in the prisons and jails. This may further perpetuate the threat of prison violence.

As states try to deal with crowding, there is likely to be more emphasis on seeking solutions in the private sector and a greater reliance on private provision of incarceration. Thus far we have not been able to build our way out of the prison crowding crisis. Prison crowding influences privatization, violence, and limited inmate participation in industry programs. Current policy choices, whatever their other consequences, appear likely to aggravate crowding in both prisons and jails.

There are growing pressures to revise and improve prison treatment programs both as something that is owed to inmates and as a means of reducing crime and future prison populations. In the past 30 years, substantial progress has been made in the development of correctional programming. Research indicates that well-planned, carefully implemented treatment programs can substantially reduce recidivism among correctional populations. Although prisons will retain their punitive and security function, there are increasing calls for the development of more therapeutic efforts in correctional institutions.

INCARCERATION IN THE TOTAL SYSTEM

Prisons and jails are inextricably linked to the larger society in which they exist. As we saw, prison researchers identified what they called a prisoner subculture, a social system with its own set of values and language set apart from the broader society. The prison subculture placed a positive value on criminality, resistance to correctional officials, and inmate solidarity (Clemmer, 1940; Sykes, 1958). Two explanations for the existence of this subculture emerged: a deprivation model and an importation model (Schwartz & Travis, 1997:127–130). The **deprivation model** suggests that the inmate code or prison subculture develops as a reaction to the losses experienced by prisoners on arrival in the prison. The loss of freedom, autonomy, goods and services, and the like supports the development of a new social order based on the values of the prison inmates born in reaction to their plight as prisoners. The importation model, however, suggests that the prison subculture is actually a product of the selection of inmates. Those persons sent to prison, according to the importation model, already subscribe to an antiauthority code of conduct. Thus, the prison imports its problems from the free society.

All of the issues examined in this chapter display links between incarceration and the wider society. Privatization reflects a broader movement to reduce governmental bureaucracy that has trickled down to prisons and jails. Prisoners' rights represent, at some level, an expansion of the general due process revolution in American criminal justice that began under the Warren Supreme Court. Violence in institutions reflects the growing violence in American society, including gang- and drug-related violence. The problem of crowding in American prisons and jails, however, is perhaps the best illustration of how the criminal justice system and its subsystems interact with their environment. Crowding in correctional institutions is a product of changes in the justice system as well as broader social changes. All components of the justice system feel its effects.

One of the explanations for prison and jail crowding is demography. The maturing of the baby-boom generation means that the raw material for the justice system—the number of potential criminals—has increased dramatically. Thus, criminal justice agencies must respond to changes in the material environment. Similarly, the war on drugs, mandatory sentencing for drunk drivers, and a general rise in the desire for punitive solutions among Americans all help to explain both the increasing rate of imprisonment and the longer sentences imposed. Here, we can see criminal justice agencies reacting to changes in the philosophical environment.

Since the war on drugs and the development of sentencing guidelines for the federal courts, the composition of the inmate population has both increased and changed. The redefinition of drug offenses as more serious crimes and the attempt to ensure greater retribution in the allocation of criminal punishments increased the number of drug offenders sentenced to federal prisons. The proportion of the federal prison population comprised of violent offenders decreased as a result. Given the large increase in the total population, however, this change reflects the increased imprisonment of those convicted of drug crimes, not a reduction in the use of prison for violent offenders. At the state level, there has been an increase in the number of violent offenders partly because of increasing lengths of terms for those convicted of violent crimes, related to truth in sentencing. The greatly increased costs of incarceration, combined with little perceived reduction in crime, have caused some people to question the emphasis we place on prison as a response to crime. This questioning of imprisonment supports an expanded role for community alternatives to incarceration, raising issues in community corrections. Thus, changes in our views of crimes and improvements in our ability to detect and apprehend offenders have ripple effects throughout the justice system that are most keenly felt in the corrections component that is "at the end of the line."

THEORY AND INCARCERATION

Once again, we see multiple explanations for criminal justice issues. Some observers contend that crowding is the product of changes in the law that led to more people being sent to prison for longer periods. Others explain crowding as a function of changes in the criminal justice environment or the outcome of people and organizations pursuing economic advantages. Crowding, in turn, is an explanation for violence and privatization.

There are varying explanations of the effects of crowding on levels of violence, the influence of prison industry on offender recidivism, and the levels of violence found in different prisons have a number of explanations. Researchers report that the characteristics of individual inmates influence levels of violence in correctional institutions. They also report that much of the variation in violence is a product of the characteristics of the prison or jail as an organization.

A former governor of Alabama, Lester Maddox, once said that we never will improve prisons in America until we get a better class of prisoner. He presented an individual theory of prisons. By this logic, prison conditions are the product of the prisoners we confine. Other explanations suggest we can change prisons by changing correctional personnel. If we increase standards for hiring, or enhance training, prison conditions will improve. Alternatively, some suggest it is the organization itself. If prisons were smaller, or larger, or friendlier, or more caring of inmates, conditions would be better.

Remember the factors we have identified that might explain differences in prisons and prison issues. Each of these factors can represent a theory of prison, and by extension, can contribute to a theory of criminal justice. What explains prisons? Do those same things also explain police and courts?

REVIEW QUESTIONS

1. Explain what is meant by privatization in institutional corrections, and identify at least two ethical issues regarding the privatization trend.

2. Explain how the crisis of crowding impacted jail capacity.

3. Identify two strategies a state may employ to deal with prison crowding, and name some factors that may help explain why a particular state selects a specific option.

4. Identify the effects of prison crowding on inmates, staff, and prison administrators.

5. Identify the ways in which prison industry has been organized throughout the years.

6. What positive benefits are expected for the state and inmates from successful prison industry programs?

7. Describe the issues related to mental health care in prisons.

8. Describe how many of the problems faced by prisons and jails are reflections of broader changes and issues in the larger society.

REFERENCES

Alarid, L. (2003). A gender comparison of prisoner selection for job assignments while incarcerated. *Journal of Crime and Justice, 26*(1), 95–116.

Ashton, P., & Petteruti, A. (2011). *Gaming the system: How the political strategies of private prison companies promote ineffective incarceration policies*. Washington, DC: Justice Policy Institute.

Astor, J.H., Fagan, T.J., & Shapiro, D. (2018). The effects of restrictive housing on the psychological functioning of inmates. *Journal of Correctional Health Care, 24*, 8–20.

Atkin-Plunk, C.A., & Sloas, L.B. (2018). Specialty courts: Funneling offenders with special needs out of the criminal justice system. In Dodson, K.D. (Ed.), *Routledge handbook on offenders with special needs* (pp. 37–53). New York: Routledge.

Auerbach, B. (1982). New prison industries legislation: The private sector re-enters the field. *The Prison Journal, 62*(2), 25–36.

Austin, J., & Irwin, J. (2012). *It's about time: America's imprisonment binge* (4th ed.). Belmont, CA: Wadsworth.

Baćak, V., & Ridgeway, G. (2018). Availability of health-related programs in private and public prisons. *Journal of Correctional Health Care, 24*, 62–70.

Bales, W., Bedard, L., Quinn, S., Ensley, D., & Holley, G. (2005). Recidivism of public and private state prison inmates in Florida. *Criminology & Public Policy, 4*(1), 57–82.

Beck, A.J. (2015). *Use of restrictive housing in U.S. prisons and jails, 2011–12*. Washington, DC: Bureau of Justice Statistics.

Beck, A.J., & Stroop, J. (2017). *PREA data collection activities, 2017*. Washington, DC: Bureau of Justice Statistics.

Beck, A.J., Rantala, R.R., & Rexroat, J. (2014). *Sexual victimization reported by adult correctional authorities, 2009–11*. Washington, DC: Bureau of Justice Statistics.

Benenson Strategy Group (2017). *ACLU National Survey*. Retrieved from: www.aclu.org/report/smart-justice-campaign-polling-americans-attitudes-criminal-justice.

Berk, R., Ladd, H., Graziano, H., & Baek, J. (2003). A randomized experiment testing inmate classification systems. *Criminology & Public Policy, 2*(2), 215–242.

Bierie, D. (2012). Is tougher better? The impact of physical prison conditions on inmate violence. *International Journal of Offender Therapy and Comparative Criminology, 56*(3), 338–355.

Blitz, C.L., Wolff, N., & Shi, J. (2008). Physical victimization in prison: The role of mental illness. *International Journal of Law and Psychiatry, 31*, 385–393.

Bowker, L.H. (1980). *Prison victimization*. New York: Elsevier.

Braswell, M., Montgomery, R., & Lombardo, L. (Eds.). (1994). *Prison violence in America*. (2nd ed.). Cincinnati, OH: Anderson.

Brinkley-Rubinstein, L., McKenzie, M., Macmadu, A., Larney, S., Zaller, N., Dauria, E., & Rich, J. (2018). A randomized, open label trial of methadone continuation versus forced withdrawal in a combined US prison and jail: Findings at 12 months post-release. *Drug and Alcohol Dependence, 184*, 57–63.

Bronson, J., & Berzofsky, M. (2017). *Indicators of mental health problems reported by prisoners and jail inmates, 2011–12*. Washington, DC: Bureau of Justice Statistics.

Bronson, J., Maruschak, L.M., & Berzofsky, M. (2015). *Disabilities among prison and jail inmates, 2011–12*. Washington, DC: Bureau of Justice Statistics.

Bronson, J., Stroop, J., Zimmer, S., & Berzofsky, M. (2017). *Drug use, dependence, and abuse among state prisoners and jail inmates, 2007–2009*. Washington, DC: Bureau of Justice Statistics.

Bryan, B. (2017). Paternal incarceration and adolescent social network disadvantage. *Demography, 54*, 1477–1501.

Bureau of Justice Assistance (2004). *Prison industry enhancement certification program*. Washington, DC: Bureau of Justice Assistance.

Bureau of Justice Statistics (1986). *Population density in state prisons*. Washington, DC: U.S. Department of Justice.

Burgess-Proctor, A., Huebner, B.M., & Durso, J.M. (2016). Comparing the effects of maternal and paternal incarceration on adult daughters' and sons' criminal justice system involvement. *Criminal Justice and Behavior, 43*(8), 1034–1055.

Burkhardt, B.C. (2017). Who is in private prisons? Demographic profiles of prisoners and workers in American private prisons. *International Journal of Law, Crime, and Justice, 51*, 24–33.

Burkhardt, B.C., & Jones, A. (2016). Judicial intervention into prisons: Comparing private and public prisons from 1990–2005. *Justice System Journal, 37*, 39–52.

Carson, E.A. (2014). *Prisoners in 2013*. Washington, DC: Bureau of Justice Statistics.

Carson, E.A. (2015). *Prisoners in 2014*. Washington, DC: Bureau of Justice Statistics.

Carson, E.A. (2018). *Prisoners in 2018*. Washington, DC: Bureau of Justice Statistics.

Carson, E.A., & Sabol, W.J. (2012). *Prisoners in 2011*. Washington, DC: Bureau of Justice Statistics.

Carson, E.A., & Golinelli, D. (2013). *Prisoners in 2012*. Washington, DC: Bureau of Justice Statistics.

Carson, E.A., & Anderson, E. (2016). *Prisoners in 2015*. Washington, DC: Bureau of Justice Statistics.

Carson, E.A., & Sabol, W.J. (2016). *Aging of the state prison population, 1993–2013*. Washington, DC: Bureau of Justice Statistics.

Chang, T.F.H., & Thompkins, D. (2002). Corporations go to prisons: The expansion of corporate power in the correctional industry. *Labor Studies Journal, 27*, 45–69.

Chari, K.A., Simon, A.E., DeFrances, C.J., & Maruschak, M.A. (2016). *National survey of prison health care: Selected findings* (Report no. 96). Retrieved from the U.S. Department of Health and Human Resources, National Center for Health Sciences website: www.cdc.gov/nchs/data/nhsr/nhsr096.pdf

Clear, T., & Austin, J. (2009). Reducing mass incarceration: Implications of the iron law of prison populations. *Harvard Law & Policy Review, 3*, 307–324.

Clear, T, & Schrantz, D. (2011). Strategies for reducing prison populations. *The Prison Journal, 91*, 38S–59S.

Clemmer, D. (1940). *The prison community*. Boston: Christopher.

Clifford, K., & White, R. (2017). *Media and crime: Content, context, and consequence*. Melbourne: Oxford University Press Australia.

Cochran, J., Siennick, S.E., & Mears, D.P. (2018). Social exclusion and paternal incarceration impacts on adolescents' networks and school engagement. *Journal of Marriage and Family, 80*, 478–498.

Conrad, J. (1982). What do the undeserving deserve? In Johnson, R. & Toch, H. (Eds.), *The pains of imprisonment* (pp. 313–330). Beverly Hills, CA: Sage.

Cooper, L. (1993). Minimizing liability with private management of correctional facilities. In Bowman, G., Hakim, S., & Seidenstat, P. (Eds.), *Privatizing correctional institutions* (pp. 131–137). New Brunswick, NJ: Transaction.

Cox, R. (2016). The effect of private sector work opportunities in prison and labor market outcomes of the formerly incarcerated. *Journal of Labor Research, 37*, 412–440.

Criminal Justice Newsletter (1985). Controversial A.C.A. policy calls for "privatization." *Criminal Justice Newsletter, 16*(3), 1–3.

Cullen, F.T., & Travis, L.F., III. (1984). Work as an avenue of prison reform. *New England Journal on Criminal and Civil Confinement, 10*(1), 45–64.

Culp, R. (2011). Prison privatization turns 25. In Ismaili, K. (Ed.), *U.S. criminal justice policy: A contemporary reader*. Sudbury, MA: Jones & Bartlett.

Davis, A. (1968). Sexual assaults in the Philadelphia prison system and sheriffs vans. *Trans-Action, 6*, 13.

DeLisi, M. (2015). Mass incarceration is the style, mass offending is the substance. *Journal of Criminal Justice, 43*, 404–405.

Durham, A. (1989). Managing the costs of modern corrections: Implications of nineteenth-century privatized prison-labor programs. *Journal of Criminal Justice, 17*(2), 441–455.

Duwe, G. (2015). An outcome evaluation of a prison work release program: Estimating its effects on recidivism, employment, and cost avoidance. *Criminal Justice Policy Review, 26*(6), 531–554.

Duwe, G., & Clark, V. (2013). The effects of private prison confinement on offender recidivism: Evidence from Minnesota. *Criminal Justice Review, 38*(3), 375–394.

Dyer, J. (2000). *The perpetual prison machine*. Boulder, CO: Westview.

Ellis, H., & Alexander, V. (2017). The mentally ill in jail: Contemporary clinical and practice perspectives for psychiatric-mental health nursing. *Archives of Psychiatric Nursing, 31*, 217–222.

Fazel, S., Hayes, A.J., Bartellas, K., Clerici, M., & Trestman, R. (2016). Mental health of prisoners: Prevalence, adverse outcomes, and interventions. *Lancet Psychiatry, 3*, 871–881.

Fowler, S., Blackburn, A., Marquart, J., & Mullings, J. (2010). Would they officially report an in-prison sexual assault? An examination of inmate perceptions. *The Prison Journal, 90*, 220–243.

Fox, V. (1983). *Correctional institutions*. Englewood Cliffs, NJ: Prentice Hall.

Gaes, G. (2008). Cost, performance studies look at prison privatization. *National Institute of Justice Journal, 259*, 32–36.

Gau, J.M. (2019). *Criminal justice policy: Origins and effectiveness*. New York: Oxford.

Geis, G., Mobley, A., & Shichor, D. (1999). Private prisons, criminological research, and conflict of interest: A case study. *Crime & Delinquency, 45*(3), 372–388.

Ghandnoosh, N. (2018). *Can we wait 75 years to cut the prison population in half?* Washington, DC: The Sentencing Project.

Greenberg Quinlan Rosner Research (2018). *The evolving landscape of crime and incarceration*. Retrieved from: https://storage.googleapis.com/vera-web-assets/inline-downloads/iob-poll-results-summary.pdf

Greenwood, P. (1981). *Private enterprise prisons? Why not? The job would be done better and at less cost*. Santa Monica, CA: RAND.

Griffin, M., & Hepburn, J. (2013). Inmate misconduct and the institutional capacity for control. *Criminal Justice and Behavior, 40*, 270–288.

Grob, G.N., & Goldman, H.H. (2006). *The dilemma of federal mental health policy: Radical reform or incremental change?* New Brunswick, NJ: Rutgers University Press.

Guerino, P., Harrison, P.M., & Sabol, W.J. (2011). *Prisoners in 2010*. Washington, DC: Bureau of Justice Statistics.

Harrison, P., & Beck, A. (2005). *Prison and jail inmates at midyear 2004*. Washington, DC: Bureau of Justice Statistics.

Harrison, P., & Beck, A. (2006). *Prisoners in 2005*. Washington, DC: Bureau of Justice Statistics.

Hassine, V. (1996). *Life without parole: Living in prison today*. Los Angeles: Roxbury.

Hopper, J. (2013). Benefits of inmate employment programs: Evidence from the prison industry enhancement certification program. *Journal of Business & Economics Research, 11*(5), 213–222.

Johnson, R. (1987). *Hard time: Understanding and reforming the prison*. Monterey, CA: Brooks/Cole.

Jones, T., & Pratt, T. (2008). The prevalence of sexual violence in prison: The state of the knowledge base and implications for evidence-based correctional policy making. *International Journal of Offender Therapy and Comparative Criminology, 52*(3), 280–295.

Kerrison, E.M. (2018). Exploring how prison-based drug rehabilitation programming shapes racial disparities in substance use disorder recovery. *Social Science & Medicine, 199*, 140–147.

Keve, P. (1986). *The history of corrections in Virginia*. Charlottesville: University of Virginia Press.

Kim, D.Y. (2016). Psychiatric deinstitutionalization and prison population growth: A critical literature review and its implication. *Criminal Justice Policy Review, 27*, 3–21.

Krajick, K. (1984). Punishment for profit. *Across the Board, 21*(3), 20–27 (March).

Labrecque, R.M. (2018). Specialized or segregated housing units: Implementing the principles of risk, needs, and responsivity. In Dodson, K.D. (Ed.), *Routledge handbook on offenders with special needs* (pp. 69–83). New York: Routledge.

Lahm, K. (2008). Inmate-on-inmate assault: A multilevel examination of prison violence. *Criminal Justice and Behavior, 35*, 120–137.

Lanza-Kaduce, L., Parker, K., & Thomas, C. (1999). A comparative recidivism analysis of releases from private and public prisons. *Crime & Delinquency, 45*(1), 28–47.

Lincoln, T., Johnson, B.D., McCarthy, P., & Alexander, E. (2018). Extended-release naltrexone for opioid use disorder stated during or following incarceration. *Journal of Substance Abuse Treatment, 85*, 97–100.

Lindsey, A.M., Mears, D.P., & Cochran, J.C. (2016). The privatization debate: A conceptual framework for improving (public and private) corrections. *Journal of Contemporary Criminal Justice, 32*(4), 308–327.

Lockwood, D. (1980). *Prison sexual violence.* New York: Elsevier.

Loeb, S.J., & Steffensmeier, D. (2011). Older inmates' pursuit of good health: A focus group study. *Research in Gerontological Nursing, 4*, 185–194.

Lonski, P.D. (1986). Illinois shatters myth—Industries boost local economy. *Corrections Today, 48*(7), 52–56.

Lundahl, B., Kunz, C., Brownell, C., Harris, N., & Van Vleet, R. (2009). Prison privatization: A meta-analysis of cost and quality of confinement indicators. *Research on Social Work Practice, 19*(4), 383–394.

Mai, C., & Subramanian, R. (2017). *The price of prisons: Examining state spending trends, 2010–2015.* New York: Vera Institute of Justice.

Marquart, J. (2008). Addicted to prisons and asking "why don't they riot?" *Criminology & Public Policy, 7*, 153–158.

Martin, J., Lichtstein, B., Jenkot, R., & Forde, D. (2012). "They can take us over any time they want": Correctional officers' responses to prison crowding. *The Prison Journal, 92*, 88–105.

Maruschak, L. (2004). *HIV in prisons, 2001.* Washington, DC: Bureau of Justice Statistics.

Maruschak, L., & Beck, A. (2001). *Medical problems of inmates, 1997.* Washington, DC: Bureau of Justice Statistics.

Maruschak, L., Berzofsky, M., & Unangst, J. (2015). *Medical problems of state and federal prisoners and jail inmates, 2011–12.* Washington, DC: Bureau of Justice Statistics.

McLaughlin, M., Pettus-Davis, C., Brown, D., Veeh, C., & Renn, T. (2016). *The economic cost of incarceration in the U.S.* St. Louis, MO: Institute for Advancing Justice Research and Innovation.

Mears, D.P., Cochran, J.C., & Cullen, F.T. (2015). Incarceration heterogeneity and its implications for assessing the effectiveness of imprisonment on recidivism. *Criminal Justice Policy Review, 26*(7), 691–712.

Miller, R., Sexton, G.E. & Jacobsen, V.J. (1991). *Making jails productive.* Washington, DC: National Institute of Justice.

Moses, M., & Smith, C. (2007). Factories behind fences: Do prison "real work" programs work? *National Institute of Justice Journal, 257*, 32–35 (June).

National Institute on Drug Abuse (2018). Principles of drug addiction treatment: A research-based guide (3rd ed.). Washington, DC: Author.

Noonan, M.E. (2016). *Mortality in state prisons, 2001–2014—Statistical tables*. Washington, DC: Bureau of Justice Statistics.

Office of the Inspector General, U.S. Department of Justice (2017). *Review of the Federal Bureau of Prisons' use of restrictive housing for inmates with mental illness*. Retrieved from: https://oig.justice.gov/reports/2017/e1705.pdf

Olson, D.E., & Lurigio, A.J. (2014). The long term effects of prison-based drug treatment and aftercare services on recidivism. *Journal of Offender Rehabilitation, 53*, 600–619.

Phelps, M.S., & Pager, D. (2016). Inequality and punishment: A turning point for mass incarceration? *Annals of the American Academy of Political and Social Sciences, 663*, 185–203.

Pitzer, P. (2013). Federal overincarceration and its impact on correctional practices: A warden's perspective. *Criminal Justice, 28*, 41–45.

Pollock, J. (2004). *Prisons and prison life*. Los Angeles: Roxbury.

Pompoco, A., Wooldredge, J., Lugo, M., Sullivan, C., & Latessa, E.J. (2017). Reducing inmate misconduct and prison returns with facility education programs. *Criminology & Public Policy, 16*(2), 515–547.

Porter, L.C., & King, R. (2015). Absent fathers or absent variables? A new look at paternal incarceration and delinquency. *Journal of Research in Crime and Delinquency, 52*(3), 414–443.

Powers, R.A., Kaukinen, C., & Jeanis, M. (2017). An examination of recidivism among inmates released from a private reentry center and public institutions in Colorado. *The Prison Journal, 97*(5), 609–627.

Pratt, T., & Maahs, J. (1999). Are private prisons more cost-effective than public prisons? A meta-analysis of evaluation research studies. *Crime & Delinquency, 45*(3), 358–371.

Primeau, A., Bowers, T.G., Harrison, M.A., & Xu, X. (2013). Deinstitutionalization of the mentally ill: Evidence for transinstitutionalization from psychiatric hospitals to penal institutions. *Comprehensive Psychology, 2*, 1–10.

Reidy, T.J., Cihan, A., & Sorensen, J.R. (2017). Women in prison: Investigating trajectories of institutional female misconduct. *Journal of Criminal Justice, 52*, 49–56.

Reisig, M., & Pratt, T. (2000). The ethics of correctional privatization: A critical examination of the delegation of coercive authority. *The Prison Journal, 80*(2), 210–222.

Rich, J.D., McKenzie, M., Wong, J.B., Tran, L., Clarke, J., Noska, A., et al. (2015). Methadone continuation versus forced withdrawal on incarceration in a combined US prison and jail: A randomized, open-label trial. *Lancet, 386*, 350–359.

Rynne, J., Harding, R., & Wortley, R. (2008). Market testing and prison riots: How public-sector commercialization contributed to a prison riot. *Criminology & Public Policy, 7*(1), 117–142.

Schaefer, L., Sams, T., & Lux, J. (2016). Saved, salvaged, or sunk: A meta-analysis of the effects of faith-based interventions on inmate adjustments. *The Prison Journal, 96*(4), 600–622.

Schwartz, M.D., & Travis, L.F., III (Eds.). (1997). *Corrections: An issues approach* (4th ed.). Cincinnati, OH: Anderson.

Schwarz, R.J. (1986). New Mexico: The anatomy of a riot. *Corrections Magazine, 6*(2), 6–24.

Seiter, R.P. (2017). *Correctional administration: Integrating theory and practice* (3rd ed.). Boston: Pearson.

Sexton, G. (1995). *Work in American prisons: Joint ventures with the private sector*. Washington, DC: National Institute of Justice.

Shover, N., & Einstadter, W. (1988). *Analyzing American corrections*. Belmont, CA: Wadsworth.

Steiner, B., Butler, H.D., & Ellison, J.M. (2014). Causes and correlates of prison inmate misconduct: A systematic review of the evidence. *Journal of Criminal Justice, 42*, 462–470.

Stephan, J., & Jankowski, L. (1991). *Jail inmates, 1990*. Washington, DC: Bureau of Justice Statistics.

Struckman-Johnson, C., & Struckman-Johnson, D. (2006). A comparison of sexual coercion experiences reported by men and women in prison. *Journal of Interpersonal Violence, 21*(12), 1591–1615.

Substance Abuse and Mental Health Services Administration (2013). *Intervention summary: Correctional therapeutic community for substance abusers*. Retrieved from: https://nrepp. samhsa.gov/Legacy/ViewIntervention.aspx?id=338

Sykes, G. (1958). *The society of captives*. Princeton, NJ: Princeton University Press.

Teasdale, B., Daigle, L.E., Hawk, S.R., & Daquin, J.C. (2016). Violent victimization in the prison context: An examination of the gendered contexts of prison. *International Journal of Offender Therapy and Comparative Criminology, 60*(9), 995–1015.

The Pew Charitable Trusts (2017). *Prison health care: Costs and quality*. Retrieved from: www. pewtrusts.org/~/media/assets/2017/10/sfh_prison_health_care_costs_and_quality_ final.pdf

The Sentencing Project (2017). *Private prisons in the United States*. Retrieved from: www.sentencingproject.org/publications/private-prisons-united-states/

Thomson, N.D., Towl, G.J., & Centifanti, L.C.M. (2016). The habitual female offender inside: How psychopathic traits predict chronic prison violence. *Law and Human Behavior, 40*(3), 257–269.

Trammell, R. (2011). Symbolic violence and prison wives: Gender roles and protective pairing in men's prisons. *The Prison Journal, 91*, 305–324.

van Womer, K., & Persson, L.E. (2010). Drug treatment within the U.S. federal prison system: Are treatment needs being met? *Journal of Offender Rehabilitation, 49*(5), 363–375.

Walker, S. (2015). *Sense and nonsense about crime, drugs, and communities* (8th ed.). Belmont, CA: West/Wadsworth.

West, H. (2010). *Prisoners at midyear 2009*. Washington, DC: Bureau of Justice Statistics.

Whitehead, J.T., Dodson, K.D., & Edwards, B.D. (2013). *Corrections: Exploring crime, punishment, and justice in America* (3rd ed.). Waltham, MA: Elsevier.

Whitehead, J.T., Edwards, B., & Griffin, H. III. (2017). Ethics and prisons: Selected issues. In Braswell, M.C., McCarthy, B.R., & McCarthy, B.J. *Justice, crime, and ethics* (pp. 315–337. New York: Routledge.

Williams, B., Stern, M., Mellow, J., Safer, M., & Greifinger, R. (2012). Aging in correctional custody: Setting a policy agenda for older prisoner health care. *American Journal of Public Health, 102*(8), 1475–1481.

Wolff, N., Shi, J., & Bachman, R. (2008). Measuring victimization inside prisons: Questioning the questions. *Journal of Interpersonal Violence, 23*(10), 1343–1362.

Wolff, N., Shi, J., & Siegel, J. (2009). Understanding physical victimization inside prisons: Factors that predict risk. *Justice Quarterly, 26*(3), 445–475.

Wolff, N., Blitz, C., Shi, J., Siegel, J., & Bachman, R. (2007). Physical violence inside prisons: Rates of victimization. *Criminal Justice and Behavior, 34*, 588–599.

Wooldredge, J. (2003). Keeping pace with evolving prison populations for effective management. *Criminology & Public Policy, 2*(2), 253–258.

Wright, J.P., & DeLisi, M. (2016). *Conservative criminology: A call to restore balance to the social sciences.* New York: Routledge.

Zeng, Z. (2018). *Jail inmates in 2016.* Washington, DC: Bureau of Justice Statistics.

Zoukis, C. (2017). *Federal prison handbook: The definitive guide to surviving the federal bureau of prisons.* Charleston, SC: Middle Street.

Zweig, J.M., Yahner, J., Visher, C.A., & Lattimore, P.K. (2015). Using general strain theory to explore the effects of prison victimization experiences on later offending and substance use. *The Prison Journal, 95*, 84–113.

IMPORTANT CASE

Brown v. Plata, 131 S. Ct. 1910 (2011).

CHAPTER 12

Probation, Parole, and Community Corrections

At any given time, there are roughly two times as many people under probation and parole supervision as there are people incarcerated in the United States. By the end of 2016, more than 6.6 million persons were under correctional authority, with more than 4.7 million under probation or parole supervision (Kaeble & Cowhig, 2018). The majority of persons under community supervision (3.7 million) were on probation. Approximately 2% of the adult population of the United States is under either probation or parole supervision (Kaeble & Cowhig, 2018). Although incarceration may be the cornerstone of American corrections, the majority of criminal offenders receive sentences to probation.

Probation and parole supervision as currently operated, like the prison, are American inventions. Both involve the conditional release of convicted offenders into the community under supervision. Probationers and parolees experience similar treatment but, as we shall see, there are important differences between the two.

Probation and parole can be considered the "bookends" of imprisonment (see Box 12.1). **Probation** is a sanction generally imposed in lieu of incarceration and, thus, it occurs before imprisonment. **Parole** involves offenders released early from incarceration sentences, so it occurs after a period of imprisonment. Therefore, probation and parole flank imprisonment on either side as criminal sanctions.

This chapter examines community supervision in the criminal justice system of the United States. We will describe probation and parole supervision as well as some other forms of community control of offenders. We will explore the history and practice of probation and parole, and we will examine other community-based sanctions and the various populations involved in these sanctions.

IMPORTANT TERMS

absconders

benefit of clergy

community service orders

discretionary release

diversion

furlough

halfway houses

home incarceration

judicial reprieve

mandatory release

mark system

"on paper"

parole

probation

reentry courts

release on recognizance (ROR)

restorative justice

special conditions

standard conditions

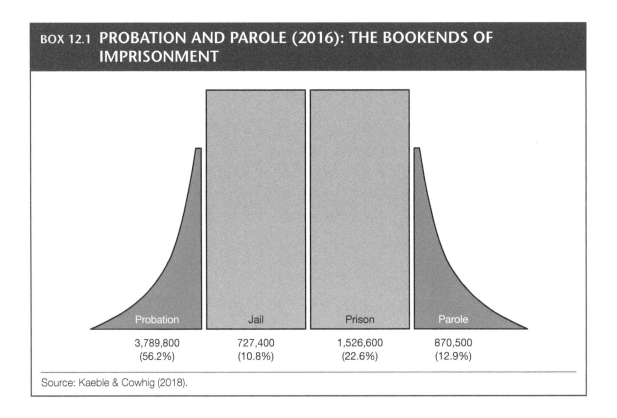

BOX 12.1 **PROBATION AND PAROLE (2016): THE BOOKENDS OF IMPRISONMENT**

Probation	Jail	Prison	Parole
3,789,800	727,400	1,526,600	870,500
(56.2%)	(10.8%)	(22.6%)	(12.9%)

Source: Kaeble & Cowhig (2018).

THE ORIGINS OF COMMUNITY SUPERVISION

supervised release

suretyship

ticket of leave

work release

Probation and parole developed in the nineteenth century in the United States, although each had precursors in Western civilization. An examination of the history of community supervision shows how the justice system has changed with the social and intellectual currents in the larger society. Shortly after the creation of the penitentiary, many people came to view incarceration as a less-than-adequate response to all offenders. Probation and parole developed as alternatives to incarceration for select groups of offenders.

Probation evolved from several prior practices in the English courts that allowed judges to grant leniency to offenders and avoid the harsh corporal and capital punishments provided in the common law. Among these were the benefit of clergy, judicial reprieve, and release on recognizance (ROR) (Allen, Eskridge, Latessa, & Vito, 1985:37–40). Each of these practices allowed the sentencing judge to postpone or avoid the execution of sentence.

The **benefit of clergy** was a practice that developed during the medieval period so that members of the clergy were accountable to ecclesiastic (church) courts rather than civil courts. The accused could claim the benefit of clergy to have his or her case moved from the civil courts to the church courts. The test for benefit of clergy came to be one of literacy, in which the court required the accused to read the text of the fifty-first Psalm (see Box 12.2). In due time, illiterate common criminals committed the psalm to memory so that they could pretend to read it and thus avoid the punishments of the king's courts (Clear & Cole, 1986:232).

BOX 12.2 FIFTY-FIRST PSALM: THE NECK VERSE

Have mercy upon me, O God,

according to thy loving kindness,

According to the multitude of thy tender mercies

blot out my transgressions.

The fifty-first Psalm, because it allowed many offenders to avoid hanging as the penalty for their crimes, came to be known as the "neck verse." After a period of expansion of the benefit of clergy (from the fourteenth through the eighteenth centuries), the practice was disallowed by statute in 1827. No longer was it possible to escape in this way to the less severe sanctions of the church courts.

Judicial reprieve was a common practice in England in the nineteenth century. Under this practice, the offender could apply to the judge for a reprieve, which required that sentencing of an offender be delayed on the condition of good behavior for a specific period. After the allotted time, the offender could ask the king for a pardon. Here, we see the addition of two components of contemporary probation: (1) a set period and (2) the requirement that the offender abide by conditions of good behavior.

Release on recognizance (ROR) was a practice (combined with peace bonds) that was a forerunner to bail. While awaiting the arrival of the circuit magistrate, an accused offender obtained release by posting a surety or by having someone vouch for him or her. This practice most directly led to the development of contemporary probation.

We generally credit John Augustus, a Boston boot maker, with being the father of probation. It was common practice in Massachusetts courts to release offenders on the recognizance of a third party. Augustus began a nearly 20-year career as a voluntary probation officer by posting bail in the Boston Police Court in 1841 for a man accused of drunkenness. Between 1841 and 1858, he supervised nearly 2,000 people. He was so successful that the state of Massachusetts passed legislation authorizing probation as a disposition and provided for the first paid probation officer. Over time, other states emulated the Massachusetts practice. Today, probation is the most common disposition of criminal cases.

Parole also developed in the mid-1800s. By the 1850s, observers of the penitentiary system grew dissatisfied with the effectiveness of incarceration in preventing further criminal behavior by offenders. These critics began to call for changes in incarceration practices that would serve to reform inmates and produce law-abiding citizens. The outcome of this reform movement was parole release and supervision. Parole has two components. The first is discretionary early release from prison. The second component is the period of supervision in the community that follows such a release.

In England and other European countries, several practices were already in place that laid the groundwork for the creation of parole. The term "parole" comes from the French phrase *parole d'honneur*, meaning word of honor. Prisoners of war gained release on their "parole" that they would not again take up arms against their captors. This term was later applied to the procedure for allowing prison inmates to return to society prior to the expiration of their prison terms. Essentially, officials expected the prisoners to vow that they would not violate the law, in return for which they were granted release.

Banishment and transportation also have been considered to be precursors to parole, in that these procedures essentially allowed an offender to avoid a more harsh penalty on the condition that the offender not return to the land of the original crime (Barnes & Teeters, 1959). Closer to modern parole practice, however, were release procedures developed by Walter Crofton and Captain Alexander Maconochie. As superintendents of penal facilities, each of these men created a system of inmate discipline that allowed the prisoners to earn early release.

Crofton devised a "ticket of leave" for inmates in the Irish prison system. Prisoners were classified into three stages of treatment, ranging from segregated confinement, through work on public projects (which was increasingly free of supervision), until final release "on license." Successful inmates earned their ticket of leave through hard work and good behavior. When Crofton believed an inmate to be ready for release, he would issue a **ticket of leave**, which authorized the inmate to leave the prison, return home, and report to the local police. There was no supervision of the released inmate.

Alexander Maconochie is often called the father of parole. His system was very similar to that of Crofton. Maconochie operated the British penal colony on Norfolk Island, in the South Pacific. There he classified offenders into three groups and instituted a **mark system**. All inmates began at the penal stage, which involved close supervision while engaging in hard labor with a large group of fellow prisoners. Good behavior and industry earned "marks" for an inmate, and upon acquiring enough marks, the prisoner moved to the next stage. The social stage involved working and living in groups of about seven prisoners, with less supervision than the penal stage. Again, prisoners earned marks leading to promotion to the individual stage, during which the prisoner was allowed a cottage and was given individual work. Prisoners were subject to being moved back to earlier stages for misconduct or laziness. Those in the individual stage who continued to demonstrate good behavior and industry eventually earned a ticket of leave or conditional pardon. Often they were apprenticed to citizens on mainland Australia (Travis, 1985).

The apparent success of these programs was noticed in the United States. In 1870, the American Prison Association in Cincinnati provided the forum for reformers to push for the creation of parole and a system of reformatory discipline in the United States (Lindsey, 1925). With growing support for the early release and reformatory discipline, New York enacted legislation creating a reformatory at Elmira, where first offenders received sentences to terms that would last "until reformation, not to exceed five years." Parole release had been born in the United States.

Simon (1993) has reviewed and assessed the development of parole. He suggested that the practice of third-party recognizance, what he calls "suretyship," was a forerunner to modern parole. Suretyship was the practice of a person of good standing in the community taking responsibility for guaranteeing the lawful behavior of another person. When the prison became the dominant form of punishment, Simon suggested, people recognized that not all offenders needed to remain in prison, or at least not for full terms. Parole release and supervision allowed authorities to select worthy offenders and release them from prison. Officials used parole to support discipline, in that well-behaved, industrious inmates who could secure employment (and thus keep busy at socially acceptable activities) received parole. When employment opportunities decreased during the Great Depression, "clinical" parole replaced disciplinary parole. In this revised model, the purpose of parole was to support the treatment and rehabilitation of inmates. In both cases, parole enabled authorities to differentiate between those convicted criminals who could be reformed and those who could not.

We can apply a similar argument to the development of probation supervision, as the functions and definitions of probation have mirrored those of parole. The task of probation evolved from diverting selected offenders from prison through control and discipline in the community, to the provision of treatment and rehabilitative services to offenders who did not need the more intensive treatment of prison.

During the first 6 or 7 decades of its existence, parole faced a number of legal challenges. The practice of discretionary early release from incarceration gained acceptance by the 1940s. In addition, during this era (called the Progressive Era), increasing attention was focused on the role of postrelease supervision of offenders (as well as probation supervision). This led eventually to the current system of parole involving both early release and supervision in the community (Rothman, 1980).

Since the mid-1990s, there has been increased concern about reentry, the return of former inmates to life in the community. Throughout the 1990s, as prison populations continued to increase, it became apparent that relatively large numbers of parole violators accounted for a substantial portion of the inmate

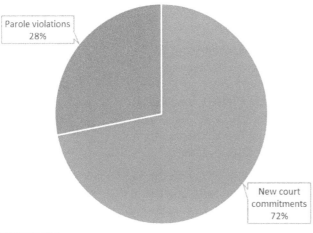

Parole violations
28%

New court
commitments
72%

FIGURE 12.1
Source of New Prison
Admissions.

Source: Carson (2018).

population (Burke, 2004). In 2016, more than one-quarter of all inmates admitted to state prisons were parole violators (see Figure 12.1).

The twin concerns of assisting released inmates to adjust to law-abiding life in the community and reducing the risk of new crime posed by released offenders led to the development of **reentry courts**, where services for and supervision of parolees are coordinated and monitored in a court environment. These operate much like drug courts in which a judge meets frequently with the offender to monitor progress and, if needed, to change conditions of supervision and release (Miller & Khey, 2017; Taylor, 2012). In many ways, the contemporary focus on prisoner reentry is similar to the original interest in developing parole (Burke & Tonry, 2006). We might say that we have rediscovered parole in the past decade.

THE ORGANIZATION OF COMMUNITY SUPERVISION: PROBATION AND PAROLE

Probation and parole, although very similar, differ in how they are organized. Whereas many states charge supervising officers with the responsibility of serving both parolees and probationers, there are other states where the two tasks are administratively separate. Probation tends to be a county and municipal function. Parole, on the other hand, usually is a state function, even in states where probation officers supervise parolees. In many states, counties can choose to provide probation supervision themselves, or may turn responsibility for the supervision of probationers over to the state.

Probationers are under the jurisdiction of the sentencing court, and held to a set of conditions imposed by the sentencing judge. Probation officers are responsible for carrying out the wishes of the sentencing judge. Parolees, on the other hand, are under the jurisdiction of the state paroling authority, and held to a set of conditions imposed by that authority. Parole officers are responsible for carrying out the wishes of the parole authority, although in most cases parole supervision is independent of the parole release authority (Bonczar, 2008). In jurisdictions in which the same officers supervise both parole and probation, the officers wear two hats, and their behavior is contingent on the legal status (probationer or parolee) of the client.

Inmates can be released from prison under a variety of terms. Several states allow prisoners to be released from a prison sentence with no parole supervision. These offenders are often said to have "maxed out" their sentence. Those inmates who enter parole supervision most often either receive a discretionary parole or achieve mandatory release. **Mandatory release** occurs at the expiration of a prison term reduced by good behavior, or when a sentencing law requires that inmates completing their prison terms receive supervision in the community for some period of time (Glaze & Bonczar, 2006:8). For the period of their supervision, they are on a conditional release and must obey the rules and conditions of parole or face a return to prison.

Discretionary release is granted by a parole board that will typically assess the inmate's perceived likelihood of successfully following the conditions of parole. The use of discretionary parole declined significantly from the 1970s until 2008. In 1977, 69% of inmates released from prison received discretionary parole release (Hughes, Wilson, & Beck, 2001). By 2008, discretionary parole represented fewer than 27% of parolees (Glaze & Bonczar, 2009). In 2016, discretionary release has again become the most common type of entry to parole (Kaeble, 2018) (see Figure 12.2).

In 1984, Congress eliminated parole at the federal level and established a post-confinement monitoring system called **supervised release**. A term of supervised release is imposed at the initial sentencing hearing, requiring the offender to be monitored by a federal probation officer and to abide by certain conditions upon release from prison. Unlike parole, supervised release does not serve as a substitute for imprisonment. If the offender violates the terms of supervised release, he or she can be required to serve the term of supervised release in prison (United States Sentencing Commission, 2010).

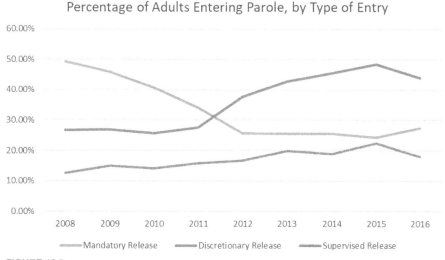

FIGURE 12.2
Percentage of Adults Entering Parole, by Type of Entry.

Source: Bureau of Justice Statistics Reports.

TABLE 12.1 Characteristics of the Probation and Parole Population, 2016		
Characteristic	Probation (%)	Parole (%)
Sex		
Male	75	87
Female	25	13
Race		
White	55	45
African American	28	38
Hispanic	14	15
Other	2	2
Most serious offense		
Violent	20	30
Property	26	21
Drug	24	31
Other	30	17

Source: Kaeble (2018).

In 2016, more than 2 million adults were sentenced to probation. At the same time, approximately 450,000 persons entered parole supervision from state prisons (Kaeble, 2018). Table 12.1 presents a description of the characteristics of the probation and parole populations in 2016. The probation population contains more females, fewer persons convicted of a violent offense, and fewer ethnic minorities than does the parole population. Given that ethnic minorities and males are more likely to receive prison sentences, it follows that the parole population (comprised of those who had been in prison) would have disproportionately fewer women and whites. Not only do parole populations generally tend to be composed of offenders with more serious criminal records, but also, largely because of incarceration, parolees tend to have greater needs in the areas of housing, employment, and personal relations than do probationers. These challenges place parolees at greater risk of recidivism. Approximately 29% of adults on parole return to prison before the end of their parole term (Kaeble, 2018).

Other Forms of Community Supervision

Although probation and parole are the major components of community supervision of criminal offenders, other programs play a role in the nonincarceration treatment of convicted offenders. These programs include halfway houses, community service, furlough, work and education, home incarceration, and various diversion programs. Very frequently, courts or parole authorities require participation in these programs as conditions of release under probation or parole. Nonetheless, we will briefly discuss each of these separately.

Halfway Houses

Halfway houses are generally small residential programs based in the community that serve populations of fewer than 30 people (Wilson, 1985). Changes in thinking about the role of the community in the development of socially acceptable behavior spurred the development of halfway houses for both criminal offenders and the mentally ill (Beha, 1975). Although these houses have a long tradition, their application to criminal corrections has experienced tremendous growth in recent decades as correctional officials have increasingly focused on successful reentry.

Halfway houses are so-named because they represent an intermediate step that is halfway between incarceration and community supervision. They are either "halfway-out" houses that deal with parolees and other ex-inmates, or "halfway-in" houses for probationers and others not imprisoned (Keller & Alper, 1970). In either case, treatment is a part of the halfway house routine.

In addition to providing room and board, halfway houses generally offer counseling services that include group and individual counseling sessions. Some halfway houses restrict their client population to special needs offenders, such as abusers of alcohol and other drugs. Other houses accept a wider range of persons and provide or contract for a wider range of services. Inasmuch as the popularity of halfway houses has increased, relatively little attention has been devoted to their effectiveness. Evidence is mixed regarding whether halfway houses reduce recidivism or provide a cost saving to states (Bell et al., 2013; Costanza, Cox, & Kilburn, 2015). In the past several years, the traditional halfway house name has been changing to "residential reentry center," though the function remains the same.

Community Service Programs

Community service orders are programs in which courts sentence convicted offenders to a number of hours of service to community organizations or governmental agencies (Allen, Latessa, & Ponder, 2013). The work takes place in the community, and the offender is generally at liberty except for the scheduled work hours. Community service is used most often for low-level offenders and delinquent juveniles. The hope is that this type of punishment can give back to the community while serving as a deterrent for low-level offenders. See Figure 12.3.

Often community service is symbolically retributive in that the court requires the offender to render some service related to the offense. For example, drunk drivers might be ordered to assist in a hospital emergency room, vandals ordered to clean and repair damaged buildings, and similar reflective penalties may be imposed on other offenders. Probation officers typically supervise those completing community service sentences. Frequently, community service is a condition of probation.

Community service is often a core component of restorative justice programs. The term **restorative**

FIGURE 12.3
Talbert House in Cincinnati, Ohio, was one of the first community-based agencies to receive accreditation from the American Correctional Association's Commission on accreditation for corrections.

Ellen S. Boyne

justice refers to efforts to repair the harm to victims and/or communities caused by crime through interventions with the offender. Karp and Drakulich (2004) reported on a statewide program of restorative justice in Vermont that involved those convicted of less serious offenses receiving probation with a condition that they participate in the restorative justice program. Nearly two-thirds of restorative justice cases included community service orders. More recently, McDowell, Braswell, and Whitehead (2017) described the possible expansion of restorative justice practice involving more serious crimes such as sex offenses and white-collar crimes. If restorative justice efforts increase, we can expect an increased reliance on community service in the future.

Furlough Programs

Furlough programs also are alternatives to the traditional sanction of incarceration. At base, a furlough program allows an inmate to leave the penal facility for a specified period to perform an identified function. Several prisons and jails operate work and educational furlough programs. Inmates are released, without escort, to participate in educational programs or to report to work. After work or school, they return to the institution.

Furloughs have a relatively long history in American corrections. Traditionally, furloughs were available to prison inmates in cases of family emergencies (e.g., to visit seriously ill relatives or to attend funerals). In these traditional furloughs, the inmate and the inmate's family were required to pay all expenses (Fox, 1983:147). In the 1960s, several states began granting furloughs to qualifying inmates for social visits by issuing weekend passes to certain inmates. Today, many states use furlough programs to allow inmates that are near the time of their release on parole to seek employment or arrange for residences. Each year thousands of prison inmates receive furloughs from prisons across the country.

Judges can also grant furloughs to help those who have been found not guilty by reason of insanity gradually reintegrate into society. Perhaps the most well-known case of insanity involved John Hinkley, who attempted to assassinate President Reagan in 1981. After his not guilty verdict, Hinkley was placed in a psychiatric hospital where he was kept until his release in 2016. Prior to his release, Hinkley was granted a series of increasingly long furlough visits to stay with his mother. In this case, the court system, in collaboration with hospital staff, used furloughs to maintain family relationships and assist in the defendant's eventual reentry to society.

WORK AND EDUCATIONAL RELEASE PROGRAMS

In a move related to furloughs, which tend to be brief releases for a specified period, correctional agencies are making use of work or study/educational release. Work or study release refers to the practice of allowing inmates of correctional facilities to leave the institution during the day to attend classes or work at a job. **Work release** programs allow inmates to secure or maintain employment while serving terms of incarceration. The opportunity for inmates released for work to maintain ties to conventional lifestyles and to their communities is a benefit of work release programs. Such inmates also can help support their families and often must contribute to the cost of their incarceration. This reduces overall correctional costs to the community. Although relatively few inmates participate in such release programs, those who do participate benefit from leaving the correctional facility and by being able to keep jobs or secure an education. See Figure 12.4.

FIGURE 12.4

A woman serving the last 6 months of a 17-month prison sentence fills out a job application as she rides the bus. She participates in the Clark County (Washington) work release program, which is a partial confinement program designed for offenders to continue their employment or transition back into society.

AP Photo/*The Columbian*, Steven Lane

Duwe (2015) reported an evaluation of work release programming in the state of Minnesota. He found that roughly 7% of all offenders released in a given year participate in work release. Most work release participants successfully completed the program, with just under a quarter (24%) of inmates failing the program due to rules violations or criminal behavior. Participants in the work release program were almost twice as likely to find employment compared to those released straight from prison without participating in the program. Further, each participant saved the state, on average, nearly $700 in cost avoidances.

Home Incarceration

As a response to crowded prisons and jails, there has been a resurgence of interest in the practice of house arrest or **home incarceration**. Offenders sentenced to home incarceration are essentially grounded. They must remain at home except for approved absences, such as attending school, going to work, or keeping medical appointments. While thus incarcerated, the offender is out of society, yet not confined in a jail. The offender retains ties to family and the community, the state faces reduced costs, and the offender generally experiences better conditions of confinement than would exist in a penal facility. House arrest can either be used pretrial to ensure that an offender does not abscond or as a criminal punishment in itself.

Proponents of home incarceration argue that this practice is less costly than incarceration and provides a more satisfactory alternative to incarceration than do the traditional practices of fines or probation. With home incarceration, someone (usually a probation officer) must be responsible for monitoring the offender to ensure that he or she stays home. As will be discussed in Chapter 13, probation officers often use electronic monitoring to confirm compliance with probation conditions. Although house arrest is clearly a less punitive punishment compared to incarceration, one study showed that nearly one-third of offenders who had experienced house arrest would have preferred to spend a month in jail rather than 6 months on house arrest (Martin, Hanrahan, & Bowers, 2009). These offenders mentioned the inconvenience, cost, and shame associated with being confined to their home while wearing an ankle bracelet.

Diversion

The general rubric of **diversion** incorporates a wide variety of programs that occur at all stages of the justice system. Diversion is included in the discussion of community supervision because its most common formal application is at some point in the court process prior to conviction or sentencing. These programs are most commonly available for first time, low-level offenders or those suffering from mental illness or substance abuse.

Diversion programs prevent some offenders from passing through the justice system, or minimize the extent of their processing. Supporters hope diverting offenders from the system will enable them to avoid the stigma of a criminal label and other negative effects of justice system processing. Critics point out that diversion programs might result in a "net-widening" effect because they target offenders who would likely have their cases dismissed altogether if the diversion programs were not available (Walker, 2015).

To the extent that net-widening occurs, this may indeed reduce the cost savings associated with the program. However, diversion also allows community-based organizations to effectively deal with underlying issues which might be related to the criminal activity. For example, diverting an offender from the criminal court to a mental health or drug court allows for a collaborative and specialized treatment plan to occur outside of a penal facility (Landess & Holoyda, 2017). As discussed in Chapter 11, prisons and jails are not ideal settings for the treatment of these issues. Overall, diverting certain types of offenders to treatment options has shown to have significant positive societal benefits (Zarkin et al., 2015).

BEING ON PAPER

We discuss additional types of community sanctions or forms of probation and parole supervision in Chapter 13. These include shock probation and parole, electronic monitoring, and day reporting programs. Just as the alternatives discussed above, each of these innovations either represents a form of community supervision or is imposed as a condition of probation or parole. Remember that probation and parole are the core components of community corrections, and that each of these is a form of conditional release from prison. The underlying assumption is that the offender would be in prison or jail if not for his or her participation in community programming.

Community supervision is the punishment of choice for most offenders. The pressures from prison crowding have enhanced the already important role community supervision plays in the justice system. Community supervision is an alternative to imprisonment. We generally hold offenders under community supervision to a higher moral standard than members of the free society, by virtue of the conditions of release. A term probationers and parolees often use to describe their status is to say they are "**on paper.**" This refers to the written probation or parole agreements that set out the conditions of release.

Supervision Conditions

In the final analysis, probation and parole represent an agreement between the offender and the state. In return for the decision not to incarcerate (or for the decision to release from incarceration), the offender agrees to abide by several conditions while at liberty in the community. Violation of any of these conditions can be the basis for revocation of liberty and subsequent incarceration of the offender.

Probation and parole conditions cover a wide variety of behavior and serve a number of purposes (Doherty, 2016). The primary goal of supervision conditions is to prevent future criminality on the part of probationers and parolees. There are two general types of conditions: **standard conditions** that are imposed on everyone under supervision in a jurisdiction, and **special conditions** imposed on individual offenders that relate directly to the offender's status and characteristics (Travis & Stacey, 2010).

Standard conditions of parole and probation vary across jurisdictions. They usually include restrictions on travel and on the freedom to change addresses or jobs, and involve instructions on reporting to the supervising officer. These conditions ensure that the probation or parole officer will be able to monitor the offender. Table 12.2 presents some of the most common standard conditions of parole in the United States during the past quarter century. Special conditions vary greatly (Doherty, 2016). They include such things as restrictions on association with particular people, requirements of attending

TABLE 12.2 **Standard Parole Conditions**	
Comply with the law	Restrictions on the use of controlled substances
Restrictions on changing residence	Restrictions on changing employment
Refrain from possessing firearm/dangerous weapons	Pay fees/restitution
Maintain employment/education program	Report arrest/questioning by law enforcement
Report regularly	Submit to drug test
Permit home/work visits and searches	Restriction on undesirable associates/locations
Restrictions on out of state travel	Obey parole officer instructions

Source: Travis and Stacey (2010:606).

treatment programs, restrictions on the consumption of alcoholic beverages, and restitution orders.

The conditions of probation and parole serve a number of purposes. Some conditions are necessary to ensure that supervision can occur. Thus, offenders may not be allowed to change their jobs or addresses without notifying the supervising officer. They may be required to report to a probation or parole officer on a regular schedule. Other supervision conditions impose punishment, such as a requirement that an offender pay restitution, write a punishment lesson, or do community service work. Still other conditions seek to ensure that offenders receive treatment for their problems by requiring participation in educational, psychological, or substance abuse programs.

Offenders who fail to submit to supervision (do not report as directed, change jobs or addresses without notifying their supervising officer, etc.) are called **absconders**, as they absconded (escaped) from supervision. The term absconder is generally reserved for offenders who actively attempt to avoid supervision. A study of probation absconders (Mayzer, Gray, & Maxwell, 2004) found that these offenders often seem to abscond in the face of probable revocation and they do not appear to differ from probationers who ultimately have their conditional liberty revoked. Belshaw (2011) found that absconding was among the most common circumstances leading to a probation revocation. A more recent study showed that employment, education level, and income status all related to the probability of an offender absconding (Stevens-Martin & Liu, 2017). It seems likely that those who abscond from supervision might either have difficulty paying the fees associated with probation or have violated other terms of their supervision.

We also use special conditions to reduce the risk of new criminality. For instance, jurisdictions across the country have placed specialized conditions on the supervision of sex offenders. These conditions often require the sex offender to register on a public database and to live outside of a specified distance of schools, day care centers, parks, and so on. Finally, some conditions do not impose restrictions on behavior, but give the offender notice. This may be the case, for example, when a condition of supervision is that the offender remains on probation or parole until discharged. What this condition does, of course, is tell the offender that supervision does not end until released by the supervising authority.

Client Perspectives

The experience of being on paper reflects the number and content of conditions as well as the characteristics of the probation or parole officer. A probationer facing conditions that include a curfew, requirements of restitution, restriction on associating with friends, and some period of incarceration will have a qualitatively different experience than will a probationer under less restrictive conditions. Similarly, a parolee whose supervising officer is rule oriented and unsympathetic will experience a different sanction than one whose officer is flexible and empathetic.

Criminal justice scholars have asked probationers and parolees for their views on community supervision. In large part, these offenders prefer probation or parole to any form of incarceration. Many feel fortunate to have an opportunity to prove themselves in the community, but most also identify problems associated with being under supervision. Probationers and parolees most often complain about what they perceive as the pettiness and unfairness of some conditions of release. They also dislike the requirement that they seek approval of their supervising officers for many minor decisions (Williams, May, & Wood, 2008).

In some cases, offenders see community supervision as a less desirable sanction than incarceration (Martin, Hanrahan, & Bowers, 2009; Wood & May, 2003). If the prison or jail term will be short and followed by unsupervised release, but probation or parole supervision will entail close supervision and restrictive conditions, many offenders would rather just "get it over with." They would prefer a short term of incarceration followed by freedom instead of what they expect will be a period of close supervision followed by prison if they are found in violation of conditions. Relatedly, some inmates who are close to "maxing out" their prison sentence will choose to forgo any scheduled parole hearings (Ostermann, 2012). Again, the prospect of completing their prison sentence might be more desirable than submitting to the various parole conditions. This suggests that, at some level, it is possible to design a community supervision sanction that is as unpleasant as incarceration. However, this also means that probation and parole officers may be very limited in their ability to control offender behavior through the threat of sanctions (Flory, May, Minor, & Wood, 2006). If offenders would prefer jail time to intensive supervision, the threat of jail is unlikely to convince them to abide by the conditions of supervision.

Despite the view that probation and parole is a punitive sanction, many offenders view community corrections favorably. The relationship between the offender and their probation or parole officer appears to be very important. Offenders who report having a positive rapport with their supervising officer have been shown to have lower rates of recidivism compared to those who have a more strained relationship (Blasko, Friedmann, Rhodes, & Taxman, 2015; Chamberlain, Gricius, Wallace, Borjas, & Ware, 2017). Fortunately, recent research has found that most probationers and parolees are satisfied with their supervising officers' job performance and feel that the officer truly cares about them (Holmstrom, Adams, Morash, Smith, & Cobbina, 2017; Zortman, Powers, Hiester, Klunk, & Antonio, 2016).

One complaint appears to be the social distance between offenders and their supervising officer (Helfgott, 1997). In other words, the offenders perceive the officers as having too little in common with them, making it difficult to appreciate the daily struggle faced by those under community supervision

(Box 12.3). This complaint likely has merit, as most probation and parole officers do not have an extensive criminal history. One study also found that offenders preferred to be supervised by an officer of their own race (Springer, Applegate, Smith, & Sitren, 2009).

BOX 12.3 POLICY DILEMMA: PEER MENTORING

In an attempt to bridge the social distance gap between offenders and their probation officer, many reentry programs are beginning to experiment with the use of peer mentors. The hope is that clients will be more receptive to advice coming from someone who has lived through a period of incarceration and has successfully reintegrated into society. Peer mentoring has been successful in helping at-risk juveniles (Fletcher, 2007), but research involving adult populations is lacking. Peer mentor programs face a number of potential barriers (Umez, Cruz, Richey, & Albis, 2017). For example, many prisons have rules prohibiting visits from those with criminal records. This rule would prevent a peer mentor from establishing a relationship with an inmate prior to release. Additionally, standard probation conditions often regulate a probationer's association with those who have a criminal record or a history of violent behavior. Traditionally, justice officials have viewed associations with these types of individuals as a risk factor for recidivism. Do you feel that we should begin to allow certain ex-offenders to provide counseling services to parolees? If so, what types of regulations should we put in place on such a service?

Officer Perspectives

Probation and parole officers lead a "schizophrenic" existence in relation to their jobs. For years, research into the role of the community supervision officer has identified conflicting dimensions of the responsibilities to help but also to control (Clear & Cole, 1986). The officer is expected to befriend and assist his or her clients, as well as to monitor and control them. Most officers think that offender success is primarily a function of rational choice and generally view the offender's complaint of social distance as an excuse (Gunnison & Helfgott, 2011).

Crean (1985:118) reported, "Within a typical work week, a line officer is usually called upon to be an investigator, a biographer, a watchdog, counselor, friend, confidant, reporter, expert witness, and broker of outside services." The line officer is a public employee and responsible to the authority that placed the offender under the officer's supervision. Further, state legislators in recent years have increased the number of tasks that community corrections officers are responsible for carrying out (Hsieh et al., 2015). Many of these new tasks are related to risk and needs assessment. Although research has shown that risk and needs assessment holds promise for improving outcomes, DeMichele and Payne (2012) found that probation officers did not place as much value on these tasks compared to their supervisors.

Traditionally, research has focused on two "styles" of community corrections: social work (emphasis on helping the offender) and law enforcement (emphasis on controlling criminal conduct). Recent scholars have questioned the continued use of this simple dichotomy. In a nationwide study of probation officers, Miller (2015:325–326) found that officers generally fit into one of four typologies:

1. **High engagers**: These officers took on both law enforcement and social work roles. Specifically, they emphasized the consequences of poor behaviors, the risks of detention, and the need to fully enforce the probation conditions. However, they also valued the establishment of trust and rapport with probationers as well as the utilization of various rehabilitative strategies. Finally, they helped their clients recognize risky activities and engaged third parties (e.g., family, community, police) to assist in the supervision process.
2. **Medium engagers—community collaboration**: These officers valued both the law enforcement and social work roles, but to a lesser degree than the high engagers. They sometimes also sought to help probationers avoid risky activities and seek help from third parties to assist in the supervision process.
3. **Medium engagers—traditional**: These officers demonstrated actions consistent with both the law enforcement and social work roles, but rarely helped probationers avoid risky activities or sought help from third parties to assist in the supervision process.
4. **Low engagers**: These officers demonstrated supervision tactics consistent with both law enforcement and social work roles. However, they used each surveillance tactic less frequently compared to the other groups.

As you can see, Miller's (2015) survey found that the differences between officers primarily involved their work output rather than adherence to either a law enforcement or social work role. In fact, these findings suggest that most probation officers embrace at least some elements of both roles. As Lutze (2014: 11) described, "in order to be successful, community corrections officers understand that both the risk offenders pose to the community and offenders' needs and behaviors most likely to sabotage their success must be addressed." De Michele (2007) also argued that punishment and rehabilitation are not mutually exclusive goals. He pointed out that one of the major problems is that political ideologies are used to promote one strategy while being dismissive of the other. In reality, probation officers are forced to work within the context of an ever-evolving political landscape. As can be seen with contemporary sex offender registration laws, these changes can directly influence the probation officer's job duties.

Community supervision involves a rather large degree of discretion and responsibility. Research has found that most of the community corrections officers' time is actually spent in the office meeting with clients (DeMichele, 2007; Matz, Conley, & Johanneson, 2017). However, the job also permits them to leave the office for various duties. For example, many probation officers are required to conduct home and employment visits as part of their supervisory duties, but may be able to choose when to conduct these visits. The amount of discretion given to a probation officer regarding his or her use of time ultimately depends on their caseload, duties expected of them, and their department's organizational culture.

Even though there are psychological, monetary, and social rewards in being a probation or parole officer, there is also the danger of dealing with offenders. In a survey of probation officers in three states, 41% of officers reported being threatened or assaulted by an offender (Lewis, Lewis, & Garby, 2013). Not surprisingly, experiencing these events led to a greater level of occupational burnout. Despite the challenges, probation and parole officers generally exhibit the characteristics of professionals, and they frequently report that the benefits outweigh the drawbacks of the job.

OFFENDER RIGHTS IN COMMUNITY SUPERVISION

Probation and parole normally are considered privileges and not rights (del Carmen, 1985). Once placed on probation or parole, however, the offender does have a constitutionally protected interest in retaining that status. To date, there have been relatively few U.S. Supreme Court decisions dealing with the acceptability of various conditions of release and supervision practices. State or federal appellate courts have decided most of the existing cases in this area. In 2006, the U.S. Supreme Court decided the case of *Samson v. California*. Samson's parole conditions required him to submit to warrantless searches by police or his parole officer. One such search uncovered criminal evidence resulting in his return to prison. The Court ruled that the search was acceptable because of the parole condition that authorized it.

Essentially, courts have supported all types of probation and parole conditions, as long as those conditions were constitutional, reasonably related to criminality, clearly written, and/or contributed to the rehabilitation of the offender. The decisions of courts vary across jurisdictions, and are indeed case specific. For example, the courts are likely to uphold a prohibition against the consumption of alcoholic beverages in a case in which the offender's criminality stemmed from drunkenness, but void it in a case in which the offender has no history of alcohol abuse.

Probation and parole conditions often serve as an arsenal of leverage for correctional authorities. Because violation of any of the conditions of parole or probation constitutes grounds for revocation and incarceration, the conditions provide the officer with tools for controlling the offender. With regard to revocation, however, the U.S. Supreme Court has set forth due process protections for probationers and parolees.

In 1972, the Court decided the case of *Morrissey v. Brewer*, in which a parolee sought relief after his parole had been revoked. The Court ruled that parolees facing revocation must be granted the following protections: (1) written notice of the claimed violations, (2) disclosure of evidence against the parolee, (3) the opportunity to be heard and to present evidence, (4) a limited right to confront and cross-examine witnesses, (5) a hearing before a neutral body, and (6) a written statement of the decision and evidence on which it was based.

The following year, the Court decided the case of *Gagnon v. Scarpelli* (1973), which dealt with probation revocation. In this case, the Court stated that, at least regarding revocation procedures, there was no substantial difference between probationers and parolees. They granted probationers the same protections as those given to parolees.

In 1983, the Court heard the case of *Bearden v. Georgia*, in which a man was placed on probation with a condition to pay $750 in fines and restitution. Shortly thereafter, the man lost his job and was unable to pay the balance owed. The trial court then revoked his probation and sentenced him to prison. However, the Supreme Court ruled that it was unfair to revoke the probation automatically due to the failure to pay. Instead, a determination must be made as to whether a defendant has the means to pay. Probation can only be revoked if it is determined that the probationer had wilfully refused to pay the required balance.

Persons under community supervision generally are less restricted and suffer fewer pains than those who are incarcerated. Nonetheless, community supervision is still a punishment for crime, and the status of conviction and being under sentence sets probationers and parolees apart from free citizens. Even with the protections afforded by the *Morrissey* and *Gagnon* decisions, probationers and parolees may have their liberty revoked and may be incarcerated for offenses that would not carry similar penalties for free citizens. In addition, incarceration may follow hearings for which the burden of proof is not as high as the trial standard of proof beyond a reasonable doubt.

Community supervision serves crime control functions by allowing the state to limit the risk of new crimes being committed through relaxed requirements for revocation (as compared to conviction). Yet, probationers and parolees have protected interests in conditional liberty, and the actions and decisions of probation and parole authorities are constrained by some due process requirements. It is instructive to observe that although a number of states and the federal government have taken steps to eliminate discretionary parole release, most have retained a period of postrelease supervision. Community supervision itself represents a balance between individual interests in avoiding incarceration and community interests in controlling potential criminality.

See Box 12.4 for a table of selected court cases on probation and parole.

BOX 12.4 SELECTED CASES ON COMMUNITY SUPERVISION

Morrissey v. Brewer 408 U.S. 471 (1972)	Parolees facing revocation must be granted: written notice of the claimed violations; disclosure of evidence against the parolee; the opportunity to be heard and to present evidence; a limited right to confront and cross-examine witnesses; a hearing before a neutral body; and a written statement of the decision and the evidence on which it was based
Gagnon v. Scarpelli 411 U.S. 778 (1973)	At least in regard to revocation procedures, there is no substantial difference between probationers and parolees; therefore, probationers are granted the same protections as parolees (for specific protections granted, see *Morrissey v. Brewer*)
Bearden v. Georgia 461 U.S. 660 (1983)	A judge cannot revoke probation for failure to pay a fine unless it is proven that the probationer had the ability to pay but made the choice not to do so
Samson v. California 547 U.S. 843 (2006)	Supreme Court affirmed the decision holding that suspicionless searches of parolees are lawful under California law and reasonable under the Fourth Amendment

Supervision conditions also represent part of the arsenal of the prosecutor. As we discussed earlier, it is common for prosecutors to dismiss new criminal charges when they can proceed to a probation or parole revocation. Kingsnorth, Macintosh, and Sutherland (2002) studied probation violation processing in eight California counties. They reported that prosecutors frequently seek probation violations rather than file new criminal charges. This is especially true when the new offense is less serious than the one for which the offender was placed on probation.

COMMUNITY SUPERVISION IN THE WHOLE SYSTEM

In previous chapters, we have discussed several areas of overlap between community supervision and other components of the justice system. Pretrial release, diversion from court, prosecutorial charging, incarceration, and the size of the prison population all involve decisions that depend (more or less) directly on the operation of community supervision. For example, if probation and parole were not available as alternatives to incarceration, absent the development of other practices, the prison and jail population would increase threefold. The growth of intermediate sanctions and other changes in community supervision practices show how the criminal justice process is an open system.

The development of intermediate sanctions and other changes in community supervision agencies and practices illustrate the open nature of the criminal justice system. The primary motivation for these new alternatives has been the need to control correctional costs, and the prison population specifically. As the prison population grew, so did community supervision populations. The same forces that produced the increase in prisoners have worked to swell community supervision populations. The war on drugs, for example, brought a large number of offenders into the justice system. Low-level drug offenders are now prime candidates for intermediate sanctions while under community supervision. A common adjustment to truth in sentencing, which requires longer prison terms for violent offenders, has been to divert property and public order offenders from prison to probation or parole.

The prison population crisis has changed the nature of the community supervision population so that there are increased numbers of felons and more serious offenders under probation and parole supervision than historically was the case. The increase in the numbers of offenders provides the impetus for changes in community supervision management, financing, and programs. Because they involve closer supervision and impose more restrictive conditions of release, some of these programs have relatively high rates of supervision failure. Those revoked from intermediate sanction programs often are imprisoned, but now come to prison as community supervision failures.

In these ways, we can see the links between the components of the justice system, and between the justice system and the larger society. A redefinition of the seriousness of drug offenses led to enhanced drug crime enforcement. Increased enforcement changed the size and composition of the correctional population, both those incarcerated and those under community supervision. Seeking ways of managing the larger populations, we have developed intermediate sanctions that have altered traditional community supervision programs. Changes in community supervision, in turn, seem to have put increased pressure on prison populations through increased numbers of supervision failures. The next chapter explores changes in community supervision in more detail.

REVIEW QUESTIONS

1. Distinguish between probation and parole.
2. Briefly trace the origins of community supervision in corrections.
3. Describe the organization of community supervision.
4. Tell what is meant by the term "reentry."
5. Identify five forms of community supervision, excluding traditional probation and parole.
6. Distinguish between standard and special conditions of probation or parole.
7. Describe the role and importance of community supervision in the criminal justice system.
8. Identify the due process rights of probationers and parolees at revocation.

REFERENCES

Allen, H.E., Latessa, E.J., & Ponder, B.S. (2013). Corrections in America: An introduction (13th ed.). Upper Saddle River, NJ: Pearson.

Allen, H.E., Eskridge, C.W., Latessa, E.J., & Vito, G.F. (1985). *Probation and parole in America.* Monterey, CA: Brooks/Cole.

Barnes, H.E., & Teeters, N.D. (1959). New horizons in criminology. Englewood Cliffs, NJ: Prentice Hall.

Beha, J.A. (1975). Halfway houses in adult corrections: The law, practice, and results. *Criminal Law Bulletin, 11*(4), 437–477.

Bell, N., Bucklen, K.B., Nakamura, K., Tomkiel, J., Santore, A., Russell, L., et al. (2013). *Recidivism report, 2013.* Mechanicsburg, PA: Pennsylvania Department of Corrections.

Belshaw, S.H. (2011). Are all probation revocations treated equal? An examination of felony probation revocations in a large Texas county. *International Journal of Punishment and Sentencing, 7,* 64–76.

Blasko, B.L., Friedmann, P.D., Rhodes, A.G., & Taxman, F.S. (2015). The parolee–parole officer relationship as mediator of criminal justice outcomes. *Criminal Justice and Behavior, 42*(7), 722–740.

Bonczar, T. (2008). *Characteristics of state parole supervising agencies, 2006.* Washington, DC: Bureau of Justice Statistics.

Burke, P. (2004). Parole violations: An important window on offender reentry. *Perspectives, 28*(1), 24–31.

Burke, P., & Tonry, M. (2006). *Successful transition and reentry for safer communities: A call to action for parole.* Silver Spring, MD: Center for Effective Public Policy.

Carson, E.A. (2018). *Prisoners in 2016.* Washington, DC: Bureau of Justice Statistics.

Chamberlain, A.W., Gricius, M., Wallace, D.M., Borjas, D., & Ware, V. (2017). Parolee-parole officer rapport: Does it impact recidivism? *International Journal of Offender Therapy and Comparative Criminology.* Advance online publication. doi: 10.1177/0306624X 17741593

Clear, T.R., & Cole, G.F. (1986). *American corrections.* Monterey, CA: Brooks/Cole.

Costanza, S.E., Cox, S.M., & Kilburn, J.C. (2015). The impact of halfway houses on parole success and recidivism. *Journal of Sociological Research, 6,* 39–55.

Crean, D.M. (1985). Community corrections: On the line. In Travis, L.F., III. (Ed.), *Probation, parole and community corrections* (pp. 109–124). Prospect Heights, IL: Waveland.

del Carmen, R.V. (1985). Legal issues and liabilities in community corrections. In Travis, L.F., III. (Ed.), *Probation, parole and community corrections* (pp. 47–70). Prospect Heights, IL: Waveland.

DeMichele, M.T. (2007). *Probation and parole's growing caseloads and working allocation: Strategies for managerial decision making.* Lexington, KY: American Probation & Parole Association.

DeMichele, M., & Payne, B. (2012). Measuring community corrections' officials perceptions of goals, strategies, and workload from a systems perspective: Differences between directors and nondirectors. *The Prison Journal, 92*(3), 388–410.

Doherty, F. (2016). Obey all laws and be good: Probation and the meaning of recidivism. *Georgetown Law Journal, 104*, 291–354.

Duwe, G. (2015). An outcome evaluation of a prison work release program: Estimating its effects on recidivism, employment, and cost avoidance. *Criminal Justice Policy Review, 26*(6), 531–554.

Fletcher, R.C. (2007). *Mentoring ex-prisoners: A guide for prisoner re-entry programs.* Washington, DC: U.S. Department of Labor.

Flory, C., May, D., Minor, K., & Wood, P. (2006). A comparison of punishment exchange rates between offenders under supervision and their supervising officers, *Journal of Criminal Justice, 34*(1), 39–50.

Fox, V. (1983). *Correctional institutions.* Englewood Cliffs, NJ: Prentice Hall.

Glaze, L., & Bonczar, T. (2006). *Probation and parole in the United States, 2005.* Washington, DC: Bureau of Justice Statistics.

Glaze, L., & Bonczar, T. (2009). *Probation and parole in the United States, 2008.* Washington, DC: Bureau of Justice Statistics.

Gunnison, E., & Helfgott, J. (2011). Factors that hinder offender reentry success: A view from community corrections officers. *International Journal of Offender Therapy and Comparative Criminology, 55*(2), 287–304.

Helfgott, J. (1997). Ex-offender needs versus community opportunity in Seattle, Washington. *Federal Probation, 61*, 12–24.

Holmstrom, A.J., Adams, E.A., Morash, M., Smith, S.W., & Cobbina, J.E. (2017). Supportive messages female offenders receive from probation and parole officers about substance avoidance: Message perceptions and effects. *Criminal Justice and Behavior, 44*(11), 1496–1517.

Hsieh, M.L., Hafoka, M., Woo, Y., Wormer, J., Stohr, M.K., & Hemmens, C. (2015). Probation officer roles: A statutory analysis. *Federal Probation, 79*, 20–37.

Hughes, T., Wilson, D., & Beck, A. (2001). *Trends in State parole, 1991–2000.* Washington, DC: Bureau of Justice Statistics.

Kaeble, D. (2018). *Probation and parole in the United States, 2016.* Washington, DC: Bureau of Justice Statistics.

Kaeble, D., & Cowhig, M. (2018). *Correctional populations in the United States, 2016.* Washington, DC: Bureau of Justice Statistics.

Karp, D., & Drakulich, K. (2004). Minor crime in a quaint setting: Practices, outcomes, and limits of Vermont reparative probation boards. *Criminology & Public Policy, 3*(4), 655–686.

Keller, O., & Alper, B. (1970). *Halfway houses: Community centered corrections and treatment.* Lexington, MA: D.C. Heath.

Kingsnorth, R., MacIntosh, R., & Sutherland, S. (2002). Criminal charge of probation violation? Prosecutorial discretion and implications for research in criminal court processing. *Criminology, 40*(3), 553–578.

Landess, J., & Holoyda, B. (2017). Mental health courts and forensic assertive community treatment teams as correctional diversion programs. *Behavioral Science & the Law, 35*, 501–511.

Lewis, K.R., Lewis, L.S., & Garby, T.M. (2013). Surviving the trenches: The personal impact of the job on probation officers. *American Journal of Criminal Justice, 38*, 67–84.

Lindsey, E. (1925). Historical sketch of the indeterminate sentence and parole system. *Journal of Criminal Law & Criminology, 16*(1925), 9–126.

Lutze, F.E. (2014). *Professional lives of community corrections officers: The invisible side of reentry.* Thousand Oaks, CA: Sage.

Martin, J., Hanrahan, K., & Bowers, J., Jr. (2009). Offenders' perceptions of house arrest and electronic monitoring. *Journal of Offender Rehabilitation, 48*, 547–570.

Matz, A.K., Conley, T.B., & Johanneson, N. (2017). What do supervision officers do? Adult probation/parole officer workloads in a rural western state. *Journal of Crime and Justice.* Advance online publication. doi: 10.1080/0735648X.2017.1386119

Mayzer, R., Gray, M., & Maxwell, S. (2004). Probation absconders: A unique risk group? *Journal of Criminal Justice, 32*(2), 137–150.

McDowell, L.A., Braswell, M.C., & Whitehead, J.T. (2017). Restorative justice and ethics: Real-world applications. In Braswell, M.C., McCarthy, B.R., & McCarthy, B.J. (Eds.), *Justice, crime, and ethics* (pp. 261–291). New York: Routledge.

Miller, J. (2015). Contemporary modes of probation officer supervision: The triumph of the "synthetic" officer? *Justice Quarterly, 32*(2), 314–336.

Miller, J., & Khey, D. (2017). Fighting America's highest incarceration rates with offender programming: Process evaluation implications from Louisiana's 22nd judicial district reentry court. *American Journal of Criminal Justice, 42*, 574–588.

Ostermann, M. (2012). Recidivism and the propensity to forgo parole release. *Justice Quarterly, 29*(4), 596–618.

Rothman, D.J. (1980). *Conscience and convenience.* Boston: Little, Brown.

Simon, J. (1993). *Poor discipline.* Chicago: University of Chicago Press.

Springer, N.F., Applegate, B.K., Smith, H.P., & Sitren, A.H. (2009). Exploring the determinants of probationers' perceptions of their supervising officers. *Journal of Offender Rehabilitation, 48*(3), 210–227.

Stevens-Martin, K., & Liu, J. (2017). Fugitives from justice: An examination of felony and misdemeanour probation absconders in a large jurisdiction. *Federal Probation, 81*, 41–51.

Taylor, C.J. (2012). Balancing act: The adaptation of traditional judicial roles in reentry court. *Journal of Offender Rehabilitation, 51*(6), 351–369.

Travis, L.F., III. (1985). The development of American parole. In Allen, H.E., Eskridge, C., Latessa, E.J., & Vito, G.F. (Eds.), *Probation and parole in America* (pp. 19–35). New York: Free Press.

Travis, L.F., III., & Stacey, J. (2010). A half century of parole rules: Conditions of parole in the United States, 2008. *Journal of Criminal Justice, 38*(4), 606.

Umez, C., Cruz, J., Richey, M., & Albis, K. (2017). *Mentoring as a component of re-entry: Practical considerations from the field.* New York: The Council of State Governments Justice Center.

United States Sentencing Commission (2010). *Federal offenders sentenced to supervised release.* Washington, DC. Retrieved from www.ussc.gov/Education_and_Training/Annual_National_Training_Seminar/2012/2_Federal_Offenders_Sentenced_to_Supervised_Release.pdf

Walker, S. (2015). *Sense and nonsense about crime, drugs, and communities* (8th ed.). Stamford, CT: Cengage.

Williams, A., May, D., & Wood, P. (2008). The lesser of two evils? A qualitative study of offenders' preferences for prison compared to alternatives. *Journal of Offender Rehabilitation*, 46(3,4), 71–90.

Wilson, G.P. (1985). Halfway house programs for offenders. In Travis, L.F., III. (Ed.), *Probation, parole and community corrections* (pp. 151–164). Prospect Heights, IL: Waveland.

Wood, P., & May, D. (2003). Racial differences in perceptions of the severity of sanctions: A comparison of prison with alternatives. *Justice Quarterly*, 20(3), 605–631.

Zarkin, G.A., Cowell, A.J., Hicks, K.A., Mills, M.J., Belenko, S., Dunlap, L.J., & Keyes, V. (2015). Lifetime benefits and costs of diverting substance-abusing offenders from state prison. *Crime & Delinquency*, 61(6), 829–850.

Zortman, J.S., Powers, T., Hiester, M., Klunk, F.R., & Antonio, M.E. (2016). Evaluating re-entry programming in Pennsylvania's board of probation & parole: An assessment of offenders' perceptions and recidivism outcomes. *Journal of Offender Rehabilitation*, 55(6), 419–442.

IMPORTANT CASES

Bearden v. Georgia, 461 U.S. 660 (1983).

Gagnon v. Scarpelli, 411 U.S. 778 (1973).

Morrissey v. Brewer, 408 U.S. 471 (1972).

Samson v. California, 547 U.S. 843 (2006).

CHAPTER 13

Issues in Community Supervision

Many important issues in community supervision center on developments in the financing, management, and technology involved in probation and parole. Largely in response to crowded institutions and budgetary concerns, there is continuing interest in community supervision as an alternative to incarceration. Recent developments are attempts to improve and enhance the effectiveness and economy of probation and parole. Following a period of time when the central practices of community supervision (probation and parole supervision) had themselves come under attack, it appears that states are again seeing the value of community corrections. This chapter examines the controversy over community supervision, and discusses specific attempts to improve the efficiency of community corrections.

DOES COMMUNITY SUPERVISION WORK?

von Hirsch and Hanrahan (1978) published a report for the U.S. Department of Justice titled "Abolish Parole?" In this brief document, they suggested that parole supervision might be useless. On both rational (practical) and philosophical grounds, the authors concluded that there was little reason to continue the practice of parole supervision. They held that a period of supervision under conditions constituted an added penalty. That is, assuming the offender served a prison sentence as punishment, there was no reason to impose conditional release in addition to incarceration. From a practical standpoint, they argued that the research to date did not support parole supervision as a crime control strategy. The authors also argued that parole officers were not very effective in providing needed services to parolees.

von Hirsch and Hanrahan were not the first to question parole supervision, but they may have been the most eloquent critics. Later reports (e.g., Gottfredson, Mitchell-Herzfeld, & Flanagan, 1982) showed that success of supervision in preventing crimes by parolees often depends on how "new crimes" are defined. If a parolee returned to prison for violating the conditions of release without committing a new crime counts as a failure, parole supervision may be less effective in preventing recidivism than no supervision at all. Because parolees must obey

IMPORTANT TERMS

boot camp

casework model

day reporting

drug testing

electronic monitoring

global positioning system (GPS)

intensive supervision

justice reinvestment

model case management system

net-widening

Probation Subsidy Act

radio frequency

restitution

shock incarceration

shock parole

shock probation

split sentence

supervision fees

team supervision

technical violations

Transition from Prison to the Community (TPC)

the conditions of release, they may be returned to prison for technical violations. **Technical violations** are infractions of the rules of supervision that do not involve any new criminality. Examples include violating curfew, failing to report to a parole officer, and other noncriminal behavior. If returns to prison based on technical violations count as failures, parolees have more chances to fail than do nonparolees, who must be convicted of new crimes to be incarcerated.

With the increased use of both probation and parole in response to prison crowding, it is important to determine whether these tools are effective. Wodahl, Ogle, and Heck (2011) observed that community-based corrections is facing a crisis of legitimacy resulting from high rates of probation and parole revocation. In fact, the most recent data available from the Bureau of Justice Statistics show that only around 60% of those who exit probation and parole have successfully completed their term (Kaeble, 2018). With large numbers of revocations, the potential cost savings and impact on prison crowding are not realized.

Other scholars suggest that the high revocation rate does not necessarily mean that community corrections is failing. Instead, Hamilton and Campbell (2013) argued that the revocation rate is largely due to the courts' overly punitive response to technical violations. By revoking probation or parole following a technical violation, we may be incarcerating more offenders than is necessary to ensure public safety. Many jurisdictions have responded to this issue by implementing various strategies including increased funding for reentry programs, stronger departmental policies to guide officers' discretion in the event of noncompliance, graduated sanctions that provide incremental responses to noncompliant behavior, and limiting the consequences offenders receive when they do not comply with supervision conditions (Wodahl et al., 2011).

In attempting to answer the question of whether community supervision works, we must first determine what we mean by the word "work." In general, we can apply three criteria to community supervision to determine "what works." Harland (1996:2–3) noted that the definition of what works depends on the perspective of the individual asking the question. Elected officials and the public often define "working" as reducing rates of recidivism. Policymakers may define "working" as reducing rates of commitment or lengths of stay for prison and jail inmates. Budget officials may define what works as managing offenders at a lower cost. For retributivists, what works is a system of sanctions that matches the pain of punishment with the harm of the crime and blameworthiness of the offender. In the following discussion, probation, parole, and other community supervision practices work to the degree that they can (1) control risk of new crime, (2) reduce incarceration and correctional costs, and/or (3) match punishments with offenders.

Controlling Risk

Much attention has been paid to the ability or inability of community correctional programs to control risk. As Bennett (1991:95) noted, "The public is upset, and perhaps rightly so, that people placed on regular probation often do not receive either help or supervision because of large caseloads and inadequate supervision." The critical question is how safe is safe enough? To be clear, all community corrections programs involve some level of risk to the public. Even if we use the best conceivable program, any offender who is supervised in the community has a certain risk of failure. Although no new crime would be the preferred outcome, achieving levels of new crime among a population diverted from prison that is less than (or no more than) the level among incarcerated offenders may be "safe enough." In other words, the goal is to manage the risk as well as possible and make sure that the overall societal benefits outweigh the risks.

One way to evaluate community corrections is the impact that offenders might have on the communities in which they remain (or return) in lieu of incarceration. Hipp and Yates (2009) reported that in one city, the return of parolees led to increases in neighborhood crime rates. Their data indicated that parolees resume their criminal ways, or otherwise produce increased rates of crime in neighborhoods in which they reside. More recently, Chamberlain (2018) also found that parolees caused several harmful effects on neighborhoods. Consistent with the results from Hipp and Yates, Chamberlain found that the number of parolees in a community was related to that community's crime rate. Specifically, she found that "just a 1% increase in parolees in a neighborhood would result in a 1.1% increase in neighborhood violent crime" (p. 191). Additionally, Chamberlain found that parolees can impact the neighborhood in fundamental ways (e.g., property sales and vacancies) aside from the increase in crime rate. Given these findings, it is no surprise that the public may be scared of the idea that offenders will return or remain in their communities instead of being incarcerated.

During the past 2 decades, substantial progress has been made in the development and use of risk assessment devices in both probation and parole. Many classification and prediction instruments exist, and several appear to offer valid predictions of the risk of recidivism among correctional populations (Harris, Rice, Quinsey, & Cormier, 2015). More so now than in any other time in history, we have the ability to identify those who are most at risk of committing a crime in the community. Most judges, and community corrections offices now use these risk assessment tools to determine which offenders receive community supervision.

Many scholars have expressed concerns with the increased use of risk assessments. One problem with these instruments is the incredible diversity of correctional populations. Most risk assessment devices do not account for specific subpopulations such as females or racial and ethnic minorities. As a result, the accuracy of risk predictions can vary across different groups (Taylor & Blanchette, 2009). Another common concern relates to how the assessments may disproportionately score racial minorities and those with low socioeconomic status as at-risk. Finally, many states refuse to allow offenders to view or challenge their score on these risk assessment tests (Rhine, Petersilia, & Reitz, 2017). Given the importance of the decision to grant or deny community supervision, the debate surrounding the use of risk assessments is likely to continue.

Surely, the safest course of action in terms of protecting the community from convicted offenders is to incarcerate the offenders. To do so, however, does not really manage risk but rather seeks to avoid it. At some point, most of these offenders will return to the community, regardless of their level of risk. Attempts to identify and manage risk through probation and parole supervision are necessary, if difficult. Baird (2009) criticized risk classification efforts as being too complicated and lacking accuracy in their conclusions. Baird warned that classification may be more accurate in assessing risk than in identifying ways to reduce risk.

Reducing Incarceration and Costs

As Jones (1991) observed, a primary factor motivating the development and spread of community supervision sanctions is a desire to reduce prison crowding and correctional costs. Clear and Byrne (1992:321) flatly said, "The frank bottom line for the intermediate sanction movement must be whether it is able to reduce overcrowding in corrections." The impetus behind the modern movement to expand community supervision sanctions has been the economic downturn of the mid-2000s and the increasingly popular view that probation or parole could serve as a better rehabilitation tool compared to incarceration.

The creation of alternative sanctions in the community can have the effect of reducing demand for prison space (Gowdy, 1993). Conversely, the development of more severe community-based sanctions may simply result in net-widening. Phelps (2013) argued that in many cases, probation has served as both an alternative to prison and a net-widener. In this way, we cast the "net" of punishment wider to catch more people who would have otherwise not been under correctional supervision. To the extent that community supervision agencies strive to identify and punish violators, the "failure" rates for probation and parole increase, and more commmunity supervision violators go to prison. If she is correct, alternative sanctions as presently developed will not meet the goal of reducing correctional costs and prison crowding.

Community supervision can work as an alternative to incarceration and many intermediate punishments developed in the past few decades seek to accomplish just that purpose. The problem arises when, because of these more restrictive community punishments, failure rates increase, leading to increased levels of imprisonment. In order for us to say that community supervision "works," we must be able to show that it lowers incarceration rates. Evaluations of the effectiveness of community supervision use different measures of outcome. The inclusion of technical violations, including absconding, with new criminality produces relatively high rates of failure but may not represent high rates of crime. Most evaluations seem to measure failure by the offenders' arrest, reconviction, or incarceration after being under community supervision.

Several studies have compared the outcomes of community-based sanctions to incarceration. Overall, this research tends to show that those sentenced to community supervision are arrested and convicted of new crimes at lower rates than those who are sentenced to incarceration (Bales & Piquero, 2012; Caudy, Tillyer, & Tillyer, 2018). However, Villettaz, Gillieron, and Killias (2015) point out that most of the existing research must be viewed with caution due to the inherent bias of comparing those who have been sentenced to incarceration to those who have been given community-based sanctions. Simply put, those who have been sentenced to incarceration are usually higher-risk offenders compared to those sentenced to probation. Even with today's advanced statistical tests used in an attempt to control for such differences, this is a difficult task. Villettaz et al. reported that the most methodologically sound studies typically find no difference in recidivism between incarceration and community-based punishments.

In attempting to answer the question of whether parole is effective, several studies have examined the outcomes of offenders who were released from prison unconditionally (maxed out) compared to those who were granted parole. In one of the few national studies of this kind, the Urban Institute found that those released under discretionary parole were somewhat less likely to be rearrested than those who were released with either no parole supervision or those whose parole was mandated at the time of sentence (Solomon, Kachnowski, & Bhati, 2005). However, we do not know if the benefit of discretionary release was due to the effectiveness of the postrelease supervision or an indicator that the parole board successfully identified the least risky offenders for release. It should be noted that this nationwide study could mask any differences in effectiveness between jurisdictions. This is a major limitation given that states vary considerably in terms of their parole decision making and supervision process.

Several other evaluations have examined the effectiveness of parole in a particular state. For example, recent studies in Kentucky (Vito, Higgins, & Tewksbury, 2017) and New Jersey (Ostermann, 2015) have found that offenders who are granted parole have better outcomes than those who max out their sentences and are released with no supervision. Ostermann and Hyatt (2016) tested the effectiveness of a short-lived program implemented in New Jersey that limited the ability of offenders to max out their sentence. Under this program, inmates who were not granted discretionary parole were released via mandatory parole 6 months before the end of their sentence. This ensured that each released offender received at least 6 months of postrelease supervision. Ostermann and Hyatt found that offenders released under the mandatory parole program were less likely to be rearrested than if they had been released unconditionally. However, this positive effect only lasted during the 6 months that they were under active supervision. The long-term effects of community supervision is still unclear.

Matching Punishments with Offenders

A third goal of the development of intermediate sanctions in the community is to create a range of penalties appropriate to the range of offenses and offenders who come before the courts. Many see the traditional choice between probation and incarceration as inadequate for responding to the wide array of crimes and criminals that exist. Morris and Tonry (1990:38) argued that a "variety of intermediate punishments, along with appropriate treatment conditions, should be part of a comprehensive, integrated system of sentencing and punishment."

Community supervision programs and practices have undergone tremendous change in recent times as we struggle to develop this range of punishments. The matching of punishments to offenders includes two dimensions as anticipated by Morris and Tonry (1990). First, the severity of the penalty should reflect more closely the seriousness of the crime. In this way, there needs to be a range of punishments. Second, punishments should be sensitive to the needs of specific offenders. This necessitates a range of appropriate treatment conditions.

The ability of probation supervision to provide for community safety remains critical. Matejkowski, Festinger, Benishek, and Dugosh (2011) advocated for policies that would encourage probation and parole officers to distinguish between offender behavior that is willfully criminal (noncompliant) and behavior that is a result of a psychiatric disorder (nonresponsive). This distinction would allow probation and parole officers to respond appropriately based on the needs of the offender. Other changes involve making both probation and parole more punitive, such as the increasing use of shock incarceration and intensive supervision, electronic monitoring, day reporting, and other sanctions that are more severe than traditional probation or parole supervision.

Taxman and Caudy (2015) argued that while we have come a long way in regards to effectively identifying an offenders' risk of recidivism, research has not yet done a good job at identifying what types of treatment each offender should receive. Instead, we too often resort to using a few evidence-based treatments to address a broad group of offenders, even when this treatment might not be appropriate. Based on analysis of criminogenic needs (e.g., antisocial personality, low self-control) and destabilizing factors (e.g., employment and educational deficiencies) among a sample of offenders in North Carolina, Taxman and Caudy (2015:90–91) identified four classes of offenders:

1. Offenders with moderate criminogenic needs but high destabilizers
2. Offenders with low criminogenic needs and few destabilizers
3. Offenders with high criminogenic needs and high destabilizers
4. Offenders with high criminogenic needs and moderate destabilizers.

Each of these classes of offenders would benefit from separate treatment plans based on their specific risk and needs. Effective matching of offenders with the proper intervention is likely to be the next development in community super-vision (Byrne & Pattavina, 2017; Gill & Wilson, 2017). To be successful, we must ensure that a true continuum of sanctions is in place so that the severity of the punishment reflects the seriousness of the crime. At the same time, we must have available a variety of evidence-based treatment options to address each offender's specific needs. Without this range of penalties and treatment options, it may be impossible to link crimes and punishments in any mean-ingful fashion.

SHOCK INCARCERATION

Programs of shock incarceration and intensive supervision have affected the operation of traditional probation and parole. They illustrate clearly how com-munity supervision serves as an alternative to incarceration. In both programs, the attempt is to use community supervision to meet sentencing goals normally associated with incarceration. Specifically, the desired result is the enhancement of both the deterrent and incapacitation effects of community supervision.

Shock probation and shock parole attempt to deter offenders from continued criminality by imposing a prison sentence that is later commuted to a period of supervision. The initial incarceration "shocks" the offender by the severity of the punishment. It also informs offenders what to expect if they continue to break the law (Vito, 1985). Thus, with **shock probation**, the judge might sentence the offender to a long prison term but, within 6 months, alter the sen-tence to a probation period. In some states, the inmate could petition the court for shock probation. In other states, shock probation is administered solely at the discretion of the judge.

Shock parole is similar to shock probation in that it involves an early release from a relatively long prison term. The difference here is that it is the parole authority rather than the judge who grants an early release from incarceration to parole supervision. For example, a convicted forger who receives a 10-year prison sentence could receive shock probation from the judge after 4 or 5 months. If there is no shock probation, in several states the parole authority is empowered to grant early or shock parole to the offender at his or her first hearing before the board.

In practice, the effectiveness of shock programs is unclear. Many of the programs provide no shock value because offenders expect release. Ideally, the incarcer-ated offender is shocked when released. In practice, however, the only shock may come if the court does not grant early release. Through a combination of a short prison term followed by community supervision, shock programs attempt to gain the benefits of both incarceration and supervision as sanctions. We hope the sentence will deter the offender. In theory, shock probation and parole provide a stern warning to the inmate. Without the kindness of the judge or parole authority, the offender would be serving a long prison term. In effect, the released inmate realizes that he or she is living "on borrowed time" and will face a long term of incarceration if supervision is unsuccessful.

An innovation to the practice of shock probation and parole has been the development of the prison **boot camp** (MacKenzie & Armstrong, 2004). These programs, sometimes called **shock incarceration** because the conditions of incarceration are very severe but limited in duration, also seek to increase punitiveness. Boot camp programs in prisons subject inmates to harsh conditions, including physical conditioning and strict discipline combined with hard labor, akin to the boot camp experience of new recruits in the military. We discussed boot camps in Chapter 9 as evidence of attempts to make punishment more physically painful for offenders (Kurlychek, 2010). These camps relate to community supervision because they serve to reduce the length of term served by participants. The boot camp provides a harsher penalty for a shorter time than traditional imprisonment. Many boot camp programs release their graduates to community supervision as probationers or parolees.

As Mackenzie and Parent (1991) noted, legislators may support boot camp programs because the conditions in these programs are more punitive than in the typical prison, and thus shorter terms can be equally tough on crime. However, the available evidence suggests that boot camps do not reduce rates of new crime by program graduates and may actually hinder inmates in making a positive adjustment (Bottcher & Ezell, 2005; Wilson, MacKenzie, & Mitchell, 2005). Benda, Toombs, and Peacock (2006) found that boot camp programs appear to have different effects for different kinds of offenders. Similar to the problem of applying general risk assessments to diverse populations, boot camp programs seem to help some offenders but are harmful to others. Improvement in boot camps (and other correctional efforts) may depend on our ability to match the right program to the right type of offender. It should be noted that the use of boot camps has declined significantly in recent years.

Another practice that is somewhat akin to shock probation is the imposition of split sentences. A **split sentence** is a penalty that is divided (split) between a period of incarceration and a period of probationary supervision. Because of the great flexibility that judges are allowed in determining the conditions of probation, split sentences are relatively widely used. This is true even in jurisdictions where there is no law that specifically allows the judge to use split sentencing.

Parisi (1980) described four historical methods of imposing split sentences, including shock probation as practiced in several states:

1. With a defendant convicted of several offenses or counts, the judge may order incarceration for some offenses and probation for others. Thus, a defendant convicted of two counts of theft receives sentences to 90 days on the first count and 2 years of probation on the second.

2. In shock probation, the same offender may receive an initial 5-year prison term, and be "shocked" to probation in 90 days.
3. Legislation that allows the judge to combine incarceration and probation in one sentence, such as a sentence of 90 days of incarceration followed by 2 years of probation, is another way to impose split sentences.
4. Finally, most states have legislation allowing the judge to use incarceration as a condition of probation. In this case, the offender might receive 2 years of probation with the condition that the offender serves the first 90 days in jail.

Regardless of the method used to impose split sentences, the outcomes are the same. Judges, offenders, and the public often see probation as a lenient sanction. The use of split or combination sentences allows judges to increase the harshness of the penalty. A judge may not wish to send a low-level offender to prison for a long period but may want the offender to spend some time in jail. Split sentences allow judges to adjust the severity of sanctions. After analysing the flow of jail inmates in California, Usta and Wein (2015) found that split sentencing was the best available option to reduce the jail population while also reducing recidivism.

INTENSIVE SUPERVISION

Intensive supervision programs (ISPs) seek to provide more control and service to specific subgroups (e.g., drug or sex offenders) or high-risk offenders who otherwise would be incarcerated (Buttars, Huss, & Brack, 2016; Travis, 1984). In practice, these programs rely on lower client-to-officer ratios and, thus, they assume a higher level of supervision and service delivery (Cullen, Wright, & Applegate, 1996; Latessa, 1980). Several states have implemented ISPs. Typically, an ISP requires the probationer to have more contacts with his or her probation officer. The net effect of these differences is to make the penalty more painful. The probationer experiences more intrusions by the officer. The penalty is also more incapacitative. The officer supervises the offender more closely and this prevents the offender from relapsing into crime.

Several strategies of intensive supervision have been developed (Drake, 2018). Many states use a surveillance-oriented model which emphasizes the closer monitoring (e.g., drug testing) without providing additional services to offenders. This model tends to respond to rule violations with incarceration. Other intensive supervision models stress the importance of providing evidence-based treatment in coordination with the decreased caseloads. It is likely safe to assume that the offender's experience with intensive supervision is quite different depending on which supervision model is used.

DeVall, Lanier, Hartmann, Williamson, and Askew (2017) described an intensive supervision model used in Michigan to supervise high-risk offenders. Based on deterrence theory, Michigan's program links intensive supervision with swift and certain consequences for failing to comply with supervision conditions. Any rule violation resulted in a hearing within 72 hours where the court would respond by sentencing the violator to a brief confinement in jail. Although this model punishes every rule violation with a jail sentence, the intent is to return the offender to the community. In other words, the court's response is designed to increase compliance rather than send the offender to a prolonged prison sentence. DeVall et al.'s evaluation of this program determined that participants had lower recidivism rates compared to those supervised by regular probation.

Reviews of intensive supervision have been mixed. Overall, the research suggests that intensive supervision does not result in a lower recidivism rate (Farrington & Welsh, 2005; Georgiou, 2014; Hyatt & Barnes, 2017). Likewise, most research has failed to show a clear benefit to reduced caseloads. However, reduced caseloads can be helpful when combined with evidence-based practices (Drake, 2018; Jalbert, Rhodes, Flygare, & Kane, 2010; Jalbert et al., 2011). ISPs in which the level of service delivery actually increases and is linked to the risk and needs of offenders appears to be moderately effective. Finally, Lowenkamp, Flores, Holsinger, Makarios, and Latessa (2010) found that not only should ISPs have a treatment philosophy, but that the staff's commitment to this philosophy was equally important.

ELECTRONIC MONITORING

In **electronic monitoring** programs, offenders are fitted with transmitting devices that allow correctional staff to monitor their whereabouts. In recent years, electronic monitoring has gained popularity as a practical tool used to supervise pretrial offenders released on bail, or used by probation and parole officers to increase supervision of certain offenders. From 2005 to 2015, the number of individuals under electronic supervision increased more than 150% (The Pew Charitable Trusts, 2016).

Although the use of surveillance technology to monitor criminal offenders is relatively new, we have long recognized the potential of such technology for crime control. As early as 1966, Ralph Schwitzgebel described a potential telemetric monitoring system for probationers and parolees, and in 1968, a prototype of the system was developed and tested. An assessment of the legal ramifications of electronic monitoring appeared in the *Harvard Law Review* in 1966. Later, Robert Schwitzgebel, Ralph's brother, experimented with telemetric monitoring with volunteers in California (Schwitzgebel, 1969). For years, a debate about the acceptability of electronic monitoring continued (Corbett & Marx, 1991; del Carmen & Vaughn, 1986), but the sheer practicality of the technology has meant that, although many important issues remain to be solved, electronic monitoring of criminal offenders is an ongoing practice. There are two types of technology available for electronic monitoring: **radio frequency** and **global positioning system (GPS)**.

Currently, two basic types of radio frequency electronic surveillance systems are in use (National Institute of Justice, 1999; The Pew Charitable Trusts, 2016). These systems are either active or passive surveillance. In an active system, the supervising agency takes positive steps to monitor the offender. Generally, this system involves fitting the offender with a transmitting device, and the transmitter sends a tone over the telephone. A computer program randomly calls offenders at times when they are supposed to be at home. The offender must answer the phone and place the transmitter in a special telephone connection, so that the transmitter sends a message to the computer. In the passive system, a transmitter attached to the offender emits a continuous signal. The offender must keep the transmitter within range of an amplifier/transmitter or the signal will not reach a monitoring computer.

The random calling system allows for both voice identification and the transmission of a monitoring signal. Because it operates over telephone lines, this system is unaffected by interruptions in transmission caused by walls, structural steel, or other radio transmissions. The active surveillance random calling system appears to have a lower false-alarm rate than the passive surveillance system. The passive-surveillance continuous-transmission system, although more prone to false alarms, provides a continuous monitoring of the offender's whereabouts.

A more recent innovation is the use of a GPS unit to maintain constant surveillance of the whereabouts of a person being electronically monitored (Waldo, 2012). GPS units are also available in active and passive formats. Similar to radio frequency systems, active GPS monitoring works through an ankle bracelet attached to the offender. The GPS connects to a satellite system, providing real-time monitoring without requiring the offender to have a landline telephone. Passive GPS systems collect and store tracking information throughout the day and then send a day's worth of data to the probation and parole officer (National Institute of Justice, 2011). Although GPS is typically considered to be a more advanced technology than radio frequency systems, there are important limitations. One of the greatest challenges of using GPS to track offenders involves the technological limitations of the GPS signal inside large buildings (Drake, 2008). Table 13.1 provides advantages and disadvantages to using a GPS monitoring system.

Most jurisdictions have adopted one or the other form of monitoring technology, and a few have used both. The random calling system allows somewhat greater freedom. For example, the offender could be next door at a neighbor's house and still be called to the telephone. The use of both systems could provide a progression in the severity of monitoring. A probationer ordered to the random calling monitor who then misses a call could be "punished" by being issued a continuous transmission monitor. The possibility of varying the level of restriction within a category of penalty called "monitoring" adds to the attractiveness of this sanction.

TABLE 13.1 Advantages and Disadvantages of GPS Monitoring	
Advantages	**Disadvantages**
• Provides real-time tracking of an offender's location	• Cannot track signal inside large buildings, subways, etc.
• Triggers an instant alarm if it is tampered with or removed	• Limited to areas with good-quality cellular telephone coverage
• Permits creation of customized inclusion zones and exclusion zones	• Initial and replacement costs are high
	• Requires timely responses to notice of violation
• Does not require the offender to have a landline telephone in his or her residence	• Requires comprehensive training
	• Active system operating costs can be relatively high
• Allows the offender to return to the workforce while still being monitored	• Notice will be sent any time the offender enters an exclusion zone, even if accidentally
• Has been shown to positively alter the behavior of those who comply with the programs	

Source: Downing (2006:44).

Renzema and Mayo-Wilson (2005) reviewed scores of studies on electronic monitoring and concluded that the best evidence available suggests that the effect of electronic monitoring on reducing recidivism is limited. However, Padgett, Bales, and Blombert (2006) found that electronic monitoring was effective at reducing the likelihood of reoffending and absconding from supervision. More recently, Bales and his colleagues (2010) conducted a comprehensive assessment of electronic monitoring outcomes in Florida. Results showed that offenders placed on electronic monitoring were 31% less likely to fail, compared to offenders participating in other forms of community supervision. Electronic monitoring was effective regardless of offense type and offender age, but GPS was shown to be more effective than radio frequency supervision. Despite the positive findings, interviews with offenders showed that electronic monitoring often had a negative impact on personal relationships and the ability to obtain employment (Bales et al., 2010).

The issues involved in electronic surveillance recall the questions raised about the development of probation and parole supervision. On the one hand are critics who suggest that monitoring is an insufficient penalty for many offenders. They contend that this leniency reduces the deterrent effect of the law. Other critics argue that the use of monitoring technology to allow the release from incarceration of "dangerous" offenders poses too great a risk to the community. In contrast are those who criticize this technology as too oppressive because it violates current standards of privacy and infringes on constitutional rights to protection against unreasonable searches and seizures. Finally, some critics fear that electronic monitoring, in practice, will increase the severity of community supervision for offenders who otherwise would receive traditional probation or parole (see Box 13.1).

BOX 13.1 POLICY DILEMMA: ELECTRONIC MONITORING

Yeh (2015) proposed a dramatic increase in the use of electronic monitoring. The proposal included releasing those convicted of a felony from prison after serving 50% of their sentence, with the remainder of the sentence to be carried out via home detention using electronic monitoring. Additionally, those who commit less serious offenses who ordinarily receive probation would instead be sentenced to the same period of home detention with electronic monitoring. Yeh estimated that this proposal would eliminate nearly 95% of the crime currently committed by probationers and parolees, while saving over $400 billion annually. Obviously, this type of proposal would require several changes to the law and a shift in public opinion. Does this type of policy proposal have merit?

Proponents of surveillance argue that the technology enhances public safety by ensuring supervision of offenders in the community. Further, surveillance itself deters offenders from committing crimes. Similarly, the proponents suggest that the ability to monitor offenders results in a lessening of penalty severity. They argue that some offenders are in prison who do not actually need to be incarcerated. Continual supervision will allow judges and parole boards to leave these offenders in the community. Whereas the current focus may be on the electronics of contemporary surveillance, at base, the questions and criticisms are the same ones that have always surrounded community supervision.

Despite the ongoing debate, electronic monitoring as a sanction, either alone or in combination with other forms of community supervision, appears to be here to stay. Most observers believe that electronic monitoring is a cost-effective alternative to incarceration for many offenders (Evans, 1996; Yeh, 2010). Learning how well electronic monitoring can meet the other goals of community supervision (fairness, a reduction in future crime, etc.) must await further study. It should be noted that electronic monitoring can only tell officials the location of an offender, not what they are doing while at that location. As DeMichele (2014) pointed out, electronic monitoring is not a "silver bullet" to replace face-to-face supervision.

DAY REPORTING AND OTHER SANCTIONS

Unlike many other intermediate sanction alternatives, **day reporting** is of relatively recent origin. Although England has used day reporting in the past, the first day reporting program in the United States was started in Massachusetts in 1986 (McDevitt, 1988). This program worked as an early release alternative for prison and jail inmates near the date of their parole. Participants in the program were required to report to the center each day, prepare an itinerary for their next day's activities, and report by telephone to the center throughout the day (Larivee, 1990). By 1994, there were 114 reporting centers operating in the United States (Parent, Byrne, Tsarfaty, Valade, & Esselman, 1995). Figure 13.1 offers a description of these early day reporting centers. As community corrections has become more popular, the use of day reporting centers has continued to increase. Contemporary day reporting centers are used in a variety of ways, including as a graduated sanction for probation and parole violators.

Day reporting clients typically make at least one in-person and several telephone contacts with center staff daily, yet they are allowed to remain in the community throughout much of the day. Most programs are limited to between 2 and 4 months in duration, followed by a period of probation or parole supervision. Often, day reporting clients work with program staff to develop and obtain substance abuse, psychological, and employment treatment and services. These programs promise to reduce prison crowding and costs, protect community safety, and provide specifically tailored services to offenders.

Evaluations of day reporting centers have generally found them to be an effective tool for reducing recidivism among probationers (Champion, Harvey, & Schanz, 2011), pretrial offenders (Martin, Lurigio, & Olson, 2003), and offenders with mental illness (Carr, Baker, & Cassidy, 2016). A study found that parolees enrolled in one of seven day reporting centers located in New Jersey performed no better (and in some cases worse) than those who instead received regular parole supervision (Boyle, Ragusa-Salerno, Langerman, & Marcus, 2013). However, Ostermann (2013) suggested that this finding could be due to the poor quality of programming in those centers. As mentioned earlier, the use of evidence-based practices and proper matching of offenders to services are essential for the success of any community corrections program.

FIGURE 13.1
Services Provided by Day Reporting Centers.

Source: Parent, Byrne, Tsarfaty, Valade, and Esselman (1995:13).

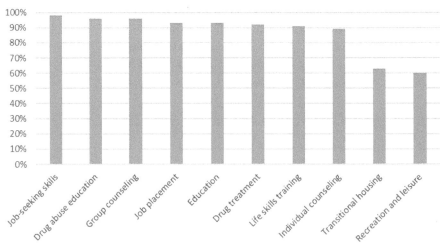

Services Provided by Day Reporting Centers

In Chapter 12 we discussed furlough programs. Work and study furloughs, as well as prerelease programs, help inmates make the adjustment from incarceration to living in the community. Placements in residential community corrections facilities also provide custodial supervision of offenders not needing or deserving jail or prison incarceration (Holsinger, Latessa, Turner, & Travis, 1997). Other changes in probation and parole have similarly bridged the gap between traditional supervision and incarceration. There has been an increasing use of community service sentencing and the imposition of monetary penalties (such as fines and restitution to crime victims). All these changes in traditional community supervision have worked to make probation and parole more severe as penalties, and to provide greater restrictions and controls over offenders in the community. Shock probation and parole, split sentencing, and ISPs blur the distinction between incarceration and community supervision. The experience of being "on paper" becomes much more like that of incarceration. In these programs, the probationer or parolee faces additional deprivations of autonomy, liberty, and the like.

ISSUES IN COMMUNITY SUPERVISION

Several issues have developed as the expectations for community corrections have continued to evolve. Broad trends, including results-driven management and the reemergence of correctional rehabilitation, have had an influence on community supervision. As with the police and courts, an increased emphasis on collaboration across agencies in communities is common to probation and parole. Specialization, in the form of focused caseloads and services, is common. Drug courts and other problem-solving courts typically work through probation departments. It is once again difficult to select issues for specific attention given the range of available topics. Perhaps the greatest demand on community supervision is that it respond to the burgeoning growth in correctional populations.

As prison populations have increased since the 1970s, so have the numbers of people under community supervision. The change in types of offenders on probation and parole supervision is an important factor in understanding changes in community supervision. The simple growth in the size of the population, regardless of its characteristics, also has led to changes. Three areas in which we observe changes are the financing, management, and technology of probation and parole. These changes merit attention, regardless of whether they involve special programs such as shock or intensive supervision.

Financing

Several states have implemented community corrections legislation that includes various funding formulas to support community supervision activities. These laws provide financial incentives to counties to reduce their prison commitments and to retain offenders in the community. The typical law either authorizes a subsidy for counties that reduce their commitment rates or provides financial support for improved and increased community corrections programs (Clear & Schrantz, 2011).

California was the first state to employ an incentive program to encourage communities to keep offenders out of the prison system. In 1965, the California legislature passed the **Probation Subsidy Act**, which paid counties for each offender not sentenced to prison in each county. The state developed a formula that estimated the number of offenders expected to be sentenced to prison, and then paid $4,000 to the county for each offender, less than that number that was not sentenced to prison. If a county was expected to commit 1,000 offenders to prison but actually committed only 900, the county received a subsidy of $400,000 ($4,000 × 100). Subsidy funds were earmarked for the improvement of local correctional services. Other states subsequently developed similar models.

The California subsidy program faced several obstacles (Clear & Cole, 1986). There was no subsidy assistance to law enforcement, although the effect of the program was to keep offenders in the community. No inflation factor was included, so that within 10 years, the purchasing power of the subsidy declined by more than one-third. The formula did not consider counties that historically had kept offenders in local correctional custody. For example, a county traditionally may have kept nonviolent offenders in the community on probation. Under the subsidy formula, that county had a lower estimate of commitments. Another county may have traditionally incarcerated nearly every felony offender. Under the subsidy program, the first county could only receive aid by keeping violent or more serious offenders in the community. The second county could begin to use probation for minor, nonviolent offenders, and could reap a large subsidy.

Later funding formulas for community corrections attempted to overcome some of the original difficulties in the California subsidy program. Minnesota, Oregon, and Colorado passed community corrections legislation that included more options for counties. These states also tried to adjust for crime and incarceration rates and included inflation factors. Other states, such as Ohio, began subsidy programs for specific practices that counties could adopt to reduce prison commitments. In each of these cases, funding relates to the development and expansion of community programs. The effects are to support probation and other community services and to assist the counties in handling their increasing caseloads.

Related to community corrections legislation is the emerging practice of **justice reinvestment**. A term first coined by Tucker and Cadora (2003), justice reinvestment seeks to not only save money by diverting offenders to community-based supervision, but also to redirect a portion of the savings to rebuild the communities that have been most impacted by high levels of incarceration. The reality is that many of those on community supervision live in disadvantaged neighborhoods that lack the infrastructure (e.g., parks, schools) and social support to provide offenders with a reasonable chance of success. Research has consistently shown that community-level factors have an impact on offender recidivism (Chamberlain & Wallace, 2016; Monteiro & Frost, 2015). Too often, offenders simply cycle back and forth from these disadvantaged neighborhoods to prison, and back again. Justice reinvestment seeks to break this cycle by improving these communities.

By 2016, nearly half of the states had enacted some type of justice-reinvestment style sentencing reforms (Harvell, Welsh-Loveman, & Love, 2017). Together, these states have reinvested more than $446 million to victims' services, law enforcement grants, probation improvements, and other types of diversion services. However, critics point out that some states have failed to reinvest the savings gained from reducing their prison population. Additionally, most of the reinvestments that have occurred appear to be directed to government-based social services rather than addressing the fundamental issues affecting the communities. Clear (2011) proposed that a portion of the reinvestments could support a voucher system that employers and community-based organizations could then use to create jobs or provide services to offenders who would otherwise be imprisoned. Although it is unclear exactly how the justice reinvestment movement will develop, it appears that the idea will continue to receive attention.

Another recent development is in the charging of **supervision fees**. Many state parole authorities and probation offices now require that the client make a monthly payment to the agency to offset the costs of supervision (Teague, 2011). This requirement further increases the cost advantage of probation and parole over incarceration. Similarly, it is common for a condition of supervision to be the payment of court costs. This requires that probation and parole officers serve (at least part time) as bill collectors for the courts.

A related financial alteration in the operation of probation and parole is the growing use of **restitution**. Offenders on probation and parole increasingly must make restitution to the victims of their crimes. Probation and parole officers are then required to manage the payment process for restitution and fines (Weisburd, Einat, & Kowalski, 2008). The combination of supervision fees and restitution can often add up to hundreds of dollars per month. Probation and parole officers must develop strategies for dealing with offenders who are unable to pay these fees.

Management

The traditional approach to probation and parole supervision consisted of the casework model. In casework, each officer was responsible for a caseload of offenders. The officer was a generalist expected to supervise a variety of persons having a variety of needs. The **casework model** expected the single officer to be capable of providing needed services to all of the offenders. In this model, each officer is assigned a certain caseload with the assumption that each offender will require the same amount of effort (DeMichele, Payne, & Matz, 2011).

Several other models of organization and caseload management have been proposed and adopted. Several jurisdictions use teams of officers responsible for large groups of offenders. In **team supervision**, officers can take advantage of their varied strengths and skills. Thus, an officer who is particularly effective with offenders that require employment services can concentrate on that type of case for the entire team (DeMichele, Payne, & Matz, 2011).

The National Institute of Corrections (1981) developed a **model case management system**. The system requires each supervising officer to complete a risk and needs assessment and then review the results with the offender. Through discussion of the instruments, the officer and the offender develop a plan for dealing with needs. They develop case objectives from that plan. This system helps identify concrete actions that the officer should take, and it provides an ability to assess case progress.

As a management system, the model system involves three components: (1) classification, (2) case planning, and (3) the assignment of workload units. Classification reflects the assessments mentioned above. Depending on the levels of need or risk, or on combinations of the two, the offender is placed in a supervision category that is somewhere between high (frequent officer attention) and low (little or no direct contact by officer) (see Box 13.2). The second component, case planning, creates case objectives based on a structured interview with the client. The officers' supervisors review the objectives and approve or modify them. Finally, cases are assigned varying degrees of difficulty or workload units. These units describe how much of an officer's time will be consumed by the case. In this manner, the officer and his or her supervisor know how much effort to expend on that case, and how much effort to spend on the total caseload.

BOX 13.2 KIOSK REPORTING

Many jurisdictions now allow low-risk offenders to complete their monitoring requirements using a kiosk instead of meeting with a probation officer. The kiosks typically use biometric identification to verify the offender's identity, are capable of receiving supervision fee payments, and ask the same types of questions that would be asked during a face-to-face meeting (Ahlin, Hagen, Harmon, & Crosse, 2016). This type of efficient caseload management was not possible prior to recent technological advances and the development of valid risk assessment tools.

From the information derived through individual case planning, it is possible for the agency or office to assess its needs, objectives, and progress. If 35% of the agency's caseload has high employment needs, administrators can see the utility in providing officers with employment development training. The agency might assign some officers as employment specialists. If a large enough percentage of the caseload has high-risk scores, administrators may want to create special surveillance units that would serve risk-control goals distinct from service delivery (Clear & O'Leary, 1983).

In 2001, the National Institute of Corrections developed the **Transition from Prison to the Community (TPC)** case management model. The TPC model reflects a renewed interest in managing an offender's successful reentry to the community. Unlike traditional case management, the TPC model approaches offender reentry as a process that begins at sentencing and continues until discharge from community supervision. Thus, case management is a continuous process consisting of three phases (Thigpen, Beauclair, Moore, & Humphries, 2010:7):

1. *The institutional phase*: The period of incarceration
2. *The release phase*: The months just before and just after release
3. *The community phase*: The period after an offender reenters and stabilizes in the community through supervision, discharge from supervision, and beyond.

The National Institute of Corrections (NIC) devoted considerable resources to the development and dissemination of this case management program. From 2001 to 2009, eight states received technical assistance and guidance from the NIC. In 2009, six states were selected for additional technical assistance, with the technical assistance period ending in 2012. To implement the TPC model effectively, participating states increased collaboration between institutional corrections, field supervision, and the parole boards (Jannetta, Neusteter, Davies, & Horvath, 2012). Box 13.3 describes a sample postrelease supervision model.

BOX 13.3 SAMPLE POSTRELEASE SUPERVISION MODEL

Extremely high-risk offenders: Surveillance and monitoring for risk management.

Medium- to high-risk offenders: Supervision and case management specifically directed toward risk reduction, along with addressing factors that can contribute to community stability—obtaining forms of identification, housing, employment, and so forth.

Low-risk offenders: Minimum supervision and correctional case management with attention to stability factors, including referral to noncorrectional resources to address other needs.

Administrative cases: The major issue for an offender is completion of financial and administrative compliance.

Source: Thigpen, Beauclair, Moore, and Humphries (2010:48).

Management innovations in community supervision include the development and adoption of automated information systems to improve case supervision and monitoring. Existing software programs designed for salespeople are also well suited to probation and parole use. This software manages scheduling, keeps tracks of contacts, and otherwise performs the information storage and retrieval functions most useful to probation and parole officers. In recent years there has been increased emphasis in two important management areas. First, community supervision agencies are moving toward "evidence-based practice" (Cullen, Jonson, & Mears, 2017; DeMichele, 2007). Coupled with automated information systems, administrators seek to assess the effects of different practices on specific types of offenders or specific problems. Increasingly, community supervision managers seek information about how to affect the behaviors of offenders. In 2001, the American Probation and Parole Association began revising accreditation standards for probation and parole agencies to include performance-based standards. As described by Taylor (2004:21), the new standards not only include compliance with policies and procedures but also include "measurement not only of what an agency does, but how well it does it." Finally, the increased development of automated case management systems allows for better information sharing among justice officials (Matz, 2012).

Technology

A final area of recent innovations in community supervision involves the application of communications and other technologies to supervision. These applications have taken a number of forms, such as the information technologies already mentioned. Developments in areas including word processing, telecommunications, and video recording have had some effects and will play a larger role in the future. We have already addressed some of the effects of improved information systems and classification and assessment procedures as a form of information technology.

Drug testing is currently common throughout all of our society. Repeated or continuing wars on drugs, increasingly tough drunk driving laws, deaths of celebrities from drug-related causes, and notorious transportation accidents attributed to drug use all have focused attention on substance abuse. Today, a number of relatively simple technologies are available for the detection of alcohol and other drug consumption. They range from the Breathalyzer test to determine if a person has been driving under the influence of alcohol, to blood tests to determine alcohol or other drug content. Urine testing is becoming an increasingly common component of community supervision conditions. Every parole supervision agency in the country reports using drug testing with parolees (Bonczar, 2008).

Regardless of the attempts to control substance abuse, probationers and parolees continue to acquire and use alcohol and other drugs. The development of easily administered detection tests has strengthened the ability of probation and parole officers to identify and control substance abuse among their clients. Many believe that merely testing for use will deter offenders. As Atmore and Bauchiero (1987) reported:

> We have noted significant behavioral improvements when we test regularly, because it serves as a deterrent. Therefore, we are returning less people to higher security for positive urines because they know they are taking a huge risk by substance use. In other words, test regularly!

FIGURE 13.2
An official attaches a SCRAM (Secure Continuous Remote Alcohol Monitor) anklet at the municipal courthouse in North Las Vegas, Nevada. The court-ordered device tests vapor in the skin for alcohol and sends the results wirelessly to a phone modem.

AP Photo/*Las Vegas Sun*/Steve Marcus

> You need to have a consistent policy for testing and sanctions for positive results, or else word will get around quickly that one should not take this seriously.

In addition to urine tests, probation and parole officers have a wide array of other testing technologies available to them (see Box 13.4). These include the Breathalyzer and blood tests, as well as some newer developments, such as a saliva test for alcohol use. With the saliva test, the subject's saliva is placed on a "blotter" that is actually a form of litmus paper. If the subject has recently used alcohol, the paper will change colors. Efforts are currently underway to develop similar saliva tests for other drug use. Finally, alcohol use can now be detected by a device that is fitted on one's wrist or ankle (Leffingwell et al., 2013) (see Figure 13.2).

BOX 13.4 IGNITION INTERLOCK DEVICE

Ignition interlock devices are increasingly used to prevent repeat drunk driving. These devices are essentially in-car Breathalyzers that prevent an automobile from starting if the driver has been drinking alcohol. Many devices also require the driver to retest at regular intervals while traveling. All states have enacted laws allowing judges to require the devices to be installed on the cars of at least some DWI offenders (Shulman-Laniel, Vernick, McGinty, Frattaroli, & Rutkow, 2017). Over half of the states require all DWI offenders to temporary install an ignition interlock device on their automobile. These devices have shown to be an effective tool to help reduce the recidivism among drunk driving offenders.

Substance-abuse testing technologies enhance the ability of officers to control the risks of crime posed by their clients. These procedures also may change the nature of the job. The officers must now test their clients, serve as medical technicians, and otherwise assume the role of "cop" instead of "helper." Some believe that testing alters the nature of the officer's job and his or her relationship with offenders. As mentioned in Chapter 12, the rapport between a probation officer and his/her client has been shown to be an important factor that influences the success of community supervision. Thus, community corrections personnel must find a way to balance the use of these new technological surveillance tools with the need to maintain a positive relationship with their clients.

Developments in geographic information systems have been applied to community supervision. Harries (2003) described a program in which the Maryland Division of Probation and Parole used geographic information to improve supervision. As described by Piquero (2003), Maryland used geographic analysis of crime to develop programs involving police, probation, community members, and other agencies to develop coordinated efforts to reduce and prevent crime. Geographic analysis can improve community supervision in a variety of ways ranging from describing "good" and "bad" environments in which to place offenders through minimizing officer travel time by assigning caseloads based on geography. As mentioned, it is now possible to use GPS units to track offenders. With GPS monitors, officers can monitor the location of probationers and parolees at all times. Mellowa, Schlager, and Caplan (2008) reported on the process of using geographic analysis to evaluate parole services. By mapping the residential locations of parolees and the locations of required services, the evaluators were able to assess the adequacy of services easily available to parolees.

THEORY AND COMMUNITY SUPERVISION

We explored community supervision in this and the preceding chapter. This assessment identified several issues in contemporary community corrections and described a range of changes during the recent past. A number of explanations for the emergence of issues or the implementation of changes exist. Each of these explanations may represent a theory of community supervision.

The impact of supervision on recidivism depends on organizational goals and structures. Recidivism also seems to reflect the characteristics of the offenders under supervision as well as the qualifications and efforts of supervising officers. Changes in the law, or in public perceptions of offense seriousness, are related to changes to community supervision populations and practices. Finally, the growth of community supervision is directly related to the fiscal burden created by overcrowded prisons (Wodahl & Garland, 2009).

Once again, we can identify individual, organizational, and social theories of community supervision. It appears the same types of explanations offered for policing, courts, and incarceration apply to community supervision. We will return to this topic in the final chapter, but it seems possible to identify theories of criminal justice.

DUE PROCESS, CRIME CONTROL, AND COMMUNITY SUPERVISION

Many changes in community supervision were grafted onto existing practices of probation and parole supervision. Thus, we added restitution orders and supervision fees as conditions of probation. Periods of confinement have been added to supervision conditions to enable courts to achieve split sentences or to create programs such as shock incarceration. Supervision has changed to create intensive monitoring programs for the more serious offenders now placed in probation and parole caseloads.

Some of the changes in community supervision represent the development of new programs that complement traditional probation and parole supervision. Day reporting centers, for example, often operated as adjuncts of the jail or as institutional components of local community corrections agencies. The growth of residential community correctional facilities similarly has been outside of probation and parole agencies. Such facilities serve as a resource to those agencies.

Indeed, one of the problems inherent in estimating the numbers of persons participating in these various programs is that jurisdictions report the participants as members of the probation and parole population. Thus, probation and parole supervision now includes thousands of cases in which offenders are under electronic monitoring, residing in halfway houses, attending day reporting centers, or receiving other forms of intermediate sanctions. As perceptions about the adequacy of sentences have changed, and as the number of offenders eligible for correctional supervision has increased, a large proportion of the burden of adapting to these changes has fallen on the community supervision agencies of the criminal justice system.

Traditionally, community supervision has taken the role of assistance to convicted offenders through the provision of services and a reduction in the severity of criminal sanctions. In this regard, throughout most of the twentieth century, probation and parole supported the interests and needs of the individual, having a due process orientation. In recent years, however, the role of community supervision in crime control has come to dominate. Lutze, Johnson, Clear, Latessa, and Slate (2012) presented the argument that criminal justice policymakers must invest in community corrections if they hope to reduce crime in the future. They contend that community supervision and alternative sanctions can support crime control efforts efficiently if we utilize evidence-based practices and view community corrections as a human service profession. Probation and parole officers should be supported through reduced caseloads, better resources, and rewards for reaching positive outcomes. At the same time, they should be held professionally accountable for what they have control over.

Although crime control advocates might argue that sanctions short of incarceration inadequately protect the public and fail to deter offenders, due process advocates fear increasing the numbers of offenders subjected to criminal justice processing (i.e., net-widening) or increasing the restrictiveness of interventions on the lives of offenders. In addition, due process proponents fear that the proliferation of intermediate sanctions has simply meant that offenders who would otherwise have received regular probation now experience programs such as electronic monitoring, intensive supervision, and day reporting (Harland, 1996). They believe that instead of providing true alternatives to incarceration, intermediate sanctions have brought more offenders under tighter control by the correctional system.

The growth of the community corrections population, especially if coupled with that of the prison and jail populations, suggests the due process proponents may be correct. Nearly 3% of the adult population of the United States is under some form of correctional control (Kaeble & Cowhig, 2018). What remains unsettled is whether the offenders or society as a whole benefits from increased correctional intervention. Whereas more offenders than ever before are in the custody of corrections officials, intermediate sanctions may have saved tens of thousands of offenders from serving prison and jail terms, and may result in improved conditions for them in the future.

Community supervision represents a compromise between due process and crime control proponents. Offenders under community supervision face restrictions and control. Still, these offenders are under less onerous control than those incarcerated. Community supervision often seeks to provide the "right" level of state control over the lives and behavior of convicted offenders. We will return to this issue in the final chapter, but growing interest in restorative justice seeks to balance the interests of the community, victim, and offender. In most restorative justice programs, community supervision plays a key role precisely because it is well suited to serve both due process and crime control ends.

REVIEW QUESTIONS

1. How have probation and parole populations changed in recent years?

2. Name and describe two innovations in probation and parole that allow community supervision to approximate the severity of incarceration more closely.

3. What are split sentences, and how may they be imposed by a judge?

4. Take a position on the question of whether community supervision is effective. Include public risk, cost, and other relevant factors in arguments for your position.

5. Describe funding approaches that serve to encourage communities to keep offenders out of state prisons, and argue either in favor of or in opposition to them.

6. How can classification of probationers and parolees aid in the use of community supervision resources?

7. Describe two ways in which emerging technologies have affected the ability of probation and parole agencies to supervise offenders in the community.

8. Are current changes in community supervision evidence of a growing dominance of the crime control perspective in American criminal justice? Explain.

REFERENCES

Ahlin, E.M., Hagen, C.A., Harmon, M.A., & Crosse, S. (2016). Kiosk reporting among probationers in the United States. *The Prison Journal, 96*(5), 688–708.

Atmore, T., & Bauchiero, E. (1987). Substance abusers: Identification and treatment. *Corrections Today, 49*(7), 22–24, et seq.

Baird, C. (2009). *A question of evidence: A critique of risk assessment models used in the justice system.* Madison, WI: National Council on Crime and Delinquency.

Bales, W., Mann, K., Blomberg, T., Gaes, G., Barrick, K., Dhungana, K., et al. (2010). *A quantitative and qualitative assessment of electronic monitoring.* Tallahassee, FL: Center for Criminology and Public Policy Research.

Bales, W.D., & Piquero, A.R. (2012). Assessing the impact of imprisonment on recidivism. *Journal of Experimental Criminology, 8,* 71–101.

Benda, B., Toombs, N., & Peacock, M. (2006). Distinguishing graduates from dropouts and dismissals: Who fails boot camp? *Journal of Criminal Justice, 34*(1), 27–38.

Bennett, L. (1991). The public wants accountability. *Corrections Today, 53*(4), 92–95.

Bonczar, T. (2008). *Characteristics of state parole supervising agencies, 2006.* Washington, DC: Bureau of Justice Statistics.

Bottcher, J., & Ezell, M. (2005). Examining the effectiveness of boot camps: A randomized experiment with a long-term follow up. *Journal of Research in Crime and Delinquency, 42,* 309–332.

Boyle, D.J., Ragusa-Salerno, L., Lanterman, J.L., & Marcus, A.F. (2013). An evaluation of day reporting centers for parolees: Outcomes of a randomized trial. *Criminology & Public Policy, 12,* 119–143.

Buttars, A., Huss, M.T., & Brack, C. (2016). An analysis of an intensive supervision program for sex offenders using propensity scores. *Journal of Offender Rehabilitation, 55,* 51–68.

Byrne, J., & Pattavina, A. (2017). Next generation assessment technology: The potential and pitfalls of integrating individual and community risk assessment. *Probation Journal, 64*(3), 242–255.

Carr, W.A., Baker, A.N., & Cassidy, J.J. (2016). Reducing criminal recidivism with an enhanced day reporting center for probationers with mental illness. *Journal of Offender Rehabilitation, 55*(2), 95–112.

Caudy, M.S., Tillyer, M.S., & Tillyer, R. (2018). Jail versus probation: A gender-specific test of differential effectiveness and moderators of sanction effects. *Criminal Justice and Behavior*. Advance online publication. doi: 10.1177/0093854818766375

Chamberlain, A.W. (2018). From prison to the community: Assessing the direct, reciprocal, and indirect effects of parolees on neighborhood structure and crime. *Crime & Delinquency*, 64(2), 166–200.

Chamberlain, A.W., & Wallace, D. (2016). Mass reentry: Neighborhood context and recidivism: Examining how the distribution of parolees within and across neighborhoods impacts recidivism. *Justice Quarterly*, 33(5), 912–941.

Champion, D.R., Harvey, P.J., & Schanz, Y.Y. (2011). Day reporting center and recidivism: Comparing offender groups in a western Pennsylvania county study. *Journal of Offender Rehabilitation*, 50(7), 433–446.

Clear, T.R. (2011). A private-sector, incentives-based model for justice reinvestment. *Criminology & Public Policy*, 20(3), 585–608.

Clear, T.R., & O'Leary, V. (1983). *Controlling the offender in the community*. Lexington, MA: Lexington Books.

Clear, T.R., & Cole, G.F. (1986). *American corrections*. Monterey, CA: Brooks/Cole.

Clear, T.R., & Byrne, J. (1992). The future of intermediate sanctions: Questions to consider. In Byrne, J., Lurigio, A., & Petersilia, J. (Eds.), *Smart sentencing: The emergence of intermediate sanctions* (pp. 319–331). Beverly Hills, CA: Sage.

Clear, T.R., & Schrantz, D. (2011). Strategies for reducing prison populations. *The Prison Journal*, 91, 138S–159S.

Corbett, R., & Marx, G. (1991). Critique: No soul in the new machine: Technofallacies in the electronic monitoring movement. *Justice Quarterly*, 8(3), 399–414.

Cullen, F., Wright, J., & Applegate, B. (1996). Control in the community: The limits of reform. In Harland, A. (Ed.), *Choosing correctional options that work* (pp. 69–116). Thousand Oaks, CA: Sage.

Cullen, F., Jonson, C., & Mears, D. (2017). Reinventing community corrections. *Crime and Justice*, 46, 27–93.

Davies, G., & Dedel, K. (2006). Violence risk screening in community corrections. *Criminology & Public Policy*, 5(4), 743–770.

del Carmen, R., & Vaughn, J. (1986). Legal issues in the use of electronic surveillance in probation. *Federal Probation*, 50(2), 60–69.

DeMichele, M. (2007). *Probation and parole's growing caseloads and workload allocation: Strategies for managerial decision making*. Lexington, KY: American Probation and Parole Association.

DeMichele, M. (2014). Electronic monitoring: It is a tool, not a silver bullet. *Criminology & Public Policy*, 13(3), 393–400.

DeMichele, M., Payne, B., & Matz, A. (2011). *Community supervision workload considerations for public safety*. Lexington, KY: American Probation and Parole Association.

DeVall, K.E., Lanier, C., Hartmann, D.J., Williamson, S.H., & Askew, L.N. (2017). Intensive supervision programs and recidivism: How Michigan successfully targets high-risk offenders. *The Prison Journal*, 97(5), 585–608.

Downing, H. (2006). The emergence of global positioning satellite (GPS) systems in correctional applications. *Corrections Today, 68*(6), 42–45.

Drake, E. (2008). *Offender tracking technologies: Where are we now?* Retrieved January 26, 2014, from www.cepprobation.org/uploaded_files/Pres%20EM09%20Dra.pdf

Drake, E.K. (2018). The monetary benefits and costs of community supervision. *Journal of Contemporary Criminal Justice, 34*, 47–68.

Evans, D. (1996). Electronic monitoring: Testimony to Ontario's Standing Committee on Administration of Justice. *APPA Perspectives*, 8–10, Fall.

Farrington, D.P., & Welsh, B.C. (2005). Randomized experiments in criminology: What have we learned in the last two decades? *Journal of Experimental Criminology, 1*, 9–38.

Georgiou, G. (2014). Does increased post-release supervision of criminal offenders reduce recidivism? Evidence from a statewide quasi-experiment. *International Review of Law and Economics, 37*, 221–243.

Gill, C., & Wilson, D.B. (2017). Improving the success of re-entry programs: Identifying the impact of service-need fit on recidivism. *Criminal Justice and Behavior, 44*(3), 336–359.

Gottfredson, M., Mitchell-Herzfeld, S., & Flanagan, T. (1982). Another look at the effectiveness of parole supervision. *Journal of Research in Crime & Delinquency, 18*(2), 277–298.

Gowdy, V. (1993). *Intermediate sanctions*. Washington, DC: National Institute of Justice.

Hamilton, Z.K., & Campbell, C.M. (2013). A dark figure of corrections: Failure by way of participation. *Criminal Justice and Behavior, 40*, 180–202.

Harland, A. (Ed.), (1996). *Choosing correctional options that work*. Thousand Oaks, CA: Sage.

Harries, K. (2003). Using geographic analysis in probation and parole. *National Institute of Justice, 249*, 32–33 (July).

Harris, G.T., Rice, M.E., Quinsey, V.L., & Cormier, C.A. (2015). *Violent offenders: Appraising and managing risk* (3rd ed.). Washington, DC: American Psychological Association.

Harvard Law Review Note (1966). Anthropotelemetry: Dr. Schwitzgebel's machine. *Harvard Law Review, 80*, 403.

Harvell, S., Welsh-Loveman, J., & Love, H. (2017). *Reforming sentencing and corrections policy: The experience of justice reinvestment initiative states*. Washington, DC: Urban Institute.

Hipp, J., & Yates, D. (2009). Do returning parolees affect neighborhood crime? A case study of Sacramento. *Criminology, 47*(3), 619–656.

Holsinger, A., Latessa, E., Turner, M., & Travis, L. (1997). *High level alternatives to incarceration: Examining community based correctional facilities*. Paper presented at the annual meeting of the Academy of Criminal Justice Science, Louisville, KY, March.

Hyatt, J.M., & Barnes, G.C. (2017). An experimental evaluation of the impact of intensive supervision on the recidivism of high-risk probationers. *Crime & Delinquency, 63*, 3–38.

Jalbert, S.K., Rhodes, W., Flygare, C., & Kane, M. (2010). Testing probation outcomes in an evidence-based practice setting: Reduced caseload size and intensive supervision effectiveness. *Journal of Offender Rehabilitation, 49*(4), 233–253.

Jalbert, S.K., Rhodes, W., Kane, M., Clawson, E., Bogue, B., Flygare, C., et al. (2011). *A multi-site evaluation of reduced probation caseload size in evidence-based practice setting.* Cambridge, MA: Abt Associates.

Jannetta, J., Neusteter, S.R., Davies, E., & Horvath, A. (2012). *Transition from prison to community initiative: Process evaluation final report.* Washington, DC: Urban Institute.

Jones, P. (1991). The risk of recidivism: Evaluating the public-safety implications of a community corrections program. *Journal of Criminal Justice, 19*(1), 49–66.

Kaeble, D. (2018). *Probation and parole in the United States, 2016.* Washington, DC: Bureau of Justice Statistics.

Kaeble, D., & Cowhig, M. (2018). *Correctional populations in the United States, 2016.* Washington, DC: Bureau of Justice Statistics.

Kurlychek, M. (2010). Transforming attitudinal change into behavioral change: The missing link. *Criminology & Public Policy, 9*(1), 119–126.

Larivee, J. (1990). Day reporting centers: Making their way from the U.K. to the U.S. *Corrections Today, 52*(6), 84.

Latessa, E.J. (1980). Intensive diversion unit: An evaluation. In Price, B., & Baunach, P.J. (Eds.), *Criminal justice research* (pp. 101–124). Beverly Hills, CA: Sage.

Leffingwell, T.R., Cooney, N.J., Murphy, J.G., Luczak, S., Rosen, G., Dougherty, D., et al. (2013). Continuous objective monitoring of alcohol use: Twenty-first century measurement using trans-dermal sensors. *Alcoholism: Clinical and Experimental Research, 37,* 16–22.

Lowenkamp, C., Flores, A., Holsinger, A., Makarios, M., & Latessa, E. (2010). Intensive supervision programs: Does program philosophy and the principles of effective intervention matter? *Journal of Criminal Justice, 38,* 368–375.

Lutze, E., Johnson, W.W., Clear, T., Latessa, E., & Slate, R. (2012). The future of community corrections is now: Stop dreaming and take action. *Journal of Contemporary Criminal Justice, 28,* 42–59.

MacKenzie, D., & Parent, D. (1991). Shock incarceration and prison crowding in Louisiana. *Journal of Criminal Justice, 19*(3), 225–237.

MacKenzie, D., & Armstrong, S. (2004). *Correctional boot camps: Military basic training or a model for corrections?* Thousand Oaks, CA: Sage.

Martin, C., Lurigio, A.J., & Olson, D.E. (2003). An examination of rearrests and reincarcerations among discharged day reporting center clients. *Federal Probation, 67,* 24–30.

Matejkowski, J., Festinger, D., Benishek, L., & Dugosh, K. (2011). Matching consequences to behavior: Implications of failing to distinguish between noncompliance and nonresponsivity. *International Journal of Law and Psychiatry, 34,* 269–274.

Matz, A.K. (2012). *Community corrections automated case management procurement guide with bid specifications.* Lexington, KY: Council of State Governments, American Probation and Parole Association.

McDevitt, J. (1988). *Evaluation of the Hampton County Day Reporting Center.* Boston: Crime and Justice Foundation.

Mellowa, J., Schlager, M., & Caplan, J. (2008). Using GIS to evaluate post-release prisoner services in Newark, New Jersey. *Journal of Criminal Justice, 36*(5), 416–425.

Monteiro, C.E., & Frost, N.A. (2015). Altering trajectories through community-based justice reinvestment. *Criminology & Public Policy*, *14*(3), 455–463.

Morris, N., & Tonry, M. (1990). *Between prison and probation: Intermediate punishments in a rational sentencing system*. Oxford: Oxford University Press.

National Institute of Corrections (1981). *Model probation and parole management project*. Washington, DC: National Institute of Corrections.

National Institute of Justice (1999). *Keeping track of electronic monitoring*. Washington, DC: National Law Enforcement and Corrections Technology Center.

National Institute of Justice (2011). *Electronic monitoring reduces recidivism*. Washington, DC: Office of Justice Programs.

Ostermann, M. (2013). Using day reporting centers to divert parolees from revocation. *Criminology & Public Policy*, *12*, 163–171.

Ostermann, M. (2015). How do former inmates perform in the community? A survival analysis of rearrests, reconvictions, and technical parole violations. *Crime & Delinquency*, *61*(2), 163–187.

Ostermann, M., & Hyatt, J.M. (2016). Is something better than nothing? The effects of short terms of mandatory parole supervision. *Justice Quarterly*, *33*(5), 785–810.

Padgett, K., Bales, W., & Blomberg, T. (2006). Under surveillance: An empirical test of the effectiveness and consequences of electronic monitoring. *Criminology & Public Policy*, *5*(1), 61–92.

Parent, D., Byrne, J., Tsarfaty, V., Valade, L., & Esselman, J. (1995). *Day reporting centers, Volume 1: Issues and practices*. Washington, DC: National Institute of Justice.

Parisi, N. (1980). Combining incarceration and probation. *Federal Probation*, *44*(2), 3–11.

Phelps, M.S. (2013). The paradox of probation: Community supervision in the age of mass incarceration. *Law & Policy*, *35*, 51–80.

Piquero, N. (2003). A recidivism analysis of Maryland's community probation program. *Journal of Criminal Justice*, *32*(4), 295–307.

Renzema, M., & Mayo-Wilson, E. (2005). Can electronic monitoring reduce crime for moderate to high-risk offenders? *Journal of Experimental Criminology*, *2*(2), 215–237.

Rhine, E.E., Petersilia, J., & Reitz, K.R. (2017). The future of parole release. *Crime and Justice*, *46*, 279–338.

Schwitzgebel, R. (1969). A belt from big brother. *Psychology Today*, *2*(11), 45–47, 65.

Shulman-Laniel, J., Vernick, J.S., McGinty, B., Frattaroli, S., & Rutkow, L. (2017). U.S. state ignition interlock laws for alcohol impaired driving prevention: A 50 state survey and analysis. *Journal of Law, Medicine, & Ethics*, *45*, 221–230.

Solomon, A., Kachnowski, V., & Bhati, A. (2005). *Does parole work?* Washington, DC: Urban Institute.

Taxman, F.S., & Caudy, M.S. (2015). Risk tells us who, but not what or how: Empirical assessment of the complexity of criminogenic needs to inform correctional programming. *Criminology & Public Policy*, *14*, 71–103.

Taylor, D. (2004). Agency accreditation: The performance-based standards experience. *Perspectives*, *28*(2), 21–23.

Taylor, K., & Blanchette, K. (2009). The women are not wrong: It is the approach that is debatable. *Criminology and Public Policy, 8*(1), 221–229.

Teague, M. (2011). Probation in America: Armed, private and unaffordable? *Probation Journal, 58*, 317–332.

The Pew Charitable Trusts (2016). *Use of electronic offender-tracking devices expands sharply.* Philadelphia, PA: Pew Charitable Trusts.

Thigpen, M., Beauclair, T., Moore, J., & Humphries, K. (2010). *TPC case management handbook: An integrated case management approach.* Washington, DC: National Institute of Corrections.

Travis, L.E., III. (1984). Intensive supervision in probation and parole. *Corrections Today, 46*(4), 34.

Tucker, S.B., & Cadora, R. (2003). Justice Reinvestment. *Open Society Institute, 3*(3), 1–7.

Usta, M., & Wein, L.M. (2015). Assessing risk-based policies for pretrial release and split sentencing in Los Angeles county jails. *PLoS One, 10*(12), 1–16.

Villettaz, P., Gillieron, G., & Killias, M. (2015). The effects of re-offending of custodial vs. non-custodial sanctions: An updated systematic review of the state of knowledge. *Campbell Systematic Review, 11*, 1–92.

Vito, G.F. (1985). Probation as punishment: New directions. In Travis, L.E., III, (Ed.), *Probation, parole and community corrections* (pp. 73–80). Prospect Heights, IL: Waveland.

Vito, G.F., Higgins, G.E., & Tewksbury, R. (2017). The effectiveness of parole supervision: Use of propensity score matching to analyse reincarceration rates in Kentucky. *Criminal Justice Policy Review, 28*(7), 627–640.

von Hirsch, A., & Hanrahan, K. (1978). *Abolish parole?* Washington, DC: U.S. Department of Justice.

Waldo, J. (2012). Implementing a GPS tracking program for community-based offenders: What correctional agencies need to know. *Journal of Offender Monitoring, 25*(2), 16–21.

Weisburd, D., Einat, T., & Kowalski, M. (2008). The miracle of the cells: An experimental study of interventions to increase payment of court-ordered financial obligations. *Criminology & Public Policy, 7*(1), 9–36.

Wilson, D.B., MacKenzie, D.L., & Mitchell, F.N. (2005). Effects of correctional boot camps on offending. *Campbell Systematic Reviews.* Retrieved from: https://campbellcollaboration.org/media/k2/attachments/Wilson_Bootcamps_review.pdf

Wodahl, E.J., & Garland, B. (2009). The evolution of community corrections: The enduring influence of the prison. *The Prison Journal, 89*, 81S–104S.

Wodahl, E.J., Ogle, R., & Heck, C. (2011). Revocation trends: A threat to the legitimacy of community-based corrections. *The Prison Journal, 91*(2), 207–266.

Yeh, S. (2010). Cost–benefit analysis of reducing crime through electronic monitoring of parolees and probationers. *Journal of Criminal Justice, 38*, 1090–1096.

Yeh, S. (2015). The electronic monitoring paradigm: A proposal for transforming criminal justice in the USA. *Laws, 4*, 60–81.

CHAPTER 14

The Juvenile Justice System

Misbehavior by juveniles poses special problems for agents of social control. Foremost among the concerns is the general societal belief that we should treat juveniles differently from adults. Indeed, an entire system of social control exists for dealing with problem youths. The juvenile justice system operates under a different set of assumptions about deviant behavior than does the adult criminal justice system, and it works somewhat independently of the adult system. This does not mean that there is no overlap between the adult and juvenile systems. In actuality, there is a great deal of similarity in the operations of the two systems. Some individuals claim that the differences are little more than semantic exercises. The aim of this chapter is to give the reader a brief overview of the problem of juvenile delinquency, the operations of the juvenile justice system, and the major issues currently facing that system.

DEFINING DELINQUENCY

Perhaps the first point of departure between the adult and juvenile systems appears in the behavior each handles. There are a number of definitions of delinquency. Many definitions reflect the same behavior outlined as criminal in the adult system. Such criminal law definitions often define a **delinquent** as a juvenile who violates the criminal laws of the jurisdiction. The emphasis is on the same behavior prohibited for adults.

Besides adult criminal acts, the juvenile justice system intervenes in a variety of specific juvenile offenses usually referred to as **status offenses** because they are applicable only to persons of a certain "status." Acts typically considered status offenses include smoking, drinking, fighting, running away, being disrespectful to parents, truancy, and various others that are allowable for adults. Although "status offense" is the most common term for these actions, some jurisdictions refer to "unruliness," "incorrigibility," "dependency," and other similar terms. Regardless of the term used, such statutes are generally very vague, leaving the interpretation of what is not acceptable behavior to the reader's discretion, making it possible to intervene with almost all youths.

Implicit in the various definitions of delinquency is a definition of "juvenile." Whereas some delinquency statutes provide a specific age, others simply refer to the age of majority defined in another statute. Clearly, juveniles are young persons who are not yet adults. The legal definition, however, varies from place to place. For example, 41 states and the District of Columbia set 18 years as the age at which youths move to adult court. Seven states move those aged 17 or older to adult court jurisdiction, and two states do so with those aged 16 and over (Office of Juvenile Justice and Delinquency Prevention, 2016). At the same time that an upper age limit is set, some states also set a lower age limit for system intervention (see Box 14.1). These lower ages typically range from age 6 to age 10, although the majority of states do not set a specific lower age of jurisdiction. In these states, a child can be held criminally responsible at the point in time that the court determines him or her to be capable of forming criminal intent. A further age consideration deals with the "waiver or transfer" of youths to adult jurisdiction. Although this issue will be dealt with later in the chapter, it is important to note that in 23 states and the District of Columbia there is no minimum age at which juveniles can be considered as adults and handled by the adult criminal justice system (see Box 14.2). Varying maximum age, minimum age, and waiver provisions mean that youths subject to the juvenile statutes in one location are adults in another jurisdiction.

BOX 14.1 LOWER AGE LIMITS FOR JUVENILE COURT JURISDICTION

Age 6	Age 7	Age 8	Age 10	
North Carolina	Connecticut	Arizona	Arkansas	Pennsylvania
	Maryland		Colorado	South Dakota
	Massachusetts		Kansas	Texas
	New York		Louisiana	Vermont
			Minnesota	Wisconsin
			Mississippi	

No Specified Lowest Age in Statute or Court Rule

Alabama	Idaho	Missouri	North Dakota	Utah
Alaska	Illinois	Montana	Ohio	Virginia
California	Indiana	Nebraska	Oklahoma	Washington
Delaware	Iowa	Nevada	Oregon	Washington, DC
Florida	Kentucky	New Hampshire	Rhode Island	West Virginia
Georgia	Maine	New Jersey	South Carolina	Wyoming
Hawaii	Michigan	New Mexico	Tennessee	

Source: Office of Juvenile Justice and Delinquency Prevention (2016).

BOX 14.2 MINIMUM AGES FOR TRANSFER TO CRIMINAL COURT

Age 10	Age 12	Age 13	Age 14		Age 15
Iowa	Colorado	Illinois	Alabama	Michigan	Connecticut
Wisconsin	Missouri	Mississippi	Arkansas	Minnesota	New Jersey
	Vermont	New Hampshire	California	North Dakota	New Mexico
		New York	Kansas	Ohio	
		North Carolina	Kentucky	Texas	
			Louisiana	Utah	
			Massachusetts	Virginia	

No Minimum Age

Alaska	Hawaii	Nebraska	South Carolina
Arizona	Idaho	Nevada	South Dakota
Delaware	Indiana	Oklahoma	Tennessee
District of Columbia	Maine	Oregon	Washington
Florida	Maryland	Pennsylvania	West Virginia
Georgia	Montana	Rhode Island	Wyoming

Source: Office of Juvenile Justice and Delinquency Prevention (2016).

MEASURING THE SCOPE OF THE PROBLEM

How large is the delinquency problem? What characterizes the typical delinquent? Answers to both of these questions can present different images. The varied responses are attributable to the range of possible considerations in defining delinquents and the various methods used for measuring delinquency. Despite the potential variability, some common features about delinquency emerge.

Official Records

The most common source of information on delinquency is the official records of the criminal and juvenile justice system. The *Uniform Crime Reports* (UCR), court records, and correctional figures are among the official crime measures that present information about the level of juvenile misbehavior. Figure 14.1 presents data from the 2016 UCR. Youths under the age of 18 accounted for just under 8% of all arrests, 10.1% of the violent Index crimes, and 13.6% of the property Index crimes (Federal Bureau of Investigation, 2016). The size of the juvenile problem appears even larger when you consider that youths between the ages of 14 and 17 make up roughly 5% of the total U.S. population (U.S. Census Bureau, 2017). Juveniles, therefore, are contributing their share to the arrest statistics.

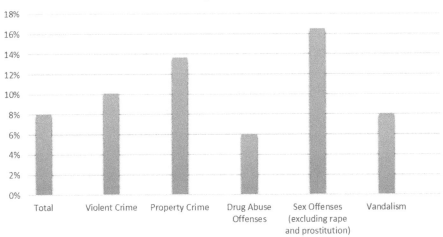

Percent of Arrests by Juveniles, by Offense Type

FIGURE 14.1
Percent of Arrests: Juveniles, by Offense Type.

Source: Federal Bureau of Investigation (2016).

Official figures also serve to provide a profile of delinquents. For example, the sex distribution of juvenile offenders is heavily skewed; males represent over 70% of all delinquency cases handled by juvenile courts. The racial breakdown in official figures shows an overrepresentation of minorities. For example, 42% of delinquency cases involving offenses against people (e.g., assault, robbery) involved black youth, even as blacks represent only 16% of the overall U.S. juvenile population (Furdella & Puzzanchera, 2015). The trend in youthful crime reveals declines over recent years. Official data show that the total number of juvenile arrests has declined 58% from 2007 to 2016 (Office of Juvenile Justice and Delinquency Prevention, 2016).

Although the official measures uncover a large amount of delinquency, these numbers probably underrepresent the actual level of juvenile misbehavior in society. First, not everyone reports all of the crimes they know about to the police; thus, any unreported deviant act is not included. Second, official records do not adequately reflect status offenses. The police may ignore or simply not record youthful misbehavior that is not also an adult criminal act. The possibility that official records underreport delinquency has led to the use of other measures of deviance. The greatest advantage of the official records lies in the fact that they are collected on an ongoing basis and in a reasonably consistent fashion.

Self-Reports of Delinquency

Researchers also measure delinquency through self-report surveys that ask the respondent what crimes he or she has committed. These measures have the potential of uncovering deviant acts not reported to the police. Self-report surveys have a long history in juvenile justice. In fact, the earliest such surveys applied specifically to the study of juvenile misbehavior. One of the best-known self-report scales was the Short–Nye Self-Report Delinquency Scale (Short & Nye, 1958). This scale was dominated by status and minor offenses such as "defied parents' authority," "took little things (less than $2)," and "bought or drank beer, wine, or liqour." Studies using these types of scales show that virtually every person is a delinquent. This probably results from the minor nature of the acts probed in the questions.

Criticisms that trivial acts dominate the scales prompted some researchers to construct scales that include more serious property and personal offenses. The National Youth Survey (NYS) included acts that qualify as felony assault, grand theft, sale of stolen items, and robbery. Unfortunately, the NYS is no longer collected. An ongoing self-report survey, the Monitoring the Future (MTF) project, questions high school students and young adults every year. As with the NYS, the MTF includes serious offenses, such as hitting a teacher, using a weapon to steal something from someone, stealing a car, and fighting in a group. The result of including these more serious items is a reduction in the number of persons claiming to have participated in deviance. In fact, the level of offending for the serious crimes is often close to that uncovered in official records.

Although self-reports can uncover crime not reported in official data, this measure of crime is not perfect. There is always a possibility that offenders will forget about an offense that they commit or an encounter with the police. Additionally, the possibility always exists that respondents could simply lie about their behavior. For example, Kirk (2006) interviewed youth about their involvement in delinquency and subsequently compared their answers to official data. He found that over 45% of youth who had been arrested failed to report the arrest and over 23% of those without an official arrest reported that they had indeed been arrested. Other researchers have found that delinquent youth tend to self-report their actions when asked (Brame, Fagan, Piquero, Schubert, & Steinberg, 2004; Piquero, Schubert, & Brame, 2014).

The demographic profile of offenders presented by self-report studies is somewhat different from the picture evident in official records. First, the peak age of offending appears to occur earlier in self-report surveys. This earlier peak may be partly due to the use of minor and status offenses in the surveys. Second, differences in the racial and social class distribution are minimal in self-report surveys. Where racial and social class differences appear, they are generally smaller than the levels observed through official data (Kirk, 2006). The greatest similarity to official records appears in the sex distribution of offenses. Males again exceed females in the level of offending (Piquero, Schubert, & Brame, 2014). Debates concerning juvenile crime policy should recognize the important differences between self-report surveys and official records.

Comparing the Delinquency Measures

The different sources of delinquency data show both similarities and differences. In general, both show that delinquency is a widespread problem. It is not restricted to any one group or type of offense. The level of offending increased throughout the 1960s and early 1970s, leveled off and showed some decreases in the late 1970s and 1980s, increased in the late 1980s and early 1990s (particularly in serious personal offenses), and has been decreasing in recent years. There is a clear diversity in offending. Youths are involved in all types of behavior—from status offenses to serious personal crimes.

No single method of measuring delinquency is better than the others. The usefulness of the measures depends entirely on the question being answered. Each method provides a different set of information about delinquency. Official records are useful for noting change in official processing and handling of youths over time. They also provide a long-term set of data that allows the inspection of changes over time. Official data are also rich in information about various demographic and offense factors not found in other measures. Self-reports provide a measure of delinquency based on the offender's viewpoint. They are capable of addressing behaviors that may not result in arrests and lead to official records. Self-reports yield information on minor crimes, the number of offenses an individual commits, offender demographics, and why an individual acts in a certain way.

GANG DELINQUENCY

The study of juvenile misbehavior often portrays delinquency as a group phenomenon (Erickson & Jensen, 1977; Hughes, 2013). Much of the interest in group delinquency revolves around the existence of juvenile gangs. One source of the public's concern about gangs may be the portrayal of gang behavior in the mass media. Movies and plays such as *The Blackboard Jungle* and *West Side Story* in the 1950s, and *Colors, Boyz N the Hood*, and *Gotti* in more recent years dramatize the lure of gangs for youths and the aggressive nature of these groups. Social and scholarly interest in gangs was particularly high in the early 1900s until the early 1960s. For a variety of reasons, including overall increases in social unrest and changes in theoretical approaches to deviance, gangs received little specific attention in the late 1960s, 1970s, and early 1980s. It is only within the past 40 years, as gang-related violence has increased, that academic and social interest in gangs reemerged.

Defining Gangs

Although there has been a great deal of interest and research in gang activity, no single, universally accepted definition of a gang has developed. However, a number of factors are common to most definitions (see Box 14.3). Gangs also are more common in poor, disorganized areas of the community. Rather than attempt to arrive at a single definition for gangs, some authors opt to identify different types of gangs. Knox (1991) offered a typology that presents gangs as developing through identifiable stages from a loose group of youths only marginally involved in criminal activity to a formal gang organized around criminal behavior for profit. Not all gangs will successfully move to the most formalized end of the development continuum. Taylor (1990) offered a typology of gangs based on the motivational factors underlying the gang behavior. Although some gangs center on protecting territory, others may exist to make money for its members. Using typologies of gangs negates the need for a single definition. By suggesting that there are different degrees or types of gangs, such typologies suggest that different types of gangs pose different problems and require different solutions.

BOX 14.3 TYPICAL ELEMENTS OF A "GANG" DEFINITION

Group Number	Usually a specified minimum number of members, most commonly defined as three or more
Identity	Clothes, hand signs, colors, etc., which serve to indicate membership
Preservation	Intent to enhance or preserve power or economic resources
Expectations	Rules for joining the group and expectation to protect its members
Turf	Territory claimed and/or controlled by the gang (not as common in many definitions)
Crime	Involvement in criminal behavior

Source: Based on Curry and Decker (1998) and National Gang Intelligence Center (2015).

The Extent of Gang Deviance

The extent of gang deviance is very difficult to gage, particularly given the varied definitions one can use to identify a gang. What was perhaps the earliest study of gangs identified 1,313 gangs with roughly 25,000 members in Chicago (Thrasher, 1936). These gangs comprised mostly adolescent males and usually ranged in size from 6 to 20 members, although some had as few as 2 to 3 members and others numbered more than 100. The great number of gangs in part reflected the fact that Thrasher considered almost any consistent grouping of youths (what he referred to as "play groups") to be a gang.

Although we do not know the exact extent of gang participation, we do know that gangs are common, and they exist throughout the United States (National Gang Intelligence Center, 2015). The most recent estimate indicates that there are nearly 30,000 gangs with about 782,000 members. Slightly less than one-third (31.6%) of local law enforcement agencies report youth gang activity in their jurisdictions (Egley & Howell, 2013) (see Figure 14.2). One study found that over 8% of students enrolled in high-risk, urban public schools were involved in a gang (Swahn, Bossarte, West, & Topalli, 2010). Overall, gang members have been shown to be disproportionately involved in delinquency

Percentage of Jurisdictions Reporting Gang Problems

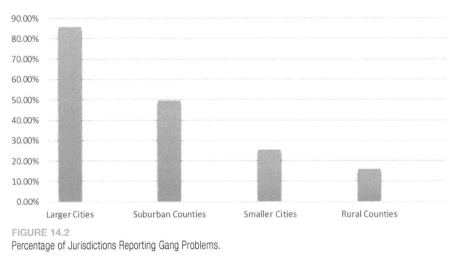

FIGURE 14.2
Percentage of Jurisdictions Reporting Gang Problems.

Source: National Youth Gang Center (2018).

and violence (Franzese, Covey, & Menard, 2016; Pyrooz, Turanovic, Decker, & Wu, 2016). In fact, one analysis found that nearly 15% of all homicides carried out by juveniles in a 37-year span were gang related (Heide, Michel, Cochran, & Khachatryan, 2017).

Why Do Youths Join Gangs?

Gang members come together and associate with one another for a wide array of reasons. Among these are the following (Taylor & Smith, 2013):

- The economic benefits offered by the gang, which can be more lucrative than a minimum-wage job
- Close relationships with friends and family who may already be involved in gang activity
- A feeling of protection where gang members may believe they are less likely to be victimized by members of the community
- To fill a need for support when the youth has experienced rejection in other areas of life, such as family or school
- An enhanced social status among peers, often symbolized by materialistic items such as clothing or jewelry
- Rebellion against traditional middle-class values and ethics.

Clearly, juveniles who find themselves faced with poverty, poor opportunities for advancement, poor school performance, lack of familial support, or other factors may find support and acceptance in a gang.

Gang Activity

The typical view of gang activity, especially as portrayed in the media, has not changed much over the years. The media portray gangs in constant violent confrontation with one another and with the public. Contrary to this media portrayal, gangs participate in a variety of different behaviors. This does not mean that the gang fights and drive-by shootings are fictional. Such confrontations have taken place in the past and continue to occur today. The media portrayal, however, distorts reality. Research reveals such violent confrontations are rare relative to other gang behavior. Joe-Laidler and Hunt (2012) pointed out that gangs provide a forum for recreation, partying, and companionship. Youth gangs are also involved in a wide variety of criminal activity aside from violence, including drug distribution, financial crime, and sex trafficking (National Gang Intelligence Center, 2015).

Even though gang violence is not as common as many believe, gang violence does occur and has changed greatly over the years. The gang fight, or rumble, was the traditional image associated with gang confrontations. The common scenario was two groups of youths in leather jackets and wielding chains, knives, or broken bottles in a prearranged fight. Such rumbles were relatively rare and resulted in few deaths. Violence in recent years does not conform to the image of a rumble. Instead, rumbles have given way to forays, in which one or two gang members attack a single rival gang member. The attack usually involves a firearm fired from a moving vehicle. The victim's gang then reciprocates against the transgressor's gang in a like fashion. From this emerges an ongoing series of small, isolated attacks between gangs. Counter to public perception, the forays target rival gang members and not the public in general. Violence against the public mainly appears in accidental injuries to bystanders. Two factors often pointed to as causes of modern gang violence are the heavy use of firearms and the role of violence in the drug trade.

The role of firearms in increased levels of violence, whether by gang members or individual youths, is very clear. Sheppard, Grant, Rowe, and Jacobs (2000) noted that the great increase in juvenile homicides in the mid- to late 1980s and early 1990s was due to the increased availability and use of firearms by youths. From 1984 to 1993, there was a 158% increase in the homicide rate involving handguns among 15–24-year-olds (Sheppard et al., 2000). According to FBI data, over 90% of gang-related homicides involved the use of a gun (Cooper & Smith, 2011).

Firearms use by gang members is a major problem. Gang members are more likely than nongang members to possess, carry, and use guns (Tigri, Reid, Turner, & Devinney, 2016; Watkins, Huebner, & Decker, 2008). Access to firearms has altered the confrontational approach of gangs from more face-to-face personal interaction to more impersonal drive-by shootings. The result of this use of lethal weapons by gangs is a much higher mortality rate among gang members when compared to the general population (Chassin, Piquero,

Losoya, Mansion, & Schubert, 2013; Teplin et al., 2014). There is evidence that possession of firearms is associated with prior exposure to or victimization by gun violence (Beardslee et al., 2018). In many cases, a juvenile might feel that owning a gun is a requirement for self-protection.

The involvement of drugs in gang activity has changed in recent years. There is no doubt that many gang members use and sell drugs, and that the sale of drugs is an integral part of some organized gangs. The degree to which gangs are involved in drugs, however, is highly variable. Drug sales in one gang may involve simple sale among its own members, really amounting to little more than a pattern of use, whereas another gang may be integrally involved in the drug trade throughout the community. Fagan (1990), in a survey of gang members in Los Angeles, San Diego, and Chicago, found that roughly 28% were rarely involved in drug use, but 35% were seriously involved in drug use and sales. In addition, he found little support for the claim that drug sales were integral to the formal organization of the gang. Conversely, a recent survey of gang members in San Antonio found that 98% reported lifetime use of marijuana and 90% reported lifetime use of cocaine (Cepeda, Onge, Nowotny, & Valdez, 2016). Over half of these gang members also reported selling drugs.

Whether drug involvement is a driving force behind gang violence is not clear. Klein, Maxson, and Cunningham (1991) reported little evidence that drugs are more prevalent in gang homicides than nongang homicides, and that violence is rare in both gang and nongang drug arrests. More recent research is mixed regarding whether gang members who sell drugs engage in more violence than those who do not sell drugs (Bellair & McNulty, 2009; Bjerregaard, 2010; Joe-Laidler & Hunt, 2012). Despite these findings, there is no doubt that conflicts over drugs and sales territories do escalate to violence. It is the extent to which such instances occur that needs further exploration.

Responding to Gangs

Responding to gangs and gang problems is an area in which much work remains. Unfortunately, the first response by many cities to an emerging gang problem is one of denial (Smith, 2017). Cities often do not want to admit that they have gangs. The outcome of such denial is the emergence of a full-blown problem before the authorities are prepared to deal with it.

A wide variety of approaches for dealing with gangs has emerged over the years. Box 14.4 lists five general gang intervention strategies identified in a survey of personnel in 245 cities (Spergel & Curry, 1990). The strategies, listed in order of prevalence at the time of the survey, show 44% of the cities reporting the use of suppression of gangs through arrest, incarceration, and supervision, closely followed by social interventions (31.5%). The least-used response was opportunities provision, although most survey respondents claimed that this was the most effective approach.

BOX 14.4 GANG INTERVENTION STRATEGIES

Suppression	Any form of social control in which the criminal justice system (police, courts, or corrections) or society attempt to impose formal or informal limits on behavior
Social Intervention	Basically a social work approach to working with gangs in the neighborhoods (such as detached worker programs)
Organizational Change and Development	An approach designed to alter the organization(s) that respond to gang problems, such as through the establishment of gang units or specialized training of personnel
Community Organization	Efforts aimed at mobilizing the community toward self-improvement and change, including both physical and social alterations
Opportunities Provision	An approach recognizing the lack of meaningful jobs and the training needed to succeed, and taking steps to change the problems; education, vocational training, and job placement are elements

Source: Adapted from Spergel and Curry (1990).

One of the interventions growing in popularity is the Gang Resistance Education and Training (GREAT) program. **GREAT** began in 1991 under a grant from the Bureau of Alcohol, Tobacco, and Firearms (known as the ATF, and now named the Bureau of Alcohol, Tobacco, Firearms, and Explosives) to the Phoenix, Arizona, Police Department. GREAT mimics the Drug Abuse Resistance Education (D.A.R.E.) program with police officers instructing middle-school youths. Lessons deal with individual rights, cultural sensitivity, conflict resolution, drugs, neighborhoods, personal responsibility, and goal setting. The program also targets self-esteem. An initial evaluation failed to find any significant impact on the level of self-reported gang participation (Esbensen & Osgood, 1997; Esbensen, Osgood, Taylor, Peterson, & Freng, 2001). However, program participants did display "more prosocial behaviors and attitudes" toward police, school, family, and peers. The promising results found in the evaluation led to the revised curriculum displayed in Box 14.5. Follow-up evaluations found the revised curriculum to be effective at preventing gang membership (Esbensen, Osgood, Peterson, Taylor, & Carson, 2013; Esbensen, Peterson, Taylor, & Osgood, 2012).

Some jurisdictions have focused on violence rather than on gangs. Where violence is linked to gangs and gang activities, interventions aimed at gangs are part of the violence control effort. These strategies fall under a general heading of "pulling levers," because officials apply any pressure they can to influence violence. Police and other justice officials use probation and parole conditions, enhanced sentencing, refusal to enter plea negotiations, and other tactics to deter or incapacitate gang members. Evaluations of these efforts indicate they are generally successful in reducing rates of gun violence and homicide (Braga, 2012).

BOX 14.5 GREAT MIDDLE-SCHOOL CURRICULUM

1. Welcome to GREAT
   - Program Introduction
   - Relationship Between Gangs, Violence, Drugs, and Crime

2. What's the Real Deal?
   - Message Analysis
   - Facts and Fiction About Gangs and Violence

3. It's About Us
   - Community
   - Roles and Responsibilities
   - What You Can Do About Gangs

4. Where Do We Go From Here?
   - Setting Realistic and Achievable Goals

5. Decisions, Decisions, Decisions
   - GREAT Decision-Making Model
   - Impact of Decisions on Goals
   - Decision-Making Model

6. Do You Hear What I Am Saying?
   - Effective Communication
   - Verbal vs. Nonverbal

7. Walk in Someone Else's Shoes
   - Active Listening
   - Identification of Different Emotions
   - Empathy for Others

8. Say It Like You Mean It
   - Body Language
   - Tone of Voice
   - Refusal-Skills Practice

9. Getting Along Without Going Along
   - Influences and Peer Pressure
   - Refusal-Skills Practice

10. Keeping Your Cool
    - GREAT Anger Management Tips
    - Practice Cooling Off

11. Keeping It Together
    - Recognizing Anger in Others
    - Tips for Calming Others

12. Working It Out
    - Consequences for Fighting
    - GREAT Tips for Conflict Resolution
    - Conflict Resolution Practice
    - Where to Go for Help

13. Looking Back
    - Program Review
    - "Making My School a GREAT Place" Project Review.

Source: Bureau of Justice Assistance (2005).

Campie, Petrosino, Fronius, and Read (2017) evaluated the effectiveness of a community-based violence prevention program used in Massachusetts. This program identified young men who had been involved in gun or gang-related crime. Workers then engaged these individuals to create a mentor relationship and provide individualized social services (e.g., workforce development, behavioral health services) as needed. Notably, this program did not include aggressive policing or intensive probation strategies. In fact, the relationship between the high-risk communities and the police had deteriorated to the point that participants indicated that they would not have participated in the program if it had included police involvement. Campie et al. found substantial reductions in violence among communities that implemented the program. This program reflects the importance of tailoring any gang prevention technique to the specific needs of the community and the juveniles involved in gang activity.

THE HISTORY OF JUVENILE JUSTICE

The history of juvenile delinquency and juvenile justice is relatively short. Although deviance on the part of young persons has always been a fact, societal intervention and participation in the handling of juvenile transgressors has gained most of its momentum in the last 150–200 years. The reasons for this are easy to see. Throughout most of history, youthful members of society did not enjoy a distinct status as "child." The young were either property or people. The very young, from birth to age 5 or 6, had much the same status as any other piece of property. The owner (parents) could buy, sell, or otherwise dispose of children as they wished. Once the individual reached the age of 5 or 6, he or she became a full-fledged adult member of society, subject to the same rules of conduct governing all "adults."

This state of indifference toward youths resulted from the health and economic conditions in society. The infant mortality rate typically exceeded 50%. The failure to develop a personal, caring attitude toward infants, therefore, was an emotional defense mechanism for reducing or eliminating any pain or sorrow attached to the death of a child. The economic conditions also meant that the birth of an infant was a financial burden on the family. Families lived from day to day on what they could produce. The very young were dependent and did not contribute to the family. A child represented a drain on the family's resources.

Over the centuries, practices developed for dealing with unwanted or burdensome children. One practice was **infanticide**, or the deliberate killing of an infant, usually by the mother. Infanticide was a common practice prior to the fourth century and appeared as late as the fourteenth century. A similar practice, which gained prominence after the fourth century, was **abandonment**. Abandoning children was less offensive than outright infanticide despite the fact that the result was the same. Parents subjected children who survived the first few years to actions such as apprenticeship and involuntary servitude. These actions were similar to the sale of youths by families. This again alleviated the need to care for the youth and brought an economic return to the family. In addition, these youths provided labor during the rise of industrialization.

Once children entered the labor force, they were viewed as adults and subject to the same rules and regulations as adults. A separate system for dealing with youthful offenders did not exist. At best, the father was responsible for controlling the child and his choices for punishment had no bounds. At the societal level, youths could be (and were) sentenced to the same penalties (including death) as were adults. Where due process protections existed, there is little evidence that the youths received them (Faust & Brantingham, 1979; Platt, 1977).

Changes in the societal view of children did not occur until the seventeenth and eighteenth centuries (the Progressive Era). During this time, medical advances were beginning to have a major impact on infant mortality and life expectancy. Additionally, scholars and religious leaders began to pay attention to the young as a means of attacking the ills of society. They saw education and protection of the young as a means of creating a moral society. Accompanying these views were alterations in beliefs about disciplining youthful offenders.

Methods for dealing with problem youths grew out of the establishment of ways to handle the poor. A key method of dealing with the poor was the removal of children from the bad influences and substandard training of poor parents. The establishment of **Houses of Refuge** in the early 1800s conformed to this idea. The first such institution opened in New York in 1825. Key features of these institutions were the use of education, skills training, hard work, and apprenticeships—all geared toward producing productive members of society. Despite the goals of the Houses of Refuge, various problems emerged. Among the concerns were the mixing of adults and juveniles, the mixing of criminals and noncriminals, overcrowding, the failure to supply intended education and training, the use of harsh physical punishment, and the exploitive use of the clients for monetary gain.

The failure of the early Houses of Refuge gave rise to the establishment of **cottage reformatories** in the second half of the 1800s. These new institutions closely paralleled a family; surrogate parents provided the education and moral training for a small number of youths. Probation and the use of foster homes also emerged at the same time as the reformatories. Unfortunately, like the earlier Houses of Refuge, these new alternatives suffered from many of the same problems.

In response to the failures of institutions to deal with problem youths and the call for new interventions, Chicago created a juvenile court in 1899. The late 1800s continued to experience great levels of immigration by lower-class Europeans, delinquency was on the rise, and there was an emergent body of sociological and psychological study attempting to explain the reasons for social ills. The new juvenile court reflected the general belief in the ability to alter youthful behavior through application of informal intervention and a desire to educate and train the child. Benevolent assistance, caring, training, and guidance were the watchwords of the new juvenile court. The mandate to help youths did not restrict the court to dealing with youths who committed criminal acts. Rather, the court could intervene in any situation in which a youth was in need of assistance. It was during this time that status offenses were included under the purview of the court. The growth of the court was phenomenal, and almost every state had at least one juvenile court by 1920 (see Box 14.6).

BOX 14.6 MILESTONES IN THE HISTORY OF JUVENILE JUSTICE

Pre-1800s	Children viewed as property or little adults Deviant youths handled in adult criminal system
1825	Establishment of Houses of Refuge—view that youths can be saved through education, moral training, hard work
1838	*Ex parte Crouse*—establishment of *parens patriae* as basis of intervention with youths
1869	Juvenile probation established in Massachusetts
Late 1880s	Move to cottage reformatories—same rationale as Houses of Refuge provided in surrogate family setup
1899	Juvenile court established in Chicago—totally separate from adult court, heavy reliance on *parens patriae* doctrine
1905	*Commonwealth v. Fisher*—court rules that *parens patriae* and good intentions are sufficient for intervention without concern for due process
1920–1960s	Various new approaches to treatment
1966	*Kent v. United States*—Justice Abe Fortas questions whether the juvenile system is providing the benevolent treatment promised by *parens patriae*; beginning of move toward due process in juvenile court
1974	The Juvenile Justice and Delinquency Prevention Act required deinstitutionalization of status offenders and protections to prevent juvenile delinquents from having contact with adult inmates
1980s–mid-2000s	Increased use of punitive-style punishments such as mandatory sentencing and waiver to adult court
Mid-2000s–Present	Decreased juvenile crime rates and new research involving juvenile brain development has led many states to repeal many of the previously enacted punitive statutes

THE PHILOSOPHY OF THE JUVENILE JUSTICE SYSTEM

The underlying philosophy of the new juvenile court was the doctrine of *parens patriae*. *Parens patriae*, or the "state as parent," was based on the actions of the English Chancery Court, which dealt with overseeing the financial affairs of orphaned juveniles. The court was to act as guardian until the child was mature enough to assume responsibility. Early interventions with juveniles also relied on the doctrine of *parens patriae*. In the case *Ex parte Crouse* (1838), the Pennsylvania Supreme Court ruled that the state had a right to intervene into a juvenile's life, against the wishes of the juvenile or the juvenile's parents, if the state thought that the parents were not capable of properly caring for the youth.

The new juvenile court borrowed this idea of guardianship for the cornerstone of its operations. Debate over the *parens patriae* doctrine was largely settled in 1905 when the Pennsylvania Supreme Court ruled in *Commonwealth v. Fisher* that intervention based on protecting, caring for, and training a youth was a duty of the state and did not violate the constitution, regardless of the youth's actions. The *parens patriae* philosophy stood largely unchallenged until the 1966 case of *Kent v. United States*. In this case, Justice Abe Fortas questioned the denial of due process for juveniles when he noted:

> There is evidence … that there may be grounds for concern that the child receives the worst of both worlds: that he gets neither the protections accorded to adults nor the solicitous care and regenerative treatment postulated for children. *Kent v. United States* (1966)

Parens patriae remained the dominant force behind interventions with juveniles through the 1970s. More recently, however, changes have been implemented that threaten the traditional *parens patriae* philosophy.

THE JUVENILE COURT PROCESS

Once a police officer takes a youth into custody, it is likely that the police will refer that youngster to juvenile court. Remaining cases are either referred to other authorities or the police handle them. When police refer a youth to juvenile court, the court personnel must then make one or more critical decisions: whether or not to detain (jail) the youth, whether or not to actually file a petition (charges) against the youth, whether to find (adjudicate) the youth a delinquent, and how to dispose of the petition. These decisions correspond to the adult court decisions of bail or jail, the filing of a formal charge versus dismissal, determination of guilt by plea or by trial, and sentencing. Several juvenile court actors—probation officers, defense attorneys, prosecutors, and judges—are involved in these important decisions. Whereas the judge is the primary decision maker, other court personnel play important roles in deciding the fate of juvenile suspects.

This section examines the critical decision points in the juvenile court process: detention, intake, waiver (transfer), adjudication, and disposition. We look at the roles the various court personnel play (and should play) in the court process. We describe what happens when a juvenile suspect goes through the juvenile court process, and compare the ideal with the reality. Finally, we examine some of the controversial issues facing juvenile court today, such as the question of how adversarial the attorneys in juvenile court should be and whether juveniles should have the right to a jury trial. Note that we do not focus on the critical question of the police and juveniles. However, many of the law enforcement issues raised earlier in the book apply to juveniles as well as to adults.

The Detention Decision

The first decision that juvenile court personnel must make is the **detention decision**. They must decide whether to keep a juvenile in custody or allow the youth to go home with his or her parents to await further court action. The detention decision is the juvenile court counterpart of the bail decision in adult court. It is very important because it concerns the freedom of the child and, therefore, resembles the disposition (sentencing) decision. In fact, children sent to detention may stay there for an extensive period, perhaps even longer than those sent to state training schools (youth prisons for juveniles determined to be delinquent). Nearly a quarter of juvenile cases result in detention (Hockenberry & Puzzanchera, 2018). As with adult bail, youths who are detained face a higher likelihood of being committed to a juvenile correctional institution (Lin, Miller, & Fukushima, 2008).

Detention workers or probation officers usually make the initial detention decision and have several options. Releasing a child to his or her parents is the most frequently used option and the preferred decision in most states. Secure detention—placing a child in the juvenile equivalent of a local jail—is another alternative. It involves placement in a locked facility that houses 10, 20, or more youths who are awaiting further court action or are awaiting transfer to a state correctional facility. Nonsecure detention is another option in some places—for youths involved in less serious crimes, youths who do not pose much threat to the community, and youths who are not a threat to themselves. Such youngsters may be placed in small group homes that are either not locked at all or at least not locked as comprehensively as a secure detention facility (hence the term "nonsecure"). Youngsters in nonsecure detention centers may even go to regular public school classes during the day. Alternatives to detention, such as home detention, have developed in recent years. These alternatives are important in light of extensive overuse of detention in the past (McCarthy, 1987).

The Intake Decision

The second major decision point in juvenile court is the intake decision, analogous to the filing or charging decision in adult court. At intake, a court official (probation officer, prosecutor, or both) decides whether to file a court petition of delinquency, status offense, neglect, abuse, or dependency in a particular case. Traditionally, a probation officer makes the intake decision. The *parens patriae* philosophy of the court dictated this approach because its treatment orientation indicated that the probation officer, ideally a trained social worker, should consider the best interests of the child as well as the legal aspects of the case (as an adult court prosecutor might). That is, an intake probation officer is supposed to consider the welfare of the child and the legal demands of the police and victim, and then attempt to resolve every case in light of those considerations.

A frequent decision of the intake officer is to refrain from filing a petition alleging delinquency or a status offense. Called "adjustment at intake" or "informal adjustment," the officer resolves the matter without resorting to a formal petition against the child. Informal adjustment practices occur approximately 25% of the time (Furdella & Puzzanchera, 2015), and have been part of juvenile court since its inception.

If an intake probation officer decides to file a petition against a child, often that decision requires the approval of an attorney, normally the prosecutor. The prosecutor's approval of the probation officer's decision to file a petition ensures that a legally trained official has reviewed the legal criteria for a properly authorized petition. The prosecutor checks the legal wording of the petition, determines that enough evidence is available for establishing the petition (finding the delinquent or status offender "guilty"), and ensures that the offense occurred in the court's jurisdiction and that the child was of proper age at the time of the offense.

Because of the importance of such legal criteria, and because of the emphasis on more punitive juvenile models, some jurisdictions have turned away from the traditional probation officer model of intake to models in which the prosecutor is either the first or the only intake decision maker. Such models are more consistent with legalistic views of juvenile court in which the state has abandoned the traditional *parens patriae* philosophy. Edwards (2011) argued that the judge should also be involved in the intake process. He noted that the lack of judicial oversight into this decision could lead to the repeated use of informal interventions for certain youth that should instead receive more formal processing This action represents a radical break with traditional juvenile court thinking and practice.

The Waiver Decision

For some youths petitioned to juvenile court, the most critical decision point is the **waiver decision** (also called the "transfer decision"). Many states allow the court to waive or transfer certain offenders (generally older offenders who commit serious crimes) to adult court. This is a crucial decision because transfer to adult court makes the youth subject to adult penalties (such as lengthy incarceration in an adult prison) as opposed to a relatively short period of incarceration in a juvenile training school. Such a decision also results in the creation of an adult criminal record, which is public and may hinder future opportunities for employment in certain occupations. In contrast, a juvenile court record is confidential.

Waivers occur through a variety of methods (see Box 14.7). The most common is in a hearing that is analogous to the preliminary hearing in adult court. At a waiver hearing, the prosecutor only must show probable cause that an offense occurred and that the juvenile committed the offense. The prosecutor does not have to prove guilt beyond a reasonable doubt. Proof of guilt is reserved for the trial in adult court if the court waives the juvenile or for the adjudication stage in juvenile court if the waiver motion fails. The juvenile transfer hearing differs from an adult court preliminary hearing in that the prosecutor must establish that the juvenile is not amenable to juvenile court intervention or that the juvenile is a threat to public safety.

BOX 14.7 FORMS OF WAIVER

Discretionary: Judge makes decision to waive youths after hearing

Mandatory: State mandates the juvenile court judge to waive jurisdiction under certain circumstances; requires a hearing

Presumptive: Statute sets presumption that certain cases are to be waived; not mandatory

Direct File: Prosecutor has right to choose whether to file in adult or juvenile court

Statutory Exclusion: State excludes certain categories of cases from juvenile court jurisdiction

Reverse Waiver: Permits a juvenile being prosecuted in adult court to be transferred back to juvenile court

Once an Adult/Always an Adult: Juvenile court jurisdiction is permanently terminated once prosecuted as an adult

Blended Sentencing: Either the juvenile court or the adult court imposes a sentence that involves either the juvenile or adult correctional systems, or both.

Source: Based on Griffin, Torbet, and Syzmanski (1998) and Office of Juvenile Justice and Delinquency Prevention (2002).

An example of nonamenability would be the case of a youth who is already on parole from a state training school for an earlier delinquent act who then commits another serious offense (e.g., armed robbery). If probable cause were established that the youth committed the robbery, then the judge would have to find that the juvenile court had a history of contacts with the youth dating back several years and that more juvenile court effort to deal with the youth's problems, either through probation or a training school placement, would be futile. An example of a case involving a threat to public safety would be a murder case or an offender with a history of violent offenses.

A number of other forms of waiver have emerged in recent years, including direct file, presumptive waiver, and statutory exclusion. Many of these are due to the more punitive attitudes toward juvenile delinquents. Table 14.1 shows the use of the varying forms of waiver found throughout the United States. Because these measures are relatively new, there are no national data available on the exact number of youths waived to adult criminal court in such ways.

Overall, approximately 1% of the juvenile cases handled formally are waived to adult court. Of waived cases in 2015, 50% involved personal crimes, and 31% involved property offenses (Hockenberry & Puzzanchera, 2018). The proportion of cases that are waived to adult court is considerably lower than the mid-1990s. One reason is that states use other methods (such as direct file or statutory exclusion) to prosecute juveniles in adult criminal courts. These cases either bypass the juvenile court altogether or restrict the judge's discretion to determine whether the statutory requirements of a waiver are met. Another reason involves the shift that has occurred since the mid-1990s back toward rehabilitative disposition of juvenile cases. Most states now reserve juvenile waivers for the most serious offenders.

TABLE 14.1 States' Use of Waiver and Blended Sentencing

	Judicial Waiver			Direct File	Statutory Exclusion	Reverse Waiver	Once/ Always	Juvenile Blended	Criminal Blended
	Discretionary	Presumptive	Mandatory						
Total States	45	15	15	15	29	24	34	14	18
Alabama	X				X		X		
Alaska	X	X			X			X	
Arizona	X			X	X	X	X		
Arkansas	X				X	X		X	X
California	X	X		X	X	X	X		X
Colorado	X	X		X		X		X	X
Connecticut			X			X		X	
Delaware	X		X		X	X	X		
DC	X	X		X			X		
Florida	X			X	X		X		X
Georgia	X		X	X	X	X			
Hawaii	X						X		
Idaho	X				X		X		X
Illinois	X	X	X		X		X	X	X
Indiana	X		X		X		X		
Iowa	X				X	X	X		X
Kansas	X	X					X	X	
Kentucky	X		X			X			X
Louisiana	X		X	X	X				
Maine	X	X					X		
Maryland	X				X	X	X		
Massachusetts					X			X	X
Michigan	X			X			X	X	X
Minnesota	X	X			X		X	X	
Mississippi	X				X	X	X		
Missouri	X						X		X
Montana				X	X	X		X	
Nebraska				X		X			X
Nevada	X	X			X	X	X		
New Hampshire	X	X					X		

	Judicial Waiver			Direct File	Statutory Exclusion	Reverse Waiver	Once/ Always	Juvenile Blended	Criminal Blended
	Discretionary	Presumptive	Mandatory						
New Jersey	X	X	X						
New Mexico					X			X	X
New York					X	X			
North Carolina	X		X				X		
North Dakota	X	X	X				X		
Ohio	X		X				X	X	
Oklahoma	X			X	X	X	X		X
Oregon	X				X	X	X		
Pennsylvania	X	X			X	X	X		
Rhode Island	X	X	X				X	X	
South Carolina	X		X		X				
South Dakota	X				X	X	X		
Tennessee	X					X	X		
Texas	X						X	X	
Utah	X	X			X		X		
Vermont	X			X	X	X			X
Virginia	X		X	X		X	X		X
Washington	X				X		X		
West Virginia	X		X						X
Wisconsin	X				X	X	X		X
Wyoming	X			X		X			

Source: Griffin, Addie, Adams, and Firestine (2011).

Research on transfer and other methods of placing youths into adult court has produced mixed results. In general, youths charged with serious offenses tend to receive harsher sentences in adult courts than young adults convicted of similar offenses (Griffin, Addie, Adams, & Firestine, 2011; Johnson & Kurlychek, 2012; Steiner, 2009). Many policymakers feel that increased punishment is needed for deterrence or to ensure public safety. However, recidivism rates among juveniles transferred to adult court tend to be similar to those who are processed through juvenile court (Loughran et al., 2010; Zane, Welsh, & Mears, 2016).

Examining juvenile court cases in Pennsylvania and Arizona, Augustyn and McGloin (2018) tested the idea that the recidivism rate among juveniles transferred to adult court might be affected by the punishment received rather than simply the waiver process itself. They evaluated recidivism rates among four groups of serious offenders: (1) those who were processed in juvenile court and received community-based sanctions, (2) those who were processed in juvenile court and were detained in a juvenile detention facility, (3) those who were waived to adult court and received community-based sanctions, and (4) those who were waived to adult court and were detained in an adult facility. They found similar recidivism rates among juveniles who were incarcerated after being waived to adult court compared to those who were detained in a juvenile detention facility. However, those who were incarcerated either in an adult or juvenile facility had higher recidivism rates compared to those who received community-based sanctions. This suggests that the waiver process might not be as inherently harmful to juveniles as once believed. However, the increased punishment that juveniles often receive after being waived to adult court might result in higher recidivism rates.

The research suggests that transfer and other means of putting juveniles into adult court are not magic solutions to some of the perceived problems in the juvenile court. Recidivism statistics do not indicate any advantage for transferred youths. In several studies, the transferred youths do worse than nontransferred youths. Further, the process of waiver (whether or not the juvenile was incarcerated) can affect one's earning potential later in life (Augustyn & Loughran, 2017). It appears that the current climate is increasingly favoring rehabilitative treatment for juvenile offenders. Indeed, the process of juvenile waiver has declined from its peak. Nevertheless, there will likely always be certain juvenile offenders who must be waived to adult court due to the nature of their conduct. Myers (2016: 933) noted: "the real issue, assuming juvenile transfer will not be eliminated, is which adolescent offenders should be waived to adult court and how they should be processed and sanctioned once they get there."

Adjudication and Disposition

For children not waived to adult criminal court, the next steps after the filing of a petition are adjudication and disposition. In the adjudication and disposition decisions, a judge determines whether there is enough evidence to establish the petition and then decides what to do if there is enough evidence. The **adjudication** decision is comparable to the conviction (plea or trial), and the **disposition** decision is like the sentencing decision in adult court.

The U.S. Supreme Court ruled certain procedural rights apply to juveniles as well as to adults (see Table 14.2 for a summary of key court decisions). In the ideal, the determination of the truth of the petition occurs in a rational fashion, with the prosecutor, defense attorney, and judge using their abilities and training to seek justice. Realistically, juvenile court sessions often are hectic and hurried, and they may reflect the self-interests of the parties involved rather than justice or the best interests of the child (Gebo, Stracuzzi, & Hurst, 2006; Newcombe, 2014).

TABLE 14.2 Selected Court Decisions on Juveniles

Kent v. United States, 383 U.S. 541 (1966)	Certain minimum safeguards apply to transfer (waiver) cases. The juvenile being considered for transfer to adult criminal court has the right to the assistance of counsel (an attorney), the right to a hearing, and a statement of the reasons for transfer if the judge decides to transfer the case to adult court
In re Gault, 387 U.S. 1 (1967)	The Fifth Amendment privilege against self-incrimination (the right to remain silent) and Sixth Amendment rights to adequate notice of charges against oneself, the right to confront and cross-examine accusers, and the right to the assistance of counsel do apply in delinquency proceedings with the possibility of confinement
In re Winship, 397 U.S. 358 (1970)	The standard proof of guilt beyond a reasonable doubt applies to juvenile delinquency proceedings as well as to adult criminal trials
McKeiver v. Pennsylvania, 403 U.S. 528 (1971)	Juveniles do *not* have a constitutional right to a jury trial
Breed v. Jones, 421 U.S. 519 (1975)	Juveniles cannot be adjudicated delinquent in juvenile court and then waived to adult court for trial without violating double jeopardy
Fare v. Michael C., 442 U.S. 707 (1979)	Trial court judges must evaluate the voluntariness of any confession obtained from a juvenile based on all circumstances of the confession. There is no rule that mandates the police consult the child's parent or an attorney before they can question a juvenile suspect. The child can waive his or her privilege against self-incrimination and the right to consult an attorney prior to interrogation
Schall v. Martin, 467 U.S. 253 (1984)	A juvenile who is awaiting court proceedings can be held in preventive detention if there is adequate concern that the juvenile would commit additional crimes while the primary case is pending further court action
Stanford v. Kentucky, 492 U.S. 361 (1989)	The constitutionality of the death penalty was upheld for youths who were either age 16 or 17 at the time they committed a murder
Roper v. Simmons, 543 U.S. 551 (2005)	The death penalty for juveniles was ruled unconstitutional
Graham v. Florida, 560 U.S. 48 (2010)	Life without parole sentences are unconstitutional for youths in nonhomicide cases
Miller v. Alabama, 132 S. Ct. 2455 (2012)	Mandatory life without parole sentences are unconstitutional for juveniles
Montgomery v. Louisiana, (577 U.S. ___) (2016)	Applied the ban on mandatory life sentences retroactively, allowing for juveniles who were sentenced prior to the *Miller* decision to have an opportunity for resentencing

Attorneys in the Juvenile Courtroom

There are several problems concerning attorneys in juvenile court. First, many communities lack access to specialized juvenile defense attorneys and resources for effective representation (National Juvenile Defender Center, 2016). Many juveniles waive their right to an attorney, often because they do not fully understand their rights, especially the importance of the right to an attorney (see e.g., Willis, 2017). A second critical problem is the burden of high caseloads for public defenders in juvenile court. Depending on the state, the caseload for the average public defender can range from 360 to 1,000 cases per defender (Jones, 2004). To place this in perspective, the American Bar Association suggests that juvenile caseloads not exceed 200 per defender (Benner, 2011). Similar to the adult justice system, large caseloads can negatively impact the quality of juvenile defense representation.

Many attorneys in juvenile court, both public defenders and private attorneys, are reluctant to utilize the zealous advocate approach that is, at least theoretically, the norm in adult criminal court (Kempf-Leonard, 2010). Attorneys in adult criminal courts justify such zealous advocacy (in which the attorney fights as hard as possible for all defendants, even defendants who have admitted that they are factually guilty) on the grounds that the system is adversarial and that the adversarial process is best for bringing out the truth. In juvenile court, some attorneys, parents, and judges believe that the adult criminal court norm of zealous advocacy is inappropriate. They may worry that strong advocacy can result in an outcome in which a child who "needs help" will not get it because failure to establish the petition leaves the court with no jurisdiction over the child. As a result, at least some attorneys act more like a concerned adult, encouraging youths to admit to petitions in cases in which an adversarial approach may have resulted in a dismissal of the petition.

The situation in America's juvenile courts appears to be that some attorneys are adversarial, some are still traditional and act as concerned adults, and some are in between the two extremes. The chief advantage of the zealous advocate model is that it is probably the best insurance that only truly guilty youths will come under court jurisdiction. Because the attorney does not pressure the child to admit to the petition (plead guilty), there is less danger that the court will attempt some type of intervention program with youths who are not guilty. An added advantage is that this approach may well generate the most respect from juveniles for the court system. Fewer youths will feel betrayed or will feel tricked into something that some adult thought was best for them, despite their own wishes.

The biggest danger of the zealous advocate approach is that it may contribute to what Fabricant (1983) called the contemporary version of benign neglect. Because many youths appearing in juvenile court come from families racked with problems, such as low income, public assistance, and/or broken homes, they indeed do need assistance. An adversarial approach may avoid forcing these children into juvenile prisons or other types of intervention owing to insufficient legal defense.

The advantage of the concerned adult model is that it seeks to address the problems of the child that presumably led the child into delinquency. The problem is that this helping philosophy has been the rationale of the juvenile court since 1899 and, as Rothman (1980) has so aptly phrased it, the rhetoric of individualized attention has always far outstripped the reality of ineffective if not abusive programs.

Although the assumption is that legal assistance will result in less severe punishments, the evidence is mixed. For example, Armstrong and Kim (2011) found that juveniles who were represented by a lawyer were more likely to receive a placement outside the home than juveniles who waived their right to counsel. This effect was true regardless of whether the juvenile was represented by a public defender or private attorney.

Jury Trials for Juveniles

In 1971, the Supreme Court ruled that juveniles do not have a constitutional right to a jury trial (*McKeiver v. Pennsylvania*, 1971). The Court noted the distinction between juvenile court and traditional criminal courts, as well as the advantages of the more informal juvenile system. Further, they concluded, "if the jury trial were to be injected into the juvenile court system as a matter of right, it would bring with it into that system the traditional delay, the formality, and the clamor of the adversary system" (p. 550). Despite this ruling, individual states are free to pass laws guaranteeing juveniles the right to a jury trial. Indeed, 10 states have specified that juveniles have a right to a jury trial, and another 8 allow juries in special circumstances (National Juvenile Defender Center, 2014).

Some maintain that it is critical for juveniles to have the right to a jury trial. For example, Feld (2017) has argued that judges require less proof than juries require and, therefore, it is easier to convict a youth in front of a judge than in front of a jury. Loveland (2017) discussed a number of potential biases to which juvenile judges are susceptible. Among these biases is the exposure to information about the juvenile that is inadmissible at trial. In making a decision, judges are only allowed to take into account evidence that has been deemed admissible and that is therefore legally relevant to the case. However, even the most well-meaning judge may implicitly use their full knowledge of the case to make a decision.

ISSUES IN JUVENILE JUSTICE

Modern juvenile justice faces a variety of concerns and issues. Many of these topics are interrelated and reflect different approaches and concerns. Among the issues are the emphasis on punitiveness and the constitutional limitations of punishment for youths.

The Emphasis on Punitiveness

Traditionally, the disposition stage of juvenile court has been the epitome of the *parens patriae* philosophy. With the advice of probation officers, social workers, psychologists, and psychiatrists, we presumed that the judge would do his or her best to act in the best interests of the child. Beginning in the

1980s, however, disposition (sentencing) in juvenile justice began to take on an increasingly punitive character.

The shift to a more punishment-oriented approach to juvenile delinquency has been referred to as the **punitive period** (Whitehead & Lab, 2015). During this time, many states took concrete measures to emphasize punishment. At least three states (Washington, New Jersey, and Texas) adopted determinate sentencing statutes with an emphasis on proportionality. The law in such states limits the discretion of judges at disposition and attempts to set penalties that are proportionate to the seriousness of the offense. Some states also enacted mandatory minimum provisions. This means that if the judge commits a child to the state youth authority, the law dictates that the youth must serve a certain minimum amount of time. Other states have adopted dispositional guidelines or suggested sentences for most adjudicated delinquents. Unless a case has some unusual factors, judges following the guidelines impose a sentence that falls within the ranges identified in the guidelines. Finally, there is the concern that the conditions of confinement during this period became more negative. For example, in 1991 almost two-thirds of all juveniles in long-term public institutions were in a facility in which the population exceeded design capacity (Office of Juvenile Justice and Delinquency Prevention, 1995).

Blended sentencing represents another punitive development. **Blended sentencing** allows either the juvenile court or the adult court to impose a sentence that can involve either the juvenile or the adult correctional system or both. The court can suspend the adult sentence pending either a violation or the commission of a new crime. The rationale is to give the judge greater flexibility in sentencing. The judge has greater discretion to adjust the sentence to the offender and the offense. The majority of states have laws authorizing some type of blended sentencing (Schaefer & Uggen, 2016).

It appears that there has been a softening of the punitive attitudes toward juvenile delinquency in recent years (Merlo & Benekos, 2010). As mentioned earlier, juvenile arrests and waivers to adult court have declined from their peaks. Similarly, juvenile detention placements have dropped dramatically in the past decade (Hockenberry, 2018). Many of these trends correspond to similar shifts in public opinion and policy response that have influenced the adult justice system.

Despite signs that the height of the punitive period may have passed, some scholars caution that remnants of this era persist. For example, Goshe (2015) points out that society's continued focus on self-sufficiency and the lack of attention to the overall well-being of youth prevents us from effectively addressing the underlying causes of juvenile delinquency. Further, juveniles continue to be placed on the sex offender registry, often for relatively minor actions (see Box 14.8). Placement of juveniles on sex offender registration has not been shown to reduce juvenile sex offending (Sandler, Letourneau, Vandiver, Shields, & Chaffin, 2017) and can significantly impair the child's rehabilitation.

Several critics have called for juveniles to be excluded from the sex offender registries (Brost & Jordon, 2017; Levenson, Grady, & Leibowitz, 2016).

BOX 14.8 POLICY DILEMMA: SEXTING

Juveniles are increasingly using cell phones to take provocative photographs of themselves and send these photographs to a classmate or dating partner. This behavior is commonly referred to as "sexting," and has become quite common. A recent study found that nearly 15% of children had sent a "sext" message and over 25% had received such a message (Madigan, Ly, Rash, Ouytsel, & Temple, 2018). Both the sending and possession of these photographs violate child pornography laws in many jurisdictions, with possible punishments including the requirement to register as a sex offender. Thomas and Cauffman (2014) argue that charging juveniles who engage in sexting with child pornography violates the original intention of such laws. Others point out the potential dangers involved in sexting, including the photograph being forwarded to third parties without the original sender's consent. Some states have passed sexting-specific legislation, which provide for less punishment and/or required counseling for juveniles who are found to be engaged in this behavior. Do you feel that states should prosecute sexting as a form of child pornography? Should legislators make a distinction between cases where a self-photograph is sent to a recipient consensually and those who then send the photograph to a third party without the consent of the original sender?

Constitutional Limitations of Punishment

In 2016, an estimated 682 juveniles (youths under 18 years of age) were arrested for murder (FBI, 2016). This continued a downward trend since the peak in 1993. Following an increase in the mid-1980s and the peak in 1993, the juvenile murder rate is currently at a historically low level. Despite this decrease, we will probably see continued media attention on juveniles who kill (see Figure 14.3).

FIGURE 14.3
Fifteen-year-old James Austin Hancock sits next to his attorney in Butler County Juvenile Court in Hamilton, Ohio. Hancock pleaded guilty to four counts of attempted murder and one count of inducing panic for a cafeteria shooting at Madison Local Schools, near Middletown, Ohio, in 2016. He was sentenced to a juvenile facility until age 21.

Greg Lynch/*Journal-News* via AP, File

Considerable discussion has taken place about the appropriateness of sentences such as the death penalty and life without the possibility of parole for juveniles accused of the most heinous crimes. Although the 2016 statistics represent a continuing decline since 1993, involvement in murder by juveniles is still alarming and prevalent enough that states will likely continue to punish these juveniles harshly. However, several recent court rulings have somewhat limited the range of punishments available for juvenile offenders.

The Supreme Court ruled that the death penalty is unconstitutional for juveniles. In *Roper v. Simmons* (2005), writing for the majority, Justice Kennedy wrote, "The Eighth and Fourteenth Amendments forbid imposition of the death penalty on offenders who were under the age of 18 when their crimes were committed." The ruling came in a case in which Christopher Simmons, age 17, with two accomplices, broke into and entered a home at 2:00 A.M., took a woman captive, drove away, and threw the woman from a railroad trestle into a river. The majority opinion reasoned that the majority of states rejected the juvenile death penalty and it is used infrequently in states that authorize it. The majority went on to cite scientific evidence that juveniles under age 18 are less mature and responsible than adults, more susceptible to peer pressure, and have character less well formed than that of an adult.

The Supreme Court has also decided a series of cases involving juveniles sentenced to life without parole. In *Graham v. Florida* (2010), the Court ruled that sentencing a juvenile to life without parole in nonhomicide cases violates the Constitution's ban on cruel and unusual punishment. The Court noted that a consensus had developed in the United States and worldwide against sentencing juveniles to life sentences without the possibility of parole.

The most recent application of the Eight Amendment in juvenile sentencing is *Miller v. Alabama* (2012). In *Miller*, the Court struck down a criminal law that mandated a life without parole sentence for juveniles convicted of murder. It is important to note that the Court did not rule that a life without parole sentence was unconstitutional per se. As stated by the Court, "Although we do not foreclose a sentencer's ability to make that judgment in homicide cases, we require it to take into account how children are different, and how those differences counsel against irrevocably sentencing them to a lifetime in prison." In 2016, the Court effectively made the *Miller* ruling retroactive to those juveniles who were sentenced to mandatory life without parole before the *Miller* decision was rendered (*Montgomery v. Louisiana*, 2016). Although this decision does not automatically reduce these sentences, it allows for the possibility that many of these cases will be resentenced. It is difficult to predict whether the Supreme Court will eventually determine that life without parole is unconstitutional in itself.

THE FUTURE OF JUVENILE JUSTICE

The juvenile justice system has a cyclical history (Lipsey, Howell, Kelly, Chapman, & Carver, 2010). As mentioned previously, many states have shifted from the traditional *parens patriae* philosophy to a more punitive model. Recent developments point to a new balance in juvenile justice policy (Bishop & Feld, 2012). We will examine several of these developments that, together, appear to be leading to a renewed focus on the rehabilitative mission of the juvenile justice system.

Developmental Approach

The past decade has seen many advances in the science of adolescent development. Specifically, empirical research shows that adolescents lack the ability to control their behavior when placed in emotional situations, are particularly sensitive to social influences such as peer pressure, and lack the ability to consider long-term consequences of their actions (Bonnie, Johnson, Chemers, & Schuck, 2013). Researchers have long studied the impact of a juvenile's social environment on delinquency. Adolescents are influenced by a variety of parental, peer group, and school factors. However, it is only recently that attention has also focused on a child's brain development. Evidence shows that the brain system influencing pleasure-seeking activity develops more rapidly than the system that supports self-control (Bonnie et al., 2013; Monahan, Steinberg, & Piquero, 2015). The advances in knowledge provide support to juvenile justice policies designed to help the juvenile make a successful transition to adulthood. In addition, they challenge the wisdom of excessively punitive punishments such as life in prison without the opportunity for parole.

Evidence-Based Practices

An important trend in the area of juvenile justice involves the increased focus on evidence-based practices. There is considerable research on "what works" in juvenile corrections. Program evaluation research has increased in rigor over the past decade, and can determine whether a particular delinquency prevention program is effective. By reviewing this research, policymakers are able to identify the types of programs that are most successful. States are increasingly using these research-based solutions to help shape their juvenile corrections policies (Greenwood & Welsh, 2012).

Several sources of information regarding the effectiveness of delinquency programs exist. For example, the Blueprints for Violence Prevention initiative reviewed more than 600 programs to identify 11 model programs that have demonstrated the greatest level of empirical support (Mihalie, Fagan, Irwin, Ballard, & Elliot, 2004). The Blueprints website (see www.blueprintsprograms.com) serves as a tool that government agencies can use to identify a program that meets the needs of their community. Another approach is to identify common characteristics of successful programs. After reviewing 548 evaluation studies, Lipsey (2009) reported that successful delinquency programs were those that were implemented well, targeted high-risk juveniles, and were therapeutic (rather than punitive) in nature. Although this research does not endorse specific programs, the information can assist communities that are interested in developing their own programs based on evidence-based principles.

Reemphasizing Juvenile Court

During the punitive era of juvenile justice, most states revised their laws to diminish the jurisdiction of the juvenile court. For example, many states removed certain crimes from the juvenile court regardless of age. Other states simplified the process to transfer youths to adult court or lowered the upper age of which a youth could be processed in juvenile court (Feld, 1998). The recent decline in juvenile crime and enhanced focus on evidence-based practices has motivated many states to reverse course. Many states have begun to reform their juvenile transfer and direct file laws, reserving the adult court system to only the most serious youth offenders. Another major trend in the past decade has been to expand the upper age of jurisdiction, in some cases back to 18. From 2001 to 2015, most states enacted legislation that in some way restored the jurisdiction of juvenile court (Brown, 2012, 2015).

Restorative Justice

Many communities have invested substantial resources into the development of restorative justice programs. Restorative justice can come in many forms (Van Ness & Strong, 2015). Often, the victim and offender will meet with a trained mediator to talk about the crime, its consequences, and how to make things right. Restorative justice can also take place as a conference in which the victim, offender, their families, and possibly even the arresting police officer are present. Another model takes the form of a circle in which members of the community and other interested parties can meet with the victim and offender. After forming a circle, participants speak one at a time about the crime and a possible resolution (Van Ness & Strong, 2015). What all of these efforts have in common is a focus that emphasizes the victim and the community. The approach is "focused less on achieving public safety by incarcerating individual offenders and more on reducing fear, building youth–adult relationships, and increasing the capacity of community groups and institutions to prevent crime and safely monitor offenders in the community" (Bazemore, 1999:98).

This represents a radical rethinking of the role of juvenile court. Instead of sanctioning and supervising offenders, the role of the court would be to build community so that neighborhoods can better respond to, but also prevent, delinquency. Communities would be more involved in sentencing through community panels, conferences, or dispute resolution programs. Communities would return to their role of being responsible for youths. Bergseth and

Bouffard (2013) reported that restorative justice programs seem to be effective with a wide range of juvenile offenders.

As noted, numerous communities already are working at restorative justice. A major question, however, is how far restorative justice can go. How willing are citizens to assume the responsibilities that restorative justice would give them in deciding cases and monitoring sanctions such as community service? If people are not available to staff the restorative justice programs, they will not work.

DUE PROCESS, CRIME CONTROL, AND THE JUVENILE JUSTICE SYSTEM

In some sense, the juvenile justice process is a separate and parallel system to the adult criminal justice system. Yet, both of these formal social control systems are designed to respond to behaviors that have been defined as criminal. Further, each is composed of subsystems of policing, courts, and corrections. What separates the two is the distinction drawn between adult and juvenile offenders.

The juvenile justice process is a product of the open-system nature of criminal justice. As attitudes and perceptions about children and childhood changed in the late 1800s, a separate system for the control and provision of assistance to youthful offenders was created. Originally designed to be helpful rather than punitive, the juvenile justice system was not seen as adversarial (where the interests of the juvenile were in conflict with the desires of the state for order). As a result, due process protections were largely absent from the juvenile justice process.

A growing recognition of the punitive aspects of the juvenile justice process led to concern about the rights of juvenile offenders. To protect the rights and interests of those accused of delinquency, due process protections were introduced into the juvenile justice system. The system increasingly came to resemble the adult process. What did not change was the goal of juvenile justice as "saving children." Inasmuch as the spread of due process protections in juvenile justice was not complete, the power of the juvenile justice system was constrained. Although juvenile offenders do not have a right to trial by jury, neither can the juvenile court impose a death sentence or life imprisonment. The juvenile process balances limits on the liberties of juveniles with limits on the power of the state.

For those deemed unsavable or otherwise inappropriate subjects for the juvenile courts, procedures to transfer their cases to the adult justice system are available. The existence of the juvenile justice system allows us, as a society, to resolve the dilemma of what should be done about the "criminality" of children. It lets us achieve a measure of social protection while requiring, in theory at least, that our protective efforts also be corrective or helpful to the offender.

The existence of a separate juvenile justice process by itself illustrates our deep ambivalence about the balance between due process and crime control. We want and need to control the dangerous and disruptive acts of youths, we also want to avoid doing long-term harm to these offenders. Rather than use a justice system that decides what interests will prevail (individual liberty or social control), we have created a compromise system that attempts to balance these interests in each case. As is usually true of compromises, the arrangement is uneasy. The juvenile justice system demonstrates the difficulty of balancing our competing interests in liberty and order. Hopefully, whatever changes continue to be implemented for court processing and correctional intervention with juveniles will retain some of the hope that the founders of juvenile court had for troubled youths.

REVIEW QUESTIONS

1. What is "delinquency"?

2. Why do juveniles join gangs?

3. How did the juvenile court come to be?

4. What is meant by *parens patriae*?

5. Briefly describe the juvenile justice system.

6. How does the juvenile justice system differ from the adult criminal system?

7. Describe how restorative justice is different from traditional punishment.

REFERENCES

Armstrong, G., & Kim, B. (2011). Juvenile penalties for "lawyering up": The role of counsel and extra-legal case characteristics. *Crime & Delinquency, 57*, 827–848.

Augustyn, M.B., & Loughran, T.A. (2017). Juvenile waiver as a mechanism of social stratification: A focus on human capital. *Criminology, 55*(2), 405–437.

Augustyn, M.B., & McGloin, J.M. (2018). Revisiting juvenile waiver: Integrating the incapacitation experience. *Criminology, 56*, 154–190.

Bazemore, G. (1999). The fork in the road to juvenile court reform. *Annals of the American Academy of Political and Social Science, 564*, 81–108.

Beardslee, J., Mulvey, E., Schubert, C., Allison, P., Infante, A., & Pardini, D. (2018). Gun and non-gun-related violence exposure and risk for subsequent gun carrying among male juvenile offenders. *Journal of the American Academy of Child & Adolescent Psychiatry, 57*(4), 274–279.

Bellair, P., & McNulty, T. (2009). Gang membership, drug selling, and violence in neighborhood context. *Justice Quarterly, 26*(4), 644–669.

Benner, L.A. (2011). Eliminating excessive public defender workloads. *Criminal Justice, 26*(2), 24–33.

Bergseth, K., & Bouffard, J. (2013). Examining the effectiveness of a restorative justice program for various types of juvenile offenders. *International Journal of Offender Therapy and Comparative Criminology, 57*(9), 1054–1075.

Bishop, D.M., & Feld, B.C. (2012). Trends in juvenile justice policy and practice. In Feld, B.C., & Bishop, D.M. (Eds.), *The Oxford handbook of juvenile crime and juvenile justice* (pp. 898–926). New York: Oxford University Press.

Bjerregaard, B. (2010). Gang membership and drug involvement: Untangling the complex relationship. *Crime & Delinquency, 56*, 3–34.

Bonnie, R., Johnson, R., Chemers, B., & Schuck, J. (2013). *Reforming juvenile justice: A developmental approach.* Washington, DC: National Academies Press.

Braga, A. (2012). Getting deterrence right? Evaluation evidence and complementary crime control mechanisms. *Criminology & Public Policy, 11*(2), 201–210.

Brame, R., Fagan, J., Piquero, A.R., Schubert, C.A., & Steinberg, L. (2004). Criminal careers of serious delinquents in two cities. *Youth Violence and Juvenile Justice, 2*(3), 256–272.

Brost, A.R., & Jordan, A.M.S. (2017). Punishment that does not fit the crime: The unconstitutional practice of placing youth on sex offender registries, *South Dakota Law Review, 62*(3), 806–831.

Brown, S.A. (2012). *Trends in juvenile justice state legislation: 2001–2011.* Denver, CO: National Conference of State Legislatures.

Brown, S.A. (2015). *Trends in juvenile justice state legislation: 2011–2015.* Denver, CO: National Conference of State Legislatures.

Bureau of Justice Assistance (2005). Gang Resistance Education and Training from http://greatonline.org

Campie, P., Petrosino, A., Fronius, T., & Read, N. (2017). *Community-based violence prevention study of the safe and successful youth initiative: An intervention to prevent urban gun violence.* Washington, DC: American Institutes for Research.

Cepeda, A., Onge, J.M.S., Nowotny, K.M., & Valdez, A. (2016). Associations between long-term gang membership and informal social control processes, drug use, and delinquent behavior among Mexican American youth. *International Journal of Offender Therapy and Comparative Criminology, 60*(13), 1532–1548.

Chassin, L., Piquero, A.R., Losoya, S.H., Mansion, A.D., & Schubert, C.A. (2013). Joint consideration of distal and proximal predictors of premature mortality among serious juvenile offenders. *Journal of Adolescent Health, 52*, 689–696.

Cooper, A., & Smith, E.L. (2011). *Homicide trends in the United States, 1980–2008.* Washington, DC: Bureau of Justice Statistics.

Curry, G.D., & Decker, S.H. (1998). *Confronting gangs: Crime and community.* Los Angeles: Roxbury.

Edwards, L. (2011, Spring). Intake decisions and the juvenile court system. *Juvenile and Family Justice Today.* Retrieved from www.judgeleonardedwards.com/docs/Spring2011intake.pdf

Egley, A., & Howell, J. (2013). *Highlights of the 2011 National Youth Gang Survey.* Washington, DC: Office of Juvenile Justice and Delinquency Prevention.

Erickson, M.L., & Jensen, G. (1977). Delinquency is still group behavior: Toward revitalizing the group premise in the sociology of deviance. *Journal of Criminal Law and Criminology, 68*, 262–273.

Esbensen, F., & Osgood, D.W. (1997). *National evaluation of GREAT. NIJ research in brief.* Washington, DC: National Institute of Justice.

Esbensen, F., Peterson, D., Taylor, T., & Osgood, D.W. (2012). Results from a multi-site evaluation of the G.R.E.A.T. program. *Justice Quarterly, 29*, 125–151.

Esbensen, F., Osgood, D.W., Taylor, T.J., Peterson, D., & Freng, A. (2001). How great is G.R.E.A.T.? Results from a longitudinal quasi-experimental design. *Criminology & Public Policy, 1*, 87–118.

Esbensen, F., Osgood, D.W., Peterson, D., Taylor, T.J., & Carson, D.C. (2013). Short- and long-term outcome results from a multisite evaluation of the G.R.E.A.T. program. *Criminology & Public Policy, 12*(3), 375–411.

Fabricant, M. (1983). *Juveniles in the family courts.* Lexington, MA: Lexington Books.

Fagan, J. (1990). Social processes of delinquency and drug use among urban gangs. In Huff, C.R. (Ed.), *Gangs in America.* Newbury Park, CA: Sage.

Faust, F.L., & Brantingham, P.J. (1979). *Juvenile justice philosophy: Readings, cases and comments.* St. Paul, MN: West.

Federal Bureau of Investigation (2016). *Crime in the United States 2016: Uniform crime reports.* Washington, DC: U.S. Department of Justice.

Feld, B. (1998). Juvenile and criminal justice systems' response to youth violence. *Crime and Justice, 24*, 189–261.

Feld, B. (2017). Competence and culpability: Delinquents in juvenile courts, youths in criminal courts. *Minnesota Law Review, 102*(2), 473–576.

Franzese, R.J., Covey, H.C., & Menard, S. (2016). *Youth gangs* (4th ed.). Springfield, IL: Charles Thomas.

Furdella, J., & Puzzanchera, C. (2015). *Delinquency cases in juvenile court, 2013.* Washington, DC: Office of Juvenile Justice and Delinquency Prevention.

Gebo, E., Stracuzzi, N.F., & Hurst, V. (2006). Juvenile justice reform and the courtroom workgroup: Issues of perception and workload. *Journal of Criminal Justice, 34*, 425–433.

Goshe, S. (2015). Moving beyond the punitive legacy: Taking stock of persistent problems in juvenile justice. *Youth Justice, 15*, 42–56.

Greenwood, P.W., & Welsh, B.C. (2012). Promoting evidence-based practice in delinquency prevention at the state level: Principles, progress, and policy directions. *Criminology & Public Policy, 11*(3), 493–513.

Griffin, P., Torbet, P., & Syzmanski, L. (1998). *Trying juveniles as adults in criminal court: An analysis of state transfer provisions.* Washington, DC: U.S. Department of Justice.

Griffin, P., Addie, S., Adams, B., & Firestine, K. (2011). *Trying juveniles as adults: An analysis of state transfer laws and reporting.* Washington, DC: Office of Juvenile Justice and Delinquency Prevention.

Heide, K.M., Michel, C., Cochran, J., & Khachatryan, N. (2017). Racial differences among juvenile homicide offenders: An empirical analysis of 37 years of U.S. arrest data. *Journal of Interpersonal Violence*. Advance online publication. doi: https://doi.org/10.1177/0886260517721173

Hockenberry, S. (2018). *Juveniles in residential placement, 2015*. Laurel, MD: Office of Juvenile Justice and Delinquency Prevention.

Hockenberry, S., & Puzzanchera, C. (2018). *Juvenile court statistics, 2015*. Pittsburgh, PA: National Center for Juvenile Justice.

Hughes, L.A. (2013). Group cohesiveness, gang member prestige, and delinquency and violence in Chicago, 1959–1962. *Criminology, 51*(4), 795–832.

Joe-Laidler, K., & Hunt, G.P. (2012). Moving beyond the gang–drug–violence connection. *Drugs, 19*(6), 442–452.

Johnson, B.D., & Kurlychek, M.C. (2012). Transferred juveniles in the era of sentencing guidelines: Examining judicial departures for juvenile offenders in adult criminal court. *Criminology, 50*(2), 525–564.

Jones, J.B. (2004). *Access to counsel*. Washington, DC: U.S. Department of Justice.

Kempf-Leonard, K. (2010). Does having an attorney provide a better outcome?: The right to counsel does not mean attorneys help youths. *Criminology & Public Policy, 9*(2), 357–364.

Kirk, D.S. (2006). Examining the divergence across self-report and official data sources on inferences about the adolescent life-course of crime. *Journal of Quantitative Criminology, 22*, 107–129.

Klein, M.W., Maxson, C.L., & Cunningham, L.C. (1991). "Crack," street gangs, and violence. *Criminology, 29*, 623–649.

Knox, G.W. (1991). *An introduction to gangs*. Berrien Springs, MI: Vande Vere.

Levenson, J.S., Grady, M.D., & Leibowitz, G. (2016). Grand challenges: Social justice and the need for evidence-based sex offender registry reform. *Journal of Sociology & Social Welfare, 43*, 3–38.

Lin, J., Miller, J., & Fukushima, M. (2008). Juvenile probation officers' dispositional recommendations: Predictive factors and their alignment with predictors of recidivism. *Journal of Crime and Justice, 31*(1), 1–34.

Lipsey, M.W. (2009). The primary factors that characterize effective interventions with juvenile offenders: A meta-analytic overview. *Victims and Offenders, 4*, 124–147.

Lipsey, M.W., Howell, J.C., Kelly, M.R., Chapman, G., & Carver, D. (2010). *Improving the effectiveness of juvenile justice programs: A new perspective on evidence-based practice*. Washington, DC: Center for Juvenile Justice Reform.

Loughran, T.A., Mulvey, E.P., Schubert, C.A., Chassin, L.A., Steinberg, L., Piquero, A.R., et al. (2010). Differential effects of adult court transfer on juvenile offender recidivism. *Law and Human Behavior, 34*, 476–488.

Loveland, P. (2017). Acknowledging and protecting against judicial bias at fact-finding in juvenile court. *Fordham Urban Law Journal, 45*, 283–319.

Madigan, S., Ly, A., Rash, C.L., Ouytsel, J.V., & Temple, J.R. (2018). Prevalence of multiple forms of sexting behavior among youth: A systematic review and meta-analysis. *JAMA Pediatrics, 172*(4), 327–335.

McCarthy, B.R. (1987). Preventive detention and pretrial custody in the juvenile court. *Journal of Criminal Justice, 15*, 185–200.

Merlo, A.V., & Benekos, P.J. (2010). Is punitive juvenile justice policy declining in the United States? A critique of emergent initiatives. *Youth Justice, 10*, 3–24.

Mihalie, S., Fagan, A., Irwin, K., Ballard, D., & Elliot, D. (2004). *Blueprints for violence prevention*. Washington, DC: Office of Juvenile Justice and Delinquency Prevention.

Monahan, K., Steinberg, L., & Piquero, A.R. (2015). Juvenile justice policy and practice: A developmental approach. *Crime and Justice, 44*, 577–619.

Myers, D.L. (2016). Juvenile transfer to adult court: Ongoing search for scientific support. *Criminology & Public Policy, 15*(3), 927–938.

National Gang Intelligence Center (2015). *National gang report, 2015*. Retrieved from www.fbi.gov/file-repository/national-gang-report-2015.pdf/view

National Juvenile Defender Center (2014). *Juvenile right to jury trial chart*. Retrieved from http://njdc.info/wp-content/uploads/2017/03/Right-to-Jury-Trial-Chart-7-18-14.pdf

National Juvenile Defender Center (2016). *Defend children: A blueprint for effective juvenile defender services*. Washington, DC: Author.

National Youth Gang Center (2018). *National youth gang survey analysis*. Retrieved from: www.nationalgangcenter.gov/Survey-Analysis/Prevalence-of-Gang-Problems

Newcombe, S. (2014). The DOJ comes to town: An argument for legislative reform when the juvenile court fails to protect due-process rights. *University of Memphis Law Review, 44*(4), 921–976.

Office of Juvenile Justice and Delinquency Prevention (1995). *Juvenile offenders and victims: A national report*. Washington, DC: U.S. Department of Justice.

Office of Juvenile Justice and Delinquency Prevention (2002). *OJJDP statistical briefing book*. Retrieved from http://ojjdp.ncjrs.org/ojstatbb/html/qa088.html

Office of Juvenile Justice and Delinquency Prevention (2016). *OJJDP statistical briefing book*. Retrieved from www.ojjdp.gov/ojstatbb/structure_process/qa04102.asp?qaDate=2016&text=no&maplink=link1

Platt, A.M. (1977). *The child savers: The invention of delinquency*. Chicago: University of Chicago Press.

Piquero, A.R., Schubert, C.A., & Brame, R. (2014). Comparing official and self-report records of offending across gender and race/ethnicity in a longitudinal study of serious youthful offenders. *Journal of Research in Crime and Delinquency, 51*(4), 526–556.

Pyrooz, D.C., Turanovic, J.J., Decker, S.H., & Wu, J. (2016). Taking stock of the relationship between gang membership and offending: A meta-analysis. *Criminal Justice and Behavior, 43*(3), 365–397.

Rothman, D. (1980). *Conscience and convenience*. Boston: Little, Brown.

Sandler, J.C., Letourneau, E.J., Vandiver, D.M., Shields, R.T., & Chaffin, M. (2017). Juvenile sexual crime reporting rates are not influenced by juvenile sex offender registration policies. *Psychology, Public Policy, and Law, 23*(2), 131–140.

Schaefer, S.S., & Uggen, C. (2016). Blended sentencing laws and the punitive turn in juvenile justice. *Law and Social Inquiry, 41*, 435–460.

Sheppard, D., Grant, H., Rowe, W., & Jacobs, N. (2000). *Fighting juvenile gun violence. OJJDP Juvenile Justice Bulletin*. Washington, DC: Office of Juvenile Justice and Delinquency Prevention.

Short, J.F., & Nye, I. (1958). Extent of unrecorded delinquency: Tentative conclusions. *Journal of Criminal Law, Criminology, and Police Science, 49*, 296–302.

Smith, C.F. (2017). *Gangs and the military: Gangsters, bikers, and terrorists with military training*. Lanham, MD: Rowman & Littlefield.

Spergel, I.A., & Curry, G.D. (1990). Strategies and perceived agency effectiveness in dealing with youth gang problems. In Huff, C.R. (Ed.), *Gangs in America*. Newbury Park, CA: Sage.

Steiner, B. (2009). The effects of juvenile transfer to criminal court on incarceration decisions. *Justice Quarterly, 26*(1), 77–106.

Swahn, M.H., Bossarte, R.M., West, B., & Topalli, V. (2010). Alcohol and drug use among gang members: Experiences of adolescents who attend school. *Journal of School Health, 80*(7), 353–360.

Taylor, C.S. (1990). Gang imperialism. In Huff, C.R. (Ed.), *Gangs in America*. Newbury Park, CA: Sage.

Taylor, C.S., & Smith, P.R. (2013). The attraction of gangs: How can we reduce it? In Simon, T., Ritter, N., & Mahendra, R. (Eds.), *Changing course: Preventing gang membership*. Washington, DC: National Institute of Justice.

Teplin, L.A., Jakubowski, J.A., Abram, K.M., Olson, N.D., Stokes, M.L., & Welty, L.J. (2014). Firearm homicide and other causes of death in delinquents: A 16-year prospective study. *Pediatrics, 134*, 63–73.

Thomas, A.G., & Cauffman, E. (2014). Youth sexting as child pornography? Developmental science supports less harsh sanctions for juvenile sexters. *New Criminal Law Review, 17*, 631–647.

Thrasher, F.M. (1936). *The gang*. Chicago: University of Chicago Press.

Tigri, H.B., Reid, S., Turner, M.G., & Devinney, J.M. (2016). Investigating the relationship between gang membership and carrying a firearm: Results from a national sample. *American Journal of Criminal Justice, 41*, 168–184.

U.S. Census Bureau (2017). *2017 national population projections tables*. Retrieved from www.census.gov/data/tables/2017/demo/popproj/2017-summary-tables.html

Van Ness, D.W., & Strong, K.H. (2015). *Restoring justice: An introduction to restorative justice* (5th ed.). Boston: Elsevier (Anderson Publishing).

Watkins, A.M., Huebner, B.M., & Decker, S.H. (2008). Patterns of gun acquisition, carrying, and use among juvenile and adult arrestees: Evidence from a high-crime city. *Justice Quarterly, 25*(4), 674–700.

Whitehead, J.T., & Lab, S.P. (2015). *Juvenile justice: An introduction* (8th ed.). New York: Routledge.

Willis, C.D. (2017). Right to counsel in juvenile court 50 years after *In re Gault. Journal of the American Academy of Psychiatry and the Law, 45*(2), 140–144.

Zane, S.N., Welsh, B.C., & Mears, D.P. (2016). Juvenile transfer and specific deterrence hypothesis: Systematic review and meta-analysis. *Criminology & Public Policy, 15*(3), 901–925.

IMPORTANT CASES

Breed v. Jones, 421 U.S. 519 (1975).

Commonwealth v. Fisher, 213 Pa. 48 (1905).

Ex parte Crouse, 4 Wheaton (Pa.) 9 (1838).

Fare v. Michael C., 442 U.S. 707 (1979).

Graham v. Florida, 560 U.S. 48 (2010).

In re Gault, 387 U.S. 1 (1967).

In re Winship, 397 U.S. 358 (1970).

Kent v. United States, 383 U.S. 541 (1966).

McKeiver v. Pennsylvania, 403 U.S. 528 (1971).

Miller v. Alabama, 132 S. Ct. 2455 (2012).

Montgomery v. Louisiana, 577 U.S. ___ (2016).

Roper v. Simmons, 543 U.S. 551 (2005).

Schall v. Martin, 467 U.S. 253 (1984).

Stanford v. Kentucky, 492 U.S. 361 (1989).

CHAPTER 15

Discharge and Developments

In 2016, approximately 3 million persons obtained release from some form of prison, probation, or parole supervision (Carson, 2018; Kaeble, 2018). Shannon et al. (2017) estimated that 3% of all Americans have been to prison at some point in their lives, and 8% of all adults had some sort of felony criminal conviction. These numbers have grown over time, and vary by racial group. For example, an estimated 33% of all African-American adult males have a felony conviction. These figures illustrate the need to understand the discharge process and the lifelong impact that an offender's involvement with the justice system will have on their lives.

This chapter examines what happens to those people who progress through the justice system to the point of discharge from custody, or to where the justice system stops. In doing so, we see that, for many, the effects of justice system processing do not stop, but start anew. We will also examine changes in the criminal justice system we can anticipate, or that seem to be beginning now.

IMPORTANT TERMS

civil death

collateral consequences

discharge

executive clemency

identity theft

intelligence gathering

recidivism

war on drugs

DISCHARGE

The last major point in the criminal justice process is the discharge of offenders. In many cases, discharge is not actually a decision point but rather an event that occurs at a point in time. For example, if you received a sentence to 5 years in prison on January 1, 2019, the sentence would expire at midnight on December 31, 2023. The state must discharge you from custody at that point, because there would no longer be a legal justification for custody. In many states, however, discharge itself is a decision.

Traditionally, we considered a convicted offender to be under the custody of the justice system until his or her sentence had expired. With the imposition of relatively long sentences, this meant that a person sentenced to a life term could be in custody for his or her entire lifetime. Most states developed mechanisms for limiting the period of custody, and empowered the parole authority to grant a discharge from sentence after a period of successful parole supervision. Further, many states have specifically empowered correctional officials to grant final releases or discharges from sentence.

Some states allow offenders to apply for **executive clemency**, or mercy. A convict could ask the governor for a commutation or pardon. If granted, the prisoner would receive a discharge from sentence. Someone serving a 20-year term, for example, might be imprisoned for 2 years, serve an additional 2 years under parole supervision, and still "owe" 16 years on the sentence. That offender could ask the governor to commute the sentence to 4 years and receive a discharge, or the governor could grant a pardon for the offense. A pardon effectively strikes the crime from the offender's record, while clemency simply reduces the punishment while keeping the record of the offense intact. Those convicted of federal crimes can seek presidential clemency. The use of presidential clemency seems to be increasing. In fact, President Obama commuted more sentences than the previous 11 presidents combined (Obama, 2017).

Regardless of the method, the fact remains that, at some point, most offenders receive a discharge from sentence. At the point of discharge, whether by the discretionary decision of a justice system official or at the expiration of sentence, the offender is free from direct intervention by the criminal justice system. The effects of conviction, however, will linger.

COLLATERAL CONSEQUENCES OF CONVICTION

Once convicted of a criminal offense, an individual experiences several negative effects. The sentence imposed for crime is a direct consequence of conviction, and is intentionally painful. Other effects that result from conviction, called **collateral consequences**, are not part of the stated penalty for the crime. In many ways, these effects can be more disturbing and painful than the imposed sentence.

A stigma attaches to a person convicted of a crime. The label of "ex-con" is difficult to overcome. If you were hiring people to work for a company and had two equally qualified candidates, one of whom was an ex-con, which applicant would you hire? At a party, how would you react when meeting someone who was an ex-con? That we can discuss a set of people simply by using the label "ex-con" illustrates the point of stigma. The Justice Department recently announced a policy to replace the terms "felon" and "convict" with less disparaging labels in an attempt to reduce the associated stigma attached to those who have been convicted of a crime (Noble, 2016). Regardless of the terminology, the fact of a prior conviction is meaningful to us in our dealings with those who have committed a crime (Blumstein & Nakamura, 2009).

In addition to the stigma of having a criminal record, other collateral consequences attend criminal conviction. Conviction closes some avenues of employment to offenders. Many jobs in our society require that the employee be bonded (insured), and ex-offenders normally are not considered good risks by bonding companies. Thus, an ex-offender may even have difficulty obtaining employment as a cashier in a convenience store. Other occupational groups, such as barbers, beauticians, teachers, physicians, nurses, and attorneys, require licensure. In many cases, a felony conviction is a bar to licensure (Davidenas, 1983).

Assuming the offender served a prison term, he or she may have a particularly difficult time in securing credit or employment because of the incarceration. How does an ex-inmate answer questions about where he or she worked for the past 2 years, or about where he or she resided? (One of the author's students, an ex-convict, routinely answered such questions by saying that he had "worked for the state" for the past 7 years.) Denver, Pickett, and Bushway (2018) reported that employers are conducting criminal history checks of employees and applicants more frequently now. They found that an estimated 31 million adults per year are asked about their criminal convictions, with most of these inquiries occurring at the application stage of the hiring process. In some cases, state laws require employers to check the criminal history of workers. The impact of a prior conviction on employment prospects continues to be negative (see Box 15.1).

BOX 15.1 POLICY DILEMMA: BAN THE BOX

One of the hurdles facing individuals convicted of a crime is obtaining a job. Studies have shown that employers are less likely to consider applications from those with even relatively minor criminal convictions (Agan & Starr, 2017; Uggen, Vuolo, Lageson, Ruhland, & Whitham, 2014). This is especially true when an applicant is forced to disclose their conviction on the initial job application. Many jurisdictions and private employers are beginning to "ban the box" that asks an applicant to disclose their criminal history. These laws do not ban the employer from conducting background checks on applicants. Rather, they simply seek to allow an applicant to get their foot in the door by delaying the background check until later in the application process. By that time, the hope is that the applicant would have had a chance to meet with their potential employer without the stigma associated with their conviction. Despite the strong momentum of such policies, some critics suggest that banning the box will not change many employers' reluctance to hire an individual with a criminal record. Instead of asking about criminal history on the application, an employer might guess about an applicant's criminal history based on demographic factors such as race (Vuolo, Lageson, & Uggen, 2017). If this occurs, the policy might unintentionally lower the chances of obtaining employment for some minorities who do not have a criminal record. Do you think that employers should be required to interview an applicant before they know the details of that applicant's criminal history?

Many states require ex-offenders to register with the local police, or maintain registries of released offenders that are available to the public. All states require registration of some offenders, especially some sexual offenders, and most restrict locations where at least some sex offenders can reside. Research into the effect of residence restrictions and registration requirements does not support the practices. Zgoba and Levenson (2012) found that sex offenders who fail to register pose no greater risk of committing new sex offenses than those who do register. Research also shows that residency restrictions for sex offenders are not effective at preventing sexual victimization by known sex offenders or at serving as a deterrent for sex crimes (Nobles, Levenson, & Youstin, 2012). Other research has reported that residency restrictions are not effective at keeping sex offenders away from protected locations (schools, parks, playgrounds, etc.). Further, the restrictions seem to result in the concentration of sex offenders in certain neighborhoods—typically poor, urban neighborhoods. Finally, the restrictions also place a heavy burden on ex-offenders in terms of their ability to find housing and employment (Hughes & Burchfield, 2008; Hughes & Kadleck, 2008; Socia & Stamatel, 2012).

As if the natural consequences of conviction were not severe enough, most states impose specific limitations on the rights of those convicted of felonies. These limitations are in addition to whatever sentences the courts order (Buckler & Travis, 2003; Burton, Cullen, & Travis, 1987). Some states have provisions allowing certain offenders (generally, those receiving life sentences) to be declared civilly dead (Rottman & Strickland, 2006). **Civil death** means that as far as civil rights (contracting, marriage, voting, etc.) are concerned, the offender is "dead." Most states restrict some civil rights for at least as long as the offender is serving his or her sentence. In many jurisdictions, conviction of a felony carries the permanent loss of some civil rights. Given the collateral consequences of conviction, for many offenders criminal justice processing never ends.

Restoration of Rights

Most states today have provisions for the restoration of all, or most, of the civil rights that are lost on conviction of a felony. In 2017, 48 states and the District of Columbia had some type of felony disenfranchisement restrictions (Box 15.2). Only two states, Maine and Vermont, do not restrict a felon's voting rights in some way (Chung, 2017). However, states vary considerably in the

---

BOX 15.2 VOTING RESTRICTIONS

Almost every state currently has voting restrictions for convicted felons. Restrictions range from only current inmates to convicted felons who are released and are no longer under any type of correctional supervision. A breakdown of restrictions is as follows:

No restriction: 2 states (Maine, Vermont)

Inmates only: 15 states (District of Columbia, Hawaii, Illinois, Indiana, Maryland, Massachusetts, Michigan, Montana, New Hampshire, North Dakota, Ohio, Oregon, Pennsylvania, Rhode Island, and Utah)

Inmates and parolees: 4 states (California, Colorado, Connecticut, and New York)

Inmates, parolees, and probationers: 18 states (Alaska, Arkansas, Georgia, Idaho, Kansas, Louisiana, Minnesota, Missouri, New Jersey, New Mexico, North Carolina, Oklahoma, South Carolina, South Dakota, Texas, Washington, West Virginia, and Wisconsin)

Inmates, parolees, probationers, and ex-felons: 12 states (Alabama, Arizona, Delaware, Florida, Iowa, Kentucky, Mississippi, Nebraska, Nevada, Tennessee, Virginia, and Wyoming).

Source: Chung (2017).

level of restriction. In 15 states, felons are only restricted from voting as long as they are in prison. However, approximately 77% of disenfranchised felons live in the community, either under probation or parole supervision or having completed their sentence entirely (Chung, 2017). Uggen, Larson, and Shannon (2016) reported that over 7% of African-American adults are disenfranchised, which is more than four times the rate of non-African Americans.

Many states automatically restore the felon's voting rights upon being eligible for restoration. Other states are much more restrictive, with some requiring the convicted felon to apply to the governor for a pardon. Several states have recently eased the restoration process. For example, Florida has traditionally had one of the most restrictive disenfranchisement laws. Until recently, felons seeking voting restoration were required to appear personally at a hearing before the governor. A new policy allows for the automatic restoration of voting rights for nonviolent offenders. Felons convicted of serious offenses such as murder or sex crimes must either petition the clemency board for an in-person hearing or wait 15 years (crime-free) to apply without a hearing (Porter, 2010).

In 2016, Virginia's governor issued an executive order restoring voting rights to felons who were living in the community and had completed their probation or parole supervision. However, the Virginia Supreme Court invalidated this order by ruling that clemency decisions can only be made on a case-by-case basis. Virginia's current governor has vowed to restore voting rights on an ongoing basis as felons complete their sentences. However, the continuation of this policy will be dependent upon the philosophy of future governors.

Those opposed to the imposition of collateral consequences of felony conviction argue restrictions may be counterproductive. Adding to penalties by limiting rights may cause ex-offenders to become bitter toward society in general and the justice system in particular. Restrictions regarding employment may be especially harmful. They limit an ex-offender's ability to lead a law-abiding life by closing opportunities for socially accepted means of earning a living. In addition, many restrictions on civil rights, such as restrictions on the voting rights of someone with a prior conviction for theft, are unrelated to the offender's crime. Finally, there are those who argue that the imposition of additional restrictions continuing after the sentence means that the offender can never pay his or her debt to society.

Those favoring the restriction of civil rights for people convicted of criminal offenses argue that such restrictions serve several purposes. First, rights are balanced by duties and, by failing to meet the duty of obeying the law, offenders have lost the privilege of exercising their rights (Vile, 1981). Others suggest that a "principle of least eligibility" applies, in that ex-offenders are least eligible of all citizens to be protected in their rights to vote, work, and so on (Simon, 1993:265–266). Others suggest that the fact of conviction evidences unacceptable character flaws, so that restricting offenders from holding public office, serving on juries, and voting actually protects law-abiding citizens from possible election fraud and malfeasance in office by ex-offenders. Other restrictions on rights are justified by proponents as pragmatic concessions. In many states, incarceration is grounds for divorce; it is justified by the argument that it is not fair to require the spouse to stand by an absent mate. Loss of parental rights has been justified because a felony conviction represents proof of being an unfit parent. Finally, civil death has been justified to enable family members to dispose of the convicted offender's debts and property.

RECIDIVISM

As a "business," the criminal justice system is not supposed to encourage repeat "customers." If the system operated at peak effectiveness, no one who committed a crime would commit a second offense. Unfortunately, the data indicate that the justice system produces many customers who return repeatedly. We call these repeat offenders recidivists.

Accurately measuring recidivism is key to understanding the operation of the criminal justice system. Throughout this text, we have cited research that has evaluated numerous crime-reduction techniques as well as the impact that certain societal factors might have on the success of these programs. Recidivism has been the most common outcome measure used to determine the success or failure of these programs. Unfortunately, there is no perfect way to measure recidivism.

Defining Recidivism

Perhaps no other concept in criminal justice has been as fully studied and debated as has recidivism. **Recidivism** is hard to define, but generally means a return to crime or other trouble with the criminal justice system. The notion of recidivism is so controversial that contemporary writers tend to use other labels for it, such as "failure" or "return." At base, recidivism means repetition of crime. The term is confusing because it is not exactly clear what "repetition" should include. For example, if a convicted robber released from custody now commits a theft, has the offender repeated? If the offender is arrested, but not convicted, is it recidivism? If the offender's parole is revoked for failing to report to the supervising officer, is the offender a recidivist? These seem to be technical distinctions, but they can be very important.

Whatever definition one adopts, the concept is crucial to evaluations of the effectiveness of criminal justice processing. Gottfredson, Mitchell-Herzfeld, and Flanagan (1982) suggested that the definition employed has an effect on the level of recidivism detected. For example, if recidivism includes technical violations of probation or parole rules, the rate of return to crime will be higher. Counting arrests yields higher recidivism rates than counting only convictions. Counting only returns to prison as repeat offenses leads to still lower rates of recidivism, and counting only repeated convictions for the same crime yields the lowest rates of all.

Harris, Lockwood, and Mengers (2009) argued that time is yet another important component of the definition. The longer the period over which we track offenders, the higher will be the total level of recidivism. For example, Jung, Spjeldnes, and Yamatani (2010) tracked the recidivism rate of jail inmates in a Pennsylvania county over a 3-year time frame. Within a year, more than 36% of inmates had been rearrested. At the end of the 3-year tracking period, nearly 60% of inmates had been rearrested for a new crime. This finding illustrates the importance of a researcher's follow-up period. A study that examines a particular program's recidivism rate should be measured against research of competing programs using the same follow-up period. Likewise, any examination of recidivism trends should use a standard tracking period.

Recidivism and Criminal Justice Policy

One problem with recidivism is that it is very difficult to measure. Knowledge of the effectiveness of correctional programs and judicial sentencing decisions based on the rate of return to crime can help us to design crime control policies. The fact that recidivism depends so heavily on what we count, and for how long, means that we must be especially careful in interpreting and using recidivism statistics as the basis for policy decisions.

A study of boot camp effectiveness illustrates this point. Duwe and Kerschner (2008) followed participants of a Minnesota boot camp for an average of more than 7 years, and compared their recidivism rates to a control group. At the end of the follow-up period, 62% of the boot camp participants had been rearrested. However, only 32% were reconvicted, and 22% were sentenced to jail or prison terms. Offenders participating in boot camps returned to prison at virtually the same rate as the control group. However, the boot camp offenders were more likely to return to prison owing to a technical violation as opposed to a new crime (Duwe & Kerschner, 2008). As a policymaker, what does this information tell you about the use of boot camps as a correctional tool?

Unfortunately, it tells us little. Paradoxically, it also tells us much. Ultimately, the decision about using boot camps as a punishment will depend on this and other information, and on the attitudes of the policymakers themselves. If they count new prison terms as recidivism, 78% of boot camp participants will succeed, and boot camps are effective for eligible offenders. If the policymakers count arrests as failures, 62% of boot camp offenders will fail, and boot camps may be an inappropriate disposition for offenders in Minnesota. What is important is that to make an informed judgment, the decision makers need to know not only the statistics (62% or 22% "failures"), but also what the statistics mean. Duwe and Kerschner (2008) calculated that the boot camp program has resulted in significant cost savings to the state of Minnesota. Even though the boot camp offenders returned to prison at the same rate as the control group, the members of the control group received a longer average sentence because they were more likely to return for a new offense as opposed to a technical violation.

In a study of drug courts, Mitchell, Wilson, Eggers, and Mackenzie (2012) reported that drug court participants generally have lower recidivism rates than nonparticipants. However, results varied by type of drug court and the measure of effectiveness. They identified three distinct categories of recidivism: general recidivism, drug-related recidivism, and drug use. Results showed that participants in adult and juvenile drug courts have significantly lower rates of general recidivism. However, drug-related recidivism was only lower for the participants of the adult court. This suggests that the effectiveness of juvenile drug courts might vary depending on the different types of recidivism. Lumping all the ways in which offenders may fail into a single category of recidivism may actually make it more difficult to understand how to intervene successfully in the lives of offenders.

DEVELOPMENTS IN CRIMINAL JUSTICE

Ritter (2006) interviewed three leading criminal justice experts to get their perspectives on what the criminal justice field will look like in 2040. All three experts agreed that global alliances would become increasingly important. Technology advances have allowed information sharing to increase across the globe, so that successful innovations occurring in one country could expand rapidly to other countries. As our country becomes more culturally diverse, the changing demographics also will help shape the operation and expectation of the criminal justice system. One expert pointed out that expected "aging" of the American populace might influence crime rates. Another expert predicted that the media would increasingly affect the public's expectations of the justice system. Finally, one of the experts suggested that the future of the criminal justice system would depend on whether we adopt an evidence-based model for the creation of new crime policies, or continue to base our policies on what seems to make sense without empirical evidence.

It is always risky to predict the future. As we have seen, criminal justice practices and policies are the products of a complex set of forces. Ideological shifts, economic changes, demographic variations, organizational goals, and the attitudes of individual agents and offenders all affect criminal justice. Attempting to predict how these factors will change and develop, and how they will interact to produce different decisions and patterns of criminal justice practice, is hazardous and prone to error. For example, it is unlikely that these experts would have predicted the emergence of Donald Trump's "America First" political philosophy and the impact that this might have on global cooperation. Nonetheless, it is incumbent on us to attempt to foresee the future so that we can try to shape it.

Looking at the future of criminal justice in the United States, we examine three trends and project possible futures. In doing so, we must pay attention to the

forces that gave rise to the trends, as well as to the powerful forces opposing any
real change in the operation of the criminal justice system. Hedging our bets, we
shall call these trends "possibilities." It should be remembered, of course, that it
is also possible that nothing much will change. The possibilities that we examine
here are increased federalism, drug policy changes, and technological justice.

Increased Federalism

We have described the American criminal justice system as fragmented, with
multiple sets of crime control apparatus operating in thousands of counties
and/or cities across the country. Traditionally, criminal justice and crime control policy have been local issues. America has not had a single national criminal justice system comprised of federal laws, police, courts, and corrections.
Rather, each state has its own criminal code, and most municipalities have
their own police and courts. During the past 70 years, however, the federal role
in criminal justice has expanded greatly. In the late 1960s, the U.S. Congress
passed the Omnibus Crime Control and Safe Streets Act, creating a federal
bureaucracy that influenced local justice policy and practice through the provision of federal aid. As Friel (2000:2) put it,

> Beginning with this Act and the Federal funds that then flowed to State
> and local criminal justice agencies, the Federal Government would
> become a major player in local crime control and justice policy. ... On
> top of the long litany of decisions by the Federal courts that have tem-
> pered almost every area of the justice system in the past 40 years, we
> now question whether there is too much federal involvement. Some
> argue for more, while others demand less.

During the last half-century, the federal government has provided funding,
training, and technical support to local criminal justice agencies on a variety of
topics. The debate over the role of the federal government in local justice policy
hinges on a disagreement about the motive for, and impact of, this assistance.
Those who believe the federal government is exerting too much influence over
local policy contend that this aid acts as an inducement, or bribe. Accepting
federal support requires accepting any "strings" attached to the support. In
essence, the federal government is paying local criminal justice agencies to
change their policies and practices.

For example, in 1998, the Office of Juvenile Justice and Delinquency Prevention began funding what it called the Juvenile Accountability Incentive Block
Grants (JAIBG) program. The program provides funding and other support to
local juvenile justice systems that agree to implement "accountability" efforts.

> Holding a juvenile offender 'accountable' in the juvenile justice system
> means that once the juvenile is determined to have committed law-
> violating behavior, by admission or adjudication, he or she is held respon-
> sible for the act through consequences of sanctions, imposed pursuant

to law, that are proportionate to the offense. Consequences or sanctions that are applied swiftly, surely, and consistently, and are graduated to provide appropriate and effective responses to varying levels of offense seriousness and offender chronicity work best in preventing, controlling, and reducing further law violations (Hurst, 1999).

To receive JAIBG funding, a state or local juvenile justice agency must demonstrate a plan to hold juveniles accountable in accordance with the JAIBG program's definitions and purposes.

Similarly, other federal assistance has been available for programs involving truth in sentencing, where federal support is available to states that propose to ensure that offenders convicted of violent crimes will serve at least 85% of their sentence in prison. Traditionally, these offenders would be incarcerated for one-half or less of their sentence. States received support for the construction of new prisons, or alternatives to prison for nonviolent offenders. They had to ensure, however, that violent offenders would serve 85% of their sentences in prison. There was also the federal initiative supporting community-oriented policing, in which the Office of Community Oriented Policing Services (COPS) disbursed funds to law enforcement agencies to support efforts to place an additional 100,000 police officers on the streets of the nation. Funding was available, over the life of the program, for hiring new officers, paying overtime to existing officers, and for other efforts (hiring civilians, implementing new technologies, etc.) aimed at making more time available for police officers to serve on the streets. Thousands of police agencies took advantage of these programs. Of course, it was not enough simply to increase the number of police officers on patrol. The increased personnel were required to be dedicated to "community policing." As of 2018, the COPS program is still operational and provides millions of dollars in funding to local police departments annually.

Critics view these funding and support programs as efforts by the federal government to standardize and direct local criminal justice practices. On the other side of the issue, however, are those who support and applaud this increased federal role. They argue that no one forces local officials to participate in the programs. Indeed, many jurisdictions did not apply to participate in each of these programs. Local justice officials are free to ignore offers of federal assistance that do not meet local needs, and to apply only for those that fit local priorities.

The impact of federal incentives on local operations is still unclear. Helms and Gutierrez (2007) reported that COPS funding is associated with important innovation and organizational change in large police agencies. Worrall and Kovandzic (2007), however, found that the influx of community policing dollars had little impact on rates of serious crime. It is also unclear how permanent any changes may be. Some cynics suggest that local agencies "follow the money" and accept federal funding without planning any lasting reforms. For example, in 2003, fewer than half of local police departments had a written plan concerning terrorist attacks (Hickman & Reaves, 2006). With the advent of the Department of Homeland Security and the availability of some federal support for counterterrorism efforts, local justice agencies began to purchase additional equipment and increase terrorism response training. However, the grants often prevented local agencies from spending the funds according to their individual needs (Eisinger, 2006).

Beginning in 2005, the federal Bureau of Justice Assistance began offering federal grant money through the Justice Assistance Grant (JAG) program, awarding more than $268.2 million in fiscal year 2016. The distribution of JAG funds is illustrated in Figure 15.1. The Bureau of Justice Statistics calculates each state's fund allocation based on its share of violent crime and the population. Funds are shared between state and local governments, to be used for one of seven purposes: (1) law enforcement, (2) prosecution and courts, (3) prevention and education, (4) corrections and community corrections, (5) drug treatment, (6) planning, evaluation, and technology, and (7) crime victim and witness programs (Cooper, 2013).

With the passage of the Adam Walsh Act in 2006, the federal government created a set of national standards for the registration and community notification of sex offenders. States that did not comply with the national standards by July 2011 faced a 10% reduction in their JAG funds. These requirements created many challenges for states. As of 2018, only 18 states have substantially implemented the new standards (SMART, 2018). Among the challenges facing

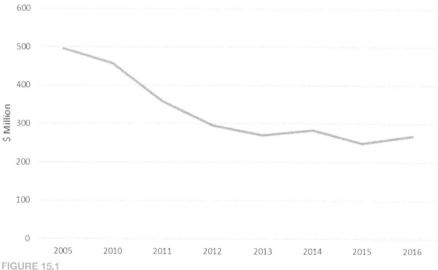

FIGURE 15.1
Allocation of Justice Assistance Grants, Selected Years.

Source: Cooper (2012a, 2012b, 2013, 2014, 2016); Cooper and Hyland (2015); Cooper and Reaves (2011); Hickman (2005).

states is the federal standard expanding sex offender registration to include many juvenile offenses. This requirement directly conflicts with many state's juvenile delinquency laws, and policymakers in these states have thus far been unwilling to implement a registration for juvenile sex offenders.

Beyond the provision of additional funds, the federal role in local justice policy and practice has also increased in two additional areas. First, the federal government has taken the lead in publicizing successful criminal justice practices by supporting a range of conferences, training, and information dissemination activities. One result of this has been to make the same information available to criminal justice officials all over the nation, creating a common knowledge of criminal justice. Second, the past 2 decades have seen greatly increased collaboration between federal and local agencies. Perhaps the most important criminal justice initiative of President George W. Bush's administration was "Project Safe Neighborhoods," which joins local and state police and prosecutors to crack down on gun crimes (Healy, 2002). And, of course, after the terrorist attacks of September 11, 2001, the provisions of the USA PATRIOT Act and the effort to prevent future acts of terrorism have combined to support increased centralization of terror prevention activities and broader information sharing between local, state, and federal law enforcement agencies (International Association of Chiefs of Police, 2004). In 2010, the Justice Reinvestment Initiative (JRI) was formed by a public–private partnership between the Department of Justice and the Pew Charitable Trusts. The initiative works directly with states to enact policies designed to safely reduce the prison population (Obama, 2017).

The past several decades have seen an increased role for the federal government in local criminal justice policy and practice. The provision of direct support for specific criminal justice policies has increased in recent years, and the federal courts have been involved in all aspects of the justice process. It remains to be seen what the ultimate impact of increased federal involvement may be, but at present, it seems certain that the federal role in local criminal justice is not likely to diminish in the near future.

Drug Policy

In earlier chapters, we discussed how changes in environmental pressures (e.g., public opinion, fiscal constraints) affects the criminal justice system. Perhaps the most contemporary illustration of this occurring has been in the ongoing war on drugs. It is important to note that most of the drugs that we now consider dangerous were once legal. Although some states enacted laws targeting certain types of drugs in the late 1800s, federal laws prohibiting drug use were virtually nonexistent until the early 1900s. In 1914, the Harrison Narcotics Act was passed as a way to regulate and tax the distribution of opium, morphine, and cocaine. The most significant federal drug law, the Comprehensive Drug Abuse and Control Act, was implemented in 1970. This Act remains the legal foundation of the federal government's war on drugs (Stolberg, 2009). Among other things, the law created five categories, or schedules, which each drug was to be placed into (see Box 15.3).

BOX 15.3 FEDERAL DRUG SCHEDULES

Schedule	Definition
Schedule I	The drug has a high potential for abuse The drug has no currently accepted medical use
Schedule II	The drug has a high potential for abuse The drug has a currently accepted medical use Abuse of the drug may lead to severe psychological or physical dependence
Schedule III	The drug has a potential for abuse less than the drugs in schedules I and II The drug has a currently accepted medical use Abuse of the drug may lead to moderate or low physical dependence or high psychological dependence
Schedule IV	The drug has a low potential for abuse relative to the drugs in schedule III The drug has a currently accepted medical use Abuse of the drug may lead to limited physical dependence or psychological dependence relative to the drugs in schedule III
Schedule V	The drug has a low potential for abuse relative to the drugs in schedule IV The drug has a currently accepted medical use Abuse of the drug may lead to limited physical dependency or psychological dependence relative to the drugs in schedule IV

As drug use increased in the 1970s and 1980s, more Americans began to see it as a serious problem (Robison, 2002a). Congress responded to these concerns by passing the Anti-Drug Abuse Act of 1988. This law significantly increased the penalties for drug abuse violations, allocated more federal dollars toward drug control efforts, and allowed for the denial of federal benefits (e.g., loans, mortgage guarantees) to certain drug offenders. Many scholars argue that this law, alongside the increasing concern regarding drug abuse in general, was largely responsible for the significant growth in incarceration rates from the 1980s until 2009. Currently, nearly half of federal prison inmates are incarcerated due to a drug offense violation (Carson, 2018).

The specific types of substances targeted by the war on drugs have evolved over time. During the time that the Anti-Drug Abuse Act was passed, the public's attention was largely focused on crack cocaine (Robison, 2002b; Vagins & McCurdy, 2006). Although cocaine usage has declined significantly from its peak, adolescent usage has risen over the past few years (Schneider, Krawczyk, Xuan, & Johnson, 2018). In recent years, however, the national focus has seemingly turned toward the opioid crisis (Blendon & Benson, 2018). The criminal justice system has responded by targeting "doctor shopping" and others who traffic prescription medication illegally. At the same time, the justice system has relied on public health agencies to increase monitoring of prescription drugs, improve physician-prescribing practices to reduce the overall supply of opioids, and treatment of those who are already addicted to these drugs (Compton, Boyle, & Wargo, 2015; Obama, 2017). The opioid crisis provides yet another example of how criminal justice is a system of interrelated parts which must work together to find effective solutions to our current problems.

While Americans seem increasingly concerned about the opioid crisis, there appears to be a softening of public animosity toward marijuana. In the 1970s, less than 20% of Americans supported legalizing marijuana. However, recent Gallup polls have showed increasing support for legalization. In 2017, the Gallup poll found that 64% of Americans support legalization of marijuana (McCarthy, 2017). Democrats and independents are historically more likely to support legalization, but the most recent poll shows that a majority of Republicans also now support legalizing marijuana. The reason behind this shift in public opinion is complex, and is likely at least partially due to the increased number of states that have voted to legalize marijuana for recreational or medicinal use.

In 1998, California and Washington were the first two states to legalize medical marijuana. A decade later, the number of states that had legalized medical marijuana had grown to 12. As of 2018, 29 states and the District of Columbia had legalized the medicinal use of marijuana. The specific laws vary considerably in terms of the amount of possession allowed, whether the patient is able to cultivate the marijuana from their home, the availability of dispensaries, and the approved conditions for which marijuana can be prescribed to treat.

These state-level initiatives are in direct conflict with federal law, which lists marijuana as a schedule 1 drug, subject to harsh punishments. However, the Justice Department has sent mixed messages regarding its enforcement of federal marijuana law in states that have legalized its use. The Obama administration generally took a hands-off approach to enforcement of such laws. The hands-off approach was reversed under the Trump administration when Attorney General Jeff Sessions issued a memo instructing federal prosecutors to apply federal law in these states. However, an amendment to the annual budget appropriations bill has prevented federal law enforcement officials from using federal funds to enforce such laws. While this amendment was still in place at the time of this writing, there is no guarantee that it will be continued in the future.

Critics of the movement to legalize medical marijuana fear that these laws will encourage drug abuse or increase crime rates. The existing research suggests that marijuana use is higher in states that have legalized medical marijuana (Cerdá, Wall, Keyes, Galea, & Hasin, 2012). One explanation could simply be that states with a higher level of preexisting marijuana usage are more likely to adopt legalization initiatives. Several studies have found that the legalization of medical marijuana does not in itself appear to increase drug use (Choo et al., 2014; Harper, Strumpf, & Kaufman, 2012). One study found that the effects of medical marijuana laws depend on the specifics of each law (Pacula, Powell, Heaton, & Sevigny, 2015). This study found that while medical marijuana laws in general were not associated with increased marijuana consumption, those states that allow dispensaries do experience an increased rate of marijuana use. Finally, Shepard and Blackley (2016) found no evidence that medical marijuana laws have led to an increase in violent or property crime.

The successful movement to legalize medical marijuana has led several states to take the additional step of legalizing marijuana for nonmedical usage. In 2012, Washington and Colorado became the first states to legalize recreational marijuana. As of 2018, nine states and the District of Columbia have adopted this policy. Thus far, the movement has largely been carried out by voter ballot initiatives instead of by the legislative process. In 2018, Vermont became the first state to approve recreational marijuana by the vote of the state legislature. Not surprisingly, recreational marijuana is much more controversial than medicinal use. Supporters and opponents of recreational marijuana advance multiple arguments to debate the merits of legalization (McGinty, Niederdeppe, Heley, & Barry, 2017). However, the public discourse has largely been limited to the states that have formally considered whether to legalize its use (McGinty et al., 2016).

Limited research is available to help determine the effects of these policy changes. Cerdá et al. (2017) compared the change in adolescent marijuana usage in Washington and Colorado following their vote to legalize recreational marijuana with states that had not voted to legalize recreational marijuana. They found that the legalization policy increased marijuana usage among adolescents in Washington. However, adolescents in Colorado did not report increased marijuana usage following legalization. It should be noted that the legalization of marijuana might affect surrounding states. For example, one report claimed that Colorado-based marijuana seized in the U.S. mail increased over 900% in the four years following the legalization of recreational marijuana (Rocky Mountain High Intensity Drug Trafficking Area, 2017). It is unclear whether the shipment of Colorado-based marijuana to other states will continue as the number of states with legalized marijuana continues to increase.

It is difficult to predict the future of drug policy in the United States. What is clear is that the public is much more accepting of marijuana than at any time in the recent past. While marijuana is still illegal at the federal level, the shift in public support will likely constrain the ability of federal law enforcement officers to enforce these laws. Thus far, the public debate regarding legalization has been focused nearly exclusively on marijuana. It seems unlikely that this discussion will expand to other drugs in the near future. However, as recently as 30 years ago it seemed unlikely that marijuana would have received such broad support for legalization. See Figure 15.2.

FIGURE 15.2
A customer shops for marijuana at The Station, a retail and medical cannabis dispensary in Boulder, Colorado.

AP Photo/Brennan Linsley

Technological Justice

In all aspects of life, technological changes have caused adjustments in behavior. We take for granted such advances as household electricity, indoor plumbing, telephones, automobiles, and other commonplace technological conveniences. Today, most Americans have a mobile phone capable of sending text messages, taking photographs, and connecting to the internet. Yet, until fairly recently, these technologies were unavailable to large segments of our population. The effects of technological advances on criminal justice have been staggering, as they have been on other areas of life. The telephone and the automobile changed the nature of policing. No longer did the neighborhood cop walk a beat and know everyone. Now a simple telephone call to an impersonal dispatcher results in the arrival, by automobile, of an equally impersonal police officer. In the past, offenders could simply give the police an alias or cross a state line to avoid prosecution. Today, computerized criminal records and fingerprint checks can lead to quick identifications. A simple traffic stop may lead to a felony arrest when a computer check yields outstanding warrants.

Crime records and court documents that once filled storage warehouses are now stored on computer files. The availability of electronic databases and powerful search engines has made the police, courts, and correctional agencies more effective in identifying and tracking known and/or wanted offenders. Television, videotapes, and computer simulations can be used to great effect in investigations and at trial to present and test evidence and to reduce delays and costs. Teleconferencing can be used to replace on-site hearings for parole or probation revocation, or even for parole release hearings; this advance reduces costs and increases efficiency.

Scientific advances (including genetic testing) have been adapted to criminal justice uses. Currently, investigators use DNA "fingerprinting" to help identify offenders conclusively, as well as to help identify innocent suspects (Rau, 1991). DNA evidence has been used in trials in most states. In 2018, California authorities identified the notorious "Golden State Killer" by sending DNA that had been found at the crime scene to a popular genealogy website, allowing police to narrow their search down to a single family (Selk, 2018). Other investigatory aids include computer-generated offender profiles to assist police in identifying likely suspects, and a variety of drug detection tests to both locate contraband and establish drug use among criminal suspects.

Modern prisons and jails are making use of a variety of technological innovations to improve custody and services. Closed-circuit television, automatic locking devices, and careful planning have reduced the need for custodial staff in prisons and jails. Joseph (2017) described cutting-edge correctional technologies including heartbeat monitors that can detect the heartbeat of those hiding in cars or cargo. Some states are experimenting with eye scanning or facial recognition to ensure the inmates' identity (Joseph, 2017). Automated video surveillance has enhanced the ability of correctional personnel to monitor inmate behavior and maintain prison safety (Turner, 2007). Prisons can also monitor inmates by fitting them with a wristband or ankle bracelet equipped with active radio frequency software, similar to the systems widely used by community corrections departments (Hickman, Davis, Wells, & Eisman, 2010). This technology allows for real-time monitoring of inmates from a central location and is currently in use by several U.S. correctional facilities (see Box 15.4).

BOX 15.4 EXPECTED BENEFITS FOR RADIO FREQUENCY IDENTIFICATION SYSTEMS IN CORRECTIONAL ENVIRONMENTS

- Improved monitoring and control of inmates
- Reduction in violence and injuries
- Reduction in escapes
- Improved investigative capabilities

- Reduction in inmate grievances, lawsuits, and disciplinary actions.

Source: Hickman, Davis, Wells, and Eisman (2010).

With technological development, the justice system now is able to detect, apprehend, and process offenders more quickly and economically. The technologies have allowed us to conduct business as usual but in a more efficient manner. The limited research available in the United States suggests that the use of surveillance cameras may have a modest effect in reducing some types of crime (Lim & Wilcox, 2017; McLean, Worden, & Kim, 2013; Welsh, Farrington, & Taheri, 2015). However, these cameras are often put in place without appropriate monitoring or the capabilities for police to quickly respond to situations that are discovered by the cameras. Piza, Caplan, Kennedy, and Gilchrist (2015) found that violent crime and social disorder were significantly reduced when video surveillance was combined with police units who were dedicated to responding to incidents uncovered using the cameras.

Advances in medical science have also affected crime. Giacopassi, Sparger, and Stein (1992) note that improved emergency medical care may have suppressed the homicide rate. Many of today's felonious assaults would have been homicides but for the fact that medical care has been improved. Saving the life of the crime victim reduces the seriousness of the crime. Indeed, Hanke and Gundlach (1995) suggested that differences in access to quality emergency medical care may explain the observed differences in seriousness of crime between Caucasian and African-American offenders. They provide evidence that lack of access to quality emergency medical care in minority, inner-city neighborhoods may help account for higher rates of homicide and more serious injuries to victims of violent crime. Other technologies have arisen that may alter the conduct of criminal justice business in more significant ways.

New technologies have affected the delicate balance between the rights of individuals and those of the community. For example, there has been an increased interest in the use of license plate-reading cameras. Mounted on patrol cars or fixed locations, license plate readers can capture images of every passing license plate. The readers are increasingly being used to identify stolen cars, unregistered vehicles, and those with stolen license plates (Baker, 2011; Willis, Koper, & Lum, 2018). In essence, the license plate readers have the capability to track the location of a vehicle's movement over a period of time. Civil liberties groups have expressed concern that few protections are in place to prevent the license plate readers' use in abusive or discriminatory targeting based on an individual's religion or ethnicity (ACLU, 2013). The Department of Homeland Security has expressed interest in creating a national database that would compile the information from license plate readers across the country (Nakashima & Hicks, 2014). The future of crime seems more easily predictable than does the outcome of the balance of state and individual interests (see Box 15.5).

BOX 15.5 TECHNOLOGY USE BY LAW ENFORCEMENT

- January 14, 2014, a North Dakota farmer was convicted in an incident involving his failure to return three cows from a neighboring property that had strayed onto his property (Peck, 2014). This case is significant not because of the crime committed, but because the conviction represents the first time in American history that a U.S. citizen had been sentenced to prison with the assistance of an unmanned drone (without a search warrant). Some commenters argue that the use of unmanned drones is an appropriate and desirable tool for domestic law enforcement (Straub, 2014). However, privacy advocates suggest that strict regulations be placed on the use of these devices (Stanley & Crump, 2011). Drones can carry sophisticated surveillance technologies, such as:

- High-power zoom lenses

- Night vision

- See-through imaging

- Video analytics with capabilities to recognize specific people, events, and objects.

- On July 6, 2016, Dallas police officers attached a bomb to the extension arm of a robot in order to kill a suspect who had murdered five police officers and wounded several others at an otherwise peaceful protest. The robot was used after the suspect barricaded himself inside a motel room and negotiations between officers and the suspect had broken down. This was the first time that a robot had been used in a show of lethal force against a criminal suspect (Thielman, 2016). Although most observers have noted that the police likely had no other way to apprehend the suspect without putting officers' lives at risk, the move has raised new questions about how to handle similar situations in the future.

Technology has also affected the types of crimes that will be the focus of criminal justice efforts. Johnson (1981) explained that changes in police organization and operation were partly owing to the changes in the nature and extent of crime. Similarly, Lundman (1980) explained the development of police, in part as a result of changes in the rates and images of crime. That is, the police (and, by extension, the entire criminal justice system) must be responsive to those acts that we view as seriously threatening public order and safety. Technological change also changes crime.

Willie Sutton, a famous bank robber, is credited with a witty response to a question about his actions. When asked why he committed bank robberies, Sutton is said to have replied, "Because that's where the money is." If we look at crimes throughout American history, we see that, very often, they take place "where the money is." Criminals such as Jesse James robbed stagecoaches because these coaches carried large sums of money (corporate payments, payrolls, etc.). These men became train robbers when cash began to be transported by rail. Bank robberies became more common when checks and fund transfers replaced cash transactions. Today's offender might well be described as a "lineman," that is, he or she often steals by computer. Fraud with computerized banking, already a problem, is likely to become more significant in the future. Counterfeiting is on the rise as more and more people gain access to sophisticated copiers, printers, and digital printing resources (Morris, Copes, & Perry-Mullis, 2009).

Not only is money itself at risk in an increasingly computerized commercial system, but information is also more often the target of theft. Trade secrets, customer lists, business records, and all sorts of other information (including national defense secrets) are stored in computer databases. Theft of such information often is more damaging (and more profitable for the thief) than fraudulent fund transfers. Detection of computer crime (and enforcement of laws against it) will require a different type of law enforcement response than that required by traditional street crime. The systems analyst at the terminal may replace the officer in the patrol car. This does not mean that street crime will disappear; it means that computer-related offenses will increase in frequency and importance, which will lead to justice system adaptations to combat them.

A survey of law enforcement officials by Burns, Whitworth, and Thompson (2004) revealed that while most believe internet fraud was being investigated by police at all levels, it would be better pursued by federal agencies. A lack of resources and inadequate organizational structure hamper local law enforcement efforts to respond to internet fraud and similar computer crimes. We can only expect that all sorts of computer crime will increase in the future. There is great interest in **identity theft**, the theft or misuse of personal information, similar to fraud or forgery (Table 15.1). Common forms of the crime involve "posing" as someone else to use their credit cards or access their financial records or accounts. The federal government and some states have taken repeated steps to enhance penalties for identity theft in hopes of deterring offenders (Copes & Vieraitis, 2009).

TABLE 15.1 Prevalence of Identity Theft, 2014		
Type of Identity Theft	Number of Victims	Percent of All Persons
Total	17,576,200	7%
Existing credit card	8,598,600	3.4
Existing bank card	8,082,600	3.2
Other existing account	1,452,300	0.6
New account	1,077,100	0.4
Personal information	713,000	0.3

Source: Harrell (2017).

Technology affects the types of crimes that can be (and are) committed. In response, the qualifications of justice system officials, the nature of cases in the criminal justice process, and responses to offenders will change. There is little doubt that, similar to the shift from stagecoach robbery to bank robbery, the practices of criminal justice will be required to adapt to high-technology crimes of computer fraud. These changes, however, do not appear likely to change the criminal justice system in any fundamental way.

Computer crimes may pit justice system investigators skilled in system analysis against offenders skilled in programming. Evidence at trials may come to be comprised of documents and disks more often than of hairs and weapons, but the process itself can continue in its present form. In this case, the type of crime and criminal may be different, but the basic response to crime remains. Schroeder and White (2009) reported that DNA evidence was used rarely in homicide investigations in New York City, largely because it was not needed. Likewise, one recent study showed that the delay in receiving ballistics imaging reports forced investigations to proceed as normal without the benefit of such analysis (King, Campbell, Matusiak, & Katz, 2017). Other effects of technology, however, pose a greater likelihood of altering the way in which criminal justice is accomplished.

In addition to the opportunities for crime that computers offer, computerized record keeping can aid the efforts of criminal justice officials. Tremendous amounts of information about almost every citizen are kept somewhere on computer records. If a suspected offender has ever applied for credit, or if he or she operates a motor vehicle, police can garner much information, quickly and easily, by requesting computer records. Credit applications generate data about employment, earnings, savings, debts, addresses, dependents, references, and demographic characteristics. Access to a person's credit report can provide a tremendous amount of personal information about him or her. Operation of a motor vehicle provides similar background data, important identifying numbers (e.g., Social Security numbers), information about possible criminal record, and even height and eye color. It is possible to conduct a relatively thorough background check on the average citizen without ever leaving a computer.

Credit card purchase records allow investigators to track travel patterns of individuals. Credit card purchasing and payment information exists on computer records that do not require a check on access (other than perhaps knowledge of a correct password). This means that personal data are potentially available to anyone. Consider what information about your private life someone could learn with access to computer records. If someone had unlimited access, could that person discover your name, address, telephone number, income, age, sex, race, general whereabouts over the past month, courses you are taking, and your grade point average? How difficult would it be for such a person to identify your friends, family, or taste in clothes or music? Even though much of this information is freely available on various social media sites, we would consider much of the available information to be personal. Our right to be protected from unreasonable "searches and seizures" is increasingly at risk, and we are frequently unaware of the risk. The sheer availability of information about suspects, witnesses, and offenders could lead to increased surveillance by the justice system. There is no doubt that much of this information would prove useful in not only criminal investigations, but also for what is called "intelligence gathering."

Intelligence gathering relates to the compilation of information that may be helpful in solving crimes, but with no specific probable cause to believe the information will lead to solving a particular crime. As the emphasis on proactive, preventive efforts of criminal justice agencies increased, justice agencies compiled larger and more diverse sets of information that could later be used to solve crime. After the 9/11 attacks, the National Security Agency (NSA) collected Americans' phone records in bulk without any suspicion of wrongdoing. This program came under scrutiny after Edward Snowden disclosed its existence in 2013. In 2015, Congress restricted (but did not eliminate) the NSA's ability to conduct this type of mass surveillance. Relatedly, the Supreme Court recently ruled that police must obtain a search warrant to access location data of cell phone users (*Carpenter v. United States*, 2018). The struggle to find an appropriate balance between privacy and security in today's digital age will surely continue into the foreseeable future.

Crime mapping and geographic analysis have become important planning tools for law enforcement agencies (Santos, 2017), and their use has spread to courts and to correctional agencies (Harries, 2003). In most applications, the criminal justice analyst ties a variety of data sources to a known location and then seeks relationships that help explain the presence or absence of crime to develop policy directed at reducing the likelihood of crime or other problems. Most contemporary developments in policing (e.g., intelligence-led policing, COMPSTAT), rely on the gathering of intelligence to assist in the determination of resource allocation. Crime reports and computer-generated patterns of police problems guide the deployment of police resources.

Previous editions of this text predicted that the use of risk and needs assessments would grow. Indeed, risk assessment tools have continued to evolve and are now commonly used in bail, sentencing, prison classification, and parole decisions. It is now safe to say that an offender's experience with the criminal justice system is largely determined with reference to risk and needs assessment tools. The problem posed by this new decision-making technology for criminal justice revolves around the question of who makes the decisions. To the extent that all these classification and prediction technologies provide information and guidance to decision makers, they do not significantly alter justice system processing. These devices simply become another factor in the equation that produces criminal justice outcomes. The problem is that these technologies may actually alter the decisions themselves.

For sentencing decisions, guidelines are often stifling. Convicted offenders receive predetermined sentences because of factors that often are beyond their control. When prior criminal record influences criminal sentences there is nothing the offender can do to alter his or her sentence. The prior record never gets better. When one considers that parole and sentencing guidelines are often based on past experience (what kinds of sentences were imposed in the past), the problem is more apparent. Arguably, one reason for the development of these models is to improve decision making, but the models are based on the very decisions they are expected to improve (Gottfredson, Cosgrove, Wilkins, Wallerstein, & Rauh, 1978).

The tendency to adopt decision-making aids will probably continue and spread throughout the justice system. Although courts and corrections personnel have typically used risk assessment tools, some have suggested that these tools could be used more broadly. For example, police departments could potentially use risk assessment tools to help guide and audit police officer decisions (Goel, Rao, & Shroff, 2016) or identify police officers who are at risk of engaging in misconduct (Worden, Harris, & McLean, 2014). Further, we can expect that such guidelines will further strengthen the justice system's ability to resist change. Finally, greater individualization in terms of treatment may occur as an increased variety of rehabilitation options become available.

Technology and Criminal Justice in the Total System

Technological advances in all areas of society continue at an accelerated pace. The effects of these changes for society are both positive and negative. Increased communication capability, for example, applies to both good news and bad. Having voicemail means the phone will be answered, but the chance of speaking with a living person is greatly decreased. Changes in technology affect the environment of the criminal justice system, and thus we can expect them to produce changes in criminal justice processing.

As illustrated in the previous discussion, most of the effects of technological change have occurred in the material environment. Types of crimes and methods of detection and investigation have changed. The relative frequency of homicide as compared to assault, or white-collar crimes as compared with street crimes, may have changed. Similarly, the characteristics of criminal justice personnel, and the equipment they need to do their jobs, may change in response to technological innovation.

The impact of technology on the operations of criminal justice indicates the open-system nature of the justice process. These changes occur in the means of criminal justice operations but probably have little impact on the ends of processing. Other technological advances, such as decision-making aids, are more subtle, but may have more far-reaching effects because they alter the ends of criminal justice operations. To the extent that information technology assists criminal justice actors to achieve effective and fair decisions, the impact of technological change in the environment will be positive. To the degree that these alterations produce unintended consequences, such as institutionalizing discrimination, their effects may be negative. Whatever the effect on the means and ends of criminal justice, one factor to be considered in assessing the future of criminal justice remains the future environment of the system.

CRIMINAL JUSTICE: MOVING FORWARD

What will the structure and practice of criminal justice in the United States be like as the twenty-first century progresses? If current developments continue on their present course, we can expect some changes. Three related developments promise to alter the fundamental nature of American criminal justice from its most recent tradition. There is a growing emphasis on crime prevention and rehabilitation (Weisburd, Farrington, & Gill, 2017) resulting in an expansion of the role of criminal justice in our society. We increasingly hold the justice system accountable for a broader quality of life than simply crime control. In part, this expansion of the role of criminal justice is a function of information suggesting that if we wish to reduce crime, we must attack problems of social order. Focusing on problems of social order has meant that the community has become an identified client and partner in criminal justice policymaking. Concern about the costs of justice system operation has risen at the same time that the criminal justice system is experiencing role expansion. This has meant that cost considerations have gained new prominence in criminal justice policy. Thus, concern about social order beyond crime has produced an expanded role for agents of the criminal justice system. This expanded role has increased costs, but also opened new avenues of funding for criminal justice operations. Finally, current assessments of criminal justice practices place greater emphasis on effectiveness than on democracy. In terms of our themes, contemporary criminal justice practices stress crime control over due process more than in the recent past.

Throughout the book we have explored community policing, community prosecution, community courts, and community corrections. All of these developments assume the existence of an identifiable community. What happens when there are no (or few) agreed-on community values? What happens if some members of the community are concerned about marijuana, but others are not? To whom do justice system officials listen?

The tradition in American criminal justice has been to let the community decide what issues were important and communicate this by their complaints to the police. Increasingly, justice system agents are becoming involved in community-organizing activities in which they bring neighborhood residents together and help them to define problems and select solutions. To the extent that it is criminal justice agents who create these communities, the justice system is being proactive. A proactive justice system, you will recall, is associated with a less democratic society.

People, in general, may be more willing to use the law—and the criminal law in particular—to resolve disputes and conflicts (Miethe, 1995). Therefore, the range of issues that reach the attention of police, court, and correctional authorities has expanded. So, too, has the range of solutions. Justice system officials increasingly employ noncriminal justice solutions to problems of crime and disorder. Prosecutors use civil court processes such as eviction and license revocation to remove offenders from communities, and to close problem businesses (Finn, 1995). "Zero-tolerance" policing, which seeks to minimize disorderly behavior, can produce negative police–community relations, as some parts of the community experience repressive policing (National Institute of Justice, 2000). Asset forfeiture, in which justice agents seize the property of criminal offenders, provides a profit motive for criminal justice and may alter the selection of enforcement targets and the establishment of enforcement priorities in justice agencies (Rothschild & Block, 2016). Rather than working for us, justice agents may be working for themselves.

The emphasis on crime prevention that is characteristic of recent changes in criminal justice operations is qualitatively different from an emphasis on crime control. In crime control, the justice system can be reactive and respond to instances of crime or potential crime. The system is effective to the extent that criminals are arrested and processed. A potential benefit may be that crime control efforts have preventive effects, such as deterrence and education. The emphasis on crime prevention, however, means that the justice system works to stop crimes that might happen, or that are predicted to happen. The difference is that these crimes have not yet occurred, and the offenders who might commit these crimes have yet to do anything wrong. What is the basis on which justice system authorities intervene in the lives of citizens who have not yet committed a crime?

At present, the balance between due process and crime control seems to have shifted in the direction of a greater emphasis on crime control. This can occur only at a cost to due process. In the foreseeable future, at least, we can expect increased criminal justice system efforts at crime prevention, accompanied by less concern about individual liberty and due process. We can expect that at some time in the future the balance will shift again to give primacy to individual liberty, due process, and the control of government power.

Criminal Justice Theory Revisited

How can we explain criminal justice? As we explored the criminal justice process in the United States, we reviewed a great deal of the research (and theory) about how the American criminal justice process works. We saw that some things influence all criminal justice decisions. Police, court, and correctional decisions reflect the seriousness of the offense and, whenever applicable, the availability of evidence. Beyond this, a number of other factors also prove to be influential in these decisions. The characteristics of criminal justice organizations, including size, rules, and supervision of personnel, seem to constrain decisions. The characteristics of individual decision makers (police officers, prosecutors, judges, etc.) are also important.

We can think of a number of theories or classes of theories of criminal justice. There is quite a bit of solid evidence for a legal theory of criminal justice. The law, both substantive and procedural, influences the operations of the criminal justice system. One explanation for what happens in criminal justice can be found in looking at the law.

There is also evidence for organizational theories of criminal justice. Police decisions vary, in part, based on the police department. Large police departments and departments with certain kinds of rules and supervision produce different outcomes than other departments. Courts and correctional agencies also reflect organizational characteristics including size, rules, and traditions.

The research supports group theories as well. In courts, in particular, the evidence indicates that group processes and shared understandings among prosecutors, defense attorneys, and judges explain rates and types of guilty pleas and sentences. Within police departments, different squads or units may behave differently. Probation and parole supervision outcomes reflect group beliefs shared among officers working in a particular office.

Individual characteristics of criminal justice officials also influence decisions. It seems the characteristics of individual police officers, prosecutors, judges, and the like are important, but that individual choices tend to be limited by legal, organizational, and group influences. The more independent the criminal justice official, the more likely the official's characteristics will explain decisions.

A large number of factors influence or explain criminal justice decisions and processes. It is probably most accurate to recognize that there are multiple "causes" of criminal justice outcomes. The most accurate theories of criminal justice are likely to be those that recognize that legal, organizational, group, and individual factors combine, often with other influences, to explain criminal justice outcomes. It is incumbent on us to work to develop and understand theories of criminal justice so we can refine and strengthen our understanding of the system.

CRIMINAL JUSTICE: A FINAL THOUGHT

The term "criminal justice" is paradoxical. Does it mean that justice is criminal, or that crime is just? We now know that it refers to how we define, detect, and react to behaviors that we deem criminal. We should also have an appreciation for how complex a topic it really is.

We have seen how present practices reflect various mixes of historical, political, economic, social, philosophical, and individual traits and factors. We can appreciate how the justice system, in balancing these many demands, is highly resistant to reform. There are so many places where reforms can be made, and so many levels at which reform can be stymied, that the justice process seems immune to change. Still, we have seen that changes have indeed occurred in criminal justice over the years. Some changes have been more fundamental than others. Some appear to have been more long lasting. Others are not yet complete.

Through our examination of criminal justice, we have come to see that no single purpose does (or can) predominate the system. The central dilemma is that of controlling behavior in a free society. The tension is between individual liberty and the need for an orderly and predictable society. The pendulum shifts over the decades from an emphasis on one to an emphasis on the other. The result is the appearance that no change has occurred. Criminal justice seems to proceed in a circular fashion, continually returning to earlier points. The effect is enough to make us despair in our hopes to achieve progress, but that is not the intent of the study of criminal justice.

An understanding of criminal justice promotes an understanding of our society and culture. Similarly, it is not possible to grasp the intricacies of the justice system without understanding its larger context. The failure of past reforms often can be traced to either or both of two mistakes. First, we must be reasonable in our expectations for change. It is probably not possible to eliminate injustice or inefficiency, and it may not even be desirable to do so. Second, change in the justice process is accomplished only by thorough planning and careful execution. As a system (or collection of separate systems), criminal justice is elastic and resists alteration. The would-be reformer must anticipate and prepare for reactions to change.

One thing that we can predict with confidence is that there will be a criminal justice process in the future. It is incumbent on us to try to understand it, and to work to improve it.

REVIEW QUESTIONS

1. Identify two ways in which an offender can receive a discharge from criminal justice custody.

2. What are collateral consequences of conviction?

3. Describe a process by which an ex-offender may have his or her constitutional rights restored.

4. What are some of the problems in arriving at a definition of recidivism?

5. What are the implications of the concept of recidivism for criminal justice policy?

6. Many states are legalizing the use of marijuana for medical use. Describe the impact, if any, that these new laws have had on the crime rate or overall drug use.

7. Has the justice system moved more toward an emphasis on crime control than due process?

REFERENCES

Agan, A., & Starr, S.B. (2017). The effect of criminal records on access to employment. *American Economic Review, 107*(5), 560–564.

American Civil Liberties Union (ACLU) (2013). *You are being tracked: How license plate readers are being used to record Americans' movements.* Retrieved from www.aclu.org/files/assets/071613-aclu-alprreport-opt-v05.pdf

Baker, A. (2011, April 11). Camera scans of car plates are reshaping police inquires. *The New York Times.* Retrieved from www.nytimes.com/2011/04/12/nyregion/12plates.html?pagewanted=all

Blendon, R.J., & Benson, J.M. (2018). The public and the opioid-abuse epidemic. *New England Journal of Medicine, 378,* 407–411.

Blumstein, A., & Nakamura, K. (2009). Redemption in the presence of widespread criminal background checks. *Criminology, 47*(2), 327–360.

Buckler, K., & Travis, L. (2003). Reanalyzing the prevalence and social context of collateral consequence statutes. *Journal of Criminal Justice, 31*(5), 435–453.

Burns, R., Whitworth, K., & Thompson, C. (2004). Assessing law enforcement preparedness to address Internet fraud. *Journal of Criminal Justice, 32*(6), 477–493.

Burton, V., Cullen, F., & Travis, L.F., III. (1987). The collateral consequences of a felony conviction: A national study of state statutes. *Federal Probation, 51*(3), 52–60.

Carson, E.A. (2018). *Prisoners in 2016.* Washington, DC: Bureau of Justice Statistics.

Cerdá, M., Wall, M., Keyes, K.M., Galea, S., & Hasin, D. (2012). Medical marijuana laws in 50 states: Investigating the relationship between state legislation of medical marijuana and marijuana use, abuse and dependence. *Drug and Alcohol Dependence, 120,* 22–27.

Cerdá, M., Wall, M., Feng, T., Keyes, K.M., Sarvet, A., Schulenberg, J., et al. (2017). Association of state recreational marijuana laws with adolescent marijuana use. *JAMA Pediatrics, 171*(2), 142–149.

Choo, E.K., Benz, M., Zaller, N., Warren, O., Rising, K.L., & McConnell, K.J. (2014). The impact of state medical marijuana legislation on adolescent marijuana use. *Journal of Adolescent Health, 55,* 160–166.

Chung, J. (2017). *Felony disenfranchisement: A primer.* Washington, DC: The Sentencing Project.

Compton, W.M., Boyle, M., & Wargo, E. (2015). Prescription opioid abuse: Problems and responses. *Preventive Medicine, 80,* 5–9.

Cooper, A.D. (2012a). *Justice Assistance Grant (JAG) program, 2011.* Washington, DC: Bureau of Justice Statistics.

Cooper, A.D. (2012b). *Justice Assistance Grant (JAG) program, 2012.* Washington, DC: Bureau of Justice Statistics.

Cooper, A.D. (2013). *Justice Assistance Grant (JAG) program, 2013.* Washington, DC: Bureau of Justice Statistics.

Cooper, A.D. (2014). *Justice Assistance Grant (JAG) program, 2014.* Washington, DC: Bureau of Justice Statistics.

Cooper, A.D. (2016). *Justice Assistance Grant (JAG) program, 2016.* Washington, DC: Bureau of Justice Statistics.

Cooper, A.D., & Reaves, B.A. (2011). *Justice Assistance Grant (JAG) program 2010.* Washington, DC: Bureau of Justice Statistics.

Cooper, A.D., & Hyland, S.S. (2015). *Justice Assistance Grant (JAG) program, 2015.* Washington, DC: Bureau of Justice Statistics.

Copes, H., & Vieraitis, L. (2009). Bounded rationality of identity thieves: Using offender-based research to inform policy. *Criminology & Public Policy, 8*(2), 237–262.

Davidenas, J. (1983). The professional license: An ex-offender's illusion. *Criminal Justice Journal, 7*(1), 61–96.

Denver, M., Pickett, J.T., & Bushway, S.D. (2018). Criminal records and employment: A survey of experiences and attitudes in the United States. *Justice Quarterly, 35*(4), 584–613.

Duwe, G., & Kerschner, D. (2008). Removing a nail from the boot camp coffin: An outcome evaluation of Minnesota's challenge incarceration program. *Crime & Delinquency*, *54*(4), 614–643.

Eisinger, P. (2006). Imperfect federalism: The intergovernmental partnership for homeland security. *Public Administration Review*, *66*(4), 537–545.

Finn, P. (1995). *The Manhattan District Attorney's Narcotics Eviction Program*. Washington, DC: National Institute of Justice.

Friel, C. (2000). A century of boundary changes. In Friel, C. (Ed.), *Criminal justice 2000: Boundary changes in criminal justice organizations: Vol. 2*. (pp. 1–17). Washington, DC: National Institute of Justice.

Giacopassi, D., Sparger, J., & Stein, P. (1992). The effects of emergency medical care on the homicide rate: Some additional evidence. *Journal of Criminal Justice*, *20*(3), 249–259.

Goel, S., Rao, J.M., & Shroff, R. (2016). Personalized risk assessments in the criminal justice system. *American Economic Review*, *106*(5), 119–123.

Gottfredson, D., Cosgrove, C., Wilkins, L., Wallerstein, J., & Rauh, C. (1978). *Classification for parole decision policy*. Washington, DC: U.S. Government Printing Office.

Gottfredson, M., Mitchell-Herzfeld, S., & Flanagan, T. (1982). Another look at the effectiveness of parole supervision. *Journal of Research in Crime and Delinquency*, *18*(2), 277–298.

Hanke, P., & Gundlach, J. (1995). Damned on arrival: A preliminary study of the relationship between homicide, emergency medical care, and race. *Journal of Criminal Justice*, *23*(4), 313–323.

Harper, S., Strumpf, E.C., & Kaufman, J. (2012). Do medical marijuana laws increase marijuana use? Replication study and extension. *Annals of Epidemiology*, *22*(3), 207–212.

Harrell, E. (2017). *Victims of identity theft, 2014*. Washington, DC: Bureau of Justice Statistics.

Harries, K. (2003). Using geographic analysis in probation and parole. *National Institute of Justice Journal*, *249*, 32–33 (July).

Harris, P.W., Lockwood, B., & Mengers, L. (2009). *A CJCA white paper: Defining and measuring recidivism*. Braintree, MA: Council of Juvenile Correctional Administrators.

Healy, G. (2002). *Policy analysis: "There goes the neighborhood: The Bush–Ashcroft Plan to 'help' localities fight gun crime"*. Washington, DC: CATO Institute.

Helms, R., & Gutierrez, R. (2007). Federal subsidies and evidence of progressive change: A quantitative assessment of the effects of targeted grants on manpower and innovation in large U.S. police agencies. *Police Quarterly*, *10*(1), 87–107.

Hickman, L.J., Davis, L.M., Wells, E., & Eisman, M. (2010). *Tracking inmates and locating staff with active radio-frequency identification (RFID): Early lessons learned in one U.S. correctional facility*. Santa Monica, CA: RAND.

Hickman, M. (2005). *Justice Assistance Grant (JAG) program, 2005*. Washington, DC: Bureau of Justice Statistics.

Hickman, M., & Reaves, B. (2006). *Local police departments, 2003*. Washington, DC: Bureau of Justice Statistics.

Hughes, L., & Burchfield, K. (2008). Sex offender residence restrictions in Chicago: An environmental injustice? *Justice Quarterly*, *25*(1), 647–673.

Hughes, L., & Kadleck, C. (2008). Sex offender community notification and community stratification. *Justice Quarterly*, *25*(3), 469–495.

Hurst, H. (1999). *Workload measurement for juvenile justice system personnel: Practices and needs.* Washington, DC: Office of Juvenile Justice and Delinquency Prevention.

International Association of Chiefs of Police (2004). *Executive summary: National criminal intelligence sharing plan.* Gaithersburg, MD: International Association of Chiefs of Police.

Johnson, D. (1981). *American law enforcement: A history.* St. Louis, MO: Forum Press.

Joseph, J. (2017). Technoprison: Technology and prisons. In Moriarty, L. (Ed.), *Criminal justice technology in the 21st century* (3rd ed.). Springfield, IL: Charles C. Thomas.

Jung, H., Spjeldnes, S., & Yamatani, H. (2010). Recidivism and survival time: Racial disparity among jail ex-inmates. *Social Work Research*, *34*(3), 181–189.

Kaeble, D. (2018). *Probation and parole in the United States, 2016.* Washington, DC: Bureau of Justice Statistics.

King, W.R., Campbell, B.A., Matusiak, M.C., & Katz, C.M. (2017). Forensic evidence and criminal investigations: The impact of ballistics information on the investigation of violent crime in nine cities. *Journal of Forensic Sciences*, *62*(4), 874–880.

Lim, H., & Wilcox, P. (2017). Crime-reduction effects of open-street CCTV: Conditionality considerations. *Justice Quarterly*, *34*(4), 597–626.

Lundman, R. (1980). *Police and policing: An introduction.* New York: Holt, Rinehart & Winston.

McCarthy, J. (2017). Record-high support for legalizing marijuana use in U.S. *Gallup.* Retrieved from http://news.gallup.com/poll/221018/record-high-support-legalizing-marijuana.aspx

McGinty, E.E., Niederdeppe, J., Heley, K., & Barry, C.L. (2017). Public perceptions of arguments supporting and opposing recreational marijuana legalization. *Preventive Medicine*, *99*, 80–86.

McGinty, E.E., Samples, H., Bandara, S.N., Saloner, B., Bachhuber, M.A., & Barry, C.L. (2016). The emerging public discourse on state legislation of marijuana for recreational use in the U.S.: Analysis of news media coverage, 2010–2014. *Preventative Medicine*, *90*, 114–120.

McLean, S.J., Warden, R.E., & Kim, M. (2013). Here's looking at you: An evaluation of public CCTV cameras and their effects on crime and disorder. *Criminal Justice Review*, *38*(3), 303–334.

Miethe, T. (1995). Predicting future litigiousness. *Justice Quarterly*, *12*(3), 563–581.

Mitchell, O., Wilson, D.B., Eggers, A., & MacKenzie, D.L. (2012). Assessing the effectiveness of drug courts on recidivism: A meta-analytic review of traditional and non-traditional drug courts. *Journal of Criminal Justice*, *40*, 60–71.

Morris, R., Copes, H., & Perry-Mullis, K. (2009). Correlates of currency counterfeiting. *Journal of Criminal Justice*, *40*, 60–71.

Nakashima, E., & Hicks, J. (2014, February 18). Homeland Security is seeking a national license plate tracking system. *The Washington Post*. Retrieved from www.washington-post.com/world/national-security/homeland-security-is-seeking-a-national-license-plate-tracking-system/2014/02/18/56474ae8–9816–11e3–9616-d367fa6ea99b_story.html

National Institute of Justice (2000). At-a-glance: Recent research findings—Effective police management affects citizen perceptions. *National Institute of Justice Journal, 244*, 24–25 (July).

Noble, A. (2016, May 4). Justice department program to no longer use 'disparaging' terms 'felons' and 'convicts'. *The Washington Times*. Retrieved from www.washingtontimes.com/news/2016/may/4/justice-dept-no-longer-use-terms-felon-convict/

Nobles, M.R., Levenson, J.S., & Youstin, T.J. (2012). Effectiveness of residence restrictions in preventing sex offense recidivism. *Crime & Delinquency, 58*(4), 491–513.

Obama, B. (2017). The president's role in advancing criminal justice reform. *Harvard Law Review, 130*(3), 811–866.

Pacula, R.L., Powell, D., Heaton, P., & Sevigny, E.L. (2015). Assessing the effects of medical marijuana laws on marijuana use: The devil is in the details. *Journal of Policy Analysis and Management, 34*, 7–31.

Peck, M. (2014, January 27). Predator drone sends North Dakota man to jail. *Forbes*. Retrieved from www.forbes.com/sites/michaelpeck/2014/01/27/predator-drone-sends-north-dakota-man-to-jail/

Piza, E.L., Caplan, J.M., Kennedy, L.W., & Gilchrist, A.M. (2015). The effects of merging proactive CCTV monitoring with directed police patrol: A randomized controlled trial. *Journal of Experimental Criminology, 11*, 43–69.

Porter, N.D. (2010). *Expanding the vote: State felony disenfranchisement reform, 1997–2010*. Washington, DC: The Sentencing Project.

Rau, R. (1991). Forensic science and criminal justice technology: High-tech tools for the 90's. *National Institute of Justice Reports, 224*, 6–10, June.

Ritter, N.M. (2006). Preparing for the future: Criminal justice in 2040. *National Institute of Justice Journal, 255*, 8–11.

Robison, J. (2002a, July 2). Decades of drug use: Data from the '60s and '70s. *Gallup*. Retrieved from http://news.gallup.com/poll/6331/decades-drug-use-data-from-60s-70s.aspx

Robison, J. (2002b, July 9). Decades of drug use: The '80s and '90s. *Gallup*. Retrieved from http://news.gallup.com/poll/6352/Decades-Drug-Use-80s-90s.aspx?g_source=link_NEWSV9&g_medium=tile_5&g_campaign=item_6331&g_content=Decades%2520of%2520Drug%2520Use%3a%2520The%2520%2780s%2520and%2520%2790s

Rocky Mountain High Intensity Drug Trafficking Area (2017). *The legalization of marijuana in Colorado: The impact*. Retrieved from www.rmhidta.org/html/FINAL%202017%20Legalization%20of%20Marijuana%20in%20Colorado%20The%20Impact.pdf

Rothschild, D.Y., & Block, W.E. (2016). Don't steal; the government hates competition: The problem with civil asset forfeiture. *Journal of Private Enterprise, 31*(1), 45–56.

Rottman, D., & Strickland, S. (2006). *State court organization, 2005*. Washington, DC: Bureau of Justice Statistics.

Santos, R.B. (2017). *Crime analysis with crime mapping*. Thousand Oaks, CA: Sage.

Schneider, K.E., Krawczyk, N., Xuan, Z., & Johnson, R.M. (2018). Past 15-year trends in lifetime cocaine use among US high school students. *Drug and Alcohol Dependence, 183*, 69–72.

Schroeder, D., & White, M. (2009). Exploring the use of DNA evidence in homicide investigations: Implications for detective work and case clearance. *Police Quarterly, 12*(3), 319–342.

Selk, A. (2018, April 28). The ingenious and "dystopian" DNA technique police used to hunt the "Golden State Killer" suspect. *Washington Post*. Retrieved from www.washingtonpost.com/news/true-crime/wp/2018/04/27/golden-state-killer-dna-website-gedmatch-was-used-to-identify-joseph-deangelo-as-suspect-police-say/?noredirect=on&utm_term=.b4e4e20bcc90

Shannon, S.K.S., Uggen, C., Schnittker, J., Thompson, M., Wakefield, S., & Massoglia, M. (2017). The growth, scope, and spatial distribution of people with felony records in the United States, 1948–2010. *Demography, 54*, 1795–1818.

Shepard, E.M., & Blackley, P.R. (2016). Medical marijuana and crime: Further evidence from the western states. *Journal of Drug Issues, 46*(2), 122–134.

Simon, J. (1993). *Poor discipline*. Chicago: University of Chicago Press.

SMART (2018). *Sex Offender Registration and Notification Act: State and territory implementation progress check*. Retrieved from www.smart.gov/pdfs/SORNA-progress-check.pdf

Socia, K.M., & Stamatel, J.P. (2012). Neighborhood characteristics and the social control of registered sex offenders. *Crime & Delinquency, 58*(4), 565–587.

Stanley, J., & Crump, C. (2011). *Protecting privacy from aerial surveillance: Recommendation for government use of drone aircraft*. New York: American Civil Liberties Union.

Stolberg, V.B. (2009). Comprehensive drug abuse prevention and control act. In Fisher, G.L., & Roget, N.A (Eds.), *Encyclopedia of substance abuse prevention, treatment, & recovery* (pp. 224–225). Thousand Oaks, CA: SAGE.

Straub, J. (2014). Unmanned aerial systems: Consideration of the use of force for law enforcement applications. *Technology in Society, 39*, 100–109.

Thielman, S. (2016, July 8). Use of police robot to kill Dallas shooting suspect believed to be first in US history. *The Guardian*. Retrieved from www.theguardian.com/technology/2016/jul/08/police-bomb-robot-explosive-killed-suspect-dallas

Turner, A. (2007). Automated video surveillance: Improving CCTV to detect and prevent incidents. *Corrections Today, 69*, 44–45 (June).

Uggen, C., Larson, R., & Shannon, S. (2016*). 6 million lost voters: State-level estimates of felony disenfranchisement, 2016*. Washington, DC: The Sentencing Project.

Uggen, C., Vuolo, M., Lageson, S., Ruhland, E., & Whitham, H.K. (2014). The edge of stigma: An experimental audit of the effects of low-level criminal records on employment. *Criminology, 52*(4), 627–654.

Vagins, D.J., & McCurdy, J. (2006). *Cracks in the system: Twenty years of the unjust federal crack cocaine law*. Washington, DC: American Civil Liberties Union.

Vile, J. (1981). The right to vote as applied to ex-felons. *Federal Probation, 45*(1), 12–16.

Vuolo, M., Lageson, S., & Uggen, C. (2017). Criminal record questions in the era of ban the box. *Criminology & Public Policy, 16*, 139–165.

Weisburd, D., Farrington, D.P., & Gill, C. (2017). What works in crime prevention and rehabilitation: An assessment of systematic reviews. *Criminology & Public Policy, 16*(2), 415–449.

Welsh, B.C., Farrington, D.P., & Taheri, S.A. (2015). Effectiveness and social costs of public area surveillance for crime prevention. *Annual Review of Law and Social Science, 11*, 111–130.

Willis, J.J., Koper, C., & Lum, C. (2018). The adaptation of license-plate readers for investigative purposes: Police technology and innovation re-invention. *Justice Quarterly, 35*(4), 614–638.

Worden, R.E., Harris, C., & McLean, S.J. (2014). Risk assessment and risk management in policing. *Policing: An International Journal of Police Strategies & Management, 37*, 239–258.

Worrall, J., & Kovandzic, T. (2007). COPS grants and crime revisited. *Criminology, 45*(1), 159–190.

Zgoba, K.M., & Levenson, J. (2012). Failure to register as a predictor of sex offense recidivism: The big bad wolf or a red herring? *Sexual Abuse, 24*(4), 328–349.

IMPORTANT CASE

Carpenter v. United States, 585 U.S. ____ (2018).

Glossary

A

abandonment: a way of disposing of unwanted or burdensome children, which gained prominence after the fourth century; the desertion of an infant or child by her or his parents.

absconders: offenders who fail to submit to supervision (do not report as directed, change jobs or addresses without notifying their supervising officer, etc.).

actus reus: the behavioral or action element of a criminal offense.

adjudication: the determination of the facts in a case by a judicial body. Specifically in the juvenile justice system, it refers to the fact-finding process that is similar to the trial in the adult system.

appellate courts: tribunals authorized to hear and settle questions of law that arise from lower courts. In most cases, appellate courts do not hear factual matters, and most often the appellate court comprises panels of judges or justices.

arraignment: a formal stage of the criminal justice process at which the accused is informed of the criminal charges against her or him, and asked to plead to those charges.

arrest: taking a person into custody; one of the decision points of the criminal justice system.

assigned counsel: a system for providing criminal defense services for the indigent in which attorneys are assigned or appointed to the case of a defendant from a list of candidates maintained by the court.

automobile searches: the Supreme Court ruled that police officers could conduct a full search of an automobile that the officers had legitimately stopped, as long as they had probable cause to believe that the automobile contained contraband.

B

bail: a monetary surety required of a defendant prior to release from custody to assure the defendant's appearance at later court hearings.

bench trial: a criminal trial held before a judge alone, without a jury. The judge in a bench trial both presides over the trial and serves as the finder of fact.

benefit of clergy: a forerunner to contemporary practices such as probation; the practice of excusing members of the clergy from state criminal responsibility in English courts, the benefit was later extended to all literate citizens.

blended sentencing: a practice that allows either the juvenile court or the adult court to impose a sentence on a juvenile offender that can involve either the juvenile or the adult correctional system, or both.

booking: the point in the criminal justice process at which the arrest of a criminal suspect is officially recorded (written in the police logbook).

boot camp: a specialized prison program in which offenders (typically young) are subjected to a regimen of physical training and strict discipline, but are granted release earlier than more traditional incarceration.

C

career criminal: a label given to repeat offenders, generally those who have a lengthy and involved history of criminality.

casework model: a model of organizing service delivery in probation, parole, and other settings in which the criminal justice official is assigned subjects (cases) and is responsible for generally serving all of their needs.

charging: the process by which the prosecutor (or state's attorney or district attorney) applies the criminal law to the facts of the case and identifies which provisions of the criminal code have been violated.

circuits: districts or territories assigned to courts in the federal system; 11 circuits cover the United States; the District of Columbia comprises an additional circuit.

civil death: loss or restriction of civil rights so that for legal purposes, the individual is "dead." Often a consequence of conviction in felony cases.

classification: the testing and assessment of inmates to determine inmate treatment needs and prison custody and security needs.

clear and present danger: conditions or behavior that pose an immediate threat to safety or order and relates to controls on the activities of inmates that pose a direct threat to the smooth operation of the facility.

closed system: a system or collection of interrelated parts that are relatively isolated or insulated from their environment.

cohort studies: research based on the longitudinal study of an identifiable group of individuals. The identified group is the cohort about which the study seeks information.

cold case squads: teams of detectives dedicated to pursuing cases (usually homicide cases) that have not been solved and lack significant leads.

collateral consequences: effects or products of a criminal conviction that are in addition to any criminal penalty, for example the loss of civil rights.

community courts: courts that attempt to resolve problems and disputes by addressing all concerned, including the community at large. Such courts are usually linked to community resources.

community prosecution: a type of prosecution in which the prosecutor is assigned to the case from initial appearance through disposition and works with the police, community, and other agencies not just to secure conviction, but to solve the problems that led to the criminal behavior.

community-oriented policing: an approach to policing that relies on community definitions of police functions and a partnership between the police and the community in the production of public safety.

community service orders: a command by the court that a convicted offender work, without compensation, at some task or job of benefit to the community.

compelling state interest: a legal criterion used to judge the reasonableness of a practice or condition. In general, the state must justify its use of intrusive practices or conditions by showing that the practice accomplishes an objective that the state must achieve.

CompStat: policing strategy giving the decision-making authority to the middle-level managers and then redirects resources as needed to assist in solving each district's specific problem.

concurrent term: a sentence for a criminal conviction that is executed ("runs") at the same time as another sentence.

conditional release: permission for an offender to remain in the community if he or she abides by certain conditions, such as reporting regularly to a supervising officer, refraining from consuming alcohol, or other.

congregate system: the silent or Auburn model of prison discipline in which inmates ate, worked, recreated, and worshiped together, but were housed in separate cells at night and prohibited from talking with each other.

consecutive term: a sentence for a criminal conviction that is delayed in execution until after the sentence for another conviction has expired.

consent decrees: decrees whereby the court and the state enter into a voluntary agreement about issues raised in court.

consent search: the suspect consents or agrees to the search so that any evidence found will be admissible.

constable: a court office in Norman England that had many administrative and public safety duties; a forerunner to the police.

contract system: the practice of leasing to the highest bidder inmate labor and the use of prison work areas and shops.

corporate gang: a group of individuals that has a well-organized formal structure and whose purpose is to make money.

cottage reformatories: institutions for juvenile offenders established in the mid- to late 1800s that attempted to parallel a family setting.

count: a specific criminal charge in an indictment so that, for example, three counts of robbery means three separate charges of robbery.

courtroom work group: the people who comprise the major actors in the court process and who generally develop common understandings and norms for how the business of the court will be conducted.

crackdown: usually short-term, an intensive police response to a perceived problem. For example, strict enforcement of traffic or parking laws, saturation patrol, and other intensive police efforts focused at specific problem areas.

crime: an act or omission in violation of a law, which is punishable by the state.

crime control model: an analytic device developed by Herbert Packer that describes how the criminal justice process would operate if the control of crime were the only (or the predominant) goal served by the system.

Crime Index: the total number of eight specific types of offenses that are known to the police in any given year, as reported by the Federal Bureau of Investigation.

crime rate: a standardized measure of the amount of crime per unit of population. Typically, the number of crimes known to the police per 100,000 members of the population.

criminal justice: the formal social institution designed to

respond to deviance defined as crime.

criminology: the scientific study of law breaking.

custodial interrogation: a suspect is only entitled to *Miranda* warnings during this type of interrogation.

cynicism: in terms of police, the perception or belief that citizens, department leaders, politicians, and other criminal justice officials are not truthful and honest in their dealings with police officers.

D

"dark figure" of crime: the amount of criminal activity that is unreported and undetected; specifically, the amount of crime that is not included in official statistics such as the Uniform Crime Reports.

day reporting: a community corrections program in which convicted offenders are required to check in with (or at) a supervising center each day, but are allowed to remain in their homes at night and to engage in approved activities (i.e., work or school) during the day.

defounding: the practice of reducing the seriousness of a crime alleged by a victim or complainant, such as recording a reported felony as if it were a misdemeanor.

delinquent: a juvenile who violates the criminal laws of the jurisdiction.

deprivation model: a model that explains the development of prisoner subculture as a reaction to the loss of freedom, goods, services, and ties to life outside the prison.

design capacity: the number of inmates that a prison was designed to house or hold.

detection: the decision point in the criminal justice process at which police officers come to believe that a crime has occurred.

detention decision: the decision whether to keep a juvenile in custody or to allow the youth to go home with parents or guardians to await further court action.

determinate sentencing: a sentencing to incarceration in which the exact length and nature of punishment is known at the time it is imposed.

deterrence: a reason for criminal punishment based on the idea that punishment of the individual offender produces benefits for the future by making the idea of criminal behavior less attractive.

deviance: behavior that violates socially accepted standards of proper conduct.

differentiated case management: programs in which prosecutors select certain types of cases that can proceed quickly through the court system.

discharge: release from sentence and custody or control by the criminal justice system.

discretionary release: release granted by a parole board that will typically assess the inmate's perceived likelihood of successfully following the conditions of parole.

disparity: inequality; especially in sentencing when two similarly situated offenders receive different penalties.

disposition: outcome; in the juvenile justice system and with probation and parole revocation, a specific hearing stage during which penalty/sentence is decided.

diversion: preventing cases from entering the criminal justice system or reducing how far cases progress into the system; avoiding criminal justice processing.

double jeopardy: occurs when a defendant faces trial or punishment more than once for a single offense.

drug courts: special courts dedicated to the processing and supervision of drug cases.

drug testing: chemical testing for the presence of drugs in the urine, blood, hair, and more of the test subject. A relatively common practice for suspected drug users in criminal justice populations, and often a requirement of employment in criminal justice agencies.

dual system: a term used to describe the existence of two sets of courts in the United States, one federal and the other state.

due process model: an analytic device created by Herbert Packer to describe a criminal justice system in which the most important goal is the protection of individual liberty.

E

education: a long-term function aimed at developing a spirit of exploration, developing the academic tools, and developing an introductory base of how the structure and process of the justice system works.

electronic monitoring: any of several systems in which criminal offenders are tracked and supervised, at least in part,

by radio and/or telephone contact.

entrapment: a defense to criminal charges that applies when the idea for the crime and the motivation to commit the crime are produced by the police and did not arise with the offender.

evidence-based policy: scientific research helps policymakers better understand the criminal justice system and thus make more informed policy decisions.

exclusionary rule: created by the appellate courts as protection of constitutional rights and a sanction against police misbehavior, this rule prohibits the use of illegally obtained evidence in criminal prosecutions.

exculpatory evidence: evidence that tends to establish the innocence of the accused or defendant.

executive clemency: the authority of the executive officer of a jurisdiction to grant mercy or forgiveness to those accused or convicted of a crime. Clemency includes pardon, commutation, and reprieve.

F

false negative: in criminal justice prediction, someone who is predicted to be safe, not to pose a threat of future criminality, but who is, in fact, dangerous and commits additional crimes.

false positive: in criminal justice prediction, someone who is predicted to be dangerous, to pose a threat of future criminality, but who is, in fact, safe and would not commit additional crimes.

family model: an analytic device developed by John Griffiths (in opposition to Packer's due process and crime control models) that contends that the criminal justice system should operate under the assumption that the interests of society and the interests of the offender are the same.

federalism: the structure of government in the United States that distinguishes between federal, state, and local governmental interests, duties, responsibilities, and powers.

felony: serious criminal offense defined by statute; usually punishable by a term of one year or longer in a state or federal prison.

focal concerns: the "focal concerns" of judges at sentencing are public protection, blameworthiness, and practical considerations.

forgetting: in regard to victim surveys, the possibility that a respondent will forget a crime that occurred during a specific time period.

formal charges: the official accusation of criminal conduct that the prosecutor must prove beyond a reasonable doubt if the case goes to trial; established by indictment or the filing of an information in court.

formal social control: sanctions that are applied by some authorized body after a public finding of fault.

frisk: a limited search (pat down) of the outer clothing of a suspect for the purpose of self-protection through the discovery of any weapons.

functions: the purposes or goals served by social institutions or practices. They can be both manifest (stated, expressed) purposes and latent (hidden) purposes.

funnel effect: the effect by which the criminal justice system operates like a giant sieve, continuously filtering the huge volume of crimes and criminals to the relatively small number of offenders who are incarcerated in the nation's prisons.

furlough: temporary release from custody; programs of short-term release of prison inmates for specific purposes such as seeking employment and housing prior to release, attending to a personal or family emergency, and the like.

G

general deterrence: a subtype of deterrence based on the notion that punishing a specific offender will frighten or warn the general population to avoid criminal behavior.

general jurisdiction: a term used to describe the authority of some trial courts that indicates that the court is empowered to settle questions of fact in almost all civil and criminal matters arising within its geographical area.

global positioning system (GPS): a technology used in electronic monitoring which can provide real-time tracking of an offender's location using a satellite system.

good faith exception: the tenet that if the police conduct a search believing in good faith that the search is permissible, then the evidence can be used at trial.

good time: reductions in the length of sentence granted to inmates as a reward for good behavior in the institution.

grand jury: a panel of citizens (usually 23, with a quorum of 16) that reviews evidence in criminal cases to determine whether sufficient evidence exists to justify trial of an individual.

GREAT: a program in which officers give lessons to middle-school youths on individual rights, cultural sensitivity, conflict resolution, drugs, neighborhoods, personal responsibility, and goal setting.

H

habitual offender statutes: legislative statutes that allow for increased penalties for repeat offenders.

halfway houses: a generic term describing residential programs operating in the community in which criminal offenders are housed and provided various treatments. The term indicates that this option is "halfway" between incarceration in prison or jail and release to the community under supervision.

hands-off doctrine: phrase describing the reluctance of appellate courts to intervene in the operation of prisons, jails, and other correctional facilities. The term implies that courts grant wide latitude to correctional administrators by keeping their (the court's) hands off questions of facility administration.

hierarchy rule: data collected though this reporting procedure mask the actual numbers of offenses and offenders.

home incarceration: a sanction of incarceration that is served in the offender's home, essentially a restriction on liberty that

requires the offender to remain in his or her residence during specified hours for a set term, and typically enforced by the use of electronic monitoring.

hot pursuit: a circumstance in which the police are closely chasing a crime suspect. Officers engaged in a hot pursuit are not required to seek a warrant to search the area in which the suspect is caught or trapped.

hot spot policing: a type of policing in which police departments have responded to "crime waves" by increasing police presence in a given area.

hot spots of crime: a term coined by Sherman, Gartin, and Buerger to refer to locations where much more crime can be found than at other places.

Houses of Refuge: institutions established in the early 1800s for youthful offenders; they were designed to use education, skills training, hard work, and apprenticeships to produce productive members of society.

hung jury: a jury that cannot reach consensus about the verdict.

I

identity theft: the theft or misuse of personal information, similar to fraud or forgery.

importation model: a model that explains the existence of a prisoner subculture as the result of inmates bringing criminal and antiauthority values with them from their lives in the community.

incapacitation: a reason for criminal punishment based on the notion that the penalty will prevent the offender from having the chance to commit a crime in the future.

incarceration rate: the number of persons incarcerated per 100,000 population eligible for incarceration.

indeterminate sentencing: a sentencing to incarceration that is stated as a range of time between some minimum and some maximum term and in which, at the point of sentencing, the exact length of confinement is unknown.

indictment: a true bill issued by a grand jury that establishes that the jury found probable cause to have a defendant respond to criminal charges in court. A mechanism of filing formal criminal charges against a defendant.

indigent: poor; a criminal defendant or convict who is unable to afford the cost of defense counsel and for whom counsel will be provided at state expense.

infanticide: literally, the killing of an infant; the practice in history of killing unwanted or burdensome children shortly after birth.

informal social control: mechanisms that influence behavior without the need for a public finding of fault or the use of group-authorized sanctions.

initial appearance: the criminal suspect's first appearance in a court hearing, at which the question of pretrial release is decided.

innocence projects: coordinated efforts to investigate claims of innocence maintained by persons convicted of criminal offenses.

institutionalization: the tendency for residents of "total institutions" (e.g., prison inmates) to become habituated to and dependent on the institutional routine so that they lose the ability to make independent decisions.

intake decision: a screening point in the juvenile justice system at which a court official, often a probation officer, decides whether a juvenile's case will be processed through the court or handled outside the formal court process.

intelligence gathering: the compilation of information that may be helpful in solving crimes, but with no specific probable cause to believe the information will lead to solving a particular crime.

intensive supervision a form of probation and parole supervision in which offenders receive an increased level of attention from supervision officers that usually includes more in-person contact and closer monitoring by the officer.

intermediate sanctions: the term given to describe a range of criminal penalties developed as alternatives to traditional probation or incarceration. These include intensive supervision, house arrest, community service, and other penalties.

interrogation: questioning; specifically, the in-custody questioning of crime suspects by the police.

inventory search: the routine check of seized property to establish what has been taken by the police. The term specifically applies to inventorying the contents of seized automobiles.

investigation: the search for and accumulation of evidence that links a particular crime to a particular person or persons.

J

jails: relatively short-term custodial facilities, typically operated at the municipal (county) level, used to house a variety of offenders and criminal suspects.

jailhouse lawyers: inmates who assist others in the preparation of court documents.

judicial reprieve: a practice in early English courts that served as a forerunner to probation. It was essentially a suspended sentence ordered by the judge.

jurisdiction: the limits of authority or interest placed on a criminal justice agent or agency and comprised generally of geographic boundaries and the identification of case or offender characteristics.

jury: a panel of citizens selected to hear evidence and render a decision in a criminal matter. Grand juries make charging decisions; petit juries render conviction decisions.

jury nullification: the power and practice of a petit jury rendering a not guilty verdict despite overwhelming evidence of wrong; this is a statement that the law involved is inappropriate, thus the jury nullifies the law by nonenforcement.

jury trial: the determination of guilt by adjudication before a jury at which the state must prove, in open court, all elements of the offense beyond a reasonable doubt, and the jury members make the final determination concerning guilt.

just deserts: a justification for criminal penalties based on the notion that criminals, by virtue of breaking the law, have earned their punishment, and noncriminals have earned the right to have criminals punished; therefore, crime deserves punishment.

justice reinvestment: seeks to not only save money by diverting offenders to community-based supervision, but also to redirect a portion of the savings to rebuild the communities that have been most impacted by high levels of incarceration.

juvenile: a person who by virtue of age, as defined in applicable statute, has not yet reached majority and thus is subject to different treatment than adults. The age limits for juvenile status vary among the states.

L

latent functions: the unstated or hidden goals of an institution.

lease system: a model of prison labor in which the state rented the labor of convicts and the use of prison shops to the highest bidder.

least restrictive alternative: the principle that requires the state use the least intrusive or least controlling practice or regulation to achieve its legitimate aims.

legalistic style: coined by James Q. Wilson, this label is attached to those police organizations in which the normal practice is to intervene frequently and formally with citizens and in which the police role and activities are more narrowly defined by law and law enforcement obligations.

life-course criminality: a person's involvement in crime over his or her entire lifetime, including both childhood and adulthood.

limited jurisdiction: used to describe courts or other judicial offices, this phrase generally refers to circumstances in which the court is authorized to hear only the early stages of serious cases, or only to hear less serious or specific types of cases.

lineups: an investigatory practice in which the crime suspect, in the company of a group of similar persons, is brought before the witness for identification.

local autonomy: a component of federalism, the freedom of local governmental units to define and respond to problems in their own way.

low visibility: lacking review; as used by Joseph Goldstein, discretionary decisions by police not to invoke the criminal law. These decisions would not be reviewed by courts or others in the police administration and thus would have "low visibility."

M

mala in se: "bad in itself," a term used to describe certain traditional crimes about which there is general agreement that the behavior is wrong.

mala prohibita: "bad because prohibited," a term used to describe those crimes about which there is more general disagreement among people concerning whether the behavior is wrong. Often applied to vice and regulatory offenses.

mandatory minimum sentences: sentences that reflect a required period of incarceration defined by statute that must be imposed as a sanction if the offender is convicted of a particular crime.

mandatory release: release based on earned good time or other statutory sentence-reduction measures; usually does not depend on the discretionary decision of a parole board.

manifest functions: the stated purposes of an institution.

mark system: the "token economy" of Alexander Maconochie at Norfolk Island by which inmates could earn marks by good behavior and labor. These marks then allowed inmates to progress to increasingly less strict conditions of confinement.

mens rea: "mental things," the mental element of a crime such as intention, voluntariness, and other cognitive conditions required by the law.

Miranda **warnings:** notification of rights (to remain silent and to have an attorney present) that are required to be given to suspects prior to interrogation; named for the 1966 Supreme Court case *Miranda v. Arizona*, which established these rights.

misdemeanor: a crime that is generally considered to be less serious than a felony and that is usually punishable by a term of no more than one year in a local jail.

Missouri Plan: a method of judicial selection by which judges are appointed from a slate of qualified candidates and are periodically reviewed by means of a retention election in which citizens decide if the judge, running unopposed, should continue to serve.

model case management system: a process for classifying offenders on probation and parole based on both risk of further crime as well as the service needs of the offender, using this information to structure and direct the activities of officers and entire field supervision agencies. The model system was developed and disseminated by the National Institute of Corrections beginning in the early 1980s.

multijurisdictional jails: jails that serve more than one municipality or jurisdiction.

N

National Crime Victimization Survey (NCVS): an annual survey of a representative sample of Americans asking their experiences as crime victims. The NCVS is the primary source of victimization data concerning crime in America.

net-widening: the term used to describe the phenomenon of increasing the number of persons touched by some aspect of the justice system. In particular, it is applied to cases where a new program or service results in more people being subjected to criminal justice intervention because the new program is seen as appropriate to them.

nolle prosequi: Latin term meaning "I do not prosecute"; a prosecutor may choose to refuse to prosecute, or "nol pros," a case.

nolo contendere: Latin term meaning "no contest," or

"I do not contest the charges"; a plea available to criminal defendants, with the consent of the court, in most jurisdictions. The effect of the *nolo contendere* plea is similar to that of a plea of guilty, except the defendant is not "proven" to have broken the law and thus is better able to defend against a civil suit.

nonsecure detention: housing options for youths involved in less serious crimes that are either not locked at all or at least not locked as comprehensively as a secure detention facility.

O

observations: as used here, a type of research that relies upon the on-site, in-person observations of criminal justice practice or the behavior of criminals by the research staff; more generally, "field research."

official statistics: any data routinely collected and reported by official agencies of criminal justice. The best known example is the Uniform Crime Reports.

on paper: serving a term of probation or parole supervision. The phrase refers to the supervision agreement or conditions of supervision that govern the behavior of the offender as the "paper" the offender is "on."

open system: a system that is relatively sensitive to its environment and thus adapts and reacts to changes in the environment.

operational capacity: a method of rating the size of a population appropriate for a given correctional institution, based on the facility's staff, existing programs, and services.

order maintenance: functions of the police that serve to maintain order, including settling disputes, dispersing crowds, keeping traffic flowing smoothly, and the like.

organized/corporate gang: a well-organized group that exists and functions for the purpose of making money through criminal activities.

P

pains of imprisonment: a term originated by Gresham Sykes to refer to the social psychological deprivations experienced by prison inmates that combine to make the experience of incarceration personally painful to inmates.

panel attorneys: assigned counsel in the federal system, named as such because defense counsel are assigned from a list or panel of approved lawyers.

paramilitary structure: military-like structure; having the characteristics of a military organization such as formal ranks, a chain of command, pyramidal organizational structure, and more.

parens patriae: "the state as parent"; a doctrine that suggests that the state or government has a parental interest in the welfare of children and can act as a parent when needed. The doctrine underlies the concept of the juvenile court as a nonadversarial, nonpunitive solution of juvenile offending.

parole: from the French term meaning "word of honor," it refers to the release of prison inmates, prior to expiration of term, on the condition that they agree to abide by certain restrictions. It also describes the process of conditional liberty, which includes supervision by a state official. Parole can mean either the decision to release, or supervised release in the community.

parole eligibility: established by the legislature, this defines which types of inmates can be paroled at what points in their sentence.

penitentiary: prison; although no longer commonly used in the names of correctional institutions, early prisons were considered to be places in which convicted offenders could "pay" for their crimes and "do penance." Thus, they were call penitentiaries.

peremptory challenges: authorization to exclude someone from a particular trial jury without the requirement that a cause for exclusion be shown.

piece price system: a method of organizing prison labor that was common in the earliest years of prison, in which the private contractor supplied raw materials and agreed to buy finished products made by inmates at fixed price for each piece of finished product.

plain view doctrine: a doctrine that states that police do not need to obtain a warrant to seize contraband or criminal evidence that is plainly visible and requires no search.

plea bargaining: the practice of exchange between the prosecution and defense in which the defendant agrees to plead guilty to criminal charges in return for some concession from the prosecution.

police specialization: divides tasks into special units or divisions.

preliminary hearing: a hearing in open court at which the prosecution introduces evidence to establish probable cause to have a defendant bound over for trial on criminal charges.

presentence investigation (PSI): a background report on a convicted criminal offender designed to provide the judge with information about the offender's social and criminal history and current status, for use in making a sentencing decision.

presumptive sentencing: suggested sanction for conviction of a particular offense; the expected or "presumed" sentence can be modified by the judge for cause.

preventive detention: a practice by which defendants suspected to be dangerous are denied bail until their cases are tried.

prison industrial complex: the relationships between governments, correctional authorities, and private corrections companies that combine to support increased use of prison.

Prison Industry Enhancement Certification Program (PIECP): created by Congress in 1979 to encourage states to provide inmates with employment opportunities that are similar to private-sector work opportunities.

prisons: typically larger institutions used to house convicted adult felons for terms of 1 year or longer.

Prison Litigation Reform Act: a 1995 law requiring inmates to exhaust all administrative

remedies (appeals through the prison administration and department of corrections) before they can file a suit in federal court.

Prison Rape Elimination Act: a 2003 law providing for the development of better information about the nature and incidence of rape and sexual assault in prisons, and providing funding to correctional authorities to reduce and control sexual violence in prisons.

prisonization: the term describing the process by which prison inmates come to learn and accept the values and norms of the inmate subculture.

private court: offices or commissions for dispute resolution that divert cases away from the formal courts, usually staffed by volunteers or paid staff whose salaries are lower than that of a judge.

privately retained counsel: counsel or attorney for the defense in a criminal matter, who is hired and paid privately by the defendant.

privatization: the movement to turn government functions over to operation by private sector (profit or not-for-profit) organizations.

proactive: self-motivated or self-initiated; specifically as related to policing, efforts to detect or respond to crimes that are motivated by the police themselves without reliance upon a formal complaint.

probable cause: evidence that leads a reasonable person to conclude that a crime has occurred and evidence of the crime may be found to support the search.

probation: a sentence of conditional and revocable release into the community, generally under supervision, usually imposed in lieu of incarceration. The process of supervising and enforcing conditions of release in lieu of incarceration.

Probation Subsidy Act: an early form of community corrections legislation in California that provided financial incentives to counties to retain convicted offenders at the local level under probation supervision as an alternative to incarceration in state prisons.

problem-oriented policing: this method of policing uses the SARA model to problem-solve using a situational crime prevention approach and focuses on individual responses to specific local problems that may arise in a community.

procedural justice: this perspective suggests that people oppose the rules when they believe their treatment is unfair.

prosecutorial case management: a system for assigning, managing, and conducting the work of a prosecutor's office, generally involving case classification and the establishment of priorities for the expenditure of prosecutorial resources.

protective custody: a method for protecting inmates from attacks by other inmates, usually a housing unit kept separate from the general inmate population.

public account system: a model of prison industry in which prison labor is used in the production of goods for sale on the open market, and the prison industry, as a public agency, operates like a private business.

public defender: common in larger and busier criminal jurisdictions, a model for the provision of defense counsel to the indigent that relies on a public office organized similarly to that of the prosecutor.

public works system: a model of prison industry in which inmate labor is used in the completion of public works such as road construction and maintenance, the building or repair of public buildings, and similar public projects.

punitive period: the shift to a more punishment-oriented approach to juvenile delinquency.

R

radio frequency: a technology used in electronic monitoring which confirms that an offender is at approved locations by sending a radio signal to an offender's landline telephone or a receiver located in their home

rated capacity: a method for determining the size of a population that is appropriate for a specific correctional institution, based on a judgment or rating of some official such as a health inspector or fire marshal.

reactive: responsive; specifically as it relates to policing, efforts of the police to detect or prevent crime, which are initiated by or in reaction to a formal complaint.

recidivism: the commission of criminal behavior by a person after release or discharge from the criminal justice process.

reentry: the return of former inmates to life in the community.

reentry courts: courts in which services for and supervision of parolees are coordinated and monitored in a court environment.

reformatory: a prison typically reserved for youthful adult felony offenders.

rehabilitation: a justification for the imposition of a criminal sentence based on the idea that crime is symptomatic of problems an offender has with regard to living within society. The punishment is imposed "for the offender's own good," and is intended to change the offender's need or proclivity to engage in crime.

release on recognizance (ROR): generally used at the pretrial stage of the process, the order allowing a defendant to remain at liberty in the community pending further court actions without posting any bond or surety.

restitution: repayment for the harm caused by criminal behavior; an increasingly common part of criminal sanctions that requires convicted offenders to repay their victims.

restorative justice: efforts to repair the harm to victims and/or communities caused by crime through interventions with the offender.

retribution: a justification for criminal penalties that is based on the principle that wrong deserves to be punished, regardless of whether the punishment produces any benefit.

revocation: recall; the cancellation of conditional liberty (usually probation or parole) upon a finding that the offender has violated the conditions of release.

role stress: many correctional officers report being troubled by stressors of the job, including a lack of clear job description, absence of support from superiors, and not being able to exercise personal judgment.

routine activities theory: a theory suggesting that crime occurs when a motivated offender and a suitable target (victim or property) come together in time and space in the absence of an effective guardian.

rumble: the traditional image of the gang fight involving two groups of youths in a prearranged fight.

S

search: an investigatory technique in which police or other officials seek physical evidence of criminality or contraband.

search incident to lawful arrest: this type of search is limited to the area within the immediate control of the officers and protects officers through discovery of any weapons, and to secure any evidence of the crime that the offender might otherwise be able to destroy.

segregate system: an early form of prison discipline, practiced in the Pennsylvania penitentiaries, in which inmates were kept isolated from each other for the duration of their confinement.

self-report studies: surveys that attempt to measure the amount of crime committed and describe the characteristics

of offenders by asking people if they have committed offenses.

sentencing: the imposition of a sanction upon conviction of a crime; the decision and process of deciding upon an appropriate penalty for a specific criminal act or acts.

sentencing commissions: formal bodies assigned to assess and oversee criminal sentencing and recommend reforms.

separation of powers: the reservation of specific functions and authority to particular branches of government that enables the system of "checks and balances" to operate.

service style: coined by James Q. Wilson, the term refers to a method of policing in which officers intervene frequently, but informally, in the lives of citizens.

shire reeve: an early Saxon political office representing the head of a large group of families (shire); forerunner to the contemporary office of sheriff.

shock incarceration: short-term prison programs in which the conditions of incarceration are much more severe but limited in duration (e.g., boot camps).

shock parole: early release to parole supervision after a comparatively brief period of incarceration. The purpose is to "shock" the offender with a small period of incarceration followed by an unexpected early release to supervision.

shock probation: see shock parole; a grant of probation to an offender previously sentenced to prison after serving a relatively small portion of the prison term.

smug hack: a brutal, calloused, authoritarian correctional officer; a prison guard who behaves as a petty tyrant.

social control: the process of directing and limiting the behavior of individual members of a social group or society.

special conditions: requirements for release and/or supervision written into the supervision agreement for probation or parole. These are not generally imposed on all probationers or parolees.

special jurisdiction: a nontraditional authorization; for example, in policing, special jurisdiction agencies may have full police powers within a park, on transportation routes or on waterways but lack general police powers throughout the municipality.

special prosecutor: an attorney appointed by a governmental authority for the purpose of conducting investigations and pursuing criminal cases arising from particularly complex or politically sensitive circumstances.

specific deterrence: a subtype of deterrence in which the purpose of the penalty is to frighten the individual offender into conformity, regardless of the effect of the sanction on the broader, general public.

Speedy Trial Act of 1974: this federal law outlined specific time limits in which the indictment and trial must begin after arrest.

split sentences: sentences combining a period of incarceration with a period of community supervision as part of the sentence order; splitting the total sentence between incarceration and community supervision.

standard conditions: requirements for release and/or supervision that are imposed on everyone under supervision in a given jurisdiction.

state-use system: a model of prison industry in which prison labor and prison factories are operated to produce goods solely for the use of governmental entities and not for sale on the open market; for example, using prison factories to produce state-issued automobile license plates.

status offenses: regarding juvenile offenders, acts prohibited by law or justifying juvenile justice system intervention that apply to youth solely by virtue of their status as juveniles (e.g., truancy, running away, etc.).

sting operation: a police decoy operation in which officers pretend to be involved in a criminal operation such as a stolen automobile "chop shop" or a fencing operation and in which criminal offenders who sell the proceeds of their crimes are "stung" or caught by these decoy officers.

street sense: intuition based on experience that enables police officers to detect criminality; a phrase describing how police officers develop hunches and suspicions in circumstances that would not attract a second thought from civilians.

strict liability: an offense in which the mental state or *mens rea* is presumed to be present so that mere behavior constitutes the elements of the crime regardless of intention.

supervised release: a type of release imposed at the initial sentencing hearing, requiring the offender to be monitored by a federal probation officer and to abide by certain conditions upon release from prison.

supervision fees: costs charged to offenders on probation and parole supervision that are used to offset the expenses of operating the supervision process. These fees are typically billed in monthly installments for the length of the supervision period.

suretyship: the practice of a person of good standing in the community taking responsibility for guaranteeing the lawful behavior of another person; the forerunner to modern parole.

system: a collection of interrelated parts working together toward a common goal.

T

team supervision: a model of organizing probation or parole offices in which a group or "team" of personnel are assigned to provide service and surveillance to offenders. In theory, this model allows officers to specialize, thus improving the efficiency of the supervision process.

technical violations: infractions of the rules of supervision that do not involve any new criminality (e.g., violating curfew).

telescoping: in regard to victim surveys, the possibility that a respondent will erroneously include an earlier event in reporting on criminal victimizations during a specific time period.

theory: a logical explanation for reality; a statement of how things work.

therapeutic community (TC): houses substance abusers together in specific areas of the prison where they receive a variety of individual and group therapies.

three-strikes laws: laws that increase prison terms for offenders having been convicted of a serious criminal offense on three or more separate occasions.

throw-downs: a process of criminal suspect identification in which a set of photographs, including a picture of the suspect, is presented (thrown down) before a witness who is asked to pick out the photograph of the person who committed the crime.

ticket of leave: a practice of early release developed by Sir Walter Crofton by which prisoners could be released early by issuance of a ticket of leave from the prison. A forerunner to modern parole.

tort: a civil action; an offense against an individual settled in a civil court.

total institution: a concept developed by Irving Goffman to refer to settings, such as prisons and mental hospitals, in which residents are completely dependent on facility staff and where virtually all decisions are made by staff rather than by the individuals.

training: instruction that prepares a criminal justice professional to face his or her daily challenges.

Transition from Prison to the Community (TPC): a parole case management model that approaches offender reentry as a process which begins at sentencing and continues until discharge from community supervision.

transportation: a practice of England in the eighteenth and nineteenth centuries in which offenders convicted of several types of crimes were banished to English colonies. Convicted offenders were "transported" to these distant locations as a punishment for crime. This practice was a forerunner to modern parole.

treatment: see rehabilitation; a justification for sentencing based on the belief that criminal offenders can be helped to prevent them from committing future crimes; the delivery of services—and the services themselves—provided to criminal offenders as a means to reduce future criminality.

trial: the fact-finding point in the criminal justice process; the court stage at which the state must present evidence sufficient to convince the judge or jury beyond a reasonable doubt that the defendant committed the crimes charged, or else the defendant is released from further processing.

trial courts: those courts in which criminal (and often civil) cases are heard to determine the facts of the case.

truth in sentencing: the concept that offenders should actually serve at least 85% of the term they receive; legislation requiring truth in sentencing has been passed in many states.

U

undetected crime: crime that is not known to the criminal justice system or the victim; crimes that are not recognized as crimes.

unfounding: the process that takes place when police decide that a citizen's complaint of a crime is not supported by available evidence. The unfounding of a complaint essentially "erases" the event as a crime known to the police.

unified court system: a system that combines general-jurisdiction and limited-jurisdiction courts into one.

Uniform Crime Reports (UCR): published annually by the Federal Bureau of Investigation, the most well-known source of official statistics on crime. Among other things, the report includes a count of all crimes known to police participating in the UCR program, as well as a description of the characteristics of persons arrested for crimes.

unofficial statistics: in regard to crime and criminal justice, statistics and data concerning the amount and distribution of crime and the activities of criminal justice agents and agencies that are collected and reported by persons who are not themselves involved in the official criminal justice processing of cases.

unreported crime: crime that victims recognize as law-breaking behavior, but is not brought to the attention of authorities.

unsolved: an adjective used to describe criminal cases in which the police have been unable to identify an individual or group as the offender.

unsupervised parole: a practice used in some jurisdictions in which active supervision of a parolee is terminated before the expiration of sentence and the discharge of the parolee from criminal justice custody.

V

venire: the panel of citizens from which a jury can be chosen.

vertical prosecution: the practice of assigning the responsibility of a case to a single prosecutor who then follows that same case throughout the entire court process.

victimization data: estimates of the rate and distribution of crime derived from survey respondents' reports of experiences of being the victims of crime.

victims' rights: a term referring to the status of crime victims in the criminal justice system and to summarize a movement designed to increase concern for crime victims and change criminal justice processing to be more sympathetic to the needs and concerns of crime victims.

violation: a minor criminal offense, such as speeding, overtime parking, or the like, typically punishable by a fine or restriction of privileges. The breaking of a condition of supervision by a probationer or parolee.

voir dire: "speak the truth"; the jury selection process in which members of a venire are interrogated by both sides (prosecution and defense, in criminal trials) to determine their suitability for service on the jury.

W

waiver decision: the decision to transfer a juvenile offender to the adult court; a determination that, though a juvenile in terms of age, an offender should be tried and punished as an adult by virtue of the seriousness of the crime or the characteristics of the offender.

war on drugs: a set of laws designed to increase the punishment and the drug enforcement efforts of the criminal justice system.

watchman style: coined by James Q. Wilson, the term refers to police agencies in which officers only infrequently intervene in the lives of citizens.

work release: programs which allow inmates to secure or maintain employment while serving terms of incarceration.

working rules: guidelines for behavior that identify the circumstances that justify certain police actions.

writ of certiorari: an order to the lower court to send its records of the case so that the Supreme Court can review them.

wrongful conviction: the result when innocent persons are convicted of criminal acts.

Index